Mathematical Methods for Physics

Mathematical Methods for Physics

Editor: Luca Bailey

STATES
ACADEMIC PRESS
www.statesacademicpress.com

Published by States Academic Press
109 South 5th Street,
Brooklyn, NY 11249, USA
www.statesacademicpress.com

Mathematical Methods for Physics
Edited by Luca Bailey

International Standard Book Number: 978-1-63989-345-4 (Hardback)

Cataloging-in-Publication Data

Mathematical methods for physics / edited by Luca Bailey.
p. cm.
Includes bibliographical references and index.
ISBN 978-1-63989-345-4
1. Mathematical physics. 2. Physics. I. Bailey, Luca.
QC20 .M63 2022
530.15--dc23

Contents

Preface VII

Chapter 1 **Strong Solutions for the Fluid-Particle Interaction Model with Non-Newtonian Potential** 1
Yukun Song, Heyuan Wang and Fengming Liu

Chapter 2 **Simultaneous Determination of the Space-Dependent Source and the Initial Distribution in a Heat Equation by Regularizing Fourier Coefficients of the Given Measurements** 12
Shufang Qiu, Wen Zhang and Jianmei Peng

Chapter 3 **Local Well-Posedness and Blow-Up for the Solutions to the Axisymmetric Inviscid Hall-MHD Equations** 27
Eunji Jeong, Junha Kim and Jihoon Lee

Chapter 4 **A Study of the Transport of Marine Pollutants using Adjoint Method of Data Assimilation with Method of Characteristics** 43
Xiaona Li, Minjie Xu, Xianqing Lv and Kai Fu

Chapter 5 **Plasma Parameters Effects on Dust Acoustic Solitary Waves in Dusty Plasmas of Four Components** 57
Abeer A. Mahmoud, Essam M. Abulwafa, Abd-alrahman F. Al-Araby and Atalla M. Elhanbaly

Chapter 6 **Loss, Gain, and Singular Points in Open Quantum Systems** 68
Hichem Eleuch and Ingrid Rotter

Chapter 7 **Magnetoacoustic Heating in Nonisentropic Plasma caused by Different Kinds of Heating-Cooling Function** 77
Anna Perelomova

Chapter 8 **Lump Solutions to the (3+1)-Dimensional Generalized B-Type Kadomtsev-Petviashvili Equation** 89
Xifang Cao

Chapter 9 **Vortex Structures inside Spherical Mesoscopic Superconductor Plus Magnetic Dipole** 94
A. Ludu

Chapter 10 **Diffusion Process and Reaction on a Surface** 113
M. E. K. Fuziki, M. K. Lenzi, M. A. Ribeiro, A. Novatski and E. K. Lenzi

Chapter 11 **CIP Method of Characteristics for the Solution of Tide Wave Equations** 124
Yafei Nie, Kai Fu and Xianqing Lv

Chapter 12 **Existence of Positive Ground State Solution for the Nonlinear Schrödinger-Poisson System with Potentials** 134
Lu Xiao, Haiqin Qian, Mengmeng Qu and Jun Wang

Chapter 13 **Evaluation of Heat Irreversibility in a Thin Film Flow of Couple Stress Fluid on a Moving Belt** 144
Samuel O. Adesanya, Ramosheuw S. Lebelo and K. C. Moloi

Chapter 14 **The Global Existence and Uniqueness of the Classical Solution with the Periodic Initial Value Problem for One-Dimension Klein-Gordon-Zakharov Equations** 150
Cong Sun and Lixia Li

Chapter 15 **Minimization of Stress State of a Hub of Friction Pair** 160
Vagif M. Mirsalimov and Parvana E. Akhundova

Chapter 16 **Some Analytical Solitary Wave Solutions for the Generalized q-Deformed Sinh-Gordon Equation: $\partial^2\theta/\partial z\partial\xi = \alpha[\sinh_q(\beta\theta^{\gamma})]^p - \delta$** 170
H. Eleuch

Chapter 17 **On the Second-Order Quantum (p, q)-Difference Equations with Separated Boundary Conditions** 177
Chanon Promsakon, Nattapong Kamsrisuk, Sotiris K. Ntouyas and Jessada Tariboon

Chapter 18 **The Application of Minimal Length in Klein-Gordon Equation with Hulthen Potential using Asymptotic Iteration Method** 186
Isnaini Lilis Elviyanti, Beta Nur Pratiwi, A. Suparmi and C. Cari

Chapter 19 **Rational Solutions of a Weakly Coupled Nonlocal Nonlinear Schrödinger Equation** 194
Huijuan Zhou, Chuanzhong Li and Yueh-Lung Lin

Chapter 20 **Dimensional Regularization Approach to the Renormalization Group Theory of the Generalized Sine-Gordon Model** 206
Takashi Yanagisawa

Chapter 21 **Expanded (G/G^2) Expansion Method to Solve Separated Variables for the 2+1-Dimensional NNV Equation** 213
Yong Meng

Chapter 22 **Abundant Lump-Type Solutions and Interaction Solutions for a Nonlinear (3+1) Dimensional Model** 219
R. Sadat, M. Kassem and Wen-Xiu Ma

Permissions

List of Contributors

Index

Preface

The main aim of this book is to educate learners and enhance their research focus by presenting diverse topics covering this vast field. This is an advanced book which compiles significant studies by distinguished experts in the area of analysis. This book addresses successive solutions to the challenges arising in the area of application, along with it; the book provides scope for future developments.

Mathematical physics is a branch of physics that is concerned with the application of mathematical concepts to problems in physics. It focuses on solving such problems by formulating a mathematically rigorous framework. It is characterized by the blending of mathematical and physics theoretical aspects. Mathematical physics primarily expands and elucidates physical theories. There are several distinct branches of mathematical physics. These are partial differential equations, classical mechanics, quantum theory, relativity and quantum relativistic theories, and statistical mechanics. This book attempts to understand the multiple branches that fall under the discipline of mathematical physics and how such concepts have practical applications. It strives to provide a fair idea about this discipline and to help develop a better understanding of the latest advances within this field. This book includes contributions of experts and scientists which will provide innovative insights into mathematical physics.

It was a great honour to edit this book, though there were challenges, as it involved a lot of communication and networking between me and the editorial team. However, the end result was this all-inclusive book covering diverse themes in the field.

Finally, it is important to acknowledge the efforts of the contributors for their excellent chapters, through which a wide variety of issues have been addressed. I would also like to thank my colleagues for their valuable feedback during the making of this book.

Editor

1

Strong Solutions for the Fluid-Particle Interaction Model with Non-Newtonian Potential

Yukun Song, **Heyuan Wang**, **and Fengming Liu**

College of Science, Liaoning University of Technology, Jinzhou 121001, China

Correspondence should be addressed to Yukun Song; songyukun8@163.com

Academic Editor: Manuel De León

This paper deals with a mathematical fluid-particle interaction model used to describing the evolution of particles dispersed in a viscous compressible non-Newtonian fluid. It is proved that the initial boundary value problems with vacuum admits a unique local strong solution in the dimensional case. The strong nonlinearity of the system brings us difficulties due to the fact that the viscosity term and non-Newtonian gravitational potential term are fully nonlinear.

1. Introduction

Fluid-particle interaction model arises in many practical applications in science and engineering [1–4] and is of primarily importance in the sedimentation analysis of disperse suspensions of particles in fluids. We focus on the fluid-particle interaction model that describes the evolution of particles dispersed in a viscous fluid.

Carrillo and Goudon first derived a fluid-particle interaction system by formal asymptotics from a mesoscopic description (see [5]). In the microscopic description, the cloud of the particles is related to its distribution function $f(t,x,\xi)$, which is the solution to a dimensionless Vlasov-Fokker-Planck equation. On the other hand, the fluid is described by its density $\rho(x,t) \geq 0$ and its velocity field $u(x,t)$. We assume that the fluid is compressible and isentropic, then (ρ,u) solves the compressible Euler equations for the inviscid case or the Navier-Stokes equations for the viscous case, respectively.

In [5], for the inviscid case, the coupling between the kinetic and the fluid equations is obtained through the friction forces that the fluid and the particles exert mutually. The friction force is assumed to follow the Stokes law and thus is proportional to the relative velocity of the fluid and the particles given by $\xi - u_\epsilon(t,x)$. Furthermore, both phases are affected by external forces, which are supposed to derive from a time independent potential $\Phi(x)$. The system was given as follows:

$$\begin{aligned}
& f_t + \theta\xi\cdot\nabla_x f - \kappa\nabla_x\Phi\cdot\nabla_\xi f \\
& \quad = \frac{1}{\epsilon}\operatorname{div}_\xi\left(\left(\xi - \frac{1}{\theta}u\right)f + \nabla_\xi f\right) \\
& \rho_t + \operatorname{div}_x(\rho u) = 0 \\
& (\rho u)_t + \operatorname{div}_x(\rho u\otimes u) + \chi\nabla_x P(\rho) + \alpha\theta\kappa\rho\nabla_x\Phi \\
& \quad = \frac{1}{\epsilon}\frac{\rho_P}{\rho_F}(J - \eta u)
\end{aligned} \tag{1}$$

where

$$\begin{aligned}
\eta(t,x) &= \int_{\mathbb{R}} f(t,x,\xi)\,d\xi, \\
J(t,x) &= \theta\int_{\mathbb{R}} \xi f(t,x,\xi)\,d\xi
\end{aligned} \tag{2}$$

$\theta,\kappa,\epsilon,\alpha,\chi$ are some related dimensionless parameters, $P(\rho)$, ρ_P, ρ_F are the pressure, the mass density of particles and fluid, respectively. Setting $\rho_P/\rho_F = 1/\theta^2, \kappa = \theta, \theta = 1/\sqrt{\epsilon}$, $\alpha = \text{sign}(\alpha)\epsilon$ with $\text{sign}(\alpha) = \pm 1$, then (1) becomes

$$\begin{aligned} &f_t^\epsilon + \frac{1}{\sqrt{\epsilon}}\left(\xi \cdot \nabla_x f^\epsilon - \nabla_x \Phi \cdot \nabla_\xi f^\epsilon\right) \\ &\quad = \frac{1}{\epsilon}\text{div}_\xi\left(\left(\xi - \sqrt{\epsilon}u^\epsilon\right) f^\epsilon + \nabla_\xi f^\epsilon\right) \\ &\rho_t^\epsilon + \text{div}_x\left(\rho^\epsilon u^\epsilon\right) = 0 \\ &\left(\rho^\epsilon u^\epsilon\right)_t + \text{div}_x\left(\rho^\epsilon u^\epsilon \otimes u^\epsilon\right) + \chi \nabla_x P\left(\rho^\epsilon\right) + \alpha\theta\kappa\rho^\epsilon \nabla_x \Phi \\ &\quad = \left(J^\epsilon - \eta^\epsilon u^\epsilon\right) \end{aligned} \tag{3}$$

Finally, letting $\epsilon \longrightarrow 0$, then (3) converges to system

$$\begin{aligned} &\eta_t + \text{div}_x\left(\eta\left(u - \nabla_x \Phi\right) - \nabla_x \eta\right) = 0 \\ &\rho_t + \text{div}_x(\rho u) = 0 \\ &(\rho u)_t + \text{div}_x(\rho u \otimes u) + \nabla_x(\eta + \chi P(\rho)) \\ &\quad + (\text{sign}(\alpha)\rho + \eta)\nabla_x \Phi = 0 \end{aligned} \tag{4}$$

For the viscous case, with the dynamic viscosity terms taken into consideration, then $(4)_3$ will be have an additional term of $\text{div}_x \mathbb{S}$. In [6], the viscous stress tensor $\mathbb{S} = \mathbb{S}(\nabla_x u)$ is assumed to satisfy Newton's Law for viscosity which requires that

$$\mathbb{S} = \mu\left(\nabla u + \nabla u^T\right) + \lambda\, \text{div}_x u\mathbb{I}, \tag{5}$$

where μ and λ are constant viscosity coefficients satisfying

$$\begin{aligned} &\mu > 0, \\ &\lambda + \frac{2}{3}\mu \geq 0. \end{aligned} \tag{6}$$

Then

$$\text{div}_x \mathbb{S}\left(\nabla_x u\right) = \mu \Delta_x u + \lambda \nabla_x \text{div}\, u \tag{7}$$

and thus the system turns into the following equation

$$\begin{aligned} &\rho_t + \text{div}_x(\rho u) = 0 \\ &(\rho u)_t + \text{div}_x(\rho u \otimes u) + \nabla_x(P(\rho) + \eta) - \mu\Delta_x u \\ &\quad - \lambda \nabla_x \text{div}_x u = -(\eta + \beta\rho)\nabla_x \Phi \\ &\eta_t + \text{div}_x\left(\eta\left(u - \nabla_x \Phi\right)\right) - \Delta_x \eta = 0 \end{aligned} \tag{8}$$

Moreover, if the influence of gravitational potential $\Psi = \Psi(x,t)$ was taken into consideration, there will be a Poisson equation coupled to the above system, as for the Navier-Stokes-Poisson equation in [7].

Carrillo et al. obtained the global existence and asymptotic behavior of the weak solutions and stability properties to (8). Subsequently, Fang et al. [8] studied the existence of global classical solutions in dimension one. In [9, 10], Balew and Trivisa obtained the existence of global weak solutions and weakly dissipative solutions by entropy method in dimension three. The two-phase flow hydrodynamic models have been proposed in [3]. For some mathematical results on the Navier-Stokes coupled equations such as the nematic liquid crystal flows models where viscous effects are included, for more details, we refer to [11–13] and references therein.

On the other hand, as we know, the viscous stress tensor $\mathbb{S}$ is depends on the rate of strain $\mathbb{E}_{ij}(\nabla_x u)$, where

$$\mathbb{E}_{ij}\left(\nabla_x u\right) = \frac{\partial u_i}{\partial x_j} + \frac{\partial u_j}{\partial x_i} \tag{9}$$

If the stress and rate of strain satisfy the following linear relation

$$\mathbb{S} = \mu \mathbb{E}_{ij}(\nabla u). \tag{10}$$

then the fluid is called Newtonian. The coefficient of proportionality μ is called the viscosity coefficient, and it is a characteristic material quantity for the fluid concerned, which in general depends on density, temperature, and pressure. The governing equations of motions of them will be the Navier-Stokes equations. If the relation is not linear, the fluid is called non-Newtonian. Examples of non-Newtonian fluids are molten plastics, polymer solutions, dyes, varnishes, suspensions, adhesives, paints, greases, paper pulp, and biological fluids like blood. The simplest model of the stress-strain relation for such fluids given by the power laws, which states that

$$\mathbb{S} = \mu\left(\frac{\partial u_i}{\partial x_j} + \frac{\partial u_j}{\partial x_i}\right)^q, \tag{11}$$

for $0 < q < 1$ (see [14]). Ladyzhenskaya (see [15]) proposed a special form for $\mathbb{S}$ on the incompressible model:

$$\mathbb{S}_{ij} = \left(\mu_0 + \mu_1 \left|\mathbb{E}\left(\nabla_x u\right)\right|^{p-2}\right) \mathbb{E}_{ij}\left(\nabla_x u\right) \tag{12}$$

These models are called

$$\begin{aligned} &\textit{Newtonian}, \quad \textit{for } \mu_0 > 0,\ \mu_1 = 0; \\ &\textit{Rabinowitsch}, \quad \textit{for } \mu_0, \mu_1 > 0,\ p = 4; \\ &\textit{Eills}, \quad \textit{for } \mu_0, \mu_1 > 0,\ p > 2; \\ &\textit{Ostwald}-\textit{deWaele}, \quad \textit{for } \mu_0 = 0,\ \mu_1 > 0,\ p > 1; \\ &\textit{Bingham}, \quad \textit{for } \mu_0, \mu_1 > 0,\ p = 1. \end{aligned} \tag{13}$$

For $\mu_0 = 0$, if $p < 2$ then it is a pseudo-plastic fluid, and if $p > 2$ then it is a dilatant fluid (see [14]). In the view of physics, the model captures the shear thinning fluid for the case of $1 < p < 2$, and captures the shear thickening fluid for the case of $p > 2$.

Followed by the Ladyzhenskaya model, in this paper, we investigate the compressible non-Newtonian fluid-particle interaction model in one-dimensional case, then system (8) changes to be

$$\begin{aligned}
&\rho_t + (\rho u)_x = 0, \\
&(\rho u)_t + (\rho u^2)_x + \rho\Psi_x - \lambda\left[\left(u_x^2+\mu_1\right)^{(p-2)/2} u_x\right]_x \\
&\quad + (P+\eta)_x = -\eta\Phi_x, \quad (x,t)\in\Omega_T \\
&\left[\left(\Psi_x^2+\mu_2\right)^{(q-2)/2}\Psi_x\right]_x = 4\pi g\left(\rho - \frac{1}{|\Omega|}\int_\Omega \rho \,\mathrm{d}x\right), \\
&\eta_t + [\eta(u-\Phi_x)]_x = \eta_{xx}
\end{aligned} \tag{14}$$

with the initial and boundary conditions

$$\begin{aligned}
(\rho,u,\eta)|_{t=0} &= (\rho_0,u_0,\eta_0), \quad x\in\Omega, \\
u|_{\partial\Omega} &= \Psi|_{\partial\Omega} = 0, \quad t\in[0,T],
\end{aligned} \tag{15}$$

and the no-flux condition for the density of particles

$$(\eta_x + \eta\Phi_x)|_{\partial\Omega} = 0, \quad t\in[0,T]. \tag{16}$$

where Ω is a bounded interval, $\Omega_T = \Omega\times[0,T]$. ρ, u, η denote the fluid density, velocity, and the density of particle in the mixture, respectively. $P(\rho) = a\rho^\gamma$ is the pressure where $a > 0, \gamma > 1$, $\Psi(x,t)$ stands for the non-Newtonian gravitational potential, and the given function $\Phi(x)$ denotes the external potential. $\lambda > 0$ is the viscosity coefficient and $\mu_1 > 0, \mu_2 > 0$, $p > 2, q > 2$ are constants.

As to the non-Newtonian fluids, there has been much research both theoretically and experimentally, see ([15–23]). Indeed, we have investigated the existence results of solutions for $p > 2$ and $1 < p < 2$ in the absence of term $\Psi(x)$ for (14) in [22, 23]. However, the influence of non-Newtonian gravitational potential for a practical model was not taken into consideration there. To our knowledge, there seems very few mathematical results for the case of the fluid-particle interaction model systems with non-Newtonian gravitational potential, even in dimension one. The existence results to problem (14)-(16) when $p, q > 2$ which describes the motion of the compressible viscous isentropic gas flow is driven by a non-Newtonian gravitational force is still open up to now. We are interested in the existence and uniqueness of strong solutions on a one-dimensional bounded domain. In fact, the strong nonlinearity of (14) brings us new difficulties in getting the upper bound of ρ and the method used in [8] does not suitable for us. Motivated by Cho *etal*'s [24, 25] work on Navier-Stokes equations, we establish local existence and uniqueness of strong solutions by the iteration techniques.

Throughout the paper we assume that $a = \lambda = 1$. In the following sections, we will use simplified notations for standard Sobolev spaces and Bochner spaces, such as $L^p = L^p(\Omega)$, $H_0^1 = H_0^1(\Omega)$, $C(0,T;H^1) = C(0,T;H^1(\Omega))$.

1.1. Main Results

Theorem 1. *Let $\mu_1 > 0$ be a positive constant and $\Phi \in C^2(\Omega)$, and assume that the initial data (ρ_0, u_0, η_0) satisfy the following conditions:*

$$\begin{aligned}
0 \le \rho_0 &\in H^1(\Omega), \\
u_0 &\in H_0^1(\Omega)\cap H^2(\Omega), \\
\eta_0 &\in H^2(\Omega)
\end{aligned} \tag{17}$$

and

$$\begin{aligned}
&-\left[\left(u_{0x}^2+\mu_1\right)^{(p-2)/2} u_{0x}\right]_x + (P(\rho_0)+\eta_0)_x + \eta_0\Phi_x \\
&= \rho_0^{1/2}(g+\beta\Phi_x),
\end{aligned} \tag{18}$$

for some $g \in L^2(\Omega)$, where $\beta \ne 0$ is a constant. Then there exist a time $T_* \in (0,+\infty)$ and a unique strong solution (ρ, u, η) to (14)-(16) such that

$$\begin{aligned}
&\rho \in L^\infty(0,T_*;H^1(\Omega)), \\
&u \in L^\infty(0,T_*;W_0^{1,p}(\Omega)\cap H^2(\Omega)), \\
&\eta \in L^\infty(0,T_*;H^2(\Omega)) \\
&\rho_t \in L^\infty(0,T_*;L^2(\Omega)), \\
&u_t \in L^2(0,T_*;H_0^1(\Omega)), \\
&\eta_t \in L^\infty(0,T_*;L^2(\Omega)) \\
&\sqrt{\rho}u_t \in L^\infty(0,T_*;L^2(\Omega)), \\
&\left(\left(u_x^2+\mu_1\right)^{(p-2)/2}u_x\right)_x \in L^2(0,T_*;L^2(\Omega)),
\end{aligned} \tag{19}$$

2. A Priori Estimates for Smooth Solutions

In this section, we will prove the local existence of strong solutions. Provided that (ρ, u, η) is a smooth solution of (14)-(16) and $\rho_0 \ge \delta$, where $0 < \delta \ll 1$ is a positive number. We denote by $M_0 = 1 + \mu_1 + \mu_1^{-1} + |\rho_0|_{H^1} + |g|_{L^2}$ and introduce an auxiliary function

$$\begin{aligned}
Z(t) = \sup_{0\le s\le t}&\left(1 + |\rho(s)|_{H^1} + |u(s)|_{W_0^{1,p}} + |\eta_t(s)|_{L^2}\right. \\
&\left. + |\eta(s)|_{H^1} + |\sqrt{\rho}u_t(s)|_{L^2}\right)
\end{aligned} \tag{20}$$

Then we estimate each term of $Z(t)$ in terms of some integrals of $Z(t)$ and apply arguments of Gronwall-type and thus prove that $Z(t)$ is locally bounded.

2.1. Estimate for $|\rho|_{H^1}$. First we need to do the following estimates. Using $(14)_1$, we rewritten $(14)_2$ as

$$\begin{aligned}
&\rho u_t + \rho u u_x + \rho\Psi_x - \left[\left(u_x^2+\mu_1\right)^{(p-2)/2}u_x\right]_x \\
&\quad + (P+\eta)_x = -\eta\Phi_x
\end{aligned} \tag{21}$$

By virtue of

$$\left|\left[\left(u_x^2+\mu_1\right)^{(p-2)/2}u_x\right]_x\right| \geq \mu_1^{(p-2)/2}\left|u_{xx}\right| \tag{22}$$

Then

$$\left|u_{xx}\right| \leq C\left|\rho u_t+\rho u u_x+\rho\Psi_x+(P+\eta)_x+\eta\Phi_x\right| \tag{23}$$

Taking the above inequality by L^2 norm, we get

$$\begin{aligned}\left|u_{xx}\right|_{L^2} &\leq C\left|\rho u_t+\rho u u_x+\rho\Psi_x+(P+\eta)_x+\eta\Phi_x\right|_{L^2}\\ &\leq C\left(|\rho|_{L^\infty}^{1/2}\left|\sqrt{\rho}u_t\right|_{L^2}+|\rho|_{L^\infty}|u|_{L^\infty}\left|u_x\right|_{L^2}\right.\\ &\quad\left.+|\rho|_{L^2}\left|\Psi_x\right|_{L^2}+\left|P_x\right|_{L^2}+\left|\eta_x\right|_{L^2}+|\eta|_{L^2}\left|\Phi_x\right|_{L^2}\right)\end{aligned} \tag{24}$$

We deal with the term of $|\Psi_x|_{L^2}$.

Multiplying $(14)_3$ by Ψ and integrating over Ω, we obtain

$$\begin{aligned}&\int_\Omega\left[\left(\Psi_x^2+\mu_2\right)^{(q-2)/2}\Psi_x\right]_x\Psi\,\mathrm{d}x\\ &\quad=4\pi g\left(\int_\Omega\rho\Psi\,\mathrm{d}x-m_0\int_\Omega\Psi\,\mathrm{d}x\right)\\ &\int_\Omega\left|\Psi_x\right|^q\mathrm{d}x\leq\int_\Omega\left(\Psi_x^2+\mu_2\right)^{(q-2)/2}\Psi_x^2\mathrm{d}x\\ &\quad=-\int_\Omega\left(\left(\Psi_x^2+\mu_2\right)^{(q-2)/2}\Psi_x\right)_x\Psi\,\mathrm{d}x\\ &\quad=-4\pi g\left(\int_\Omega\rho\Psi\,\mathrm{d}x-m_0\int_\Omega\Psi\,\mathrm{d}x\right)\\ &\quad\leq|\Psi|_{L^\infty}8\pi g m_0\leq\left|\Psi_x\right|_{L^q}8\pi g m_0\\ &\quad\leq\frac{1}{q}\left|\Phi_x\right|_{L^q}^q+\frac{1}{p}\left(8\pi g m_0\right)^p\end{aligned} \tag{25}$$

then we have

$$\int_\Omega\left|\Psi_x\right|^q\mathrm{d}x\leq C\left(m_0\right),\quad q>2. \tag{26}$$

Hence, we deduce that

$$\left|u_{xx}\right|_{L^2}\leq CZ^{\gamma+2}(t) \tag{27}$$

From $(14)_3$, taking it by L^2 norm, we get

$$\begin{aligned}\left|\eta_{xx}\right|_{L^2}&\leq\left|\eta_t+\left(\eta\left(u-\Phi_x\right)\right)_x\right|_{L^2}\\ &\leq\left|\eta_t\right|_{L^2}+\left|\eta_x\right|_{L^2}|u|_{L^\infty}+C\left|\eta_x\right|_{L^2}+|\eta|_{H^1}\left|u_x\right|_{L^p}\\ &\quad+C|\eta|_{H^1}\leq CZ^2(t)\end{aligned} \tag{28}$$

Multiplying $(14)_1$ by ρ, integrating over Ω, we have

$$\frac{1}{2}\frac{\mathrm{d}}{\mathrm{d}t}\int_\Omega|\rho|^2\mathrm{d}s+\int_\Omega(\rho u)_x\rho\,\mathrm{d}x=0 \tag{29}$$

Integrating by parts, using Sobolev inequality, we deduce that

$$\frac{\mathrm{d}}{\mathrm{d}t}|\rho(t)|_{L^2}^2\leq\int_\Omega\left|u_x\right||\rho|^2\mathrm{d}x\leq\left|u_{xx}\right|_{L^2}|\rho|_{L^2}^2 \tag{30}$$

Differentiating $(14)_1$ with respect to x, and multiplying it by ρ_x, integrating over Ω, using Sobolev inequality, we have

$$\begin{aligned}\frac{\mathrm{d}}{\mathrm{d}t}\int_\Omega\left|\rho_x\right|^2\mathrm{d}x&=-\int_\Omega\left[\frac{3}{2}u_x\left(\rho_x\right)^2+\rho\rho_x u_{xx}\right](t)\,\mathrm{d}x\\ &\leq C\left[\left|u_x\right|_{L^\infty}\left|\rho_x\right|_{L^2}^2+|\rho|_{L^\infty}\left|\rho_x\right|_{L^2}\left|u_{xx}\right|_{L^2}\right]\\ &\leq C|\rho|_{H^1}^2\left|u_{xx}\right|_{L^2}\end{aligned} \tag{31}$$

From (30) and (31), by Gronwall's inequality, it follows that

$$\begin{aligned}\sup_{0\leq t\leq T}|\rho(t)|_{H^1}^2&\leq\left|\rho_0\right|_{H^1}^2\exp\left\{C\int_0^t\left|u_{xx}\right|_{L^2}\mathrm{d}s\right\}\\ &\leq C\exp\left(C\int_0^t Z^{\gamma+2}(s)\,\mathrm{d}s\right)\end{aligned} \tag{32}$$

Besides, we can also get the following estimates. By using $(14)_1$,

$$\begin{aligned}\left|\rho_t(t)\right|_{L^2}&\leq\left|\rho_x(t)\right|_{L^2}|u(t)|_{L^\infty}+|\rho(t)|_{L^\infty}\left|u_x(t)\right|_{L^2}\\ &\leq CZ^2(t)\end{aligned} \tag{33}$$

From $(14)_3$, we have

$$\mu_1^{(q-2)/2}\Psi_{xx}\leq\left[\left(\Psi_x^2+\mu_1\right)^{(q-2)/2}\Psi_x\right]_x=4\pi g\left(\rho-m_0\right) \tag{34}$$

then

$$\left|\Psi_{xx}\right|_{L^2}\leq C\left(m_0\right) \tag{35}$$

Differentiating $(14)_3$ with respect to time t, multiplying it by Ψ_t, integrating over Ω to x, and using Young's inequality, we have

$$\begin{aligned}\mu_1^{(q-2)/2}\int_\Omega\Psi_{xt}^2\mathrm{d}x&\leq\int_\Omega\left[\left(\Psi_x^2+\mu_1\right)^{(q-2)/2}\Psi_x\right]_t\Psi_{xt}\mathrm{d}x\\ &=-4\pi g\int_\Omega\rho_t\Psi_t\mathrm{d}x\\ &\leq\int_\Omega\left|\rho_t\right|_{L^2}\left|\Psi_{xt}\right|_{L^2}\\ &\leq C\left|\rho_t\right|_{L^2}^2+\eta\left|\Psi_{xt}\right|_{L^2}^2\end{aligned} \tag{36}$$

thus, we get

$$\left|\Psi_{xt}\right|_{L^2}^2\leq CZ^4(t) \tag{37}$$

where C is a positive constant, depending only on M_0.

2.2. Estimate for $|u|_{W_0^{1,p}}$. Multiplying (21) by u_t, integrating (by parts) over Ω_T, we have

$$\begin{aligned}&\iint_{\Omega_T}\rho\left|u_t\right|^2\mathrm{d}x\mathrm{d}s\\ &\quad+\iint_{\Omega_T}\left(u_x^2+\mu_1\right)^{(p-2)/2}u_x u_{xt}\mathrm{d}x\mathrm{d}s\\ &\quad=-\iint_{\Omega_T}\left(\rho u u_x+\rho\Psi_x+P_x+\eta_x+\eta\Phi_x\right)u_t\mathrm{d}x\mathrm{d}s\end{aligned} \tag{38}$$

We deal with each term as follows:

$$\int_\Omega \left(u_x^2+\mu_1\right)^{(p-2)/2} u_x u_{xt}\mathrm{d}x = \frac{1}{2}\int_\Omega \left(u_x^2+\mu_1\right)^{(p-2)/2}\left(u_x^2\right)_t \mathrm{d}x = \frac{1}{2}\frac{\mathrm{d}}{\mathrm{d}t}\int_\Omega\left(\int_0^{u_x^2}(s+\mu_1)^{(p-2)/2}\mathrm{d}s\right)\mathrm{d}x$$

$$\int_0^{u_x^2}(s+\mu_1)^{(p-2)/2}\mathrm{d}s = \int_{\mu_1}^{u_x^2+\mu_1} t^{(p-2)/2}\mathrm{d}t = \frac{2}{p}\left[\left(u_x^2+\mu_1\right)^{p/2}-\mu_1^{p/2}\right] \geq \frac{2}{p}\left|u_x\right|^p - \frac{2}{p}\mu_1^{p/2} \tag{39}$$

$$-\iint_{\Omega_T} P_x u_t \mathrm{d}x\mathrm{d}s = \iint_{\Omega_T} P u_{xt}\mathrm{d}x\mathrm{d}s = \frac{\mathrm{d}}{\mathrm{d}t}\iint_{\Omega_T} P u_x \mathrm{d}x\mathrm{d}s - \iint_{\Omega_T} P_t u_x \mathrm{d}x\mathrm{d}s$$

From $(14)_1$ we get

$$P_t = -\gamma P u_x - P_x u \tag{40}$$

$$-\iint_{\Omega_T}\left(\eta_x+\eta\Phi_x\right)u_t\mathrm{d}x\mathrm{d}s = \frac{\mathrm{d}}{\mathrm{d}t}\iint_{\Omega_T}\left(\eta_x+\eta\Phi_x\right)u\,\mathrm{d}x\mathrm{d}s - \iint_{\Omega_T}\left(\eta_x+\eta\Phi_x\right)_t u\,\mathrm{d}x\mathrm{d}s \tag{41}$$

$$-\iint_{\Omega_T}\left(\eta_x+\eta\Phi_x\right)_t u\,\mathrm{d}x\mathrm{d}s = \iint_{\Omega_T}\eta_t\left(u_x-\Phi_x u\right)\mathrm{d}x\mathrm{d}s = -\iint_{\Omega_T}\left[\eta_x-\eta\left(u-\Phi_x\right)\right]\left(u_x-\Phi_x u\right)_x\mathrm{d}x\mathrm{d}s$$

Substituting the above into (38), and using Young's inequality, we obtain

$$\begin{aligned}\int_0^t\left|\sqrt{\rho}u_t(s)\right|_{L^2}^2\mathrm{d}s + \left|u_x(t)\right|_{L^p}^p &\leq C + \int_\Omega\left|Pu_x\right|\mathrm{d}x + \iint_{\Omega_T}\left(\left|\rho u u_x u_t\right| + \left|\rho\Psi_x u_t\right| + \left|\gamma P u_x^2\right| + \left|P_x u u_x\right|\right)\mathrm{d}x\mathrm{d}s \\ &+ \iint_{\Omega_T}\left(\left|\eta_x u_{xx}\right| + \left|\eta_x\Phi_x u_x\right| + \left|\eta_x\Phi_{xx}u\right| + \left|\eta u u_{xx}\right| + \left|\eta u^2\Phi_{xx}\right| + \left|\eta u\Phi_x u_x\right| + \left|\eta\Phi_x u_{xx}\right| + \left|\eta\Phi_x\Phi_{xx}u\right| + \left|\eta\Phi_x^2 u_x\right|\right)\mathrm{d}x\mathrm{d}s \\ &\leq C + C\int_0^t\left(|\rho|_{L^\infty}|u|_{L^\infty}^2 \cdot \left|u_x\right|_{L^p}^2 + |\rho|_{L^\infty}\left|\Psi_x\right|_{L^2}^2 + |P|_{L^\infty}\left|u_x\right|_{L^p}^2 + \left|P_x\right|_{L^2}|u|_{L^\infty} \cdot \left|u_x\right|_{L^p} + \left|\eta_x\right|_{L^2}\left|u_{xx}\right|_{L^2} + \left|\eta_x\right|_{L^2}\left|u_x\right|_{L^p} + \left|\eta_x\right|_{L^2} \right. \\ &\left. \cdot |u|_{L^\infty} + |\eta|_{L^\infty}|u|_{L^\infty}\left|u_{xx}\right|_{L^2} + |\eta|_{L^\infty}|u|_{L^\infty}^2 + |\eta|_{L^\infty} \cdot |u|_{L^\infty}\left|u_x\right|_{L^p} + |\eta|_{L^\infty}\left|u_{xx}\right|_{L^2} + |\eta|_{L^\infty}|u|_{L^\infty} + |\eta|_{L^\infty} \cdot \left|u_x\right|_{L^p}\right)\mathrm{d}s + C\left|P(t)\right|_{L^2}^2 + \frac{1}{2}\int_0^t\left|\sqrt{\rho}u_t\right|_{L^2}^2\mathrm{d}s \\ &+ \frac{1}{2}\left|u_x(t)\right|_{L^p}^p \leq C\left(1+\int_0^t Z^{\gamma+4}(s)\,\mathrm{d}s + \left|P(t)\right|_{L^2}^2\right)\end{aligned} \tag{42}$$

Using $(14)_1$, we have

$$\begin{aligned}\int_\Omega\left|P(t)\right|^2\mathrm{d}x &= \int_\Omega\left|P(0)\right|^2\mathrm{d}x + \int_0^t\frac{\partial}{\partial s}\left(\int_\Omega P(s)^2\mathrm{d}x\right)\mathrm{d}s \\ &\leq \int_\Omega\left|P(0)\right|^2\mathrm{d}x + 2\int_0^t\int_\Omega P(s)\,\gamma\rho^{\gamma-1}\left(-\rho_x u - \rho u_x\right)\mathrm{d}x\mathrm{d}s \\ &\leq C + C\int_0^t|P|_{L^\infty}|\rho|_{L^\infty}^{\gamma-1}|\rho|_{H^1}\left|u_x\right|_{L^p}\mathrm{d}s \\ &\leq C\left(1+\int_0^t Z^{2\gamma+1}(s)\,\mathrm{d}s\right)\end{aligned} \tag{43}$$

Combining (42)-(43), yields

$$\int_0^t\left|\sqrt{\rho}u_t(s)\right|_{L^2}^2(s)\,\mathrm{d}s + \left|u_x(t)\right|_{L^p}^p \leq C\left(1+\int_0^t Z^{2\gamma+3}(s)\,\mathrm{d}s\right) \tag{44}$$

where C is a positive constant, depending only on M_0.

2.3. Estimate for $|\eta_t|_{L^2}$ and $|\eta|_{H^1}$. Multiplying $(14)_4$ by η, integrating the resulting equation over Ω_T, and using the boundary conditions (16), Young's inequality, we have

$$\begin{aligned}\int_0^t\left|\eta_x(s)\right|_{L^2}^2\mathrm{d}s + \frac{1}{2}\left|\eta(t)\right|_{L^2}^2 &\leq \iint_{\Omega_T}\left(\left|\eta u\eta_x\right| + \left|\eta\Phi_x\eta_x\right|\right)\mathrm{d}x\mathrm{d}s \\ &\leq \frac{1}{4}\int_0^t\left|\eta_x(s)\right|_{L^2}^2\mathrm{d}s + C\int_0^t\left|u_x\right|_{L^p}^2|\eta|_{H^1}^2\mathrm{d}s + C\int_0^t|\eta|_{H^1}^2 + C \\ &\leq \frac{1}{4}\int_0^t\left|\eta_x(s)\right|_{L^2}^2\mathrm{d}s + C\left(1+\int_0^t Z^4(t)\,\mathrm{d}s\right)\end{aligned} \tag{45}$$

Multiplying $(14)_4$ by η_t, integrating (by parts) over Ω_T, and using the boundary conditions (16), Young's inequality, we have

$$\begin{aligned}
&\int_0^t |\eta_t(s)|_{L^2}^2 \,ds + \frac{1}{2}|\eta_x(t)|_{L^2}^2 \\
&\le \int\int_{\Omega_T} |\eta(u-\Phi_x)\eta_{xt}|\,dxds \\
&\le \frac{1}{4}\int_0^t |\eta_{xt}(s)|_{L^2}^2\,ds + C\int_0^t |\eta|_{H^1}^2 |u_x|_{L^p}^2\,ds \\
&\quad + C\int_0^t |\eta|_{H^1}^2\,ds + C \\
&\le \frac{1}{4}\int_0^t |\eta_{xt}(s)|_{L^2}^2\,ds + C\left(1+\int_0^t Z^4(t)\,ds\right)
\end{aligned} \tag{46}$$

Differentiating $(14)_4$ with respect to t, multiplying the resulting equation by η_t, and integrating (by parts) over Ω_T, we get

$$\begin{aligned}
&\int_0^t |\eta_{xt}(s)|_{L^2}^2\,ds + \frac{1}{2}|\eta_t(t)|_{L^2}^2 = \int\int_{\Omega_T} (\eta(u-\Phi_x))_t \\
&\cdot \eta_{xt}\,dxds \le C + \int\int_{\Omega_T} (|\eta_t u\eta_{xt}| + |\eta_t\Phi_x\eta_{xt}| \\
&+ |\eta_x u_t\eta_t| + |\eta u_{xt}\eta_t|)\,dxds \le C\Big(1 \\
&+ \int_0^t \big(|\eta_t|_{L^2}^2|u_x|_{L^p}^2 + |\eta_t|_{L^2}^2 + |\eta_x|_{L^2}^2|\eta_t|_{L^2}^2 \\
&+ |\eta|_{H^1}^2|\eta_t|_{L^2}^2\big)\,dx\Big) + \frac{1}{2}\int_0^t |\eta_{xt}|_{L^2}^2 + \frac{1}{2} \\
&\cdot \int_0^t |u_{xt}|_{L^2}^2 \le C\left(1+\int_0^t Z^{2\gamma+6}(s)\,ds\right)
\end{aligned} \tag{47}$$

Combining (45)-(47), we get

$$\begin{aligned}
&|\eta|_{H^1}^2 + |\eta_t|_{L^2}^2 + \int_0^t \left(|\eta_x|_{L^2}^2 + |\eta_t|_{L^2}^2 + |\eta_{xt}|_{L^2}^2\right)(s)\,ds \\
&\le C\left(1+\int_0^t Z^{2\gamma+6}(s)\,ds\right)
\end{aligned} \tag{48}$$

2.4. Estimate for $|\sqrt{\rho}u_t|_{L^2}$. Differentiating (21) with respect to t, multiplying the result equation by u_t, and integrating it over Ω with respect to x, we have

$$\begin{aligned}
&\frac{1}{2}\frac{d}{dt}\int_\Omega \rho|u_t|^2\,dx + \int_\Omega \left[\left(u_x^2+\mu_1\right)^{(p-2)/2} u_x\right]_t u_{xt}\,dx \\
&= \int_\Omega \Big[(\rho u)_x\left(u_t^2 + uu_xu_t + \Psi_x u_t\right) - \rho u_x u_t^2 \\
&\quad - \rho\Psi_{xt}u_t - (P+\eta)_t u_{xt} - \eta_t\Phi_x u_t\Big]\,dx
\end{aligned} \tag{49}$$

Note that

$$\begin{aligned}
&\left[\left(u_x^2+\mu_1\right)^{(p-2)/2} u_x\right]_t u_{xt} \\
&= \left(u_x^2+\mu_1\right)^{(p-4)/2}(p-1)\left(u_x^2+\mu_1\right)u_{xt}^2 \\
&\ge \mu_1^{(p-2)/2} u_{xt}^2
\end{aligned} \tag{50}$$

Combining (40), (49) can be rewritten into

$$\begin{aligned}
&\frac{1}{2}\frac{d}{dt}\int_\Omega \rho|u_t|^2\,dx + \int_\Omega |u_{xt}|^2\,dx \\
&\le 2\int_\Omega \rho|u||u_t||u_{xt}|\,dx + \int_\Omega \rho|u||u_x|^2|u_t|\,dx \\
&+ \int_\Omega \rho|u|^2|u_{xx}||u_t|\,dx + \int_\Omega \rho|u|^2|u_x||u_{xt}|\,dx \\
&+ \int_\Omega \rho|u||\Psi_{xx}||u_t|\,dx + \int_\Omega \rho|u||\Psi_x||u_{xt}|\,dx \\
&+ \int_\Omega \rho|u_x||u_t|^2\,dx + \int_\Omega \gamma|P||u_x||u_{xt}|\,dx \\
&+ \int_\Omega |P_x||u||u_{xt}|\,dx + \int_\Omega |\eta_t||u_{xt}|\,dx \\
&+ \int_\Omega |\eta_t||\Phi_x||u_t|\,dx + \int_\Omega \rho|\Psi_{xt}||u_t|\,dx = \sum_{j=1}^{12} I_j
\end{aligned} \tag{51}$$

By using Sobolev inequality, Hölder inequality and Young's inequality, (35), (37), we estimate each term of I_j as follows:

$$\begin{aligned}
I_1 &= 2\int_\Omega \rho|u||u_t||u_{xt}|\,dx \\
&\le 2|\rho|_{L^\infty}^{1/2}|u|_{L^\infty}|\sqrt{\rho}u_t|_{L^2}|u_{xt}|_{L^2} \\
&\le CZ^5(t) + \frac{1}{7}|u_{xt}|_{L^2}^2 \\
I_2 &= \int_\Omega \rho|u||u_x|^2|u_t|\,dx \\
&\le |\rho|_{L^\infty}^{1/2}|u|_{L^2}^2|u_x|_{L^\infty}^2|\sqrt{\rho}u_t|_{L^2} \le CZ^{2\gamma+7}(t) \\
I_3 &= \int_\Omega \rho|u|^2|u_{xx}||u_t|\,dx \\
&\le |\rho|_{L^\infty}^{1/2}|u|_{L^\infty}^2|u_{xx}|_{L^2}|\sqrt{\rho}u_t|_{L^2} \le CZ^{\gamma+6}(t) \\
I_4 &= \int_\Omega \rho|u|^2|u_x||u_{xt}|\,dx \\
&\le |\rho|_{L^\infty}|u|_{L^\infty}^2|u_x|_{L^2}|u_{xt}|_{L^2} \le CZ^8(t) + \frac{1}{7}|u_{xt}|_{L^2}^2 \\
I_5 &= \int_\Omega \rho|u||\Psi_{xx}||u_t|\,dx \\
&\le |\rho|_{L^\infty}^{1/2}|u|_{L^2}|\Psi_{xx}|_{L^2}|\sqrt{\rho}u_t|_{L^2} \le CZ^3(t)
\end{aligned}$$

$$\begin{aligned}
I_6 &= \int_\Omega \rho |u| |\Psi_x| |u_{xt}| \, dx \le |\rho|_{L^\infty} |u|_{L^\infty} |\Psi_x|_{L^2} |u_{xt}|_{L^2} \\
&\le CZ^4(t) + \frac{1}{7} |u_{xt}|_{L^2}^2 \\
I_7 &= \int_\Omega \rho |u_x| |u_t|^2 \, dx \le |u_x|_{L^\infty} |\sqrt{\rho} u_t|_{L^2}^2 \le CZ^{\gamma+4}(t) \\
I_8 &= \int_\Omega \gamma |P| |u_x| |u_{xt}| \, dx \le C |P|_{L^\infty} |u_x|_{L^p} |u_{xt}|_{L^2} \\
&\le CZ^{2\gamma+2}(t) + \frac{1}{7} |u_{xt}|_{L^2}^2 \\
I_9 &= \int_\Omega |P_x| |u| |u_{xt}| \, dx \le |P_x|_{L^2} |u|_{L^\infty} |u_{xt}|_{L^2} \\
&\le CZ^{2\gamma+2}(t) + \frac{1}{7} |u_{xt}|_{L^2}^2 \\
I_{10} &= \int_\Omega |\eta_t| |u_{xt}| \, dx \le |\eta_t|_{L^2} |u_{xt}|_{L^2} \\
&\le CZ^2(t) + \frac{1}{7} |u_{xt}|_{L^2}^2 \\
I_{11} &= \int_\Omega |\eta_t| |\Phi_x| |u_t| \, dx \le CZ^2(t) + \frac{1}{7} |u_{xt}|_{L^2}^2 \\
I_{12} &= \int_\Omega \rho |\Psi_{xt}| |u_t| \, dx \le C |\rho|_{L^\infty}^{1/2} |\Psi_{xt}|_{L^2} |\sqrt{\rho} u_t|_{L^2} \\
&\le CZ^6(t)
\end{aligned} \tag{52}$$

Substituting I_j $(j = 1, 2, \dots, 12)$ into (51), and integrating over $(\tau, t) \subset (0, T)$ on the time variable, we have

$$\begin{aligned}
&|\sqrt{\rho} u_t(t)|_{L^2}^2 + \int_\tau^t |u_{xt}|_{L^2}^2(s) \, ds \\
&\quad \le |\sqrt{\rho} u_t(\tau)|_{L^2}^2 + C \int_\tau^t Z^{2\gamma+7}(s) \, ds.
\end{aligned} \tag{53}$$

To obtain the estimate of $|\sqrt{\rho} u_t(t)|_{L^2}^2$, we need to estimate $\lim_{\tau \to 0} |\sqrt{\rho} u_t(\tau)|_{L^2}^2$. Multiplying (21) by u_t and integrating over Ω, we get

$$\begin{aligned}
\int_\Omega \rho |u_t|^2 \, dx &\le 2 \int_\Omega \Big(\rho |u|^2 |u_x|^2 + \rho |\Psi_x|^2 \\
&+ \rho^{-1} \Big| - \Big[\left(u_x^2 + \mu_1\right)^{(p-2)/2} u_x \Big]_x + (P + \eta)_x \\
&+ \eta \Phi_x \Big|^2 \Big) \, dx.
\end{aligned} \tag{54}$$

According to the smoothness of (ρ, u, η), we obtain

$$\begin{aligned}
&\lim_{\tau \to 0} \int_\Omega \Big(\rho |u|^2 |u_x|^2 + \rho |\Psi_x|^2 \\
&+ \rho^{-1} \Big| - \Big[\left(u_x^2 + \mu_1\right)^{(p-2)/2} u_x \Big]_x + (P + \eta)_x + \eta \Phi_x \Big|^2 \Big) \, dx \\
&= \int_\Omega \Big(\rho |u_0|^2 |u_{0x}|^2 + \rho_0 |\Phi_x|^2 \\
&+ \rho_0^{-1} \Big| - \Big[\left(u_{0x}^2 + \mu_1\right)^{(p-2)/2} u_{0x} \Big]_x + (P_0 + \eta_0)_x \\
&+ \eta_0 \Phi_x \Big|^2 \Big) \, dx \le |\rho_0|_{L^\infty} |u_0|_{L^\infty}^2 |u_{0x}|_{L^2}^2 + |\rho_0|_{L^\infty} |\Phi_x|_{L^2}^2 \\
&+ |g|_{L^2}^2 + \beta |\Phi_x|_{L^2}^2 \le C.
\end{aligned} \tag{55}$$

Therefore, taking a limit on τ in (53), as $\tau \to 0$, we conclude that

$$\begin{aligned}
&|\sqrt{\rho} u_t(t)|_{L^2}^2 + \int_0^t |u_{xt}|_{L^2}^2(s) \, ds \\
&\quad \le C \left(1 + \int_0^t Z^{2\gamma+7}(s) \, ds \right),
\end{aligned} \tag{56}$$

where C is a positive constant, depending only on M_0.

Combining the estimates of (27), (28), (32), (33), (44), (48), (56) and the definition of $Z(t)$, we conclude that

$$Z(t) \le \widetilde{C} \exp \left(\widetilde{\widetilde{C}} \int_0^t Z^{2\gamma+7}(s) \, ds \right) \tag{57}$$

where $\widetilde{C}, \widetilde{\widetilde{C}}$ are positive constants, depending only on M_0. This means that there exist a time $T_1 > 0$ and a constant $C > 0$, such that

$$\begin{aligned}
&\operatorname*{ess\,sup}_{0 \le t \le T_1} \Big(|\rho|_{H^1} + |u|_{W_0^{1,p} \cap H^2} + |\eta|_{H^2} + |\eta_t|_{L^2} \\
&+ |\sqrt{\rho} u_t|_{L^2} + |\rho_t|_{L^2} \Big) + \int_0^{T_1} \Big(|\sqrt{\rho} u_t|_{L^2}^2 + |u_{xt}|_{L^2}^2 \\
&+ |\eta_x|_{L^2}^2 + |\eta_t|_{L^2}^2 + |\eta_{xt}|_{L^2}^2 \Big) \, ds \le C
\end{aligned} \tag{58}$$

3. Proof of the Main Theorem

In this section, our proof will be based on the usual iteration argument and some ideas developed in [24, 25]. We construct the approximate solutions, by using the iterative scheme, inductively, as follows: first define $u^0 = 0$ and assuming that u^{k-1} was defined for $k \ge 1$, let ρ^k, u^k, η^k be the unique smooth solution to the following problems:

$$\rho_t^k + \rho_x^k u^{k-1} + \rho^k u_x^{k-1} = 0 \tag{59}$$

$$\rho^k u_t^k + \rho^k u^{k-1} u_x^k + \rho^k \Psi_x^k + L_p u^k + P_x^k + \eta_x^k = -\eta^k \Phi_x \tag{60}$$

$$L_q \Psi^k = -4\pi g \left(\rho^k - m_0\right) \tag{61}$$

$$\eta_t^k + \left(\eta^k \left(u^{k-1} - \Phi_x\right)\right)_x = \eta_{xx}^k \tag{62}$$

with the initial and boundary conditions

$$\begin{aligned}
\left(\rho^k, u^k, \eta^k\right)\big|_{t=0} &= (\rho_0, u_0, \eta_0) \\
u^k\big|_{\partial\Omega} &= \Psi^k\big|_{\partial\Omega} = 0 \\
\left(\eta_x^k + \eta^k \Phi_x\right)\big|_{\partial\Omega} &= 0
\end{aligned} \tag{63}$$

where

$$L_p u^k = -\left[\left(\left(u_x^k\right)^2 + \mu_1\right)^{(p-2)/2} u_x^k\right]_x$$
$$L_q \Psi^k = -\left[\left(\left(\Psi_x^k\right)^2 + \mu_2\right)^{(q-2)/2} \Psi_x^k\right]_x \tag{64}$$

We directly construct approximate solutions of the problem (59)–(63). More precisely, we first find ρ^k from (59) and (63) with smooth function u^{k-1}, i.e.,

$$\rho_t^k + \rho_x^k u^{k-1} + \rho^k u_x^{k-1} = 0, \tag{65}$$

and

$$\rho^k\big|_{t=0} = \rho_0, \quad \rho_0 \geq \delta. \tag{66}$$

It follows from the classical linear hyperbolic theory that there is a unique solution ρ^k on this above initial problem. Using the method of characteristics, we have

$$\frac{d\rho^k}{dt} = -\rho^k u_x^{k-1}, \tag{67}$$

$$\frac{dx}{dt} = u^{k-1}(x,t), \tag{68}$$

$$x|_{t=0} = x_0, \qquad \rho^k\big|_{t=0} = \rho_0. \tag{69}$$

By (68) and (69), we have

$$x(t) = x_0 + \int_0^t u^{k-1}(x(s), s)\, ds = U(x_0, t) \tag{70}$$

Using (67), then

$$d\left(\ln \rho^k\right) = -u_x^{k-1} dt \tag{71}$$

which means

$$\rho^k(x,t) \geq \delta \exp\left[-\int_0^{T_1} \left|u_x^{k-1}(\cdot, s)\right|_{L^\infty} ds\right] > 0, \quad \text{for all } t \in (0, T_1). \tag{72}$$

Next, combining the classical stableness results of the elliptic equation, the existence of Ψ^k can be obtained by (61) and (63), then by (62) and (63) we get η^k. The last, with ρ^k, Ψ^k, η^k being given, by virtue of (72), from (60) and (63), according to the classical theorem of quasi-linear parabolic equation (see [17], Chapter VI, Theorem 5.2), there exists a unique smooth solution u^k. With the process, the nonlinear coupled system has been deduced into a sequence of decoupled problems and each problem admits a smooth solution. And the following estimates hold:

$$\begin{aligned}
&\operatorname*{ess\,sup}_{0 \leq t \leq T_1} \left(\left|\rho^k\right|_{H^1} + \left|u^k\right|_{W_0^{1,p} \cap H^2} + \left|\eta^k\right|_{H^2} + \left|\eta_t^k\right|_{L^2}\right. \\
&\left. + \left|\sqrt{\rho^k} u_t^k\right|_{L^2} + \left|\rho_t^k\right|_{L^2}\right) + \int_0^{T_1} \left(\left|\sqrt{\rho^k} u_t^k\right|_{L^2}^2 + \left|u_{xt}^k\right|_{L^2}^2\right. \\
&\left. + \left|\eta_x^k\right|_{L^2}^2 + \left|\eta_t^k\right|_{L^2}^2 + \left|\eta_{xt}^k\right|_{L^2}^2\right) ds \leq C
\end{aligned} \tag{73}$$

where C is a generic constant depending only on M_0, but independent of k.

Next, we have to prove that the approximate solution (ρ^k, u^k, η^k) converges to a solution to the original problem (14) in a strong sense. To this end, let us define

$$\begin{aligned}
\overline{\rho}^{k+1} &= \rho^{k+1} - \rho^k, \\
\overline{u}^{k+1} &= u^{k+1} - u^k, \\
\overline{\Psi}^{k+1} &= \Psi^{k+1} - \Psi^k, \\
\overline{\eta}^{k+1} &= \eta^{k+1} - \eta^k
\end{aligned} \tag{74}$$

then we can verify that the functions $\overline{\rho}^{k+1}, \overline{u}^{k+1}, \overline{\eta}^{k+1}$ satisfy the system of equations

$$\overline{\rho}_t^{k+1} + \left(\overline{\rho}^{k+1} u^k\right)_x + \left(\rho^k \overline{u}^k\right)_x = 0 \tag{75}$$

$$\begin{aligned}
&\rho^{k+1} \overline{u}_t^{k+1} + \rho^{k+1} u^k \overline{u}_x^{k+1} + \left(L_p u^{k+1} - L_p u^k\right) \\
&= -\overline{\rho}^{k+1}\left(u_t^k + u^k u_x^k + \Psi_x^{k+1}\right) - \left(P_x^{k+1} - P_x^k\right) \\
&+ \rho^k \left(\overline{u}^k u_x^k - \overline{\Psi}_x^{k+1}\right) - \overline{\eta}_x^{k+1} - \overline{\eta}^{k+1} \Phi_x
\end{aligned} \tag{76}$$

$$L_q \Psi^{k+1} - L_q \Psi^k = -4\pi g \overline{\rho}^{k+1} \tag{77}$$

$$\overline{\eta}_t^{k+1} + \left(\eta^k \overline{u}^k\right)_x + \left(\overline{\eta}^{k+1}\left(u^k - \Phi_x\right)\right)_x = \overline{\eta}_{xx}^{k+1} \tag{78}$$

Multiplying (75) by $\overline{\rho}^{k+1}$, integrating over Ω, and using Young's inequality, we obtain

$$\begin{aligned}
\frac{d}{dt}\left|\overline{\rho}^{k+1}\right|_{L^2}^2 &\leq C\left|\overline{\rho}^{k+1}\right|_{L^2}^2 \left|u_x^k\right|_{L^\infty} + \left|\rho^k\right|_{H^1} \left|\overline{u}_x^k\right|_{L^2} \left|\overline{\rho}^{k+1}\right|_{L^2} \\
&\leq C\left|u_{xx}^k\right|_{L^2} \left|\overline{\rho}^{k+1}\right|_{L^2}^2 + C_\zeta \left|\rho^k\right|_{H^1}^2 \left|\overline{\rho}^{k+1}\right|_{L^2}^2 \\
&+ \zeta \left|\overline{u}_x^k\right|_{L^2}^2 \leq C_\zeta \left|\overline{\rho}^{k+1}\right|_{L^2}^2 + \zeta \left|\overline{u}_x^k\right|_{L^2}^2
\end{aligned} \tag{79}$$

where C_ζ is a positive constant, depending on M_0 and ζ for all $t < T_1$ and $k \geq 1$.

Multiplying (76) by $\overline{u}^{k+1}$, integrating over Ω, and using Young's inequality, we obtain

$$\begin{aligned}
&\frac{1}{2}\frac{d}{dt}\int_\Omega \rho^{k+1} \left|\overline{u}^{k+1}\right|^2 dx + \int_\Omega \left(L_p u^{k+1} - L_p u^k\right) \overline{u}^{k+1} dx \\
&\leq C \int_\Omega \left(\left|\overline{\rho}^{k+1}\right| \left(\left|u_t^k\right| + \left|u^k u_x^k\right| + \left|\Psi_x^{k+1}\right|\right) \left|\overline{u}^{k+1}\right|\right. \\
&+ \left|P_x^{k+1} - P_x^k\right| \left|\overline{u}^{k+1}\right| + \left|\rho^k \overline{u}^k\right| \left|u_x^k\right| \left|\overline{u}^{k+1}\right| \\
&+ \left|\rho^k\right| \left|\overline{\Psi}_x^{k+1}\right| \left|\overline{u}^{k+1}\right| + \left|\overline{\eta}_x^{k+1}\right| \left|\overline{u}^{k+1}\right| \\
&\left. + \left|\overline{\eta}^{k+1} \Phi_x\right| \left|\overline{u}^{k+1}\right|\right) dx
\end{aligned} \tag{80}$$

Let

$$\sigma(s) = \left(s^2 + \mu_1\right)^{(p-2)/2} s \tag{81}$$

and then

$$\sigma'(s) = \left(\left(s^2+\mu_1\right)^{(p-2)/2} s\right)' = \left(s^2+\mu_1\right)^{(p-4)/2}\left((p-1)s^2+\mu_1\right) \geq \mu_1^{(p-2)/2} \tag{82}$$

We estimate the second term of (80) as follows:

$$\begin{aligned}\int_\Omega \left(L_p u^{k+1} - L_p u^k\right)\overline{u}^{k+1} dx &= \int_\Omega \int_0^1 \sigma'\left(\theta u_x^{k+1} + (1-\theta)u_x^k\right) d\theta \left|\overline{u}_x^{k+1}\right|^2 dx \\ &\geq \mu_1^{(p-2)/2}\int_\Omega \left|\overline{u}_x^{k+1}\right|^2 dx\end{aligned} \tag{83}$$

Similarly, we have

$$\begin{aligned}\mu_2^{(q-2)/2}\int_\Omega \left(\overline{\Psi}_x^{k+1}\right)^2 dx &\leq \int_\Omega \left(L_q\Psi^{k+1} - L_q\Psi^k\right)\overline{\Psi}^{k+1} dx \\ &= 4\pi g\int_\Omega \overline{\rho}^{k+1}\overline{\Psi}^{k+1} dx\end{aligned} \tag{84}$$

and then we have

$$\left|\overline{\Psi}_x^{k+1}\right|_{L^2}^2 \leq C\left|\overline{\rho}^{k+1}\right|_{L^2}^2 \tag{85}$$

Substituting (83) into (80) and using Young's inequality, we have

$$\begin{aligned}&\frac{d}{dt}\int_\Omega \rho^{k+1}\left|\overline{u}^{k+1}\right|^2 dx + \int_\Omega \left|\overline{u}_x^{k+1}\right|^2 dx \\ &\leq C\Big(\left|\overline{\rho}^{k+1}\right|_{L^2}\left|u_{xt}^k\right|_{L^2}\left|\overline{u}_x^{k+1}\right|_{L^2} \\ &+ \left|\overline{\rho}^{k+1}\right|_{L^2}\left|u_x^k\right|_{L^p}\left|u_{xx}^k\right|_{L^2}\left|\overline{u}_x^{k+1}\right|_{L^2} \\ &+ \left|\overline{\rho}^{k+1}\right|_{L^2}\left|\Psi_x^{k+1}\right|_{L^2}\left|\overline{u}_x^{k+1}\right|_{L^2} + \left|P^{k+1}-P^k\right|_{L^2}\left|\overline{u}_x^{k+1}\right|_{L^2} \\ &+ \left|\rho^k\right|_{L^2}^{1/2}\left|\sqrt{\rho^k}\overline{u}^k\right|_{L^2}\left|u_{xx}^k\right|_{L^2}\left|\overline{u}_x^{k+1}\right|_{L^2} \\ &+ \left|\rho^k\right|_{L^\infty}\left|\overline{\Psi}_x^{k+1}\right|_{L^2}\left|\overline{u}_x^{k+1}\right|_{L^2} + \left|\overline{\eta}^{k+1}\right|_{L^2}\left|\overline{u}_x^{k+1}\right|_{L^2} \\ &+ \left|\overline{\eta}^{k+1}\right|_{L^2}\left|\overline{u}_x^{k+1}\right|_{L^2}\Big) \leq B_\zeta(t)\left|\overline{\rho}^{k+1}\right|_{L^2}^2 \\ &+ C\left(\left|\sqrt{\rho^k}\overline{u}^k\right|_{L^2}^2 + \left|\overline{\eta}^{k+1}\right|_{L^2}^2\right) + \zeta\left|\overline{u}_x^{k+1}\right|_{L^2}^2\end{aligned} \tag{86}$$

where $B_\zeta(t) = C(1+|u_{xt}^k(t)|_{L^2}^2)$, for all $t \leq T_1$ and $k \geq 1$. Using (73) we derive

$$\int_0^t B_\zeta(s)\, ds \leq C + Ct \tag{87}$$

Multiplying (78) by $\overline{\eta}^{k+1}$, integrating over Ω, and using (73) and Young's inequality, we have

$$\begin{aligned}&\frac{1}{2}\frac{d}{dt}\int_\Omega \left|\overline{\eta}^{k+1}\right|^2 dx + \int_\Omega \left|\overline{\eta}_x^{k+1}\right|^2 dx \\ &\leq \int_\Omega \left|\overline{\eta}^{k+1}\right|\left|u^k - \Phi_x\right|\left|\overline{\eta}_x^{k+1}\right| dx \\ &+ \int_\Omega \left(\left|\eta^k\right|\left|\overline{u}^k\right|\right)_x\left|\overline{\eta}^{k+1}\right| dx \\ &\leq \left|\overline{\eta}^{k+1}\right|_{L^2}\left|u^k - \Phi_x\right|_{L^\infty}\left|\overline{\eta}_x^{k+1}\right|_{L^2} \\ &+ \left|\eta_x^k\right|_{L^2}\left|\overline{u}^k\right|_{L^\infty}\left|\overline{\eta}^{k+1}\right|_{L^2} + \left|\eta^k\right|_{L^\infty}\left|\overline{u}_x^k\right|_{L^2}\left|\overline{\eta}^{k+1}\right|_{L^2} \\ &\leq C_\zeta\left|\overline{\eta}^{k+1}\right|_{L^2}^2 + \zeta\left|\overline{\eta}_x^{k+1}\right|_{L^2}^2 + \zeta\left|\overline{u}_x^k\right|_{L^2}^2\end{aligned} \tag{88}$$

Collecting (79), (86), and (88), we obtain

$$\begin{aligned}&\frac{d}{dt}\left(\left|\overline{\rho}^{k+1}(t)\right|_{L^2}^2 + \left|\sqrt{\rho^{k+1}}\overline{u}^{k+1}(t)\right|_{L^2}^2 + \left|\overline{\eta}^{k+1}(t)\right|_{L^2}^2\right) \\ &+ \left|\overline{u}_x^{k+1}(t)\right|_{L^2}^2 + \left|\overline{\eta}_x^{k+1}\right|_{L^2}^2 \\ &\leq E_\zeta(t)\left|\overline{\rho}^{k+1}(t)\right|_{L^2}^2 + C\left|\sqrt{\rho^k}\overline{u}^k\right|_{L^2}^2 + C_\zeta\left|\overline{\eta}^{k+1}\right|_{L^2}^2 \\ &+ \zeta\left|\overline{u}_x^k\right|_{L^2}^2\end{aligned} \tag{89}$$

where $E_\zeta(t)$ depends only on $B_\zeta(t)$ and C_ζ, for all $t \leq T_1$ and $k \geq 1$. Using (73), we have

$$\int_0^t E_\zeta(s)\, ds \leq C + C_\zeta t \tag{90}$$

Integrating (89) over $(0,t) \subset (0,T_1)$ with respect to t, using Gronwall's inequality, we have

$$\begin{aligned}&\left|\overline{\rho}^{k+1}(t)\right|_{L^2}^2 + \left|\sqrt{\rho^{k+1}}\overline{u}^{k+1}(t)\right|_{L^2}^2 + \left|\overline{\eta}^{k+1}(t)\right|_{L^2}^2 \\ &+ \int_0^t \left|\overline{u}_x^{k+1}(t)\right|_{L^2}^2 ds + \int_0^t \left|\overline{\eta}_x^{k+1}\right|_{L^2}^2 ds \\ &\leq C\exp\left(C_\zeta t\right)\int_0^t \left(\left|\sqrt{\rho^k}\overline{u}^k(s)\right|_{L^2}^2 + \left|\overline{u}_x^k(s)\right|_{L^2}^2\right) ds\end{aligned} \tag{91}$$

from the above recursive relation, choose $\zeta > 0$ and $0 < T_* < T_1$ such that $C\exp(C_\zeta T_*) < 1/2$, using Gronwall's inequality, we deduce that

$$\begin{aligned}&\sum_{k=1}^K \Bigg[\sup_{0\leq t\leq T_*}\left(\left|\overline{\rho}^{k+1}(t)\right|_{L^2}^2 + \left|\sqrt{\rho^{k+1}}\overline{u}^{k+1}(t)\right|_{L^2}^2 + \left|\overline{\eta}^{k+1}(t)\right|_{L^2}^2\right) \\ &+ \int_0^{T_*}\left|\overline{u}_x^{k+1}(t)\right|_{L^2}^2 dt + \int_0^{T_*}\left|\overline{\eta}_x^{k+1}(t)\right|_{L^2}^2 dt\Bigg] < C\end{aligned} \tag{92}$$

Since all of the constants do not depend on δ, as $k \longrightarrow \infty$, we conclude that sequence (ρ^k, u^k, η^k) converges to a limit $(\rho^\delta, u^\delta, \eta^\delta)$ in the following convergence:

$$\rho \longrightarrow \rho^\delta \text{ in } L^\infty\left(0, T_*; L^2(\Omega)\right), \quad (93)$$

$$u \longrightarrow u^\delta \text{ in } L^\infty\left(0, T_*; L^2(\Omega)\right) \cap L^2\left(0, T_*; H_0^1(\Omega)\right), \quad (94)$$

$$\eta \longrightarrow \eta^\delta \text{ in } L^\infty\left(0, T_*; L^2(\Omega)\right) \cap L^2\left(0, T_*; H^1(\Omega)\right), \quad (95)$$

and there also holds

$$\begin{aligned} &\operatorname*{ess\,sup}_{0\le t\le T_1} \left(\left|\rho^\delta\right|_{H^1} + \left|u^\delta\right|_{W_0^{1,p}\cap H^2} + \left|\eta^\delta\right|_{H^2} + \left|\eta_t^\delta\right|_{L^2} \right. \\ &\quad \left. + \left|\sqrt{\rho^\delta} u_t^\delta\right|_{L^2} + \left|\rho_t^\delta\right|_{L^2} \right) + \int_0^{T_*} \left(\left|\sqrt{\rho^\delta} u_t^\delta\right|_{L^2}^2 + \left|u_{xt}^\delta\right|_{L^2}^2 \right. \\ &\quad \left. + \left|\eta_x^\delta\right|_{L^2}^2 + \left|\eta_t^\delta\right|_{L^2}^2 + \left|\eta_{xt}^\delta\right|_{L^2}^2 \right) ds \le C. \end{aligned} \quad (96)$$

For each small $\delta > 0$, let $\rho_0^\delta = J_\delta * \rho_0 + \delta$, J_δ is a mollifier on Ω, and $u_0^\delta \in H_0^1(\Omega) \cap H^2(\Omega)$ is a smooth solution of the boundary value problem:

$$\begin{aligned} &L_p u_0^\delta + \left(P\left(\rho_0^\delta\right) + \eta_0^\delta\right)_x + \eta_0^\delta \Phi_x \\ &\quad = \left(\rho_0^\delta\right)^{1/2} \left(g^\delta + \beta \Phi_x\right), \\ &u_0^\delta(0) = u_0^\delta(1) = 0, \end{aligned} \quad (97)$$

where $g^\delta \in C_0^\infty$ and satisfies $|g^\delta|_{L^2} \le |g|_{L^2}$, $\lim_{\delta \longrightarrow 0^+} |g^\delta - g|_{L^2} = 0$.

We deduce that $(\rho^\delta, u^\delta, \eta^\delta)$ is a solution of the following initial boundary value problem:

$$\begin{aligned} &\rho_t + (\rho u)_x = 0, \\ &(\rho u)_t + \left(\rho u^2\right)_x + \rho \Psi_x - \lambda \left[\left(u_x^2 + \mu_1\right)^{(p-2)/2} u_x\right]_x \\ &\quad + (P + \eta)_x = -\eta \Phi_x, \\ &\left[\left(\Psi_x^2 + \mu_2\right)^{(q-2)/2} \Psi_x\right]_x = 4\pi g \left(\rho - \frac{1}{|\Omega|} \int_\Omega \rho \, dx\right), \\ &\eta_t + \left(\eta \left(u - \Phi_x\right)\right)_x = \eta_{xx}, \\ &(\rho, u, \eta)|_{t=0} = \left(\rho_0^\delta, u_0^\delta, \eta_0^\delta\right), \\ &u|_{\partial\Omega} = \Psi|_{\partial\Omega} = \left(\eta_x + \eta \Phi_x\right)|_{\partial\Omega} = 0. \end{aligned} \quad (98)$$

where $\rho_0^\delta \ge \delta$, $p, q > 2$.

By the proof of Lemma 2.3 in [20], there exists a subsequence $\{u_0^{\delta_j}\}$ of $\{u_0^\delta\}$, as $\delta_j \longrightarrow 0^+$, $u_0^\delta \longrightarrow u_0$ in $H_0^1(\Omega) \cap H^2(\Omega)$, $-(|u_{0x}^{\delta_j}|^{p-2} u_{0x}^{\delta_j})_x \longrightarrow -(|u_{0x}|^{p-2} u_{0x})_x$ in $L^2(\Omega)$, Hence, u_0 satisfies (18) of Theorem 1. By virtue of the lower semi-continuity of various norms, we deduce that (ρ, u, η) satisfies the following uniform estimate:

$$\begin{aligned} &\operatorname*{ess\,sup}_{0\le t\le T_1} \left(|\rho|_{H^1} + |u|_{W_0^{1,p}\cap H^2} + |\eta|_{H^2} + |\eta_t|_{L^2} \right. \\ &\quad \left. + \left|\sqrt{\rho} u_t\right|_{L^2} + |\rho_t|_{L^2} \right) + \int_0^{T_*} \left(\left|\sqrt{\rho} u_t\right|_{L^2}^2 + |u_{xt}|_{L^2}^2 \right. \\ &\quad \left. + |\eta_x|_{L^2}^2 + |\eta_t|_{L^2}^2 + |\eta_{xt}|_{L^2}^2 \right) ds \le C, \end{aligned} \quad (99)$$

where C is a positive constant, depending only on M_0.

The uniqueness of solution can be obtained by the same method as the above proof of convergence; we omit the details here. This completes the proof.

Conflicts of Interest

The authors declare that there are no conflicts of interest regarding the publication of this paper.

Acknowledgments

This work was supported by the National Natural Science Foundation of China (nos. 11526105; 11572146) and the Funds of Education Department of Liaoning Province (JQL201715411).

References

[1] A. A. Anthony, "Kiva-3V Release2, Improvements to Kiva-3V," Tech. Rep., Los Alamos National Laboratory, 1999.

[2] S. Berres, R. Bürger, K. H. Karlsen, and E. M. Tory, "Strongly degenerate parabolic-hyperbolic systems modeling polydisperse sedimentation with compression," *SIAM Journal on Applied Mathematics*, vol. 64, no. 1, pp. 41–80, 2003.

[3] L. Boudin, L. Desvillettes, and R. Motte, "A modeling of compressible droplets in a fluid," *Communications in Mathematical Sciences*, vol. 1, no. 4, pp. 657–669, 2003.

[4] I. Vinkovic, C. Aguirre, S. Simoëns, and M. Gorokhovski, "Large eddy simulation of droplet dispersion for inhomogeneous turbulent wall flow," *International Journal of Multiphase Flow*, vol. 32, no. 3, pp. 344–364, 2006.

[5] J. A. Carrillo and T. Goudon, "Stability and asymptotic analysis of a fluid-particle interaction model," *Communications in Partial Differential Equations*, vol. 31, no. 7-9, pp. 1349–1379, 2006.

[6] J. A. Carrillo, T. Karper, and K. Trivisa, "On the dynamics of a fluid-particle interaction model: the bubbling regime," *Nonlinear Analysis. Theory, Methods & Applications. An International Multidisciplinary Journal*, vol. 74, no. 8, pp. 2778–2801, 2011.

[7] J. Yin and Z. Tan, "Global existence of strong solutions of Navier-Stokes-Poisson equations for one-dimensional isentropic compressible fluids," *Chinese Annals of Mathematics, Series B*, vol. 29, no. 4, pp. 441–458, 2008.

[8] D. Fang, R. Zi, and T. Zhang, "Global classical large solutions to a 1D fluid-particle interaction model: the bubbling regime," *Journal of Mathematical Physics*, vol. 53, no. 3, 033706, 21 pages, 2012.

[9] J. Ballew and K. Trivisa, "Suitable weak solutions and low stratification singular limit for a fluid particle interaction model," *Quarterly of Applied Mathematics*, vol. 70, no. 3, pp. 469–494, 2012.

[10] J. Ballew and K. Trivisa, "Weakly dissipative solutions and weak-strong uniqueness for the Navier-Stokes-SMOluchowski system," *Nonlinear Analysis. Theory, Methods & Applications. An International Multidisciplinary Journal*, vol. 91, pp. 1–19, 2013.

[11] T. Huang, C. Wang, and H. Wen, "Strong solutions of the compressible nematic liquid crystal flow," *Journal of Differential Equations*, vol. 252, no. 3, pp. 2222–2265, 2012.

[12] H. Wen and C. Zhu, "Blow-up criterions of strong solutions to 3D compressible Navier-Stokes equations with vacuum," *Advances in Mathematics*, vol. 248, pp. 534–572, 2013.

[13] S. Jiang and Y. Ou, "A blow-up criterion for compressible viscous heat-conductive flows," *Acta Mathematica Scientia B*, vol. 30, no. 6, pp. 1851–1864, 2010.

[14] G. Bohme, *Non-Newtonian Fluid Mechanics*, vol. 31 of *North-Holland Series in Applied Mathematics and Mechanics*, North-Holland, Amsterdam, The Netherlands, 1987.

[15] O. A. Ladyzhenskaya, "New equations for the description of viscous incompressible fluids and solvability in the large of the boundary value problems for them," in *Boundary Value Problems of Mathematical Physics*, vol. V, American Mathematical Society, Providence, RI, USA, 1970.

[16] O. A. Ladyzhenskaya, *The Mathematical Theory of Viscous Incompressible Flow*, Gordon and Breach, New York, NY, USA, 1969.

[17] O. A. Ladyzenskaja, V. A. Solonnikov, and N. N. Uralceva, *Linear and Quasilinear Equations of Parabolic Type*, Translations of Mathematical Monographs, American Mathematical Society, Providence, RI, USA, 1968.

[18] J. Málek, J. Necas, Rokyta. M., and M. Ružicka, *Weak and Measure-v alued Solutions to Evolutionary PDEs*, Chapman-Hall, New York, NY, USA, 1996.

[19] A. E. Mamontov, "Global regularity estimates for multidimensional equations of compressible non-Newtonian fluids," *Annali dell'Università di Ferrara. Nuova Serie. Sezione VII. Scienze Matematiche*, vol. 46, pp. 139–160, 2000.

[20] H. Yuan and X. Xu, "Existence and uniqueness of solutions for a class of non-Newtonian fluids with singularity and vacuum," *Journal of Differential Equations*, vol. 245, no. 10, pp. 2871–2916, 2008.

[21] O. Rozanova, "Nonexistence results for a compressible non-Newtonian fluid with magnetic effects in the whole space," *Journal of Mathematical Analysis and Applications*, vol. 371, no. 1, pp. 190–194, 2010.

[22] Y. Song, H. Yuan, Y. Chen, and Z. Guo, "Strong solutions for a 1D fluid-particle interaction non-newtonian model: the bubbling regime," *Journal of Mathematical Physics*, vol. 54, no. 9, Article ID 091501, 12 pages, 2013.

[23] Y. Song, H. Wang, Y. Chen, and Y. Zhang, "The strong solutions for a class of fluid-particle interaction non-Newtonian models," *Boundary Value Problems*, Paper No. 108, 17 pages, 2016.

[24] Y. Cho, H. J. Choe, and H. Kim, "Unique solvability of the initial boundary value problems for compressible viscous fluids," *Journal de Mathématiques Pures et Appliquées*, vol. 83, no. 2, pp. 243–275, 2004.

[25] Y. Cho and H. Kim, "Existence results for viscous polytropic fluids with vacuum," *Journal of Differential Equations*, vol. 228, no. 2, pp. 377–411, 2006.

2

Simultaneous Determination of the Space-Dependent Source and the Initial Distribution in a Heat Equation by Regularizing Fourier Coefficients of the Given Measurements

Shufang Qiu,[1,2] Wen Zhang,[1,2] and Jianmei Peng[1]

[1]*School of Science, East China University of Technology, Nanchang, Jiangxi 330013, China*
[2]*Institute of Science and Engineering Computing, East China University of Technology, Nanchang 330013, China*

Correspondence should be addressed to Shufang Qiu; shfqiu@ecit.cn

Academic Editor: Pavel Kurasov

We consider an inverse problem for simultaneously determining the space-dependent source and the initial distribution in heat conduction equation. First, we study the ill-posedness of the inverse problem. Then, we construct a regularization problem to approximate the originally inverse problem and obtain the regularization solutions with their stability and convergence results. Furthermore, convergence rates of the regularized solutions are presented under a prior and a posteriori strategies for selecting regularization parameters. Results of numerical examples show that the proposed regularization method is stable and effective for the considered inverse problem.

1. Introduction

In the past decades, various classes of inverse heat conduction equation problems have been studied by many scholars including recovery of the initial temperature [1–5], reconstruction of the heat source [6–12], and identification of thermal diffusion coefficients [13, 14]. The inverse problems of heat equations such as the backward problems and the source reconstruction problems arise from various scientific and engineering fields, including heat conduction, hydrology, environmental controlling. It is worth noting that most of the existing literature considers recovery of only one unknown term or parameter. However, in many applications we hope to simultaneously reconstruct more than one unknown term from some overspecified conditions, which makes inverse problems very complicated. To the authors' knowledge, papers devoted to the simultaneous recovery problems are very limited. In [15], a numerical algorithm based on the fundamental solutions method is proposed to reconstruct the space-dependent heat source and the initial value simultaneously in an inverse heat conduction problem, which is transformed into a homogeneous backward problem and a Dirichlet boundary value problem for Poisson's equation. In [16], an iterative algorithm is proposed for reconstructing both the space-dependent source term and the initial value based on solving a sequence of well-posed direct problems for the heat equation. In [17, 18], the unknown initial temperature and heat source are reconstructed simultaneously from the temperature data at the final time and at a fixed internal location over the time interval. In [19], the identification of the space-dependent heat source and the heat flux at the left endpoint is studied by the Tikhonov regularization method with generalized cross validation criterion for one-dimensional inverse heat conduction problem. In [20], the authors studied the inverse problem of reconstructing the time- and space-dependent heat source and the Robin boundary condition from the measured final data. In [21], the authors considered an inverse problem to simultaneously reconstruct the time- and space-dependent heat source and the initial temperature distribution and established the conditional stability and uniqueness of the inverse problem, which is solved by the variational regularization method. In [22], the inverse problem of simultaneous determination of the time-dependent source term and the time-dependent coefficients in the heat equation is studied by the overspecified conditions of integral type. Recently, there has been a growing interest in inverse problems with fractional derivatives. In [23], the authors studied the inverse problem of the time-fractional diffusion equation in one-dimensional spatial space for determining the initial value and the heat flux on the boundary simultaneously and proved the uniqueness of the inverse problem by using the Laplace transform and the unique extension technique.

Motivated by the idea of [3, 5] for solving the backward problem of heat equation, we consider the inverse problem of heat equation to simultaneously determine the space-dependent source and the initial distribution from two final temperature measurements at two terminal times. This paper is organized as follows. The inverse problem is formulated in Section 2 with its ill-posedness. A regularization approximation problem is constructed to approximate the inverse problem, and regularized solutions are obtained with their stability and convergence rates in Section 3. Numerical examples are given in Section 4 to show the feasibility and validity of the proposed method. Finally, some conclusions are drawn in Section 5.

2. Problem Formulation and Its Ill-Posedness

2.1. Problem Formulation. Consider the initial-boundary problem of a nonhomogeneous heat conduction equation:

$$\begin{aligned} u_t - u_{xx} &= f(x), \quad (x,t) \in (0,\pi) \times [0,T], \\ u(0,t) &= u(\pi,t) = 0, \quad t \in [0,T], \\ u(x,0) &= \varphi(x), \quad x \in (0,\pi), \end{aligned} \tag{1}$$

where the initial distribution $\varphi(x)$ satisfies the compatibility condition: $\varphi(0) = \varphi(\pi) = 0$. As we all know, the direct problem is solving problem (1) for yielding $u(x,t)$ when the source $f(x)$ and the initial distribution $\varphi(x)$ are known. However, the inverse problem considered in this paper is to reconstruct $f(x)$ and $\varphi(x)$ from problem (1) and two additional final measurements

$$u(x,T_1) = g(x), \quad x \in (0,\pi), \tag{2}$$

and

$$u(x,T_2) = h(x), \quad x \in (0,\pi), \tag{3}$$

where $T_1 < T_2$.

2.2. Ill-Posedness. By the method of separation of variables, we obtain the solution to problem (1):

$$u(x,t) = \sum_{k=1}^{\infty} \left(e^{-k^2 t} \varphi_k + \frac{1 - e^{-k^2 t}}{k^2} f_k \right) \sin(kx), \tag{4}$$

where φ_k and f_k are Fourier coefficients of $\varphi(x)$ and $f(x)$, respectively. And they are defined by

$$\begin{aligned} \varphi_k &= \frac{2}{\pi} \int_0^{\pi} \varphi(x) \sin(kx)\, dx, \\ f_k &= \frac{2}{\pi} \int_0^{\pi} f(x) \sin(kx)\, dx. \end{aligned} \tag{5}$$

From the Fourier expansions of $g(x)$ and $h(x)$ with respect to the eigenfunctions $\sin(kx)$, it follows that

$$\begin{aligned} e^{-k^2 T_1} \varphi_k + \frac{1 - e^{-k^2 T_1}}{k^2} f_k &= g_k, \\ e^{-k^2 T_2} \varphi_k + \frac{1 - e^{-k^2 T_2}}{k^2} f_k &= h_k, \end{aligned} \tag{6}$$

where

$$\begin{aligned} g_k &= \frac{2}{\pi} \int_0^{\pi} g(x) \sin(kx)\, dx, \\ h_k &= \frac{2}{\pi} \int_0^{\pi} h(x) \sin(kx)\, dx. \end{aligned} \tag{7}$$

By simple calculations, from system (6) we obtain that

$$\begin{aligned} f_k &= \frac{k^2 \left(h_k - g_k e^{-k^2 (T_2 - T_1)} \right)}{1 - e^{-k^2 (T_2 - T_1)}}, \\ \varphi_k &= \frac{e^{k^2 T_1} \left[h_k \left(1 - e^{-k^2 T_1} \right) - g_k \left(1 - e^{-k^2 T_2} \right) \right]}{e^{-k^2 (T_2 - T_1)} - 1}. \end{aligned} \tag{8}$$

Thus, the solution to the inverse problem is obtained in the following form:

$$f(x) = \sum_{k=1}^{\infty} \frac{k^2 \left(h_k - g_k e^{-k^2 (T_2 - T_1)} \right)}{1 - e^{-k^2 (T_2 - T_1)}} \sin(kx), \tag{9}$$

$$\varphi(x) = \sum_{k=1}^{\infty} \frac{e^{k^2 T_1} \left[h_k \left(1 - e^{-k^2 T_1} \right) - g_k \left(1 - e^{-k^2 T_2} \right) \right]}{e^{-k^2 (T_2 - T_1)} - 1} \cdot \sin(kx). \tag{10}$$

Hence, the inverse problem has a unique solution $(f(x), \varphi(x))$ in $L^2(0,\pi)$ if and only if the Fourier coefficients of $g(x)$ and $h(x)$ satisfy that

$$\sum_{k=1}^{\infty} \left| \frac{k^2 \left(h_k - g_k e^{-k^2 (T_2 - T_1)} \right)}{1 - e^{-k^2 (T_2 - T_1)}} \right|^2 < \infty \tag{11}$$

and

$$\sum_{k=1}^{\infty} \left| \frac{e^{k^2 T_1} \left[h_k \left(1 - e^{-k^2 T_1} \right) - g_k \left(1 - e^{-k^2 T_2} \right) \right]}{e^{-k^2 (T_2 - T_1)} - 1} \right|^2 < \infty. \tag{12}$$

In other words, g_k and h_k must decay faster than $e^{-k^2T_1}$ as $k \to +\infty$ for guaranteeing the existence of the inverse solution.

Unfortunately, the two additional measurements are always contaminated by noise in practical applications. In the context, we denote the contaminated measurements of $g(x)$ and $h(x)$ by $g^\delta(x)$ and $h^\delta(x)$, which satisfy that

$$\left\|g^\delta(x) - g(x)\right\| \le \delta, \qquad \left\|h^\delta(x) - h(x)\right\| \le \delta. \tag{13}$$

Here, $\|\cdot\|$ represents the L^2-norm on the interval $(0,\pi)$, and δ is the noise level. Although $g^\delta(x)$ and $h^\delta(x)$ belong to the space $L^2(0,\pi)$, the Fourier coefficients $\{g_k^\delta\}$ and $\{h_k^\delta\}$ no longer satisfy inequalities (11) and (12) since the random noise does not decay and will be dramatically amplified by k^2 and $e^{k^2T_1}$. Therefore, the inverse problem is ill-posed for determining the heat source and the initial distribution simultaneously in $L^2(0,\pi)$.

3. Regularization Method

First, we construct a regularization approximation problem for simultaneously determining the heat source term and the initial distribution:

$$\begin{aligned} &u_t^\alpha - u_{xx}^\alpha = f^\alpha(x), \quad (x,t) \in (0,\pi)\times[0,T], \\ &u^\alpha(0,t) = u^\alpha(\pi,t) = 0, \quad t \in [0,T], \\ &u^\alpha(x,0) = \varphi^\alpha(x), \quad x \in (0,\pi), \end{aligned} \tag{14}$$

and $u^\alpha(x,t)$ satisfies that

$$u^\alpha(x,T_1) = \sum_{k=1}^{\infty} \frac{e^{-k^2T_1}}{\alpha e^{k^2} + e^{-k^2T_1}} g_k \sin(kx), \quad x \in (0,\pi), \tag{15}$$

$$u^\alpha(x,T_2) = \sum_{k=1}^{\infty} \frac{e^{-k^2T_2}}{\alpha e^{k^2} + e^{-k^2T_2}} h_k \sin(kx), \quad x \in (0,\pi), \tag{16}$$

where $\alpha \in (0,+\infty)$ is a regularization parameter. The superscript α is used to denote the dependency of the approximate solution on the regularization parameter. By using the method of separation of variables, we can easily derive the solution to the approximate problem (14):

$$u^\alpha(x,t) = \sum_{k=1}^{\infty}\left(\frac{1-e^{-k^2t}}{k^2} f_k^\alpha + \varphi_k^\alpha e^{-k^2t}\right)\sin(kx), \tag{17}$$

where

$$\begin{aligned} f_k^\alpha &= \frac{2}{\pi}\int_0^\pi f^\alpha(x)\sin(kx)\,dx, \\ \varphi_k^\alpha &= \frac{2}{\pi}\int_0^\pi \varphi^\alpha(x)\sin(kx)\,dx. \end{aligned} \tag{18}$$

From (15) and (16), we have

$$\begin{aligned} \frac{1-e^{-k^2T_1}}{k^2} f_k^\alpha + \varphi_k^\alpha e^{-k^2T_1} &= \frac{e^{-k^2T_1}}{\alpha e^{k^2} + e^{-k^2T_1}} g_k, \\ \frac{1-e^{-k^2T_2}}{k^2} f_k^\alpha + \varphi_k^\alpha e^{-k^2T_2} &= \frac{e^{-k^2T_2}}{\alpha e^{k^2} + e^{-k^2T_2}} h_k. \end{aligned} \tag{19}$$

Thus, the approximate solutions of the inverse problem are

$$f^\alpha(x) = \sum_{k=1}^{\infty} \frac{k^2\left(\left(e^{-k^2T_2}/\left(\alpha e^{k^2} + e^{-k^2T_2}\right)\right)h_k - \left(e^{-k^2T_1}/\left(\alpha e^{k^2} + e^{-k^2T_1}\right)\right)g_k e^{-k^2(T_2-T_1)}\right)}{1-e^{-k^2(T_2-T_1)}}\sin(kx), \tag{20}$$

$$\varphi^\alpha(x) = \sum_{k=1}^{\infty} \frac{e^{k^2T_1}\left[\left(e^{-k^2T_2}/\left(\alpha e^{k^2} + e^{-k^2T_2}\right)\right)h_k\left(1-e^{-k^2T_1}\right) - \left(e^{-k^2T_1}/\left(\alpha e^{k^2} + e^{-k^2T_1}\right)\right)g_k\left(1-e^{-k^2T_2}\right)\right]}{e^{-k^2(T_2-T_1)}-1}\sin(kx). \tag{21}$$

We firstly prove the posedness of problem (14)–(16). Since we have obtained the solution to problem (14)–(16), the existence and uniqueness of the solution are obvious. Therefore, we only need to prove the stability of the solution.

Before we prove the main theoretical results, we introduce a lemma.

Lemma 1. *For $T > 0$, $0 < T_1 < T_2$, and $\alpha \in (0,+\infty)$, the following inequalities are valid:*

(a) $1/(\alpha e^\lambda + e^{-T\lambda}) \le T/\alpha(1 + T(T/\alpha)^{1/(1+T)})$, $\lambda \in (-\infty,+\infty)$.

(b) $\lambda/(e^{\lambda T_2} - e^{\lambda T_1}) \le 1/(T_2 - T_1)$, $\lambda \ge 1$.

(c) $e^{\lambda T_2}/(e^{\lambda T_2} - e^{\lambda T_1}) \le e^{T_2}/(e^{T_2} - e^{T_1})$, $\lambda \ge 1$.

Proof. (a) We define the function $h(\lambda)$ by the formula

$$h(\lambda) = \frac{1}{\alpha e^\lambda + e^{-T\lambda}}, \quad \lambda > 0. \tag{22}$$

By simple computations, the function $h(\lambda)$ gets the maximum $h(\ln(T/\alpha)/(1+T))$ at the point $\lambda = \ln(T/\alpha)/(1+T)$; that is

$$h(\lambda) \le h\left(\frac{\ln(T/\alpha)}{1+T}\right) \le \frac{T}{\alpha\left(1+T(T/\alpha)^{1/(1+T)}\right)}. \tag{23}$$

(b)

$$\frac{\lambda}{e^{\lambda T_2}-e^{\lambda T_1}}=\frac{\lambda}{e^{\lambda(T_2-T_1)}-1}e^{-\lambda T_1}\le\frac{\lambda}{(\lambda(T_2-T_1)+1)-1}\le\frac{1}{T_2-T_1}. \tag{24}$$

(c) Set $F(\lambda)=e^{\lambda T_2}/(e^{\lambda T_2}-e^{\lambda T_1})$. After a simple calculation we can say $F'(\lambda)<0$, so the conclusion is true. □

Theorem 2. *Let $g(x), h(x), g^\delta(x), h^\delta(x) \in L^2(0,\pi)$ and satisfy (13), and then*

$$\left\|f^\alpha(x)-f^{\alpha,\delta}(x)\right\|_{L^2(0,\pi)}\le\frac{\sqrt{2\pi}}{T_2-T_1}\frac{T_2}{\alpha\left(1+T_2(T_2/\alpha)^{1/(1+T_2)}\right)}\delta, \tag{25}$$

and

$$\left\|\varphi^\alpha(x)-\varphi^{\alpha,\delta}(x)\right\|_{L^2(0,\pi)}\le\frac{\sqrt{2\pi}e^{T_2}}{e^{T_2}-e^{T_1}}\frac{T_2}{\alpha\left(1+T_2(T_2/\alpha)^{1/(1+T_2)}\right)}\delta; \tag{26}$$

i.e., regularized solutions $f^\alpha(x)$ and $\varphi^\alpha(x)$ are consecutively dependent on the final value data $g(x)$ and $h(x)$, where $f^{\alpha,\delta}(x)$ and $\varphi^{\alpha,\delta}(x)$ are regularized solutions corresponding to $g^\delta(x)$ and $h^\delta(x)$, respectively.

Proof. From the perturbation data $g^\delta(x)$ and $h^\delta(x)$, by simple calculations we obtain

$$f^{\alpha,\delta}(x)=\sum_{k=1}^{\infty}\frac{k^2\left(\left(e^{-k^2T_2}/\left(\alpha e^{k^2}+e^{-k^2T_2}\right)\right)h_k^\delta-\left(e^{-k^2T_1}/\left(\alpha e^{k^2}+e^{-k^2T_1}\right)\right)g_k^\delta e^{-k^2(T_2-T_1)}\right)}{1-e^{-k^2(T_2-T_1)}}\sin(kx), \tag{27}$$

$$\varphi^{\alpha,\delta}(x)=\sum_{k=1}^{\infty}\frac{e^{k^2T_1}\left[\left(e^{-k^2T_2}/\left(\alpha e^{k^2}+e^{-k^2T_2}\right)\right)h_k^\delta\left(1-e^{-k^2T_1}\right)-\left(e^{-k^2T_1}/\left(\alpha e^{k^2}+e^{-k^2T_1}\right)\right)g_k^\delta\left(1-e^{-k^2T_2}\right)\right]}{e^{-k^2(T_2-T_1)}-1}\sin(kx), \tag{28}$$

where

$$g_k^\delta=\frac{2}{\pi}\int_0^\pi g^\delta(x)\sin(kx)\,dx,\qquad h_k^\delta=\frac{2}{\pi}\int_0^\pi h^\delta(x)\sin(kx)\,dx. \tag{29}$$

From formulas (20) and (27) and Lemma 1, we can get

$$\begin{aligned}\left\|f^\alpha(x)-f^{\alpha,\delta}(x)\right\|_{L^2(0,\pi)}^2&=\left|\sum_{k=1}^{\infty}\frac{k^2}{1-e^{-k^2(T_2-T_1)}}\left[\frac{e^{-k^2T_2}}{\alpha e^{k^2}+e^{-k^2T_2}}\left(h_k-h_k^\delta\right)-\frac{e^{-k^2T_1}}{\alpha e^{k^2}+e^{-k^2T_1}}e^{-k^2(T_2-T_1)}\left(g_k-g_k^\delta\right)\right]\sin(kx)\right|^2\\&=\frac{\pi}{2}\sum_{k=1}^{\infty}\left|\frac{k^2}{e^{k^2T_2}-e^{k^2T_1}}\left[\frac{1}{\alpha e^{k^2}+e^{-k^2T_2}}\left(h_k-h_k^\delta\right)-\frac{1}{\alpha e^{k^2}+e^{-k^2T_1}}\left(g_k-g_k^\delta\right)\right]\right|^2\le\pi\sum_{k=1}^{\infty}\left[\frac{k^2}{e^{k^2T_2}-e^{k^2T_1}}\cdot\frac{1}{\alpha e^{k^2}+e^{-k^2T_2}}\right]^2\left[\left(h_k-h_k^\delta\right)^2+\left(g_k-g_k^\delta\right)^2\right]\\&\le\pi\left[\frac{1}{T_2-T_1}\frac{T_2}{\alpha\left(1+T_2(T_2/\alpha)^{1/(1+T_2)}\right)}\right]^2\left(\left\|h-h^\delta\right\|_{L^2(0,\pi)}^2+\left\|g-g^\delta\right\|_{L^2(0,\pi)}^2\right);\end{aligned} \tag{30}$$

i.e.,

$$\left\|f^\alpha(x)-f^{\alpha,\delta}(x)\right\|_{L^2(0,\pi)}\le\frac{\sqrt{2\pi}}{T_2-T_1}\frac{T_2}{\alpha\left(1+T_2(T_2/\alpha)^{1/(1+T_2)}\right)}\delta. \tag{31}$$

From formulas (21) and (28) and Lemma 1, we obtain

$$\begin{aligned}\left\|\varphi^\alpha(x)-\varphi^{\alpha,\delta}(x)\right\|_{L^2(0,\pi)}^2&=\left|\sum_{k=1}^{\infty}\frac{e^{k^2T_1}}{e^{-k^2(T_2-T_1)}-1}\left[\frac{e^{-k^2T_2}\left(1-e^{-k^2T_1}\right)}{\alpha e^{k^2}+e^{-k^2T_2}}\left(h_k-h_k^\delta\right)-\frac{e^{-k^2T_1}\left(1-e^{-k^2T_2}\right)}{\alpha e^{k^2}+e^{-k^2T_1}}\left(g_k-g_k^\delta\right)\right]\sin(kx)\right|^2=\frac{\pi}{2}\\&\cdot\sum_{k=1}^{\infty}\left|\frac{1}{e^{k^2T_2}-e^{k^2T_1}}\left[\frac{e^{k^2T_1}-1}{\alpha e^{k^2}+e^{-k^2T_2}}\left(h_k-h_k^\delta\right)\right.\right.\end{aligned}$$

$$- \frac{e^{k^2T_2} - 1}{\alpha e^{k^2} + e^{-k^2T_1}} \left(g_k - g_k^\delta\right)\Bigg]\Bigg|^2 \le \pi \sum_{k=1}^{\infty} \left[\frac{e^{k^2T_2}}{e^{k^2T_2} - e^{k^2T_1}} \cdot \frac{1}{\alpha e^{k^2} + e^{-k^2T_2}} \right]^2 \left[\left(h_k - h_k^\delta\right)^2 + \left(g_k - g_k^\delta\right)^2 \right]$$

$$\le \pi \left[\frac{e^{T_2}}{e^{T_2} - e^{T_1}} \frac{T_2}{\alpha \left(1 + T_2 \left(T_2/\alpha\right)^{1/(1+T_2)}\right)} \right]^2 \left(\left\|h - h^\delta\right\|_{L^2(0,\pi)}^2 + \left\|g - g^\delta\right\|_{L^2(0,\pi)}^2 \right); \tag{32}$$

i.e.,

$$\left\|\varphi^\alpha(x) - \varphi^{\alpha,\delta}(x)\right\|_{L^2(0,\pi)} \le \frac{\sqrt{2\pi}e^{T_2}}{e^{T_2} - e^{T_1}} \frac{T_2}{\alpha \left(1 + T_2 \left(T_2/\alpha\right)^{1/(1+T_2)}\right)} \delta. \tag{33}$$

□

Theorem 3. *Let $g(x), h(x) \in L^2(0,\pi)$, and $\alpha \in (0,T)$. $f(x)$ and $\varphi(x)$ are the true solutions of the inverse problem (1). If there exists a constant A such that*

$$\sum_{k=1}^{\infty} \left[\left(e^{k^2(1+T_2)} h_k\right)^2 + \left(e^{k^2(1+T_1)} g_k\right)^2 \right] \le A^2, \tag{34}$$

then

$$\left\|f(x) - f^\alpha(x)\right\|_{L^2(0,\pi)} \le \frac{\sqrt{\pi}}{T_2 - T_1} \frac{T_2 A}{1 + T_2 \left(T_2/\alpha\right)^{1/(1+T_2)}}. \tag{35}$$

$$\left\|\varphi(x) - \varphi^\alpha(x)\right\|_{L^2(0,\pi)} \le \frac{\sqrt{\pi}e^{T_2}}{e^{T_2} - e^{T_1}} \frac{T_2 A}{1 + T_2 \left(T_2/\alpha\right)^{1/(1+T_2)}}. \tag{36}$$

Proof. From formulas (9) and (20) and Lemma 1, we derive that

$$\left\|f(x) - f^\alpha(x)\right\|_{L^2(0,\pi)}^2 = \left| \sum_{k=1}^{\infty} \frac{k^2}{1 - e^{-k^2(T_2 - T_1)}} \left[\left(1 - \frac{e^{-k^2T_2}}{\alpha e^{k^2} + e^{-k^2T_2}}\right) h_k - \left(1 - \frac{e^{-k^2T_1}}{\alpha e^{k^2} + e^{-k^2T_1}}\right) e^{-k^2(T_2 - T_1)} g_k \right] \sin(kx) \right|^2$$

$$= \frac{\pi}{2} \sum_{k=1}^{\infty} \left| \frac{k^2}{e^{k^2T_2} - e^{k^2T_1}} \left[\frac{\alpha}{\alpha e^{k^2} + e^{-k^2T_2}} e^{k^2(1+T_2)} h_k - \frac{\alpha}{\alpha e^{k^2} + e^{-k^2T_1}} e^{k^2(1+T_1)} g_k \right] \right|^2 \le \frac{\pi}{2} \sum_{k=1}^{\infty} \left| \frac{k^2}{e^{k^2T_2} - e^{k^2T_1}} \cdot \frac{\alpha}{\alpha e^{k^2} + e^{-k^2T_2}} \left(e^{k^2(1+T_2)} h_k - e^{k^2(1+T_1)} g_k\right) \right|^2 \le \frac{\pi}{2}$$

$$\cdot \sum_{k=1}^{\infty} \left| \frac{k^2}{e^{k^2T_2} - e^{k^2T_1}} \frac{\alpha}{\alpha e^{k^2} + e^{-k^2T_2}} \left(e^{k^2(1+T_2)} h_k - e^{k^2(1+T_1)} g_k\right) \right|^2 \le \pi \left(\frac{1}{T_2 - T_1} \cdot \frac{T_2}{1 + T_2 \left(T_2/\alpha\right)^{1/(1+T_2)}} \right)^2 A^2. \tag{37}$$

i.e.,

$$\left\|f(x) - f^\alpha(x)\right\|_{L^2(0,\pi)} \le \frac{\sqrt{\pi}}{T_2 - T_1} \frac{T_2 A}{1 + T_2 \left(T_2/\alpha\right)^{1/(1+T_2)}}. \tag{38}$$

From formulas (10) and (21) and Lemma 1, we can obtain

$$\left\|\varphi(x) - \varphi^\alpha(x)\right\|_{L^2(0,\pi)}^2 = \left| \sum_{k=1}^{\infty} \frac{e^{k^2T_1}}{e^{-k^2(T_2 - T_1)} - 1} \left[\frac{\alpha e^{k^2}\left(e^{-k^2T_1} - 1\right)}{\alpha e^{k^2} + e^{-k^2T_2}} h_k + \frac{\alpha e^{k^2}\left(1 - e^{-k^2T_2}\right)}{\alpha e^{k^2} + e^{-k^2T_1}} g_k \right] \sin(kx) \right|^2 = \frac{\pi}{2}$$

$$\cdot \sum_{k=1}^{\infty} \left| \frac{e^{k^2(T_1+T_2)}}{e^{k^2T_2} - e^{k^2T_1}} \left[\frac{\alpha e^{k^2}\left(e^{-k^2T_1} - 1\right)}{\alpha e^{k^2} + e^{-k^2T_2}} h_k + \frac{\alpha e^{k^2}\left(1 - e^{-k^2T_2}\right)}{\alpha e^{k^2} + e^{-k^2T_1}} g_k \right] \right|^2 \le \pi \sum_{k=1}^{\infty} \left| \frac{e^{k^2T_2}}{e^{k^2T_2} - e^{k^2T_1}} \cdot \frac{\alpha}{\alpha e^{k^2} + e^{-k^2T_2}} \right|^2 \left[\left(e^{k^2(T_1+1)} h_k\right)^2 + \left(e^{k^2(T_1+1)} g_k\right)^2 \right]$$

$$\le \pi \left(\frac{e^{T_2}}{e^{T_2} - e^{T_1}} \frac{T_2}{1 + T_2 \left(T_2/\alpha\right)^{1/(1+T_2)}} \right)^2 A^2; \tag{39}$$

i.e.,

$$\left\|\varphi(x) - \varphi^\alpha(x)\right\|_{L^2(0,\pi)} \le \frac{\sqrt{\pi}e^{T_2}}{e^{T_2} - e^{T_1}} \frac{T_2 A}{1 + T_2 \left(T_2/\alpha\right)^{1/(1+T_2)}}. \tag{40}$$

□

Theorem 3 illustrates that $f^\alpha(x)$ and $\varphi^\alpha(x)$ are really regularized solutions, since $f^\alpha(x)$ and $\varphi^\alpha(x)$ approximate the true solutions $f(x)$ and $\varphi(x)$ as $\alpha \to 0$, respectively.

Theorem 4. *Under the conditions of Theorems 2 and 3, the choice $\alpha = \sqrt{2}\delta/A$ leads to asymptotically optimal estimates of the regularization solutions:*

$$\left\| f(x) - f^{\alpha,\delta}(x) \right\|_{L^2(0,\pi)} = O\left(\delta^{1/(1+T_2)}\right) \quad (41)$$

and

$$\left\| \varphi(x) - \varphi^{\alpha,\delta}(x) \right\|_{L^2((0,\pi))} = O\left(\delta^{1/(1+T_2)}\right). \quad (42)$$

Proof. From conclusions of Theorems 2 and 3, we have

$$\begin{aligned} &\left\| f(x) - f^{\alpha,\delta}(x) \right\|_{L^2(0,\pi)} \\ &\quad \le \left\| f(x) - f^{\alpha}(x) \right\|_{L^2(0,\pi)} \\ &\qquad + \left\| f^{\alpha}(x) - f^{\alpha,\delta}(x) \right\|_{L^2(0,\pi)} \\ &\quad \le \frac{\sqrt{\pi}}{T_2 - T_1} \frac{T_2 A}{1 + T_2 (T_2/\alpha)^{1/(1+T_2)}} \\ &\qquad + \frac{\sqrt{2\pi}}{T_2 - T_1} \frac{\delta T_2}{\alpha\left(1 + T_2 (T_2/\alpha)^{1/(1+T_2)}\right)} \end{aligned} \quad (43)$$

and

$$\begin{aligned} &\left\| \varphi(x) - \varphi^{\alpha,\delta}(x) \right\|_{L^2(0,\pi)} \\ &\quad \le \left\| \varphi(x) - \varphi^{\alpha}(x) \right\|_{L^2(0,\pi)} \\ &\qquad + \left\| \varphi^{\alpha}(x) - \varphi^{\alpha,\delta}(x) \right\|_{L^2(0,\pi)} \\ &\quad \le \frac{\sqrt{\pi} e^{T_2}}{e^{T_2} - e^{T_1}} \frac{T_2 A}{1 + T_2 (T_2/\alpha)^{1/(1+T_2)}} \\ &\qquad + \frac{\sqrt{2\pi} e^{T_2}}{e^{T_2} - e^{T_1}} \frac{\delta T_2}{\alpha\left(1 + T_2 (T_2/\alpha)^{1/(1+T_2)}\right)}. \end{aligned} \quad (44)$$

From the above inequalities, the choice $\alpha = \sqrt{2}\delta/A$ leads to the asymptotically optimal estimates:

$$\begin{aligned} \left\| f(x) - f^{\alpha,\delta}(x) \right\|_{L^2(0,\pi)} &= O\left(\delta^{1/(1+T)}\right), \\ \left\| \varphi(x) - \varphi^{\alpha,\delta}(x) \right\|_{L^2(0,\pi)} &= O\left(\delta^{1/(1+T_2)}\right). \end{aligned} \quad (45)$$

□

Theorem 4 gives a priori strategy for selecting regularization parameter. However, it is often undesirable to have the value of constant A in practical applications. It is well known that regularization parameter plays a very important role in solving ill-posed problems, and the effectiveness of a regularization method depends strongly on the choice of the regularization parameter. Here, we adopt the Morozov discrepancy principle [24–26], a posteriori strategy, to select the regularization parameter; that is, selecting the regularization parameter α makes

$$\begin{aligned} \max \Big\{ &\left\| u^{\alpha,\delta}(x, T_1) - g^{\delta}(x) \right\|^2, \\ &\left\| u^{\alpha,\delta}(x, T_2) - h^{\delta}(x) \right\|^2 \Big\} = C\delta^2, \end{aligned} \quad (46)$$

where C is a given constant. For regularization solutions $f^{\alpha,\delta}(x)$ and $\varphi^{\alpha,\delta}(x)$, the deviation equation (46) means that

$$\begin{aligned} \max \Bigg\{ &\sum_{k=1}^{\infty} \left(\frac{e^{-k^2 T_1}}{\alpha e^{k^2} + e^{-k^2 T_1}} g_k^{\delta} - g_k^{\delta} \right)^2, \\ &\sum_{k=1}^{\infty} \left(\frac{e^{-k^2 T_2}}{\alpha e^{k^2} + e^{-k^2 T_2}} h_k^{\delta} - h_k^{\delta} \right)^2 \Bigg\} = C\delta^2. \end{aligned} \quad (47)$$

Theorem 5. *Let $f(x)$ and $\varphi(x)$ be the true solutions of (1)–(3) with their Fourier coefficients f_k and g_k, respectively. If there exist constants $P > 0$ and E such that*

$$\begin{aligned} \sum_{k=1}^{\infty} \left(e^{k^2 T_1}\right)^P f_k^2 &\le E, \\ \sum_{k=1}^{\infty} \left(e^{k^2 T_1}\right)^P \varphi_k^2 &\le E, \end{aligned} \quad (48)$$

and $PT_1 < 2$, the a posteriori strategy (47) for selecting the regularization parameter leads to the following estimates:

$$\left\| f(x) - f^{\alpha,\delta}(x) \right\|_{L^2(0,\pi)}^2 = O\left(\delta^{\min\{2P/(P+2),2\}}\right), \quad (49)$$

$$\left\| \varphi(x) - \varphi^{\alpha,\delta}(x) \right\|_{L^2(0,\pi)}^2 = O\left(\delta^{\min\{2P/(P+2),2\}}\right), \quad (50)$$

where $f^{\alpha,\delta}(x)$ and $\varphi^{\alpha,\delta}(x)$ are regularization solutions with respect to a regularization parameter selected by the a posteriori strategy (47).

Proof. (1) By careful computations and the Hölder inequality, we can get

$$\begin{aligned} \left\| f(x) - f^{\alpha,\delta}(x) \right\|_{L^2(0,\pi)}^2 &= \sum_{k=1}^{\infty} \left| \frac{k^2}{1 - e^{-k^2(T_2 - T_1)}} \left[\left(h_k - \frac{e^{-k^2 T_2}}{\alpha e^{k^2} + e^{-k^2 T_2}} h_k^{\delta} \right) - e^{-k^2(T_2 - T_1)} \left(g_k - \frac{e^{-k^2 T_1}}{\alpha e^{k^2} + e^{-k^2 T_1}} g_k^{\delta} \right) \right] \right|^2 \\ &= \sum_{k=1}^{\infty} \left(\frac{k^2}{1 - e^{-k^2(T_2 - T_1)}} \right)^2 \left[\left(h_k - \frac{e^{-k^2 T_2}}{\alpha e^{k^2} + e^{-k^2 T_2}} h_k^{\delta} \right) - e^{-k^2(T_2 - T_1)} \left(g_k - \frac{e^{-k^2 T_1}}{\alpha e^{k^2} + e^{-k^2 T_1}} g_k^{\delta} \right) \right]^{4/(P+2)} \end{aligned}$$

$$
\begin{aligned}
&\cdot\left[\left(h_k-\frac{e^{-k^2T_2}}{\alpha e^{k^2}+e^{-k^2T_2}}h_k^\delta\right)-e^{-k^2(T_2-T_1)}\left(g_k-\frac{e^{-k^2T_1}}{\alpha e^{k^2}+e^{-k^2T_1}}g_k^\delta\right)\right]^{2-4/(P+2)}\\
&\le\left\{\sum_{k=1}^{\infty}\left[\left(\frac{k^2}{1-e^{-k^2(T_2-T_1)}}\right)^2\left[\left(h_k-\frac{e^{-k^2T_2}}{\alpha e^{k^2}+e^{-k^2T_2}}h_k^\delta\right)-e^{-k^2(T_2-T_1)}\left(g_k-\frac{e^{-k^2T_1}}{\alpha e^{k^2}+e^{-k^2T_1}}g_k^\delta\right)\right]^{4/(P+2)}\right]^{(P+2)/2}\right\}^{2/(P+2)}\\
&\cdot\left\{\left[\sum_{k=1}^{\infty}\left[\left(h_k-\frac{e^{-k^2T_2}}{\alpha e^{k^2}+e^{-k^2T_2}}h_k^\delta\right)-e^{-k^2(T_2-T_1)}\left(g_k-\frac{e^{-k^2T_1}}{\alpha e^{k^2}+e^{-k^2T_1}}g_k^\delta\right)\right]^{2-4/(P+2)}\right]^{(P+2)/P}\right\}^{P/(P+2)}.
\end{aligned}
\tag{51}
$$

From (13), we can get

$$
\begin{aligned}
&\left\{\sum_{k=1}^{\infty}\left[\left(\frac{k^2}{1-e^{-k^2(T_2-T_1)}}\right)^2\left[\left(h_k-\frac{e^{-k^2T_2}}{\alpha e^{k^2}+e^{-k^2T_2}}h_k^\delta\right)-e^{-k^2(T_2-T_1)}\left(g_k-\frac{e^{-k^2T_1}}{\alpha e^{k^2}+e^{-k^2T_1}}g_k^\delta\right)\right]^{4/(P+2)}\right]^{(P+2)/2}\right\}^{2/(P+2)}\\
&=\left\{\sum_{k=1}^{\infty}\left(\frac{k^2}{1-e^{-k^2(T_2-T_1)}}\right)^{P+2}\left[\left(h_k-\frac{e^{-k^2T_2}}{\alpha e^{k^2}+e^{-k^2T_2}}h_k^\delta\right)-e^{-k^2(T_2-T_1)}\left(g_k-\frac{e^{-k^2T_1}}{\alpha e^{k^2}+e^{-k^2T_1}}g_k^\delta\right)\right]^2\right\}^{2/(P+2)}\\
&=\left\{\sum_{k=1}^{\infty}\left(\frac{k^2}{1-e^{-k^2(T_2-T_1)}}\right)^{P+2}\left[\left(1-\frac{e^{-k^2T_2}}{\alpha e^{k^2}+e^{-k^2T_2}}\right)h_k+\frac{e^{-k^2T_2}}{\alpha e^{k^2}+e^{-k^2T_2}}\left(h_k-h_k^\delta\right)-e^{-k^2(T_2-T_1)}g_k\left(1-\frac{e^{-k^2T_1}}{\alpha e^{k^2}+e^{-k^2T_1}}\right)\right.\right.\\
&\left.\left.-e^{-k^2(T_2-T_1)}\frac{e^{-k^2T_1}}{\alpha e^{k^2}+e^{-k^2T_1}}\left(g_k-g_k^\delta\right)\right]^2\right\}^{2/(P+2)}\le\left\{2\sum_{k=1}^{\infty}\left(\frac{k^2}{1-e^{-k^2(T_2-T_1)}}\right)^{P+2}\left[\left(h_k-e^{-k^2(T_2-T_1)}g_k\right)^2\right.\right.\\
&\left.\left.+\left(\frac{e^{-k^2T_2}}{\alpha e^{k^2}+e^{-k^2T_2}}\left(h_k-h_k^\delta\right)+\frac{e^{-k^2T_2}}{\alpha e^{k^2}+e^{-k^2T_1}}\left(g_k-g_k^\delta\right)\right)^2\right]\right\}^{2/(P+2)}\le\left\{2\sum_{k=1}^{\infty}\left(\frac{k^2}{1-e^{-k^2(T_2-T_1)}}\right)^{P}f_k^2+4\sum_{k=1}^{\infty}\left(k^2\right)^{P+2}\right.\\
&\left.\cdot e^{-2k^2T_2}\delta^2\right\}^{2/(P+2)}\le\left(2^{P+1}E+4C_1\delta^2\right)^{2/(P+2)}
\end{aligned}
\tag{52}
$$

and

$$
\begin{aligned}
&\sum_{k=1}^{\infty}\left[\left(h_k-\frac{e^{-k^2T_2}}{\alpha e^{k^2}+e^{-k^2T_2}}h_k^\delta\right)-e^{-k^2(T_2-T_1)}\left(g_k-\frac{e^{-k^2T_1}}{\alpha e^{k^2}+e^{-k^2T_1}}g_k^\delta\right)\right]^{2-4/(P+2)}\\
&\quad\le\left\{\left[\sum_{k=1}^{\infty}\left[\left(h_k-\frac{e^{-k^2T_2}}{\alpha e^{k^2}+e^{-k^2T_2}}h_k^\delta\right)-e^{-k^2(T_2-T_1)}\left(g_k-\frac{e^{-k^2T_1}}{\alpha e^{k^2}+e^{-k^2T_1}}g_k^\delta\right)\right]^{2-4/(P+2)}\right]^{(P+2)/P}\right\}^{P/(P+2)}\\
&\quad\le\left\{2\sum_{k=1}^{\infty}\left[\left(h_k-h_k^\delta\right)-\left(\frac{e^{-k^2T_2}}{\alpha e^{k^2}+e^{-k^2T_2}}h_k^\delta-h_k^\delta\right)\right]^2\right.\\
&\quad\left.+2\sum_{k=1}^{\infty}\left[e^{-k^2(T_2-T_1)}\left(\left(g_k-g_k^\delta\right)-\left(\frac{e^{-k^2T_1}}{\alpha e^{k^2}+e^{-k^2T_1}}g_k^\delta-g_k^\delta\right)\right)\right]^2\right\}^{P/(P+2)}\le\left(8C\delta^2+8\delta^2\right)^{P/(P+2)}=C_2\delta^{2P/(P+2)}.
\end{aligned}
\tag{53}
$$

In summary,

$$\begin{aligned}&\left\|f(x)-f^{\alpha,\delta}(x)\right\|^2_{L^2(0,\pi)}\\&\le\left(2^{P+1}E+4C_1\delta^2\right)^{2/(P+2)}C_2\delta^{2P/(P+2)}\\&\le C_3E\delta^{2P/(P+2)}+C_4\delta^2=O\left(\delta^{\min\{2P/(P+2),2\}}\right).\end{aligned}\tag{54}$$

(2) Based on the Hölder inequality, we can similarly obtain that

$$\begin{aligned}\left\|\varphi(x)-\varphi^{\alpha,\delta}(x)\right\|^2_{L^2(0,\pi)}&=\sum_{k=1}^{\infty}\left|\frac{e^{k^2T_1}}{1-e^{-k^2(T_2-T_1)}}\left[\left(1-e^{-k^2T_1}\right)\left(\frac{e^{-k^2T_2}}{\alpha e^{k^2}+e^{-k^2T_2}}h_k^\delta-h_k\right)-\left(1-e^{-k^2T_2}\right)\right.\right.\\&\cdot\left.\left.\left(\frac{e^{-k^2T_1}}{\alpha e^{k^2}+e^{-k^2T_1}}g_k^\delta-g_k\right)\right]\right|^2=\sum_{k=1}^{\infty}\left(\frac{e^{k^2T_1}}{1-e^{-k^2(T_2-T_1)}}\right)^2\left[\left(1-e^{-k^2T_1}\right)\left(\frac{e^{-k^2T_2}}{\alpha e^{k^2}+e^{-k^2T_2}}h_k^\delta-h_k\right)-\left(1-e^{-k^2T_2}\right)\right.\\&\cdot\left.\left(\frac{e^{-k^2T_1}}{\alpha e^{k^2}+e^{-k^2T_1}}g_k^\delta-g_k\right)\right]^{4/(P+2)}\left[\left(1-e^{-k^2T_1}\right)\left(\frac{e^{-k^2T_2}}{\alpha e^{k^2}+e^{-k^2T_2}}h_k^\delta-h_k\right)-\left(1-e^{-k^2T_2}\right)\left(\frac{e^{-k^2T_1}}{\alpha e^{k^2}+e^{-k^2T_1}}g_k^\delta\right.\right.\\&\left.\left.-g_k\right)\right]^{2-4/(P+2)}\le\left\{\sum_{k=1}^{\infty}\left[\left(\frac{e^{k^2T_1}}{1-e^{-k^2(T_2-T_1)}}\right)^2\right.\right.\\&\cdot\left.\left.\left[\left(1-e^{-k^2T_1}\right)\left(\frac{e^{-k^2T_2}}{\alpha e^{k^2}+e^{-k^2T_2}}h_k^\delta-h_k\right)-\left(1-e^{-k^2T_2}\right)\left(\frac{e^{-k^2T_1}}{\alpha e^{k^2}+e^{-k^2T_1}}g_k^\delta-g_k\right)\right]^{4/(P+2)}\right]^{(P+2)/2}\right\}^{2/(P+2)}\\&\cdot\left\{\left[\sum_{k=1}^{\infty}\left[\left(1-e^{-k^2T_1}\right)\left(\frac{e^{-k^2T_2}}{\alpha e^{k^2}+e^{-k^2T_2}}h_k^\delta-h_k\right)-\left(1-e^{-k^2T_2}\right)\left(\frac{e^{-k^2T_1}}{\alpha e^{k^2}+e^{-k^2T_1}}g_k^\delta-g_k\right)\right]^{2-4/(P+2)}\right]^{2/(P+2)}\right\}^{P/(P+2)}\end{aligned}\tag{55}$$

From inequalities (13), we get

$$\begin{aligned}&\left\{\sum_{k=1}^{\infty}\left[\left(\frac{e^{k^2T_1}}{1-e^{-k^2(T_2-T_1)}}\right)^2\right.\right.\\&\cdot\left.\left.\left[\left(1-e^{-k^2T_1}\right)\left(\frac{e^{-k^2T_2}}{\alpha e^{k^2}+e^{-k^2T_2}}h_k^\delta-h_k\right)-\left(1-e^{-k^2T_2}\right)\left(\frac{e^{-k^2T_1}}{\alpha e^{k^2}+e^{-k^2T_1}}g_k^\delta-g_k\right)\right]^{4/(P+2)}\right]^{(P+2)/2}\right\}^{2/(P+2)}\\&\le2\left\{\sum_{k=1}^{\infty}\left(\frac{e^{k^2T_1}}{1-e^{-k^2(T_2-T_1)}}\right)^{P+2}\left[\left(\frac{e^{-k^2T_2}}{\alpha e^{k^2}+e^{-k^2T_2}}h_k^\delta-h_k\right)^2+\left(\frac{e^{-k^2T_1}}{\alpha e^{k^2}+e^{-k^2T_1}}g_k^\delta-g_k\right)^2\right]\right\}^{2/(P+2)}\\&\le2\left\{\sum_{k=1}^{\infty}\left(\frac{e^{k^2T_1}}{1-e^{-k^2(T_2-T_1)}}\right)^{P+2}\left[\left(\frac{e^{-k^2T_1}}{\alpha e^{k^2}+e^{-k^2T_1}}\left(g_k^\delta-g_k\right)-\left(1-\frac{e^{-k^2T_1}}{\alpha e^{k^2}+e^{-k^2T_1}}\right)g_k\right)^2\right.\right.\\&\left.\left.+\left(\frac{e^{-k^2T_2}}{\alpha e^{k^2}+e^{-k^2T_2}}\left(h_k^\delta-h_k\right)-\left(1-\frac{e^{-k^2T_2}}{\alpha e^{k^2}+e^{-k^2T_2}}\right)h_k\right)^2\right]\right\}^{2/(P+2)}\le4\left\{\sum_{k=1}^{\infty}\left(\frac{e^{k^2T_2}}{e^{k^2T_2}-e^{k^2T_1}}\right)^{P+2}\left[\frac{e^{k^2(PT_1-2)}}{\alpha^2}\delta^2\right.\right.\end{aligned}$$

$$+ e^{k^2T_1(P+2)}g_k^2 + \frac{e^{k^2(PT_1-2)}}{\alpha^2}\delta^2 + e^{k^2T_1(P+2)}h_k^2\Bigg]\Bigg\}^{2/(P+2)} \le 4\left\{\sum_{k=1}^{\infty}\left(\frac{e^{k^2T_2}}{e^{k^2T_2}-e^{k^2T_1}}\right)^{P+2}\left[2\frac{e^{k^2(PT_1-2)}}{\alpha^2}\delta^2 + e^{k^2T_1(P+2)}g_k^2\right.\right.$$

$$\left.\left. + e^{k^2T_1(P+2)}h_k^2\right]\right\}^{2/(P+2)} \le 4\left\{\sum_{k=1}^{\infty}\left(\frac{e^{k^2T_2}}{e^{k^2T_2}-e^{k^2T_1}}\right)^{P+2}\left[2C_1\delta^2 + e^{k^2T_1(P+2)}\left(e^{-k^2T_1}\varphi_k + \frac{1-e^{-k^2T_1}}{k^2}f_k\right)^2\right.\right.$$

$$\left.\left. + e^{k^2T_1(P+2)}\left(e^{-k^2T_2}\varphi_k + \frac{1-e^{-k^2T_2}}{k^2}f_k\right)^2\right]\right\}^{2/(P+2)} \le 4\left\{\sum_{k=1}^{\infty}\left(\frac{e^{k^2T_2}}{e^{k^2T_2}-e^{k^2T_1}}\right)^{P+2}\left[2C_1\delta^2\right.\right.$$

$$\left.\left. + 2e^{k^2T_1(P+2)}\left(2e^{-2k^2T_1}\varphi_k^2 + \frac{2e^{2k^2(T_2-T_1)}}{k^4}f_k^2\right)\right]\right\}^{2/(P+2)} \le 4\left\{\sum_{k=1}^{\infty}\left(\frac{e^{k^2T_2}}{e^{k^2T_2}-e^{k^2T_1}}\right)^{P+2}\left[2C_1\delta^2 + 4e^{k^2T_1P}\varphi_k^2 + 4\right.\right.$$

$$\left.\left. \cdot\frac{e^{k^2(T_1P+2T_2)}}{k^4}f_k^2\right]\right\}^{2/(P+2)} \le 4\left\{\sum_{k=1}^{\infty}\left(\frac{e^{T_2}}{e^{T_2}-e^{T_1}}\right)^{P+2}\left(2C_1\delta^2 + 4E + 4C_2E\right)\right\}^{2/(P+2)} \le 4\left(\frac{e^{T_2}}{e^{T_2}-e^{T_1}}\right)^2\left(2C_1\delta^2\right.$$

$$\left. + C_3E\right)^{2/(P+2)}, \tag{56}$$

and

$$\sum_{k=1}^{\infty}\left[\left(1-e^{-k^2T_1}\right)\left(\frac{e^{-k^2T_2}}{\alpha e^{k^2}+e^{-k^2T_2}}h_k^\delta - h_k\right) - \left(1-e^{-k^2T_2}\right)\left(\frac{e^{-k^2T_1}}{\alpha e^{k^2}+e^{-k^2T_1}}g_k^\delta - g_k\right)\right]^{2-4/(P+2)}$$

$$\le \left\{\left[\sum_{k=1}^{\infty}\left[\left(1-e^{-k^2T_1}\right)\left(\frac{e^{-k^2T_2}}{\alpha e^{k^2}+e^{-k^2T_2}}h_k^\delta - h_k\right) - \left(1-e^{-k^2T_2}\right)\left(\frac{e^{-k^2T_1}}{\alpha e^{k^2}+e^{-k^2T_1}}g_k^\delta - g_k\right)\right]^{2-4/(P+2)}\right]^{2/(P+2)}\right\}^{P/(P+2)}$$

$$\le \left\{2\sum_{k=1}^{\infty}\left[\left(1-e^{-k^2T_1}\right)\left(\left(\frac{e^{-k^2T_2}}{\alpha e^{k^2}+e^{-k^2T_2}}h_k^\delta - h_k^\delta\right) - \left(h_k - h_k^\delta\right)\right)\right]^2\right.$$

$$\left. + 2\sum_{k=1}^{\infty}\left[\left(1-e^{-k^2T_2}\right)\left(\left(\frac{e^{-k^2T_1}}{\alpha e^{k^2}+e^{-k^2T_1}}g_k^\delta - g_k^\delta\right) - \left(g_k - g_k^\delta\right)\right)\right]^2\right\}^{P/(P+2)} \le \left(8C\delta^2 + 8\delta^2\right)^{P/(P+2)} = C_4\delta^{2P/(P+2)}. \tag{57}$$

To sum up, we have

$$\left\|\varphi(x) - \varphi^{\alpha,\delta}(x)\right\|_{L^2(0,\pi)}^2 \le 4\left(\frac{e^{T_2}}{e^{T_2}-e^{T_1}}\right)^2\left(2C_1\delta^2 + C_3E\right)^{2/(P+2)}C_4\delta^{2P/(P+2)} \le C_5E\delta^{2P/(P+2)} + C_6\delta^2 = O\left(\delta^{\min\{2P/(P+2),2\}}\right). \tag{58}$$

□

4. Numerical Examples

In this section, we give two numerical examples to show effectiveness and stability of the proposed method: the first one has an analytical expression of the temperature concentration for the direct problem of heat conduction equation; the second one does not have an explicit solution of the direct problem, which must be solved by numerical methods such as the finite element method and the finite difference method.

In addition, the infinite summation in (27) and (28) must be truncated in practical computations. In the absence of confusion, we still denote the truncated solutions of regularization solutions by $f^{\alpha,\delta}(x)$ and $\varphi^{\alpha,\delta}(x)$; that is,

$$f^{\alpha,\delta}(x) = \sum_{k=1}^{m} \frac{k^2\left(\left(e^{-k^2T_2}/\left(\alpha e^{k^2}+e^{-k^2T_2}\right)\right)h_k^{\delta} - \left(e^{-k^2T_1}/\left(\alpha e^{k^2}+e^{-k^2T_1}\right)\right)g_k^{\delta}e^{-k^2(T_2-T_1)}\right)}{1-e^{-k^2(T_2-T_1)}}\sin(kx), \quad (59)$$

$$\varphi^{\alpha,\delta}(x) = \sum_{k=1}^{m} \frac{e^{k^2T_1}\left[\left(e^{-k^2T_2}/\left(\alpha e^{k^2}+e^{-k^2T_2}\right)\right)h_k^{\delta}\left(1-e^{-k^2T_1}\right) - \left(e^{-k^2T_1}/\left(\alpha e^{k^2}+e^{-k^2T_1}\right)\right)g_k^{\delta}\left(1-e^{-k^2T_2}\right)\right]}{e^{-k^2(T_2-T_1)}-1}\sin(kx). \quad (60)$$

Here, the Fourier coefficients g_k^{δ} and h_k^{δ} are computed by the trapezoid rule of numerical integral as follows:

$$\begin{aligned} g_k^{\delta} &= \frac{2}{\pi}\int_0^{\pi} g^{\delta}(x)\sin(kx)\,dx \\ &\approx \frac{2h}{\pi}\sum_{i=1}^{n-1} g^{\delta}(x_i)\sin(kx_i), \end{aligned} \quad (61)$$

and

$$\begin{aligned} h_k^{\delta} &= \frac{2}{\pi}\int_0^{\pi} h^{\delta}(x)\sin(kx)\,dx \\ &\approx \frac{2h}{\pi}\sum_{i=1}^{n-1} h^{\delta}(x_i)\sin(kx_i), \end{aligned} \quad (62)$$

where $g^{\delta}(x_i) = g(x_i) + \rho(2\mathrm{rand}(x_i) - 1)g(x_i)$, $h^{\delta}(x_i) = g(x_i) + \rho(2\mathrm{rand}(x_i) - 1)h(x_i)$, ρ is a relative error level, and rand is a random function in MATLAB that generates a uniformly distributed number over $(0,1)$. In all numerical experiments, we execute the Morozov discrepancy principle (46) for selecting regularization parameters as follows.

Let $r \in (0,1)$ be a fixed constant and $\alpha_0 > 0$ be an initial guess of the regularized parameter. Here, we consider regularization parameters as a geometric sequence form:

$$\alpha_k = \alpha_0 r^k, \quad k = 0,1,2,\cdots. \quad (63)$$

The a posteriori strategy for selecting regularization parameters is to select the regularization parameter α_{k^*-1} so that in the first step k^* has

$$\begin{aligned} \left\|u^{\alpha_{k^*},\delta}(x,T_1) - g^{\delta}(x)\right\|_{L^2(0,\pi)}^2 &\le C\delta^2, \\ \left\|u^{\alpha_{k^*},\delta}(x,T_2) - h^{\delta}(x)\right\|_{L^2(0,\pi)}^2 &\le C\delta^2 \end{aligned} \quad (64)$$

where $C \ge 1$ is a constant and $u^{\alpha_k,\delta}(x,T_1)$ and $u^{\alpha_k,\delta}(x,T_2)$ are also the truncated solutions as the following forms:

$$\begin{aligned} u^{\alpha_k,\delta}(x,T_1) &= \sum_{k=1}^{m}\frac{e^{-k^2T_1}}{\alpha_k e^{k^2}+e^{-k^2T_1}}g_k^{\delta}\sin(kx), \\ u^{\alpha_k,\delta}(x,T_2) &= \sum_{k=1}^{m}\frac{e^{-k^2T_2}}{\alpha_k e^{k^2}+e^{-k^2T_2}}h_k^{\delta}\sin(kx). \end{aligned} \quad (65)$$

We always take $n = 100$ and $m = 20$ and use two strategies for selecting regularization parameters in all numerical examples: (a) the a priori selection strategy given in Theorem 4 and (b) the a posteriori selection strategy 3.17 in which $\alpha_0 = 1$, $r = 1/2$, and $C = 1.0$. In order to compare the truncated regularization solutions with the true solutions, we introduce E_1 and E_2 representing the relative errors of regularization solutions and defined by

$$\begin{aligned} E_1 &= \frac{\left(\sum_{i=0}^{n}\left|f^{\alpha,\delta}(x_i) - f(x_i)\right|^2\right)^{1/2}}{\left(\sum_{i=0}^{n}\left|f(x_i)\right|^2\right)^{1/2}}, \\ E_2 &= \frac{\left(\sum_{i=0}^{n}\left|\varphi^{\alpha,\delta}(x_i) - \varphi(x_i)\right|^2\right)^{1/2}}{\left(\sum_{i=0}^{n}\left|\varphi(x_i)\right|^2\right)^{1/2}}. \end{aligned} \quad (66)$$

Example 1. Consider simultaneous determination of the source function and the initial distribution in the problem of heat conduction equation:

$$\begin{aligned} u_t - u_{xx} &= f(x), \quad (x,t)\in(0,\pi)\times[0,1], \\ u(0,t) &= u(\pi,t) = 0, \quad (x,t)\in(0,\pi)\times[0,1], \\ u(x,0) &= \varphi(x), \quad x\in(0,\pi), \\ u(x,T_1) &= g(x), \quad x\in(0,\pi), \\ u(x,T_2) &= h(x), \quad x\in(0,\pi), \end{aligned} \quad (67)$$

where $T_1 = 0.5$, $T_2 = 1$, $g(x) = (2 - e^{-0.5})\sin x$, and $h(x) = (2 - e^{-1})\sin x$. In this case, $f(x) = 2\sin x$ and $\varphi(x) = \sin x$. Numerical results of this example are shown in Table 1. Behaviors of the true solutions and the regularization solutions are shown in Figures 1 and 2.

Example 2. Consider simultaneous determination of the source function and the initial distribution in the problem of heat conduction equation:

$$\begin{aligned} u_t - u_{xx} &= f(x), \quad (x,t)\in(0,\pi)\times[0,1], \\ u(0,t) &= u(\pi,t) = 0, \quad (x,t)\in(0,\pi)\times[0,1], \quad (68) \\ u(x,0) &= \varphi(x), \quad x\in(0,\pi), \end{aligned}$$

TABLE 1: Numerical results of Example 1.

	δ	α	E_1	E_2
a priori selection strategy	0.05	4.4678e-05	2.9108e-03	1.0786e-02
	0.10	8.9357e-05	5.1605e-03	1.1633e-02
a posteriori selection strategy	0.05	$(1/2)^{12}$	2.3366e-03	4.5157e-03
	0.10	$(1/2)^{11}$	3.8406e-03	1.0433e-02

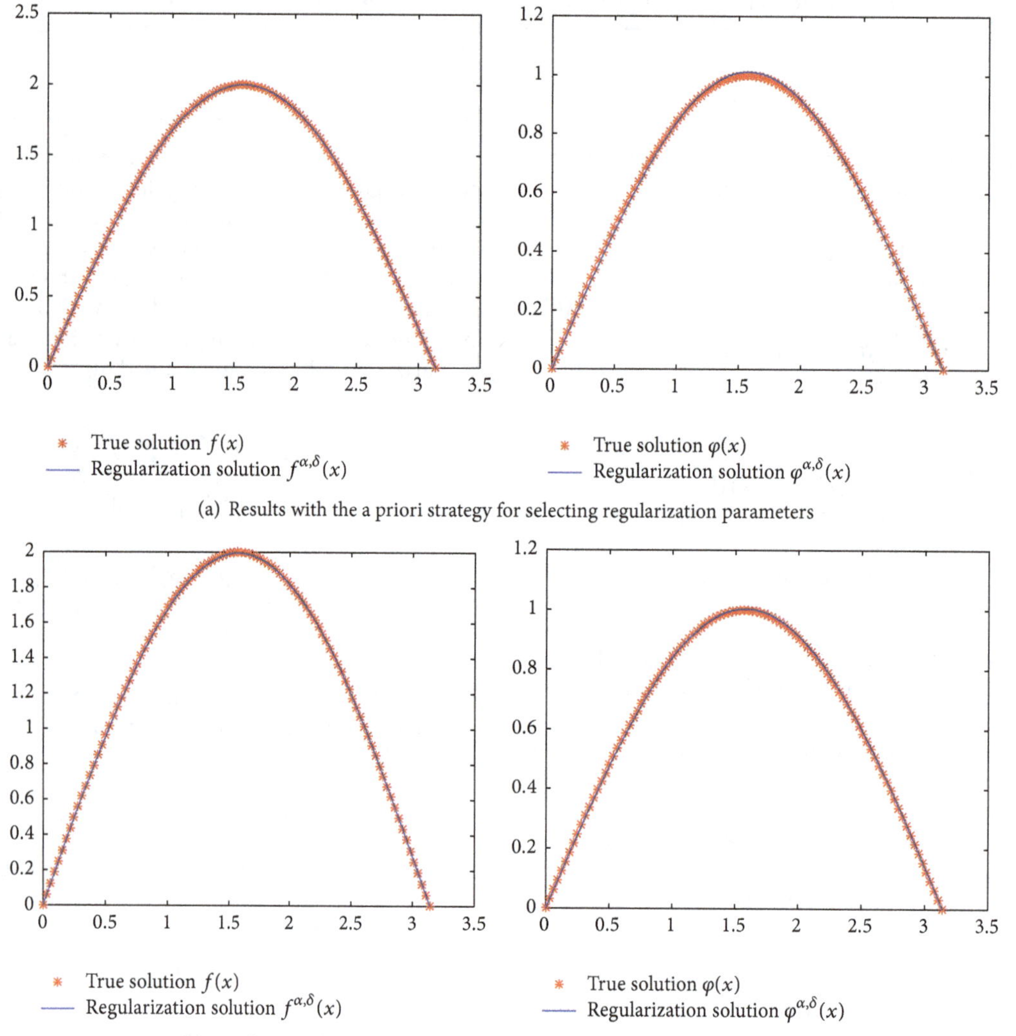

(a) Results with the a priori strategy for selecting regularization parameters

(b) Results with the a posteriori strategy for selecting regularization parameters

FIGURE 1: Behaviors of the true and the regularization solution when $\rho = 0.05$.

where we take $f(x) = x(\pi - x)$ and $\varphi(x) = x\sin(x)$. Obviously, the direct problem does not have an analytical solution. So the temperature distribution functions $u(x, T_1)$ and $u(x, T_2)$ are obtained by using the inner function "pdepe" in MATLAB to solve the direct problem. Numerical results of this example are shown in Table 2. Behaviors of the true solutions and the regularization solutions are shown in Figures 3 and 4.

From Tables 1 and 2 and Figures 1–4, we can see that the proposed method is effective and stable for the noise data especially with large noise level. And the relative errors generated by using the a posteriori strategy for selecting regularization parameters are almost equivalent to the ones generated by using the a priori strategy for selecting regularization parameters, which also indicates that regularization parameters selected by the a posteriori strategy are reasonable for obtaining the meaningful inversion solutions.

Remark 6. It is well known that the backward heat conduction problem is a severely ill-posed problem. To show the influence of the final time values T_1 and T_2 on the numerical

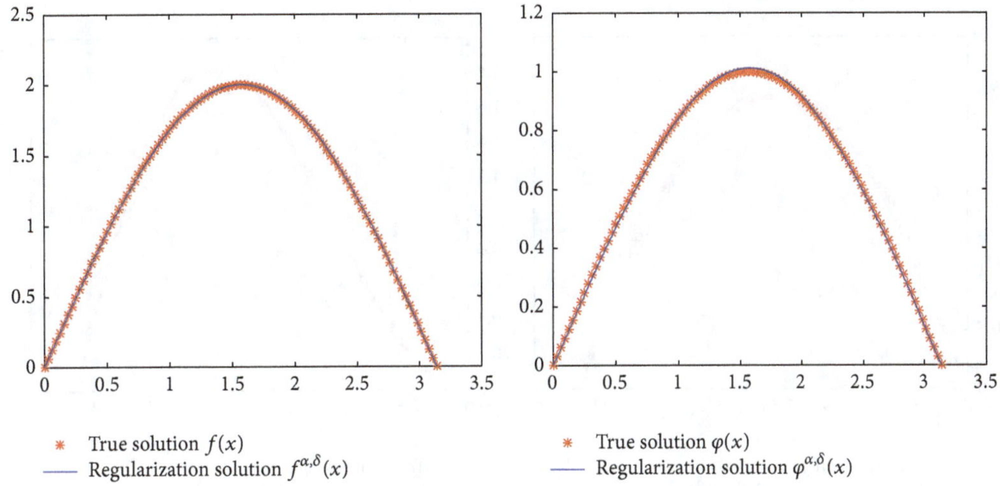

(a) Results with the a priori strategy for selecting regularization parameters

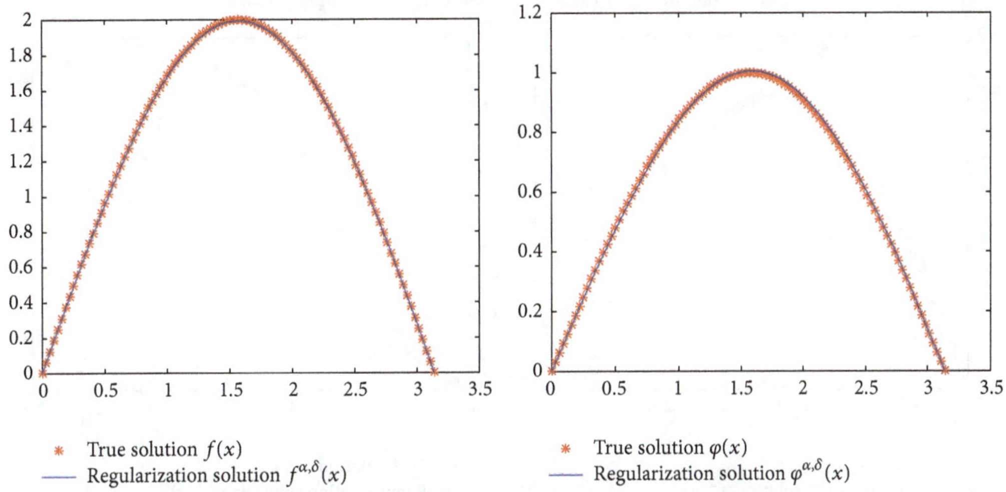

(b) Results with the a posteriori strategy for selecting regularization parameters

FIGURE 2: Behaviors of the true and the regularization solution when $\rho = 0.10$.

TABLE 2: Numerical results of Example 2.

	δ	α	E_1	E_2
a priori selection	0.05	5.9942e-05	3.8019e-02	2.8939e-02
	0.10	1.1988e-04	3.8028e-02	2.8573e-02
a posteriori selection	0.05	$(1/2)^{12}$	3.8293e-02	3.1005e-02
	0.10	$(1/2)^{10}$	3.9655e-02	7.2344e-02

inversion results, we solve the inverse problem in Examples 1 and 2 by our proposed method with different large final time values and fixed values $n = 200$, $m = 20$, and $\delta = 0.10$. Since the numerical results for Example 2 have the same behavior as Example 1, we only give the numerical results of Example 1 in Table 3, which show that the inversion results of the source term are better than that of the initial value with the increase of final time values T_1 and T_2. That is to say, the heat source can be reconstructed very well from the data measured at large final times, but the initial value is not available. It shows that the ill-posedness of the backward heat conduction problem is stronger than that of the inverse source problem.

5. Conclusions and Discussion

In this paper, we study an inverse problem for simultaneously determining the space-dependent source and the initial distribution in heat conduction equation. For obtaining stable solutions of the inverse problem, we construct a regularization problem to approximate the originally inverse problem. Stability and convergence of the regularization solutions are

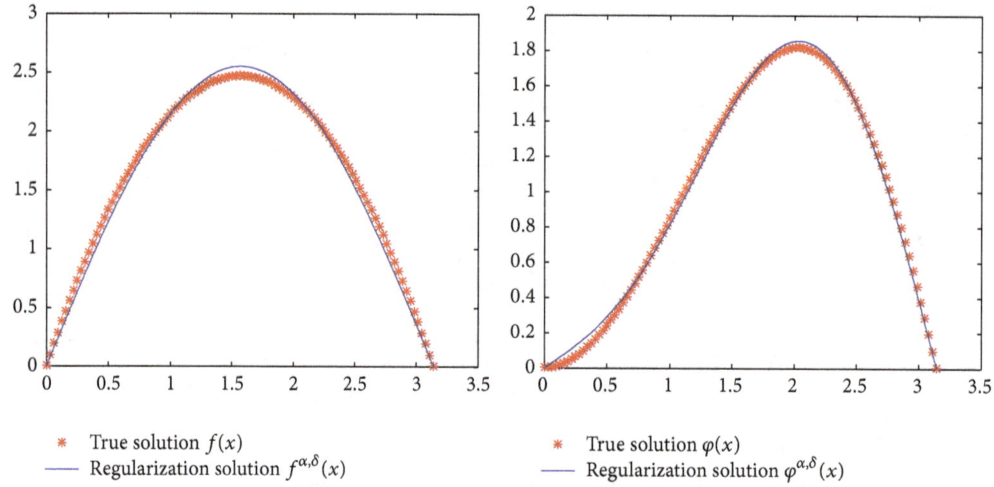

(a) Results with the a priori strategy for selecting regularization parameters

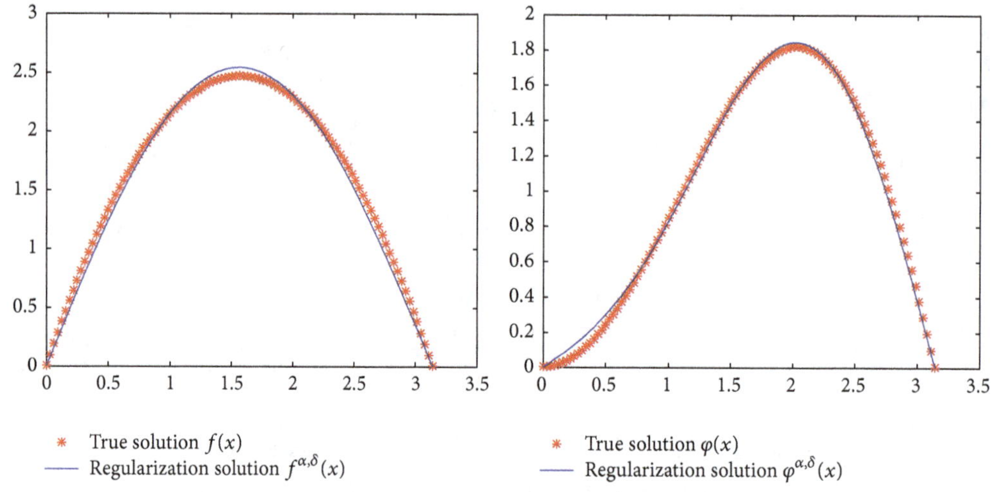

(b) Results with the a posteriori strategy for selecting regularization parameters

FIGURE 3: Behaviors of the true and the regularization solution when $\rho = 0.05$.

TABLE 3: Numerical results of Example 1 with different final time values.

	T_1	T_2	α	E_1	E_2
	1.0	1.5	7.3143e-05	1.7461e-03	2.3633e-03
	2.0	2.5	7.8949e-05	4.4939e-03	3.1173e-02
	3.0	3.5	8.1085e-05	1.1924e-02	2.6646e-01
a priori selection strategy	0.5	5.0	8.2050e-05	3.2691e-02	4.0627e-02
	1.0	5.0	8.2050e-05	3.2922e-02	1.0862e-01
	3.5	5.0	8.2050e-05	3.9430e-02	2.0275e+00
	5.0	6.0	8.2226e-05	1.1243e-01	2.3531e+01
	1.0	1.5	$(1/2)^{14}$	5262e-03	2.2503e-03
	2.0	2.5	$(1/2)^{16}$	1.2060e-03	3.4426e-02
	3.0	3.5	$(1/2)^{17}$	1.5073e-03	1.0573e-01
a posteriori selection strategy	0.5	5.0	$(1/2)^{19}$	1.1903e-03	2.4252e-02
	1.0	5.0	$(1/2)^{19}$	1.1958e-03	4.8508e-02
	3.5	5.0	$(1/2)^{19}$	1.3532e-03	4.2210e-01
	5.0	6.0	$(1/2)^{21}$	1.1297e-03	1.7172e+00

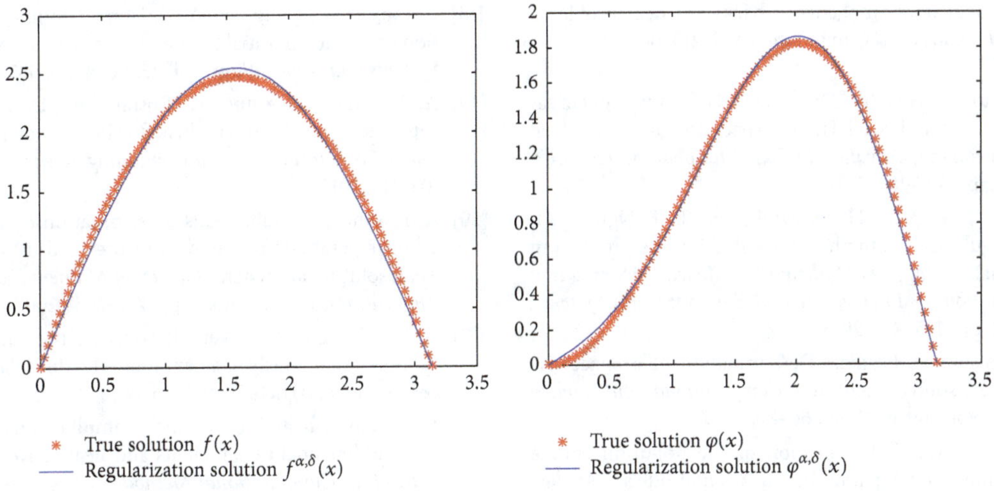

(a) Results with the a priori strategy for selecting regularization parameters

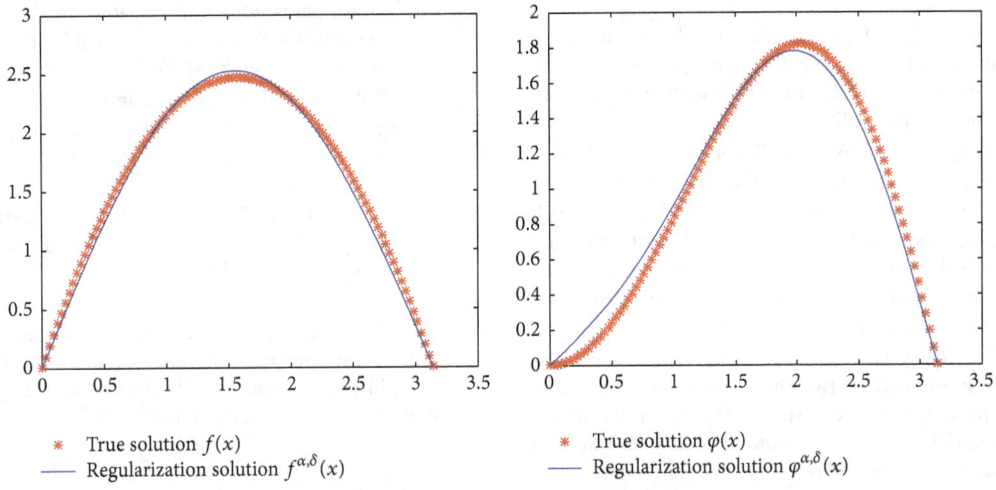

(b) Results with the a posteriori strategy for selecting regularization parameters

FIGURE 4: Behaviors of the true and the regularization solution when $\rho = 0.10$.

proved, and convergence rates of the regularized solutions are also obtained under the a priori and a posteriori strategies for selecting regularization parameters, respectively. Numerical results show that the proposed regularization method is stable and effective for simultaneously determining the space-dependent source and the initial distribution from the measurements at suitable final time values.

Conflicts of Interest

The authors declare that there are no conflicts of interest regarding the publication of this paper.

Acknowledgments

This work was supported by the National Natural Science Foundation of China [Grant nos. 11561003 and 11761007]; the Foundation of Academic and Technical Leaders Program for Major Subjects in Jiangxi Province [Grant no. 20172BCB22019]; the Natural Science Foundation of Jiangxi Province [Grant no. 20161BAB201034]; the Ground Project of Science and Technology of Jiangxi Universities [Grant no. KJLD14051].

References

[1] J. Cheng and J. J. Liu, "A quasi Tikhonov regularization for a two-dimensional backward heat problem by a fundamental solution," *Inverse Problems*, vol. 24, no. 6, Article ID 065012, pp. 1–18, 2008.

[2] D. N. Hào, N. Van, and D. Lesnic, "Regularization of parabolic equations backward in time by a non-local boundary value problem method," *IMA Journal of Applied Mathematics*, vol. 75, no. 2, pp. 291–315, 2010.

[3] N. H. Tuan and D. D. Trong, "A new regularized method for

two dimensional nonhomogeneous backward heat problem," *Applied Mathematics and Computation*, vol. 215, no. 3, pp. 873–880, 2009.

[4] Y. C. Hon and T. Takeuchi, "Discretized Tikhonov regularization by reproducing kernel Hilbert space for backward heat conduction problem," *Advances in Computational Mathematics*, vol. 34, no. 2, pp. 167–183, 2011.

[5] N. H. Tuan, T. T. Binh, N. D. Minh, and T. T. Nghia, "An improved regularization method for initial inverse problem in 2-D heat equation," *Applied Mathematical Modelling: Simulation and Computation for Engineering and Environmental Systems*, vol. 39, no. 2, pp. 425–437, 2015.

[6] T. Johansson and D. Lesnic, "Determination of a spacewise dependent heat source," *Journal of Computational and Applied Mathematics*, vol. 209, no. 1, pp. 66–80, 2007.

[7] Z. Wang and J. Liu, "Identification of the pollution source from one-dimensional parabolic equation models," *Applied Mathematics and Computation*, vol. 219, no. 8, pp. 3403–3413, 2012.

[8] C. Sun, Q. Liu, and G. Li, "Conditional well-posedness for an inverse source problem in the diffusion equation using the variational adjoint method," *Advances in Mathematical Physics*, Art. ID 6801260, 6 pages, 2017.

[9] L. Yang, M. Dehghan, J. N. Yu et al., "Inverse problem of time-dependent heat sources numerical reconstruction," *Mathematics & Computers in Simulation*, vol. 81, no. 8, pp. 1656–1672, 2011.

[10] Z.-C. Deng, K. Qian, X.-B. Rao, L. Yang, and G.-W. Luo, "An inverse problem of identifying the source coefficient in a degenerate heat equation," *Inverse Problems in Science and Engineering*, vol. 23, no. 3, pp. 498–517, 2015.

[11] L. Yang, J.-N. Yu, G.-W. Luo, and Z.-C. Deng, "Numerical identification of source terms for a two dimensional heat conduction problem in polar coordinate system," *Applied Mathematical Modelling: Simulation and Computation for Engineering and Environmental Systems*, vol. 37, no. 3, pp. 939–957, 2013.

[12] L. Yang, J.-N. Yu, G.-W. Luo, and Z.-C. Deng, "Reconstruction of a space and time dependent heat source from finite measurement data," *International Journal of Heat and Mass Transfer*, vol. 55, no. 23-24, pp. 6573–6581, 2012.

[13] Q. Chen and J. J. Liu, "Solving an inverse parabolic problem by optimization from final measurement data," *Journal of Computational and Applied Mathematics*, vol. 193, no. 1, pp. 183–203, 2006.

[14] Z.-C. Deng, Y.-C. Hon, and L. Yang, "An optimal control method for nonlinear inverse diffusion coefficient problem," *Journal of Optimization Theory and Applications*, vol. 160, no. 3, pp. 890–910, 2014.

[15] T. Wei and J. C. Wang, "Simultaneous determination for a space-dependent heat source and the initial data by the MFS," *Engineering Analysis with Boundary Elements*, vol. 36, no. 12, pp. 1848–1855, 2012.

[16] B. T. Johansson and D. Lesnic, "A procedure for determining a spacewise dependent heat source and the initial temperature," *Applicable Analysis*, vol. 87, no. 3, pp. 265–276, 2008.

[17] J. Wen, M. Yamamoto, and T. Wei, "Simultaneous determination of a time-dependent heat source and the initial temperature in an inverse heat conduction problem," *Inverse Problems in Science and Engineering*, vol. 21, no. 3, pp. 485–499, 2013.

[18] L. Yang, Z.-C. Deng, and Y.-C. Hon, "Simultaneous identification of unknown initial temperature and heat source," *Dynamic Systems and Applications*, vol. 25, no. 4, pp. 583–602, 2016.

[19] A. M. Shahrezaee and M. Rostamian, "Determination of a source term and boundary heat flux in an inverse heat equation," *Journal of Information and Computing Science*, vol. 8, no. 2, pp. 103–114, 2013.

[20] A. Hasanov, "Simultaneous determination of source terms in a linear parabolic problem from the final overdetermination: weak solution approach," *Journal of Mathematical Analysis and Applications*, vol. 330, no. 2, pp. 766–779, 2007.

[21] G.-H. Zheng and T. Wei, "Recovering the source and initial value simultaneously in a parabolic equation," *Inverse Problems*, vol. 30, no. 6, Article ID 065013, pp. 1–35, 2014.

[22] M. S. Hussein and D. Lesnic, "Simultaneous determination of time-dependent coefficients and heat source," *International Journal for Computational Methods in Engineering Science and Mechanics*, vol. 17, no. 5-6, pp. 401–411, 2016.

[23] J. J. Liu, M. Yamamoto, and L. Yan, "On the uniqueness and reconstruction for an inverse problem of the fractional diffusion process," *Applied Numerical Mathematics*, vol. 87, pp. 1–19, 2015.

[24] A. I. Prilepko, D. G. Orlovsky, and I. A. Vasin, *Methods for Solving Inverse Problems in Mathematical Physics*, vol. 231 of *Monographs and Textbooks in Pure and Applied Mathematics*, Marcel Dekker, New York, NY, USA, 2000.

[25] Z. Wang and J. Liu, "New model function methods for determining regularization parameters in linear inverse problems," *Applied Numerical Mathematics*, vol. 59, no. 10, pp. 2489–2506, 2009.

[26] Z.-w. Wang and D.-h. Xu, "On the linear model function method for choosing Tikhonov regularization parameters in linear ill-posed problems," *Chinese Journal of Engineering Mathematics*, vol. 30, no. 3, pp. 451–466, 2013.

3

Local Well-Posedness and Blow-Up for the Solutions to the Axisymmetric Inviscid Hall-MHD Equations

Eunji Jeong, Junha Kim, and Jihoon Lee

Department of Mathematics, Chung-Ang University, Seoul 06974, Republic of Korea

Correspondence should be addressed to Jihoon Lee; jhleepde@cau.ac.kr

Academic Editor: Carlo Bianca

In this paper, we consider the regularity problem of the solutions to the axisymmetric, inviscid, and incompressible Hall-magnetohydrodynamics (Hall-MHD) equations. First, we obtain the local-in-time existence of sufficiently regular solutions to the axisymmetric inviscid Hall-MHD equations without resistivity. Second, we consider the inviscid axisymmetric Hall equations without fluids and prove that there exists a finite time blow-up of a classical solution due to the Hall term. Finally, we obtain some blow-up criteria for the axisymmetric resistive and inviscid Hall-MHD equations.

1. Introduction

Magnetohydrodynamics is the study of the dynamics of the electrically conducting fluids. The dynamics of the fluids can be described by the Navier-Stokes equations and the dynamics of the magnetic field can be described by the Maxwell equations for a perfect conductor. The Hall-magnetohydrodynamics (Hall-MHD) equations differ from the standard incompressible MHD equations by the Hall term $\nabla \times ((\nabla \times B) \times B)$, which plays an important role in the study of the magnetic reconnection in the case of the large magnetic shear (see [1, 2]). In [3], Hall-MHD equations have been formally derived from using the generalized Ohm's law instead of the usual simplified Ohm's law. The Cauchy problem for three-dimensional incompressible Hall-MHD equations reads as follows:

$$\begin{aligned} &\partial_t u + (u \cdot \nabla) u + \nabla p = (\nabla \times B) \times B + \nu \Delta u, \\ &\nabla \cdot u = 0, \quad \text{in } \mathbb{R}^3 \times (0, \infty), \\ &\partial_t B - \nabla \times (u \times B) + \nabla \times ((\nabla \times B) \times B) = \kappa \Delta B, \\ &(u(0,x), B(0,x)) = (u_0(x), B_0(x)), \quad \text{in } \mathbb{R}^3, \end{aligned} \tag{1}$$

where u, B, and p represent three-dimensional velocity vector field, the magnetic field, and scalar pressure, respectively. The initial data u_0 and B_0 satisfy

$$\nabla \cdot u_0 = \nabla \cdot B_0 = 0. \tag{2}$$

Note that if $\nabla \cdot B_0 = 0$, then the divergence free condition is propagated by $(1)_3$. We only consider $\mathbb{R}^3$ for a spatial domain with vanishing at infinity condition for simplicity.

The Hall magnetohydrodynamics were studied systematically by Lighthill [2]. The Hall-MHD is important, describing many physical phenomena, e.g., space plasmas, star formation, neutron stars, and geo-dynamo (see [1, 4–8] and references therein).

The Hall-MHD equations have been mathematically investigated in several works. In [9], Acheritogaray, Degond, Frouvelle, and Liu derived the Hall-MHD equations from either two fluids' model or kinetic models in a mathematically more rigorous way. In [10], the global existence of weak solutions to (1) and the local well-posedness of classical solution are established when $\nu, \kappa > 0$. Also, a blow-up criterion for smooth solution to (1) and the global existence of smooth solution for small initial data are obtained (see [10, Theorem 2.2 and 2.3]). Some of the results have been refined by many authors (see [11–13] and references therein). Recently, temporal decay for the weak solution and smooth solution with small data to Hall-MHD are also established in [14]. Spatial and temporal decays of solutions to (1) have been investigated in [15].

Using vector identity, we can rewrite (1) as follows:

$$\partial_t u + (u\cdot\nabla)u + \nabla\left(p + \frac{|B|^2}{2}\right) = (B\cdot\nabla)B + \nu\Delta u,$$

$$\nabla\cdot u = 0,$$

$$\text{in } \mathbb{R}^3\times(0,\infty), \quad (3)$$

$$\partial_t B + (u\cdot\nabla)B + \nabla\times((\nabla\times B)\times B) = (B\cdot\nabla)u + \kappa\Delta B,$$

$$(u(0,x), B(0,x)) = (u_0(x), B_0(x)),$$

$$\text{in } \mathbb{R}^3.$$

Note that a weak solution (u, B) to (1) satisfies the following energy inequality (see [10]):

$$\begin{aligned}&\|u(t)\|_{L^2}^2 + \|B(t)\|_{L^2}^2\\ &\quad + 2\int_0^t \nu\|\nabla u(\cdot,\tau)\|_{L^2}^2 + \kappa\|\nabla B(\cdot,\tau)\|_{L^2}^2\,d\tau \\ &\le \|u_0\|_{L^2}^2 + \|B_0\|_{L^2}^2,\end{aligned} \quad (4)$$

for almost every $t\in[0,\infty)$.

Next we consider the mathematical setting for the axisymmetric Hall-MHD equations. Introducing the cylindrical coordinates

$$\begin{aligned} r &= \sqrt{x_1^2 + x_2^2},\\ \theta &= \arctan\frac{x_2}{x_1},\\ z &= x_3,\end{aligned} \quad (5)$$

and standard basis vectors for the cylindrical coordinates

$$\begin{aligned} e_r &= (\cos\theta, \sin\theta, 0),\\ e_\theta &= (-\sin\theta, \cos\theta, 0),\\ e_z &= (0,0,1),\end{aligned} \quad (6)$$

we set

$$\begin{aligned} u &= u^r(r,z,t)\,e_r + u^z(r,z,t)\,e_z,\\ B &= B^\theta(r,z,t)\,e_\theta.\end{aligned} \quad (7)$$

It is well-known that the local-in-time classical solutions to axisymmetric Navier-Stokes equations without swirl persist to any time (see [16, 17]). But the global well-posedness for the axisymmetric Navier-Stokes equations with swirl component is widely open and has been one of the most fundamental open problems in the Navier-Stokes equations.

The axisymmetric MHD equations can be written as follows:

$$\begin{aligned} &\partial_t u^r + (u^r\partial_r + u^z\partial_z)u^r - \nu\left(\partial_r^2 + \frac{1}{r}\partial_r + \partial_z^2 - \frac{1}{r^2}\right)u^r\\ &\quad + \partial_r P = -\frac{(B^\theta)^2}{r},\\ &\partial_t u^z + (u^r\partial_r + u^z\partial_z)u^z - \nu\left(\partial_r^2 + \frac{1}{r}\partial_r + \partial_z^2\right)u^z\\ &\quad + \partial_z P = 0,\\ &\partial_r u^r + \frac{u^r}{r} + \partial_z u^z = 0,\\ &\partial_t B^\theta + (u^r\partial_r + u^z\partial_z)B^\theta - \frac{u^r B^\theta}{r}\\ &\quad - \kappa\left(\partial_r^2 + \frac{1}{r}\partial_r + \partial_z^2 - \frac{1}{r^2}\right)B^\theta = 0.\end{aligned} \quad (8)$$

Lei [18] proved the global well-posedness of classical solutions to system (8) when $\kappa \ge 0$.

Then axisymmetric Hall-MHD equations are reduced to the following:

$$\begin{aligned} &\partial_t u^r + (u^r\partial_r + u^z\partial_z)u^r - \nu\left(\partial_r^2 + \frac{1}{r}\partial_r + \partial_z^2 - \frac{1}{r^2}\right)u^r\\ &\quad + \partial_r P = -\frac{(B^\theta)^2}{r},\\ &\partial_t u^z + (u^r\partial_r + u^z\partial_z)u^z - \nu\left(\partial_r^2 + \frac{1}{r}\partial_r + \partial_z^2\right)u^z\\ &\quad + \partial_z P = 0,\\ &\partial_r u^r + \frac{u^r}{r} + \partial_z u^z = 0,\\ &\partial_t B^\theta + (u^r\partial_r + u^z\partial_z)B^\theta - \frac{u^r B^\theta}{r}\\ &\quad - \kappa\left(\partial_r^2 + \frac{1}{r}\partial_r + \partial_z^2 - \frac{1}{r^2}\right)B^\theta = \partial_z\left(\frac{(B^\theta)^2}{r}\right).\end{aligned} \quad (9)$$

For axisymmetric Hall-MHD equations, the global well-posedness of the axisymmetric solutions to the viscous case $(\nu, \kappa > 0)$ was first established by Fan, Huang, and Nakamura [19]. Recently, Chae and Weng [20] showed that the incompressible Hall-MHD system without resistivity is not globally in time well-posed in any Sobolev space $H^m(\mathbb{R}^3)$ with $m > 7/2$. But local-in-time existence of smooth solution to (1) is totally open when $\kappa \equiv 0$. Compared with the work in [18], it seems very surprising that Hall term plays a dominant role for the occurrence of the singularity and even for the local well-posedness of the partially viscous Hall-MHD problems. In this paper, we intend to investigate the blow-up problem for the solutions to the partially viscous axisymmetric Hall-MHD equations and local-in-time existence of solutions to such solution with the axisymmetry. Setting $\omega^\theta = \partial_z u^r - \partial_r u^z$, $\Omega = \omega^\theta / r$, and $\Pi = B^\theta / r$, (9) are equivalent to the following equations:

$$\begin{aligned} &\frac{\partial \Omega}{\partial t} + (u^r \partial_r + u^z \partial_z)\,\Omega + 2\Pi \partial_z \Pi \\ &\quad = \nu \left(\partial_r^2 + \frac{3}{r}\partial_r + \partial_z^2 \right) \Omega, \\ &\frac{\partial \Pi}{\partial t} + (u^r \partial_r + u^z \partial_z)\,\Pi - 2\Pi \partial_z \Pi \\ &\quad = \kappa \left(\partial_r^2 + \frac{3}{r}\partial_r + \partial_z^2 \right) \Pi. \end{aligned} \tag{10}$$

First, we consider the local well-posedness of the axisymmetric Hall MHD equations with $\nu \equiv 0$ and $\kappa \equiv 0$, and (10) can be rewritten as the equations

$$\partial_t \Pi + (u^r \partial_r + u^z \partial_z)\,\Pi = 2\Pi \partial_z \Pi \tag{11}$$

$$\partial_t \Omega + (u^r \partial_r + u^z \partial_z)\,\Omega = -2\Pi \partial_z \Pi. \tag{12}$$

Theorem 1. *Let $(u_0, B_0, B_0/r, \operatorname{curl} u_0/r) \in H^m(\mathbb{R}^3) \times H^m(\mathbb{R}^3) \times H^m(\mathbb{R}^3) \times H^{m-1}(\mathbb{R}^3)$ with integer $m > 9/2$ be axisymmetric initial data. There exist $T_0 > 0$ and classical and axisymmetric solution (Π, Ω) to (11)–(12) such that*

$$\begin{aligned} &\frac{\operatorname{curl} u}{r} = \Omega, \\ &\frac{B}{r} = \Pi, \\ &\Omega(t) \in C([0,T_0); H^m) \cap \operatorname{Lip}\left([0,T_0); H^{m-1}\right) \text{ and } u(t), B(t), \Pi(t) \in C([0,T_0); H^m) \cap \operatorname{Lip}\left([0,T_0); H^{m-1}\right). \end{aligned} \tag{13}$$

Remark 2. Since the local-in-time regularity of solution to (1) is necessary to preserve the axisymmetry of the Hall-MHD equations locally in time, Theorem 1 cannot resolve the open question raised from [20]. We remark that the relation between (11)–(12) and (1) cannot be justified without local well-posedness of solution to (1) ($\kappa = \nu \equiv 0$).

Next, we consider the local well-posedness/blow-up problem for the axisymmetric Hall equations with zero fluid velocity and $\kappa = 0$. We rewrite the Hall equation for $\Pi = B^\theta / r$:

$$\begin{aligned} &\partial_t \Pi = -2\Pi \partial_z \Pi, \\ &\Pi(x,0) = \Pi_0(x). \end{aligned} \tag{14}$$

The above equation has similar features to the inviscid Burgers equation.

Theorem 3. *Assume $\Pi_0 \in H^m(\mathbb{R}^3)$ for any integer $m > 5/2$. Then there exist $T_0 > 0$ and a classical solution to (14) such that*

$$\begin{aligned} \Pi(t) &\in C\left([0,T_0); H^m\left(\mathbb{R}^3\right)\right) \\ &\cap \operatorname{Lip}\left([0,T_0); H^{m-1}\left(\mathbb{R}^3\right)\right). \end{aligned} \tag{15}$$

Furthermore, for any $\Pi_0 \neq 0$, there exists $T^* > 0$ such that the above local solution $\Pi(t)$ has singularity at a finite time $t = T^*$.

Remark 4. In [20], the authors showed that if the initial data Π_0 satisfies $\partial_z \Pi_0(0,0) \geq 10^4 C_*^2$ for some constant C_* and $\Pi_0(0,0) > 0$, then the singularity of Π and Ω to axisymmetric inviscid Hall-MHD equations happens in a finite time. Theorem 3 implies that the singularity of Π which is a solution to (14) happens in a finite time without any restriction of the initial data.

Finally, we consider the incompressible Hall-MHD equations with zero fluid viscosity, for simplicity, assuming that $\nu \equiv 0$ and $\kappa \equiv 1$.

For the solutions to (10), global a priori bounds can be obtained; that is,

$$\|\Pi\|_{L^\infty(0,T;L^\infty)} + \|\Omega\|_{L^\infty(0,T;L^2)} < \infty \quad \text{for all } T > 0. \tag{16}$$

We assume that our initial data (u_0, B_0) is axisymmetric and satisfies

$$\begin{aligned} &(u_0, B_0, \Omega_0, \Pi_0) \in H^m\left(\mathbb{R}^3\right)^4 \quad \text{with } m > \frac{5}{2}, \\ &\nabla \cdot u_0 = \nabla \cdot B_0 = 0. \end{aligned} \tag{17}$$

The local-in-time existence of a smooth solution to (1) was already obtained by Chae, Wan, and Wu [21]. We obtain the following blow-up criterion for the local-in-time solutions to the Hall-MHD equations with $\nu \equiv 0$ and $\kappa \equiv 1$.

Theorem 5. *Let (u, B, p) be a local-in-time classical solution to the axisymmetric Hall-MHD equations (9) with $\nu = 0, \kappa = 1$. Then, for the first blow-up time $T^* < \infty$ of the classical solution to (9), it holds that*

$$\begin{aligned} \limsup_{t \nearrow T^*} &\left(\|u(t)\|_{H^m}^2 + \|B(t)\|_{H^m}^2 + \|\Omega(t)\|_{H^m}^2 \right. \\ &\left. + \|\Pi(t)\|_{H^m}^2 \right) = \infty, \end{aligned} \tag{18}$$

if and only if one of the following conditions holds:

(i)

$$\int_0^{T^*} \left\| \frac{u_+^r}{r} \right\|_{L^\infty(\mathbb{R}^3 \setminus C_R)} dt + \int_0^{T^*} \|r u_+^r\|_{L^\infty(C_R)}\, dt = \infty. \tag{19}$$

(ii)

$$\int_0^{T^*} \left\|\omega^\theta\right\|^4_{L^2(\mathbb{R}^3\setminus C_R)} dt = \infty. \quad (20)$$

In the above, C_R denotes the inside of infinite cylinder such that $C_R = \{(x,y,z) \mid x^2 + y^2 < R^2\}$ for any $R > 0$ and $f_+(x)$ is defined by $\max\{f(x), 0\}$.

Remark 6. For the usual MHD equations, Lei [18] proved the global well-posedness for the axisymmetric MHD equations even for the case that $\nu \equiv 1$ and $\kappa \equiv 0$. For Hall-MHD equations, even local well-posedness is widely open for this zero resisitivity case due to the Hall term (see [20]). Theorem 5 indicates that if there exists a finite time singularity to the axisymmetric equations with $\nu \equiv 0$ and $\kappa \equiv 1$, then some norms of velocity and vorticity should approach infinity even for the outside of any infinite cylinder.

For simplicity, we denote C for the harmless constant which changes from line to line, and $\|\cdot\|_m$ for H^m-norm.

2. Proof of Theorem 1: Local-in-Time Existence

In this section, we consider the local-in-time existence of regular solution to (11)–(12). Even if this problem does not seem complicated, we have a few technical difficulties raised from the axisymmetry; e.g., mollifying equations do not preserve the axial symmetry. We briefly explain some steps to prove Theorem 1: First, we consider system (21) without giving any symmetry. We can obtain the regularized system (25) by using standard mollifier. Then we can obtain various estimates and local-in-time existence of a solution for (21). Finally, we consider the initial data which is axial symmetry and axisymmetry is also preserved by (21) and this argument gives a proof of local-in-time existence of solution to (11)–(12).

We consider the equations

$$\begin{aligned}
&\partial_t\omega + (\mathbf{u}\cdot\nabla)\,\omega = u^r\Omega - 2B\partial_z\Pi,\\
&\partial_t B + (\mathbf{u}\cdot\nabla)\,B = u^r\Pi + 2B\partial_z\Pi,\\
&\partial_t\Pi + (\mathbf{u}\cdot\nabla)\,\Pi = 2\Pi\partial_z\Pi,\\
&\partial_t\Omega + (\mathbf{u}\cdot\nabla)\,\Omega = -2\Pi\partial_z\Pi,
\end{aligned} \quad (21)$$

where ω, B, Π, and Ω are assumed to be independent scalar valued functions without assuming symmetry for a while, and the divergence free velocity field $\mathbf{u} = u^r(r,\theta,z)e_r + u^z(r,\theta,z)e_z$ is assumed to be obtained from the equation

$$-\Delta\mathbf{u} = \nabla\times(\omega e_\theta). \quad (22)$$

Thus, we have

$$\mathbf{u}(x) = \Phi * (\nabla\times(\omega e_\theta)), \quad \text{where } \Phi(x) = \frac{1}{4\pi|x|}. \quad (23)$$

If $\omega \in H^1(\mathbb{R}^3)\cap C(\mathbb{R}^3)$, then the divergence theorem and trace theorem induce the following estimates:

$$\begin{aligned}
|\mathbf{u}(x)| &\le \|\Phi\|_{L^2(B(0;1))}\|\nabla\omega\|_{L^2(\mathbb{R}^3)}\\
&\quad + \|\nabla\Phi\|_{L^2(\mathbb{R}^3\setminus B(0;1))}\|\omega\|_{L^2(\mathbb{R}^3)}\\
&\quad + \int_{\partial B(0;1)} \Phi(y)\,\omega(x-y)\,dS\\
&\le \|\Phi\|_{L^2(B(0;1))}\|\nabla\omega\|_{L^2(\mathbb{R}^3)}\\
&\quad + \|\nabla\Phi\|_{L^2(\mathbb{R}^3\setminus B(0;1))}\|\omega\|_{L^2(\mathbb{R}^3)}\\
&\quad + C\left(\int_{\partial B(0;1)} |\omega(x-y)|^2\,dS\right)^{1/2}\\
&\le \|\Phi\|_{L^2(B(0;1))}\|\nabla\omega\|_{L^2(\mathbb{R}^3)}\\
&\quad + \|\nabla\Phi\|_{L^2(\mathbb{R}^3\setminus B(0;1))}\|\omega\|_{L^2(\mathbb{R}^3)} + C\|\omega\|_{H^1(\mathbb{R}^3)}\\
&\le C\|\omega\|_{H^1(\mathbb{R}^3)}.
\end{aligned} \quad (24)$$

We define a regularized system of (21) as follows:

$$\begin{aligned}
&\partial_t\omega_\epsilon + \mathcal{J}_\epsilon(\mathcal{J}_\epsilon\mathbf{u}\cdot\nabla)\,\mathcal{J}_\epsilon\omega_\epsilon\\
&\quad = \mathcal{J}_\epsilon(\mathcal{J}_\epsilon u^r_\epsilon\mathcal{J}_\epsilon\Omega_\epsilon) - 2\mathcal{J}_\epsilon(\mathcal{J}_\epsilon B_\epsilon\partial_z\mathcal{J}_\epsilon\Pi_\epsilon),\\
&\partial_t B_\epsilon + \mathcal{J}_\epsilon(\mathcal{J}_\epsilon\mathbf{u}\cdot\nabla)\,\mathcal{J}_\epsilon B_\epsilon\\
&\quad = \mathcal{J}_\epsilon(\mathcal{J}_\epsilon u^r_\epsilon\mathcal{J}_\epsilon\Pi_\epsilon) + 2\mathcal{J}_\epsilon(\mathcal{J}_\epsilon B_\epsilon\partial_z\mathcal{J}_\epsilon\Pi_\epsilon),\\
&\partial_t\Pi_\epsilon + \mathcal{J}_\epsilon(\mathcal{J}_\epsilon\mathbf{u}\cdot\nabla)\,\mathcal{J}_\epsilon\Pi_\epsilon = 2\mathcal{J}_\epsilon(\mathcal{J}_\epsilon\Pi_\epsilon\partial_z\mathcal{J}_\epsilon\Pi_\epsilon),\\
&\partial_t\Omega_\epsilon + \mathcal{J}_\epsilon(\mathcal{J}_\epsilon\mathbf{u}\cdot\nabla)\,\mathcal{J}_\epsilon\Omega_\epsilon = -2\mathcal{J}_\epsilon(\mathcal{J}_\epsilon\Pi_\epsilon\partial_z\mathcal{J}_\epsilon\Pi_\epsilon),
\end{aligned} \quad (25)$$

where $\mathcal{J}_\epsilon$ is a standard mollifier as in [22]. Next, we obtain apriori estimates to derive a time T_0 which does not depend on $\epsilon > 0$. Then we prove that (25) have a local-in-time solution $S_\epsilon = (\omega_\epsilon, B_\epsilon, \Pi_\epsilon, \Omega_\epsilon) \in C^1(0,T_0;H^m) \times C^1(0,T_0;H^m)\times C^1(0,T_0;H^{m+1})\times C^1(0,T_0;H^m)$ space for each $\epsilon > 0$.

Proposition 7. *Let*

$$\left\|S_\epsilon\right\|_{X^m}(t) := \left\|\omega_\epsilon\right\|_m + \left\|B_\epsilon\right\|_m + \left\|\Pi_\epsilon\right\|_{m+1} + \left\|\Omega_\epsilon\right\|_m \quad (26)$$

where $X^m = H^m(\mathbb{R}^3)\times H^m(\mathbb{R}^3)\times H^{m+1}(\mathbb{R}^3)\times H^m(\mathbb{R}^3)$ with an integer $m > 5/2$. Then, for some positive constant C_0 and $T < 1/C_0S_0$ with $S_0 = \|S_\epsilon(0)\|_{X^m}$, we have

$$\begin{aligned}
&\sup_{0<t<T}\left\|S_\epsilon\right\|_{X^m}(t) \le \frac{S_0}{1-C_0S_0T},\\
&\sup_{0<t<T}\left\|\partial_t S_\epsilon\right\|_{X^{m-1}}(t) \le \left(\frac{S_0}{1-C_0S_0T}\right)^2.
\end{aligned} \quad (27)$$

Proof. For $m > 5/2$ and $1 \le s \le m$, we have the following inequality by the calculus inequality and Hölder's inequality:

$$\begin{aligned}\frac{1}{2}\frac{d}{dt}\|\nabla^s\omega_\epsilon\|_{L^2}^2 &\le \int_{\mathbb{R}^3}|\mathcal{J}_\epsilon\nabla^s\omega_\epsilon|\left|\nabla^{s-1}(\mathcal{J}_\epsilon\nabla\mathbf{u}_\epsilon\cdot\mathcal{J}_\epsilon\nabla\omega_\epsilon)\right|dx \\ &+\int_{\mathbb{R}^3}|\mathcal{J}_\epsilon\nabla^s\omega_\epsilon|\left|\nabla^s(\mathcal{J}_\epsilon u_\epsilon^r\mathcal{J}_\epsilon\Omega_\epsilon)\right|dx \\ &+\int_{\mathbb{R}^3}|\mathcal{J}_\epsilon\nabla^s\omega_\epsilon|\left|\nabla^s(\mathcal{J}_\epsilon B_\epsilon\partial_z\mathcal{J}_\epsilon\Pi_\epsilon)\right|dx \le \|\omega_\epsilon\|_m \\ &\cdot\left(\|\nabla\mathbf{u}_\epsilon\|_{L^\infty}\|\nabla^s\omega_\epsilon\|_{L^2}+\|\nabla^s\mathbf{u}_\epsilon\|_{L^2}\|\nabla\omega_\epsilon\|_{L^\infty}\right) \\ &+\|\omega_\epsilon\|_m\left(\|u_\epsilon^r\|_{L^\infty}\|\nabla^s\Omega_\epsilon\|_{L^2}+\|\Omega_\epsilon\|_{L^\infty}\|\nabla^s u_\epsilon^r\|_{L^2}\right) \\ &+2\|\omega_\epsilon\|_m\left(\|B_\epsilon\|_{L^\infty}\left\|\nabla^{s+1}\Pi_\epsilon\right\|_{L^2}\right. \\ &\left.+\|\nabla\Pi_\epsilon\|_{L^\infty}\|\nabla^s B_\epsilon\|_{L^2}\right) \le C\|\omega_\epsilon\|_m\left(\|\omega_\epsilon\|_{m-1}\|\omega_\epsilon\|_m\right. \\ &\left.+\|\Omega_\epsilon\|_m\|\omega_\epsilon\|_m+\|\Pi_\epsilon\|_{m+1}\|B_\epsilon\|_m\right).\end{aligned} \tag{28}$$

Similarly, for $m > 5/2$ and $1 \le s \le m$, we can obtain

$$\begin{aligned}\frac{1}{2}\frac{d}{dt}\|\nabla^s B_\epsilon\|_{L^2}^2 &\le \|B_\epsilon\|_m\left(\|\omega_\epsilon\|_{m-1}\|B_\epsilon\|_m\right. \\ &\left.+\|\omega_\epsilon\|_m\|\Pi_\epsilon\|_m+\|\Pi_\epsilon\|_{m+1}\|B_\epsilon\|_m\right),\end{aligned} \tag{29}$$

and $1 \le s \le m+1$,

$$\begin{aligned}&\frac{1}{2}\frac{d}{dt}\|\nabla^s\Pi_\epsilon\|_{L^2}^2 \\ &\le \|\Pi_\epsilon\|_{m+1}\left(\|\omega_\epsilon\|_m\|\Pi_\epsilon\|_{m+1}+\|\Pi_\epsilon\|_{m+1}^2\right).\end{aligned} \tag{30}$$

For an integer $m > 5/2$ and all integers $1 \le s \le m$, we have

$$\frac{1}{2}\frac{d}{dt}\|\nabla^s(\Pi_\epsilon+\Omega_\epsilon)\|_{L^2}^2 \le \|\Pi_\epsilon+\Omega_\epsilon\|_m^2\|\omega_\epsilon\|_{m-1}. \tag{31}$$

Hence, we conclude that, for $m > 5/2$,

$$\begin{aligned}&\frac{d}{dt}\|S_\epsilon\|_{X^m}(t) \\ &\le \frac{1}{2}C_0\left(\|\omega_\epsilon\|_m^2+\|B_\epsilon\|_m^2+\|\Pi_\epsilon\|_{m+1}^2+\|\Pi_\epsilon+\Omega_\epsilon\|_m^2\right) \\ &\le C_0\|S_\epsilon\|_{X^m}^2(t).\end{aligned} \tag{32}$$

Inequality (32) implies that (27) holds true. □

Proposition 8. *Assume C_0 and S_0 are the same as in Proposition 7. Let $T_0 = 3/4C_0S_0$. Then for given initial data $\omega_0, B_0, \Omega_0 \in H^m(\mathbb{R}^3)$, and $\Pi_0 \in H^{m+1}(\mathbb{R}^3)$ with an integer $m > 5/2$, there exists a unique solution $(\omega_\epsilon, B_\epsilon, \Pi_\epsilon, \Omega_\epsilon)$ to regularized system (25) such that $\omega_\epsilon, B_\epsilon, \Omega_\epsilon \in C^1(0, T_0; H^m)$, $\Pi_\epsilon \in C^1(0, T_0; H^{m+1})$.*

Proof. We set

$$\begin{aligned}F_\epsilon^1(\omega_\epsilon,B_\epsilon,\Pi_\epsilon,\Omega_\epsilon) &= -\mathcal{J}_\epsilon(\mathcal{J}_\epsilon\mathbf{u}_\epsilon\cdot\nabla)\mathcal{J}_\epsilon\omega_\epsilon \\ &+\mathcal{J}_\epsilon(\mathcal{J}_\epsilon u_\epsilon^r\mathcal{J}_\epsilon\Omega_\epsilon) \\ &-2\mathcal{J}_\epsilon(\mathcal{J}_\epsilon B_\epsilon\partial_z\mathcal{J}_\epsilon\Pi_\epsilon), \\ F_\epsilon^2(\omega_\epsilon,B_\epsilon,\Pi_\epsilon,\Omega_\epsilon) &= -\mathcal{J}_\epsilon(\mathcal{J}_\epsilon\mathbf{u}_\epsilon\cdot\nabla)\mathcal{J}_\epsilon B_\epsilon \\ &+\mathcal{J}_\epsilon(\mathcal{J}_\epsilon u_\epsilon^r\mathcal{J}_\epsilon\Pi_\epsilon) \\ &+2\mathcal{J}_\epsilon(\mathcal{J}_\epsilon B_\epsilon\partial_z\mathcal{J}_\epsilon\Pi_\epsilon), \\ F_\epsilon^3(\omega_\epsilon,\Pi_\epsilon) &= -\mathcal{J}_\epsilon(\mathcal{J}_\epsilon\mathbf{u}_\epsilon\cdot\nabla)\mathcal{J}_\epsilon\Pi_\epsilon \\ &+2\mathcal{J}_\epsilon(\mathcal{J}_\epsilon\Pi_\epsilon\partial_z\mathcal{J}_\epsilon\Pi_\epsilon), \\ F_\epsilon^4(\omega_\epsilon,\Pi_\epsilon,\Omega_\epsilon) &= -\mathcal{J}_\epsilon(\mathcal{J}_\epsilon\mathbf{u}_\epsilon\cdot\nabla)\mathcal{J}_\epsilon\Omega_\epsilon \\ &-2\mathcal{J}_\epsilon(\mathcal{J}_\epsilon\Pi_\epsilon\partial_z\mathcal{J}_\epsilon\Pi_\epsilon).\end{aligned} \tag{33}$$

First, we show that F_ϵ^1 is Lipschitz continuous on H^m space. We estimate for $\epsilon < 1, m \ge 3$

$$\begin{aligned}&\left\|F_\epsilon^1(\omega_\epsilon,B_\epsilon,\Pi_\epsilon,\Omega_\epsilon)-F_\epsilon^1(\tilde{\omega}_\epsilon,B_\epsilon,\Pi_\epsilon,\Omega_\epsilon)\right\|_m \\ &\le \|(\mathcal{J}_\epsilon(\mathbf{u}_\epsilon-\tilde{\mathbf{u}}_\epsilon)\cdot\nabla)\mathcal{J}_\epsilon\omega_\epsilon\|_m \\ &+\|(\mathcal{J}_\epsilon\tilde{\mathbf{u}}_\epsilon\cdot\nabla)(\mathcal{J}_\epsilon\omega_\epsilon-\mathcal{J}_\epsilon\tilde{\omega}_\epsilon)\|_m \\ &+\|\mathcal{J}_\epsilon(u_\epsilon^r-\tilde{u}_\epsilon^r)\mathcal{J}_\epsilon\Omega_\epsilon\|_m \\ &\le \frac{C}{\epsilon}\left(\|\mathbf{u}_\epsilon-\tilde{\mathbf{u}}_\epsilon\|_{L^\infty}+\|\omega_\epsilon-\tilde{\omega}_\epsilon\|_{m-1}\right) \\ &\cdot\left(\|\omega_\epsilon\|_m+\epsilon\|\Omega_\epsilon\|_m\right)+\frac{C}{\epsilon}\left(\|\tilde{\mathbf{u}}_\epsilon\|_{L^\infty}+\|\tilde{\omega}_\epsilon\|_{m-1}\right) \\ &\cdot\|\omega_\epsilon-\tilde{\omega}_\epsilon\|_m \le \frac{C}{\epsilon}\left(\|\omega_\epsilon\|_m+\|\Omega_\epsilon\|_m+\|\tilde{\omega}_\epsilon\|_{m-1}\right) \\ &\cdot\|\omega_\epsilon-\tilde{\omega}_\epsilon\|_m.\end{aligned} \tag{34}$$

By the similar estimates as in (34), we obtain

$$\begin{aligned}&\left\|F_\epsilon^1(\omega_\epsilon,B_\epsilon,\Pi_\epsilon,\Omega_\epsilon)-F_\epsilon^1(\omega_\epsilon,\widetilde{B}_\epsilon,\Pi_\epsilon,\Omega_\epsilon)\right\|_m \\ &\le 2\left\|\left(\mathcal{J}_\epsilon B_\epsilon-\mathcal{J}_\epsilon\widetilde{B}_\epsilon\right)\partial_z\mathcal{J}_\epsilon\Pi_\epsilon\right\|_m \\ &\le C\|\Pi_\epsilon\|_{m+1}\left\|B_\epsilon-\widetilde{B}_\epsilon\right\|_m,\end{aligned} \tag{35}$$

$$\begin{aligned}&\left\|F_\epsilon^1(\omega_\epsilon,B_\epsilon,\Pi_\epsilon,\Omega_\epsilon)-F_\epsilon^1(\omega_\epsilon,B_\epsilon,\widetilde{\Pi}_\epsilon,\Omega_\epsilon)\right\|_m \\ &\le 2\left\|\mathcal{J}_\epsilon B_\epsilon\left(\partial_z\mathcal{J}_\epsilon\Pi_\epsilon-\partial_z\mathcal{J}_\epsilon\widetilde{\Pi}_\epsilon\right)\right\|_m \\ &\le C\|B_\epsilon\|_m\left\|\Pi_\epsilon-\widetilde{\Pi}_\epsilon\right\|_{m+1},\end{aligned} \tag{36}$$

$$\begin{aligned}&\left\|F_\epsilon^1(\omega_\epsilon,B_\epsilon,\Pi_\epsilon,\Omega_\epsilon)-F_\epsilon^1(\omega_\epsilon,B_\epsilon,\Pi_\epsilon,\widetilde{\Omega}_\epsilon)\right\|_m \\ &\le \left\|\mathcal{J}_\epsilon u_\epsilon^r\left(\mathcal{J}_\epsilon\Omega_\epsilon-\mathcal{J}_\epsilon\widetilde{\Omega}_\epsilon\right)\right\|_m \\ &\le C\|\omega_\epsilon\|_{m-1}\left\|\Omega_\epsilon-\widetilde{\Omega}_\epsilon\right\|_m.\end{aligned} \tag{37}$$

By the virtue of properties of mollifier, Lipschitz continuity of the remaining functions F_ϵ^j, $j = 2,3,4$, can be obtained with constant C/ϵ. Thus, we can deduce the following for $\|\cdot\|_{X^m}, \|\tilde{\cdot}\|_{X^m} \le R$,

$$\left\|F_\epsilon(\cdot) - F_\epsilon(\tilde{\cdot})\right\|_{X^m} \le \frac{C}{\epsilon} R \|\cdot - \tilde{\cdot}\|_{X^m} \tag{38}$$

with $m \ge 3$ and $F_\epsilon = (F_\epsilon^1, F_\epsilon^2, F_\epsilon^3, F_\epsilon^4)$. Now we use the Picard theorem with domain X^m. By picking any initial data $S_\epsilon(0) \in X^m$ and choosing $T_0 = 3/4C_0S_0$, we have, for $R = S_0/(1 - C_0T_0S_0) = 4S_0$,

$$\begin{gathered}\frac{d}{dt}\left\|S_\epsilon\right\|_{X^m}(t) = F_\epsilon(S_\epsilon(t)), \\ \left\|F_\epsilon(S_\epsilon) - F_\epsilon(\tilde{S}_\epsilon)\right\|_{X^m} \le \frac{4CS_0}{\epsilon}\left\|S_\epsilon - \tilde{S}_\epsilon\right\|_{X^m}\end{gathered} \tag{39}$$

where $\|S_\epsilon\|_{X^m}, \|\tilde{S}_\epsilon\|_{X^m} \le 4S_0$. Therefore, the Picard theorem implies that, for each $0 < \epsilon < 1$, there exists a unique solution $S_\epsilon(t) \in C^1(0, T_\epsilon; X^m)$ for a fixed time $T_\epsilon > 0$. For simplicity, let T_ϵ be the maximal existence time of such solution. Suppose that, for some $0 < \epsilon < 1$, we have $T_\epsilon < T_0$. Then by Proposition 7, for arbitrarily small $\delta > 0$, we have

$$\left\|S_\epsilon\right\|_{X^m}(T_\epsilon - \delta) \le \frac{S_0}{1 - C_0S_0T_\epsilon} < 4S_0. \tag{40}$$

If we apply the standard continuation argument, then we can have local-in-time solution S_ϵ at least until T_0. This contradicts the assumption that $T_\epsilon < T_0$. Hence we prove that, for any $0 < \epsilon < 1$, there is a unique solution $S_\epsilon(t)$ with a uniform time T_0, such that $S_\epsilon(t) \in C^1(0, T_0; X^m)$. This completes the proof. □

Proposition 9. *For an integer $m > 7/2$, the solutions obtained in Proposition 8 form the Cauchy sequences in the following spaces:*

$$\begin{gathered}\{S_\epsilon\} \in C\left(0, T_0; X^1\right), \\ \{\partial_t S_\epsilon\} \in C\left(0, T_0; X^0\right).\end{gathered} \tag{41}$$

Proof. Taking ∂_i operator ($i = 1,2,3$) on both sides of $(25)_1$ and multiplying $\partial_i\omega_\epsilon$, we deduce that

$$\begin{aligned}&\frac{1}{2}\frac{\partial}{\partial t}\left|\partial_i(\omega_\epsilon - \omega_{\epsilon'})\right|^2 \\ &\quad = \partial_i\left(F_\epsilon^1(\omega_\epsilon, B_\epsilon, \Pi_\epsilon, \Omega_\epsilon) - F_{\epsilon'}^1(\omega_{\epsilon'}, B_{\epsilon'}, \Pi_{\epsilon'}, \Omega_{\epsilon'})\right) \\ &\quad \cdot \partial_i(\omega_\epsilon - \omega_{\epsilon'}) = A_1 + A_2 + A_3 + A_4 + A_5 + A_6,\end{aligned} \tag{42}$$

where

$$\begin{aligned}A_1 &= ((\mathcal{J}_\epsilon - \mathcal{J}_{\epsilon'})\partial_i(\omega_\epsilon - \omega_{\epsilon'}))(\mathcal{J}_\epsilon\partial_i\mathbf{u}_\epsilon \cdot \nabla)\mathcal{J}_\epsilon\omega_\epsilon \\ &+ (\mathcal{J}_{\epsilon'}\partial_i(\omega_\epsilon - \omega_{\epsilon'}))((\mathcal{J}_\epsilon - \mathcal{J}_{\epsilon'})\partial_i\mathbf{u}_\epsilon \cdot \nabla)\mathcal{J}_\epsilon\omega_\epsilon \\ &+ (\mathcal{J}_{\epsilon'}\partial_i(\omega_\epsilon - \omega_{\epsilon'}))(\mathcal{J}_{\epsilon'}\partial_i(\mathbf{u}_\epsilon - \mathbf{u}_{\epsilon'}) \cdot \nabla)\mathcal{J}_\epsilon\omega_\epsilon \\ &+ (\mathcal{J}_{\epsilon'}\partial_i(\omega_\epsilon - \omega_{\epsilon'}))(\mathcal{J}_{\epsilon'}\partial_i\mathbf{u}_{\epsilon'} \cdot \nabla)(\mathcal{J}_\epsilon - \mathcal{J}_{\epsilon'})\omega_\epsilon \\ &+ (\mathcal{J}_{\epsilon'}\partial_i(\omega_\epsilon - \omega_{\epsilon'}))(\mathcal{J}_{\epsilon'}\partial_i\mathbf{u}_{\epsilon'} \cdot \nabla)\mathcal{J}_{\epsilon'}(\omega_\epsilon - \omega_{\epsilon'}),\end{aligned} \tag{43}$$

$$\begin{aligned}A_2 &= ((\mathcal{J}_\epsilon - \mathcal{J}_{\epsilon'})\partial_i(\omega_\epsilon - \omega_{\epsilon'}))(\mathcal{J}_\epsilon\mathbf{u}_\epsilon \cdot \nabla)\mathcal{J}_\epsilon\partial_i\omega_\epsilon \\ &+ (\mathcal{J}_{\epsilon'}\partial_i(\omega_\epsilon - \omega_{\epsilon'}))((\mathcal{J}_\epsilon - \mathcal{J}_{\epsilon'})\mathbf{u}_\epsilon \cdot \nabla)\mathcal{J}_\epsilon\partial_i\omega_\epsilon \\ &+ (\mathcal{J}_{\epsilon'}\partial_i(\omega_\epsilon - \omega_{\epsilon'}))(\mathcal{J}_{\epsilon'}(\mathbf{u}_\epsilon - \mathbf{u}_{\epsilon'}) \cdot \nabla)\mathcal{J}_\epsilon\partial_i\omega_\epsilon \\ &+ (\mathcal{J}_{\epsilon'}\partial_i(\omega_\epsilon - \omega_{\epsilon'}))(\mathcal{J}_{\epsilon'}\mathbf{u}_{\epsilon'} \cdot \nabla)(\mathcal{J}_\epsilon - \mathcal{J}_{\epsilon'})\partial_i\omega_\epsilon \\ &+ (\mathcal{J}_{\epsilon'}\partial_i(\omega_\epsilon - \omega_{\epsilon'}))(\mathcal{J}_{\epsilon'}\mathbf{u}_{\epsilon'} \cdot \nabla)\mathcal{J}_{\epsilon'}\partial_i(\omega_\epsilon - \omega_{\epsilon'}),\end{aligned} \tag{44}$$

$$\begin{aligned}A_3 &= ((\mathcal{J}_\epsilon - \mathcal{J}_{\epsilon'})\partial_i(\omega_\epsilon - \omega_{\epsilon'}))\mathcal{J}_\epsilon\partial_i u_\epsilon^r\mathcal{J}_\epsilon\Omega_\epsilon \\ &+ (\mathcal{J}_{\epsilon'}\partial_i(\omega_\epsilon - \omega_{\epsilon'}))(\mathcal{J}_\epsilon - \mathcal{J}_{\epsilon'})\partial_i u_\epsilon^r\mathcal{J}_\epsilon\Omega_\epsilon \\ &+ (\mathcal{J}_{\epsilon'}\partial_i(\omega_\epsilon - \omega_{\epsilon'}))\mathcal{J}_{\epsilon'}\partial_i(u_\epsilon^r - u_{\epsilon'}^r)\mathcal{J}_\epsilon\Omega_\epsilon \\ &+ (\mathcal{J}_{\epsilon'}\partial_i(\omega_\epsilon - \omega_{\epsilon'}))\mathcal{J}_{\epsilon'}\partial_i u_{\epsilon'}^r(\mathcal{J}_\epsilon - \mathcal{J}_{\epsilon'})\Omega_\epsilon \\ &+ (\mathcal{J}_{\epsilon'}\partial_i(\omega_\epsilon - \omega_{\epsilon'}))\mathcal{J}_{\epsilon'}\partial_i u_{\epsilon'}^r\mathcal{J}_{\epsilon'}(\Omega_\epsilon - \Omega_{\epsilon'}),\end{aligned} \tag{45}$$

$$\begin{aligned}A_4 &= ((\mathcal{J}_\epsilon - \mathcal{J}_{\epsilon'})\partial_i(\omega_\epsilon - \omega_{\epsilon'}))\mathcal{J}_\epsilon u_\epsilon^r\mathcal{J}_\epsilon\partial_i\Omega_\epsilon \\ &+ (\mathcal{J}_{\epsilon'}\partial_i(\omega_\epsilon - \omega_{\epsilon'}))(\mathcal{J}_\epsilon - \mathcal{J}_{\epsilon'})u_\epsilon^r\mathcal{J}_\epsilon\partial_i\Omega_\epsilon \\ &+ (\mathcal{J}_{\epsilon'}\partial_i(\omega_\epsilon - \omega_{\epsilon'}))\mathcal{J}_{\epsilon'}(u_\epsilon^r - u_{\epsilon'}^r)\mathcal{J}_\epsilon\partial_i\Omega_\epsilon \\ &+ (\mathcal{J}_{\epsilon'}\partial_i(\omega_\epsilon - \omega_{\epsilon'}))\mathcal{J}_{\epsilon'}u_{\epsilon'}^r(\mathcal{J}_\epsilon - \mathcal{J}_{\epsilon'})\partial_i\Omega_\epsilon \\ &+ (\mathcal{J}_{\epsilon'}\partial_i(\omega_\epsilon - \omega_{\epsilon'}))\mathcal{J}_{\epsilon'}u_{\epsilon'}^r\mathcal{J}_{\epsilon'}\partial_i(\Omega_\epsilon - \Omega_{\epsilon'}),\end{aligned} \tag{46}$$

$$\begin{aligned}A_5 &= ((\mathcal{J}_\epsilon - \mathcal{J}_{\epsilon'})\partial_i(\omega_\epsilon - \omega_{\epsilon'}))\mathcal{J}_\epsilon\partial_i B_\epsilon\mathcal{J}_\epsilon\partial_z\Pi_\epsilon \\ &+ (\mathcal{J}_{\epsilon'}\partial_i(\omega_\epsilon - \omega_{\epsilon'}))(\mathcal{J}_\epsilon - \mathcal{J}_{\epsilon'})\partial_i B_\epsilon\mathcal{J}_\epsilon\partial_z\Pi_\epsilon \\ &+ (\mathcal{J}_{\epsilon'}\partial_i(\omega_\epsilon - \omega_{\epsilon'}))\mathcal{J}_{\epsilon'}\partial_i(B_\epsilon - B_{\epsilon'})\mathcal{J}_\epsilon\partial_z\Pi_\epsilon \\ &+ (\mathcal{J}_{\epsilon'}\partial_i(\omega_\epsilon - \omega_{\epsilon'}))\mathcal{J}_{\epsilon'}\partial_i B_{\epsilon'}(\mathcal{J}_\epsilon - \mathcal{J}_{\epsilon'})\partial_z\Pi_\epsilon \\ &+ (\mathcal{J}_{\epsilon'}\partial_i(\omega_\epsilon - \omega_{\epsilon'}))\mathcal{J}_{\epsilon'}\partial_i B_{\epsilon'}\mathcal{J}_{\epsilon'}\partial_z(\Pi_\epsilon - \Pi_{\epsilon'}),\end{aligned} \tag{47}$$

$$\begin{aligned} A_6 &= ((\mathcal{J}_\epsilon - \mathcal{J}_{\epsilon'}) \partial_i (\omega_\epsilon - \omega_{\epsilon'})) \mathcal{J}_\epsilon B_\epsilon \mathcal{J}_\epsilon \partial_i \partial_z \Pi_\epsilon \\ &+ (\mathcal{J}_{\epsilon'} \partial_i (\omega_\epsilon - \omega_{\epsilon'})) (\mathcal{J}_\epsilon - \mathcal{J}_{\epsilon'}) B_\epsilon \mathcal{J}_\epsilon \partial_i \partial_z \Pi_\epsilon \\ &+ (\mathcal{J}_{\epsilon'} \partial_i (\omega_\epsilon - \omega_{\epsilon'})) \mathcal{J}_{\epsilon'} (B_\epsilon - B_{\epsilon'}) \mathcal{J}_\epsilon \partial_i \partial_z \Pi_\epsilon \\ &+ (\mathcal{J}_{\epsilon'} \partial_i (\omega_\epsilon - \omega_{\epsilon'})) \mathcal{J}_{\epsilon'} B_{\epsilon'} (\mathcal{J}_\epsilon - \mathcal{J}_{\epsilon'}) \partial_i \partial_z \Pi_\epsilon \\ &+ (\mathcal{J}_{\epsilon'} \partial_i (\omega_\epsilon - \omega_{\epsilon'})) \mathcal{J}_{\epsilon'} B_{\epsilon'} \mathcal{J}_{\epsilon'} \partial_i \partial_z (\Pi_\epsilon - \Pi_{\epsilon'}) . \end{aligned} \tag{48}$$

$A_1, \ldots, A_6$ can be estimated as follows:

$$\begin{aligned} \int_{\mathbb{R}^3} |A_1|\, dx &\le (\epsilon + \epsilon') \|\nabla (\omega_\epsilon - \omega_{\epsilon'})\|_{L^2} \|(\mathcal{J}_\epsilon \partial_i \mathbf{u}_\epsilon \cdot \nabla) \mathcal{J}_\epsilon \omega_\epsilon\|_1 \\ &+ \|\nabla (\omega_\epsilon - \omega_{\epsilon'})\|_{L^2} (\epsilon + \epsilon') \|\nabla \mathbf{u}_\epsilon\|_1 \|\nabla \omega_\epsilon\|_{L^\infty} \\ &+ \|\nabla (\omega_\epsilon - \omega_{\epsilon'})\|_{L^2}^2 \|\nabla \omega_\epsilon\|_{L^\infty} \\ &+ \|\nabla (\omega_\epsilon - \omega_{\epsilon'})\|_{L^2} \|\nabla \mathbf{u}_{\epsilon'}\|_{L^\infty} (\epsilon + \epsilon') \|\nabla \omega_\epsilon\|_1 \\ &+ \|\nabla (\omega_\epsilon - \omega_{\epsilon'})\|_{L^2}^2 \|\nabla \mathbf{u}_{\epsilon'}\|_{L^\infty} , \end{aligned} \tag{49}$$

$$\begin{aligned} \int_{\mathbb{R}^3} |A_2|\, dx &\le (\epsilon + \epsilon') \|\nabla (\omega_\epsilon - \omega_{\epsilon'})\|_{L^2} \|(\mathcal{J}_\epsilon \mathbf{u}_\epsilon \cdot \nabla) \mathcal{J}_\epsilon \partial_i \omega_\epsilon\|_1 \\ &+ \|\nabla (\omega_\epsilon - \omega_{\epsilon'})\|_{L^2} (\epsilon + \epsilon') \|\nabla \mathbf{u}_\epsilon\|_{L^2} \|\nabla^2 \omega_\epsilon\|_{L^\infty} \\ &+ 2 \|\omega_\epsilon - \omega_{\epsilon'}\|_1^2 \|\nabla^2 \omega_\epsilon\|_{L^2} \\ &+ \|\nabla (\omega_\epsilon - \omega_{\epsilon'})\|_{L^2} \|\mathbf{u}_{\epsilon'}\|_{L^\infty} (\epsilon + \epsilon') \|\nabla^2 \omega_\epsilon\|_1 , \end{aligned} \tag{50}$$

$$\begin{aligned} \int_{\mathbb{R}^3} |A_3|\, dx &\le (\epsilon + \epsilon') \|\nabla (\omega_\epsilon - \omega_{\epsilon'})\|_{L^2} \|\mathcal{J}_\epsilon \partial_i u_\epsilon^r \mathcal{J}_\epsilon \Omega_\epsilon\|_1 \\ &+ \|\nabla (\omega_\epsilon - \omega_{\epsilon'})\|_{L^2} (\epsilon + \epsilon') \|\nabla \mathbf{u}_\epsilon\|_1 \|\Omega_\epsilon\|_{L^\infty} \\ &+ \|\nabla (\omega_\epsilon - \omega_{\epsilon'})\|_{L^2}^2 \|\Omega_\epsilon\|_{L^\infty} \\ &+ \|\nabla (\omega_\epsilon - \omega_{\epsilon'})\|_{L^2} \|\nabla \mathbf{u}_{\epsilon'}\|_{L^\infty} (\epsilon + \epsilon') \|\Omega_\epsilon\|_1 \\ &+ \|\nabla (\omega_\epsilon - \omega_{\epsilon'})\|_{L^2} \|\nabla \mathbf{u}_{\epsilon'}\|_{L^\infty} \|\Omega_\epsilon - \Omega_{\epsilon'}\|_{L^2} , \end{aligned} \tag{51}$$

$$\begin{aligned} \int_{\mathbb{R}^3} |A_4|\, dx &\le (\epsilon + \epsilon') \|\nabla (\omega_\epsilon - \omega_{\epsilon'})\|_{L^2} \|\mathcal{J}_\epsilon u_\epsilon^r \mathcal{J}_\epsilon \partial_i \Omega_\epsilon\|_1 \\ &+ \|\nabla (\omega_\epsilon - \omega_{\epsilon'})\|_{L^2} (\epsilon + \epsilon') \|\nabla \mathbf{u}_\epsilon\|_{L^2} \|\nabla \Omega_\epsilon\|_{L^\infty} \\ &+ \|\omega_\epsilon - \omega_{\epsilon'}\|_1^2 \|\nabla \Omega_\epsilon\|_{L^2} \\ &+ \|\nabla (\omega_\epsilon - \omega_{\epsilon'})\|_{L^2} \|\mathbf{u}_{\epsilon'}\|_{L^\infty} (\epsilon + \epsilon') \|\nabla \Omega_\epsilon\|_1 \\ &+ \|\nabla (\omega_\epsilon - \omega_{\epsilon'})\|_{L^2} \|\mathbf{u}_{\epsilon'}\|_{L^\infty} \|\nabla (\Omega_\epsilon - \Omega_{\epsilon'})\|_{L^2} , \end{aligned} \tag{52}$$

$$\begin{aligned} \int_{\mathbb{R}^3} |A_5|\, dx &\le (\epsilon + \epsilon') \|\nabla (\omega_\epsilon - \omega_{\epsilon'})\|_{L^2} \|\mathcal{J}_\epsilon \partial_i B_\epsilon \mathcal{J}_\epsilon \partial_z \Pi_\epsilon\|_1 \\ &+ \|\nabla (\omega_\epsilon - \omega_{\epsilon'})\|_{L^2} (\epsilon + \epsilon') \|\nabla B_\epsilon\|_1 \|\nabla \Pi_\epsilon\|_{L^\infty} \\ &+ \|\nabla (\omega_\epsilon - \omega_{\epsilon'})\|_{L^2} \|\nabla (B_\epsilon - B_{\epsilon'})\|_{L^2} \|\nabla \Pi_\epsilon\|_{L^\infty} \\ &+ \|\nabla (\omega_\epsilon - \omega_{\epsilon'})\|_{L^2} \|\nabla B_{\epsilon'}\|_{L^\infty} (\epsilon + \epsilon') \|\nabla \Pi_\epsilon\|_1 \\ &+ \|\nabla (\omega_\epsilon - \omega_{\epsilon'})\|_{L^2} \|\nabla B_{\epsilon'}\|_{L^\infty} \|\nabla (\Pi_\epsilon - \Pi_{\epsilon'})\|_{L^2} , \end{aligned} \tag{53}$$

$$\begin{aligned} \int_{\mathbb{R}^3} |A_6|\, dx &\le (\epsilon + \epsilon') \|\nabla (\omega_\epsilon - \omega_{\epsilon'})\|_{L^2} \|\mathcal{J}_\epsilon B_\epsilon \mathcal{J}_\epsilon \partial_i \partial_z \Pi_\epsilon\|_1 \\ &+ \|\nabla (\omega_\epsilon - \omega_{\epsilon'})\|_{L^2} (\epsilon + \epsilon') \|B_\epsilon\|_1 \|\nabla^2 \Pi_\epsilon\|_{L^\infty} \\ &+ \|\nabla (\omega_\epsilon - \omega_{\epsilon'})\|_{L^2} \|B_\epsilon - B_{\epsilon'}\|_{L^2} \|\nabla^2 \Pi_\epsilon\|_{L^\infty} \\ &+ \|\nabla (\omega_\epsilon - \omega_{\epsilon'})\|_{L^2} \|B_{\epsilon'}\|_{L^\infty} (\epsilon + \epsilon') \|\nabla^2 \Pi_\epsilon\|_1 \\ &+ \|\nabla (\omega_\epsilon - \omega_{\epsilon'})\|_{L^2} \|B_{\epsilon'}\|_{L^\infty} \|\nabla^2 (\Pi_\epsilon - \Pi_{\epsilon'})\|_{L^2} . \end{aligned} \tag{54}$$

Similarly, we can obtain the estimates for Π.

$$\begin{aligned} &\frac{1}{2} \frac{d}{dt} |\partial_i \partial_j (\Pi_\epsilon - \Pi_{\epsilon'})|^2 \\ &= \partial_i \partial_j (F_\epsilon^3 (\omega_\epsilon, \Pi_\epsilon) - F_{\epsilon'}^3 (\omega_{\epsilon'}, \Pi_{\epsilon'})) \partial_i \partial_j (\Pi_\epsilon - \Pi_{\epsilon'}) \\ &= B_1 + B_2 + B_3 + B_4 + B_5 + B_6 \end{aligned} \tag{55}$$

where

$$\begin{aligned} B_1 &= ((\mathcal{J}_\epsilon - \mathcal{J}_{\epsilon'}) \partial_i \partial_j (\Pi_\epsilon - \Pi_{\epsilon'})) (\mathcal{J}_\epsilon \partial_i \partial_j \mathbf{u}_\epsilon \cdot \nabla) \\ &\cdot \mathcal{J}_\epsilon \Pi_\epsilon + (\mathcal{J}_{\epsilon'} \partial_i \partial_j (\Pi_\epsilon - \Pi_{\epsilon'})) \\ &\cdot ((\mathcal{J}_\epsilon - \mathcal{J}_{\epsilon'}) \partial_i \partial_j \mathbf{u}_\epsilon \cdot \nabla) \mathcal{J}_\epsilon \Pi_\epsilon \\ &+ (\mathcal{J}_{\epsilon'} \partial_i \partial_j (\Pi_\epsilon - \Pi_{\epsilon'})) (\mathcal{J}_{\epsilon'} \partial_i \partial_j (\mathbf{u}_\epsilon - \mathbf{u}_{\epsilon'}) \cdot \nabla) \\ &\cdot \mathcal{J}_\epsilon \Pi_\epsilon + (\mathcal{J}_{\epsilon'} \partial_i \partial_j (\Pi_\epsilon - \Pi_{\epsilon'})) (\mathcal{J}_{\epsilon'} \partial_i \partial_j \mathbf{u}_{\epsilon'} \cdot \nabla) \\ &\cdot (\mathcal{J}_\epsilon - \mathcal{J}_{\epsilon'}) \Pi_\epsilon + (\mathcal{J}_{\epsilon'} \partial_i \partial_j (\Pi_\epsilon - \Pi_{\epsilon'})) \\ &\cdot (\mathcal{J}_{\epsilon'} \partial_i \partial_j \mathbf{u}_{\epsilon'} \cdot \nabla) \mathcal{J}_{\epsilon'} (\Pi_\epsilon - \Pi_{\epsilon'}) , \end{aligned} \tag{56}$$

$$\begin{aligned} B_2 &= ((\mathcal{J}_\epsilon - \mathcal{J}_{\epsilon'}) \partial_i \partial_j (\Pi_\epsilon - \Pi_{\epsilon'})) (\mathcal{J}_\epsilon \partial_i \mathbf{u}_\epsilon \cdot \nabla) \\ &\cdot \mathcal{J}_\epsilon \partial_j \Pi_\epsilon + (\mathcal{J}_{\epsilon'} \partial_i \partial_j (\Pi_\epsilon - \Pi_{\epsilon'})) \\ &\cdot ((\mathcal{J}_\epsilon - \mathcal{J}_{\epsilon'}) \partial_i \mathbf{u}_\epsilon \cdot \nabla) \mathcal{J}_\epsilon \partial_j \Pi_\epsilon \end{aligned}$$

$$
\begin{aligned}
&+\left(\mathcal{J}_{\epsilon'}\partial_i\partial_j\left(\Pi_\epsilon-\Pi_{\epsilon'}\right)\right)\left(\mathcal{J}_{\epsilon'}\partial_i\left(\mathbf{u}_\epsilon-\mathbf{u}_{\epsilon'}\right)\cdot\nabla\right)\\
&\cdot\mathcal{J}_\epsilon\partial_j\Pi_\epsilon+\left(\mathcal{J}_{\epsilon'}\partial_i\partial_j\left(\Pi_\epsilon-\Pi_{\epsilon'}\right)\right)\left(\mathcal{J}_{\epsilon'}\partial_i\mathbf{u}_{\epsilon'}\cdot\nabla\right)\\
&\cdot\left(\mathcal{J}_\epsilon-\mathcal{J}_{\epsilon'}\right)\partial_j\Pi_\epsilon+\left(\mathcal{J}_{\epsilon'}\partial_i\partial_j\left(\Pi_\epsilon-\Pi_{\epsilon'}\right)\right)\\
&\cdot\left(\mathcal{J}_{\epsilon'}\partial_i\mathbf{u}_{\epsilon'}\cdot\nabla\right)\mathcal{J}_{\epsilon'}\partial_j\left(\Pi_\epsilon-\Pi_{\epsilon'}\right),
\end{aligned}
\tag{57}
$$

$$
\begin{aligned}
B_3=&\left(\left(\mathcal{J}_\epsilon-\mathcal{J}_{\epsilon'}\right)\partial_i\partial_j\left(\Pi_\epsilon-\Pi_{\epsilon'}\right)\right)\left(\mathcal{J}_\epsilon\mathbf{u}_\epsilon\cdot\nabla\right)\\
&\cdot\mathcal{J}_\epsilon\partial_i\partial_j\Pi_\epsilon+\left(\mathcal{J}_{\epsilon'}\partial_i\partial_j\left(\Pi_\epsilon-\Pi_{\epsilon'}\right)\right)\\
&\cdot\left(\left(\mathcal{J}_\epsilon-\mathcal{J}_{\epsilon'}\right)\mathbf{u}_\epsilon\cdot\nabla\right)\mathcal{J}_\epsilon\partial_i\partial_j\Pi_\epsilon\\
&+\left(\mathcal{J}_{\epsilon'}\partial_i\partial_j\left(\Pi_\epsilon-\Pi_{\epsilon'}\right)\right)\left(\mathcal{J}_{\epsilon'}\left(\mathbf{u}_\epsilon-\mathbf{u}_{\epsilon'}\right)\cdot\nabla\right)\\
&\cdot\mathcal{J}_\epsilon\partial_i\partial_j\Pi_\epsilon+\left(\mathcal{J}_{\epsilon'}\partial_i\partial_j\left(\Pi_\epsilon-\Pi_{\epsilon'}\right)\right)\left(\mathcal{J}_{\epsilon'}\mathbf{u}_{\epsilon'}\cdot\nabla\right)\\
&\cdot\left(\mathcal{J}_\epsilon-\mathcal{J}_{\epsilon'}\right)\partial_i\partial_j\Pi_\epsilon+\left(\mathcal{J}_{\epsilon'}\partial_i\partial_j\left(\Pi_\epsilon-\Pi_{\epsilon'}\right)\right)\\
&\cdot\left(\mathcal{J}_{\epsilon'}\mathbf{u}_{\epsilon'}\cdot\nabla\right)\mathcal{J}_{\epsilon'}\partial_i\partial_j\left(\Pi_\epsilon-\Pi_{\epsilon'}\right),
\end{aligned}
\tag{58}
$$

$$
\begin{aligned}
B_4=&\left(\left(\mathcal{J}_\epsilon-\mathcal{J}_{\epsilon'}\right)\partial_i\partial_j\left(\Pi_\epsilon-\Pi_{\epsilon'}\right)\right)\mathcal{J}_\epsilon\partial_i\partial_j\Pi_\epsilon\mathcal{J}_\epsilon\partial_z\Pi_\epsilon\\
&+\left(\mathcal{J}_{\epsilon'}\partial_i\partial_j\left(\Pi_\epsilon-\Pi_{\epsilon'}\right)\right)\left(\mathcal{J}_\epsilon-\mathcal{J}_{\epsilon'}\right)\partial_i\partial_j\Pi_\epsilon\mathcal{J}_\epsilon\partial_z\Pi_\epsilon\\
&+\left(\mathcal{J}_{\epsilon'}\partial_i\partial_j\left(\Pi_\epsilon-\Pi_{\epsilon'}\right)\right)\mathcal{J}_{\epsilon'}\partial_i\partial_j\left(\Pi_\epsilon-\Pi_{\epsilon'}\right)\mathcal{J}_\epsilon\partial_z\Pi_\epsilon\\
&+\left(\mathcal{J}_{\epsilon'}\partial_i\partial_j\left(\Pi_\epsilon-\Pi_{\epsilon'}\right)\right)\mathcal{J}_{\epsilon'}\partial_i\partial_j\Pi_{\epsilon'}\left(\mathcal{J}_\epsilon-\mathcal{J}_{\epsilon'}\right)\partial_z\Pi_\epsilon\\
&+\left(\mathcal{J}_{\epsilon'}\partial_i\partial_j\left(\Pi_\epsilon-\Pi_{\epsilon'}\right)\right)\\
&\cdot\mathcal{J}_{\epsilon'}\partial_i\partial_j\Pi_{\epsilon'}\mathcal{J}_{\epsilon'}\partial_z\left(\Pi_\epsilon-\Pi_{\epsilon'}\right),
\end{aligned}
\tag{59}
$$

$$
\begin{aligned}
B_5=&\left(\left(\mathcal{J}_\epsilon-\mathcal{J}_{\epsilon'}\right)\partial_i\partial_j\left(\Pi_\epsilon-\Pi_{\epsilon'}\right)\right)\mathcal{J}_\epsilon\partial_i\Pi_\epsilon\mathcal{J}_\epsilon\partial_j\partial_z\Pi_\epsilon\\
&+\left(\mathcal{J}_{\epsilon'}\partial_i\partial_j\left(\Pi_\epsilon-\Pi_{\epsilon'}\right)\right)\left(\mathcal{J}_\epsilon-\mathcal{J}_{\epsilon'}\right)\partial_i\Pi_\epsilon\mathcal{J}_\epsilon\partial_j\partial_z\Pi_\epsilon\\
&+\left(\mathcal{J}_{\epsilon'}\partial_i\partial_j\left(\Pi_\epsilon-\Pi_{\epsilon'}\right)\right)\mathcal{J}_{\epsilon'}\partial_i\left(\Pi_\epsilon-\Pi_{\epsilon'}\right)\mathcal{J}_\epsilon\partial_j\partial_z\Pi_\epsilon\\
&+\left(\mathcal{J}_{\epsilon'}\partial_i\partial_j\left(\Pi_\epsilon-\Pi_{\epsilon'}\right)\right)\mathcal{J}_{\epsilon'}\partial_i\Pi_{\epsilon'}\left(\mathcal{J}_\epsilon-\mathcal{J}_{\epsilon'}\right)\partial_j\partial_z\Pi_\epsilon\\
&+\left(\mathcal{J}_{\epsilon'}\partial_i\partial_j\left(\Pi_\epsilon-\Pi_{\epsilon'}\right)\right)\\
&\cdot\mathcal{J}_{\epsilon'}\partial_i\Pi_{\epsilon'}\mathcal{J}_{\epsilon'}\partial_j\partial_z\left(\Pi_\epsilon-\Pi_{\epsilon'}\right),
\end{aligned}
\tag{60}
$$

$$
\begin{aligned}
B_6=&\left(\left(\mathcal{J}_\epsilon-\mathcal{J}_{\epsilon'}\right)\partial_i\partial_j\left(\Pi_\epsilon-\Pi_{\epsilon'}\right)\right)\mathcal{J}_\epsilon\Pi_\epsilon\mathcal{J}_\epsilon\partial_i\partial_j\partial_z\Pi_\epsilon\\
&+\left(\mathcal{J}_{\epsilon'}\partial_i\partial_j\left(\Pi_\epsilon-\Pi_{\epsilon'}\right)\right)\left(\mathcal{J}_\epsilon-\mathcal{J}_{\epsilon'}\right)\Pi_\epsilon\mathcal{J}_\epsilon\partial_i\partial_j\partial_z\Pi_\epsilon\\
&+\left(\mathcal{J}_{\epsilon'}\partial_i\partial_j\left(\Pi_\epsilon-\Pi_{\epsilon'}\right)\right)\mathcal{J}_{\epsilon'}\left(\Pi_\epsilon-\Pi_{\epsilon'}\right)\mathcal{J}_\epsilon\partial_i\partial_j\partial_z\Pi_\epsilon\\
&+\left(\mathcal{J}_{\epsilon'}\partial_i\partial_j\left(\Pi_\epsilon-\Pi_{\epsilon'}\right)\right)\mathcal{J}_{\epsilon'}\Pi_{\epsilon'}\left(\mathcal{J}_\epsilon-\mathcal{J}_{\epsilon'}\right)\partial_i\partial_j\partial_z\Pi_\epsilon\\
&+\left(\mathcal{J}_{\epsilon'}\partial_i\partial_j\left(\Pi_\epsilon-\Pi_{\epsilon'}\right)\right)\\
&\cdot\mathcal{J}_{\epsilon'}\partial_i\Pi_{\epsilon'}\mathcal{J}_{\epsilon'}\partial_i\partial_j\partial_z\left(\Pi_\epsilon-\Pi_{\epsilon'}\right).
\end{aligned}
\tag{61}
$$

These $B_1,\dots,B_6$ can be estimated similarly.

$$
\begin{aligned}
&\int_{\mathbb{R}^3}\left|B_1\right|dx\\
&\le\left(\epsilon+\epsilon'\right)\left\|\nabla^2\left(\Pi_\epsilon-\Pi_{\epsilon'}\right)\right\|_{L^2}\left\|\left(\mathcal{J}_\epsilon\partial_i\partial_j\mathbf{u}_\epsilon\cdot\nabla\right)\mathcal{J}_\epsilon\Pi_\epsilon\right\|_1\\
&\quad+\left\|\nabla^2\left(\Pi_\epsilon-\Pi_{\epsilon'}\right)\right\|_{L^2}\left(\epsilon+\epsilon'\right)\left\|\nabla^2\mathbf{u}_\epsilon\right\|_1\left\|\nabla\Pi_\epsilon\right\|_{L^\infty}\\
&\quad+\left\|\nabla^2\left(\Pi_\epsilon-\Pi_{\epsilon'}\right)\right\|_{L^2}\left\|\nabla\left(\omega_\epsilon-\omega_{\epsilon'}\right)\right\|_{L^2}\left\|\nabla\Pi_\epsilon\right\|_{L^\infty}\\
&\quad+\left\|\nabla^2\left(\Pi_\epsilon-\Pi_{\epsilon'}\right)\right\|_{L^2}\left\|\nabla^2\mathbf{u}_{\epsilon'}\right\|_{L^\infty}\left(\epsilon+\epsilon'\right)\left\|\nabla\Pi_\epsilon\right\|_1\\
&\quad+\left\|\Pi_\epsilon-\Pi_{\epsilon'}\right\|_2^2\left\|\nabla^2\mathbf{u}_{\epsilon'}\right\|_{L^\infty},
\end{aligned}
\tag{62}
$$

$$
\begin{aligned}
&\int_{\mathbb{R}^3}\left|B_2\right|dx\\
&\le\left(\epsilon+\epsilon'\right)\left\|\nabla^2\left(\Pi_\epsilon-\Pi_{\epsilon'}\right)\right\|_{L^2}\left\|\left(\mathcal{J}_\epsilon\partial_i\mathbf{u}_\epsilon\cdot\nabla\right)\mathcal{J}_\epsilon\partial_j\Pi_\epsilon\right\|_1\\
&\quad+\left\|\nabla^2\left(\Pi_\epsilon-\Pi_{\epsilon'}\right)\right\|_{L^2}\left(\epsilon+\epsilon'\right)\left\|\nabla\mathbf{u}_\epsilon\right\|_1\left\|\nabla^2\Pi_\epsilon\right\|_{L^\infty}\\
&\quad+\left\|\nabla^2\left(\Pi_\epsilon-\Pi_{\epsilon'}\right)\right\|_{L^2}\left\|\omega_\epsilon-\omega_{\epsilon'}\right\|_{L^2}\left\|\nabla^2\Pi_\epsilon\right\|_{L^\infty}\\
&\quad+\left\|\nabla^2\left(\Pi_\epsilon-\Pi_{\epsilon'}\right)\right\|_{L^2}\left\|\nabla\mathbf{u}_{\epsilon'}\right\|_{L^\infty}\left(\epsilon+\epsilon'\right)\left\|\nabla^2\Pi_\epsilon\right\|_1\\
&\quad+\left\|\nabla^2\left(\Pi_\epsilon-\Pi_{\epsilon'}\right)\right\|_{L^2}^2\left\|\nabla\mathbf{u}_{\epsilon'}\right\|_{L^\infty},
\end{aligned}
\tag{63}
$$

$$
\begin{aligned}
&\int_{\mathbb{R}^3}\left|B_3\right|dx\\
&\le\left(\epsilon+\epsilon'\right)\left\|\nabla^2\left(\Pi_\epsilon-\Pi_{\epsilon'}\right)\right\|_{L^2}\left\|\left(\mathcal{J}_\epsilon\mathbf{u}_\epsilon\cdot\nabla\right)\mathcal{J}_\epsilon\partial_i\partial_j\Pi_\epsilon\right\|_1\\
&\quad+\left\|\nabla^2\left(\Pi_\epsilon-\Pi_{\epsilon'}\right)\right\|_{L^2}\left(\epsilon+\epsilon'\right)\left\|\nabla\mathbf{u}_\epsilon\right\|_{L^2}\left\|\nabla^3\Pi_\epsilon\right\|_{L^\infty}\\
&\quad+\left\|\nabla^2\left(\Pi_\epsilon-\Pi_{\epsilon'}\right)\right\|_{L^2}\left\|\omega_\epsilon-\omega_{\epsilon'}\right\|_1\left\|\nabla^3\Pi_\epsilon\right\|_{L^2}\\
&\quad+\left\|\nabla^2\left(\Pi_\epsilon-\Pi_{\epsilon'}\right)\right\|_{L^2}\left\|\mathbf{u}_{\epsilon'}\right\|_{L^\infty}\left(\epsilon+\epsilon'\right)\left\|\nabla^3\Pi_\epsilon\right\|_1,
\end{aligned}
\tag{64}
$$

$$
\begin{aligned}
&\int_{\mathbb{R}^3}\left|B_4\right|dx\\
&\le\left(\epsilon+\epsilon'\right)\left\|\nabla^2\left(\Pi_\epsilon-\Pi_{\epsilon'}\right)\right\|_{L^2}\left\|\mathcal{J}_\epsilon\partial_i\partial_j\Pi_\epsilon\mathcal{J}_\epsilon\partial_z\Pi_\epsilon\right\|_1\\
&\quad+\left\|\nabla^2\left(\Pi_\epsilon-\Pi_{\epsilon'}\right)\right\|_{L^2}\left(\epsilon+\epsilon'\right)\left\|\nabla^2\Pi_\epsilon\right\|_1\left\|\nabla\Pi_\epsilon\right\|_{L^\infty}\\
&\quad+\left\|\nabla^2\left(\Pi_\epsilon-\Pi_{\epsilon'}\right)\right\|_{L^2}^2\left\|\nabla\Pi_\epsilon\right\|_{L^\infty}\\
&\quad+\left\|\nabla^2\left(\Pi_\epsilon-\Pi_{\epsilon'}\right)\right\|_{L^2}\left\|\nabla^2\Pi_{\epsilon'}\right\|_{L^\infty}\left(\epsilon+\epsilon'\right)\left\|\nabla\Pi_\epsilon\right\|_1\\
&\quad+\left\|\left(\Pi_\epsilon-\Pi_{\epsilon'}\right)\right\|_2^2\left\|\nabla^2\Pi_{\epsilon'}\right\|_{L^\infty},
\end{aligned}
\tag{65}
$$

$$
\begin{aligned}
&\int_{\mathbb{R}^3}\left|B_5\right|dx\\
&\le\left(\epsilon+\epsilon'\right)\left\|\nabla^2\left(\Pi_\epsilon-\Pi_{\epsilon'}\right)\right\|_{L^2}\left\|\mathcal{J}_\epsilon\partial_i\Pi_\epsilon\mathcal{J}_\epsilon\partial_j\partial_z\Pi_\epsilon\right\|_1\\
&\quad+\left\|\nabla^2\left(\Pi_\epsilon-\Pi_{\epsilon'}\right)\right\|_{L^2}\left(\epsilon+\epsilon'\right)\left\|\nabla\Pi_\epsilon\right\|_1\left\|\nabla^2\Pi_\epsilon\right\|_{L^\infty}\\
&\quad+\left\|\left(\Pi_\epsilon-\Pi_{\epsilon'}\right)\right\|_2^2\left\|\nabla^2\Pi_\epsilon\right\|_{L^\infty}
\end{aligned}
$$

$$+\left\|\nabla^2\left(\Pi_\epsilon-\Pi_{\epsilon'}\right)\right\|_{L^2}\left\|\nabla\Pi_{\epsilon'}\right\|_{L^\infty}\left(\epsilon+\epsilon'\right)\left\|\nabla^2\Pi_\epsilon\right\|_1 + \left\|\nabla^2\left(\Pi_\epsilon-\Pi_{\epsilon'}\right)\right\|_{L^2}^2\left\|\nabla\Pi_{\epsilon'}\right\|_{L^\infty}, \tag{66}$$

$$\begin{aligned}\int_{\mathbb{R}^3}\left|B_6\right|dx &\le \left(\epsilon+\epsilon'\right)\left\|\nabla^2\left(\Pi_\epsilon-\Pi_{\epsilon'}\right)\right\|_{L^2}\left\|\mathcal{J}_\epsilon\Pi_\epsilon\mathcal{J}_\epsilon\partial_i\partial_j\partial_z\Pi_\epsilon\right\|_1 \\ &+ \left\|\nabla^2\left(\Pi_\epsilon-\Pi_{\epsilon'}\right)\right\|_{L^2}\left(\epsilon+\epsilon'\right)\left\|\Pi_\epsilon\right\|_1\left\|\nabla^3\Pi_\epsilon\right\|_{L^\infty} \\ &+ \left\|\nabla^2\left(\Pi_\epsilon-\Pi_{\epsilon'}\right)\right\|_2^2\left\|\nabla^3\Pi_\epsilon\right\|_{L^\infty} \\ &+ \left\|\nabla^2\left(\Pi_\epsilon-\Pi_{\epsilon'}\right)\right\|_{L^2}\left\|\Pi_{\epsilon'}\right\|_{L^\infty}\left(\epsilon+\epsilon'\right)\left\|\nabla^3\Pi_\epsilon\right\|_1.\end{aligned} \tag{67}$$

The other terms B and $\Pi+\Omega$ can be estimated similarly, so we omit the details. Then we have

$$\begin{aligned}\frac{1}{2}\frac{d}{dt}\|S\|_{X^1}(t) &:= \frac{1}{2}\frac{d}{dt}\left(\left\|\omega_\epsilon-\omega_{\epsilon'}\right\|_1+\left\|B_\epsilon-B_{\epsilon'}\right\|_1 + \left\|\Pi_\epsilon-\Pi_{\epsilon'}\right\|_2+\left\|\left(\Pi_\epsilon+\Omega_\epsilon\right)-\left(\Pi_{\epsilon'}+\Omega_{\epsilon'}\right)\right\|_1\right) \\ &\le \left(\left(\epsilon+\epsilon'\right)+\|S\|_{X^1}(t)\right)\left(\left\|\nabla^3\Pi_\epsilon\right\|_{L^\infty}+\left\|\nabla^2\omega_\epsilon\right\|_{L^\infty} + \left\|\nabla^2B_\epsilon\right\|_{L^\infty}+\left\|\nabla^2\left(\Pi_\epsilon+\Omega_\epsilon\right)\right\|_{L^\infty}\right)\le C\left(\left(\epsilon+\epsilon'\right) + \|S\|_{X^1}(t)\right),\end{aligned} \tag{68}$$

for $m>7/2$. Gronwall's inequality gives us

$$\sup_{0\le t\le T_0}\|S\|_{X^1}(t)\le\left(\epsilon+\epsilon'\right)e^{CT_0}, \tag{69}$$

which implies that $\{S_\epsilon\}\in C([0,T_0],X^1)$ and this information completes the proof. □

Proof of Theorem 1. With the bounds in Proposition 7, if we use the Sobolev inequality, then we can obtain the higher order convergence, i.e., $S\in C(0,T_0;X^s)$ for all $s<m$ by the following inequality

$$\sup_{0\le t\le T_0}\left\|S_{\epsilon'}-S_\epsilon\right\|_{X^s}\le C\left\|S_{\epsilon'}-S_\epsilon\right\|_{X^1}^{1-s/m}\left\|S_{\epsilon'}-S_\epsilon\right\|_m^{s/m}. \tag{70}$$

Now, to show $S\in C(0,T_0;X^m)\cap\mathrm{Lip}(0,T_0;X^{m-1})$ where S satisfies our equations in classical sense almost every time, we begin the process of obtaining the right continuity at $t=0$ first. Because X^m is a reflexive Banach space, by Proposition 7, there exist a subsequence and limit functions $S(t)\in L^\infty(0,T_0;H^m)$ which satisfies $\partial_tS(t)\in L^\infty(0,T_0;H^{m-1})$ for any $0\le t\le T_0$, $S_{\epsilon_n}(t)\rightharpoonup S(t)$ in X^m and $\partial_tS_{\epsilon_n}(t)\rightharpoonup\partial_tS(t)$ in X^{m-1}. This implies that $\|S(t)\|_{X^m}\le S_0/(1-C_0S_0t)$, $\|\partial_tS(t)\|_{X^{m-1}}\le(S_0/(1-C_0S_0t))^2$. Thus we have

$$\limsup_{t\longrightarrow0+}\|S\|_{X^m}(t)\le S_0. \tag{71}$$

If we use the above result, $S(t)\in C(0,T_0;X^s)$ for any $s<m$, then $S\in C_w(0,T_0;X^m)$ is obtained by the following estimate. For arbitrary $\delta>0$ and $\varphi\in X^{-m}$, there exists $\tilde\varphi\in X^{-s}$ such that $\|\varphi-\tilde\varphi\|_{X^{-m}}\le\delta/8S_0$,

$$\begin{aligned}\left|\left[\varphi,S_\epsilon-S\right]_{X^m}\right| &\le\|\varphi-\tilde\varphi\|_{X^{-m}}\left\|S_\epsilon-S\right\|_{X^m} + \left|\left[\tilde\varphi,S_\epsilon-S\right]_{X^s}\right| \\ &\le\delta+\left|\left[\tilde\varphi,S_\epsilon-S\right]_{X^s}\right|,\end{aligned} \tag{72}$$

where $[\cdot,\cdot]_{X^k}$ is a dual pairing on $X^{-k}\times X^k$. If we choose $\varphi=\nabla^{2m}\omega_0\in H^{-m}$, then by the weak continuity,

$$\begin{aligned}\left|\left[\varphi,\omega_0\right]_{H^m}\right| &\le\left|\left[\varphi,\omega(t)-\omega_0\right]_{H^m}\right|+\left|\left[\varphi,\omega(t)\right]_{H^m}\right| \\ &\le\delta+\left|\left[\varphi,\omega(t)\right]_{H^m}\right|,\end{aligned} \tag{73}$$

$$\begin{aligned}\left\|\nabla^m\omega_0\right\|_{L^2}^2 &\le\delta+\left\|\nabla^m\omega(t)\right\|_{L^2}\left\|\nabla^m\omega_0\right\|_{L^2} \\ &\le\delta+\frac{\left\|\nabla^m\omega(t)\right\|_{L^2}^2}{2}+\frac{\left\|\nabla^m\omega_0\right\|_{L^2}^2}{2}.\end{aligned} \tag{74}$$

Similarly, we have

$$\liminf_{t\longrightarrow0+}\|S\|_{X^m}(t)\ge S_0. \tag{75}$$

By inequalities (71) and (75), we have the right continuity of $S(t)$ at $t=0$. If we apply the standard time translation invariant property and the time reversal techique, we also have $S(t)\in C(0,T_0;H^m)$ without any difficulty. Lipschitz continuity also can be obtained by

$$\begin{aligned}\left\|S(t_1)-S(t_2)\right\|_{X^{m-1}} &\le\delta+\left\|S_\epsilon(t_1)-S_\epsilon(t_2)\right\|_{X^{m-1}} \\ &\le\delta+\left\|\int_{t_2}^{t_1}\partial_tS_\epsilon(\tau)\,d\tau\right\|_{X^{m-1}} \\ &\le\delta + \left|t_1-t_2\right|\sup_{0\le\tau\le T_0}\left\|\partial_tS_\epsilon(\tau)\right\|_{X^{m-1}} \\ &\le\delta+\left(4S_0\right)^2\left|t_1-t_2\right|,\end{aligned} \tag{76}$$

which means $S\in C(0,T_0;X^m)\cap\mathrm{Lip}(0,T_0;X^{m-1})$. Hence it is a classical solution to (21) almost every time.

Next we assume that the initial data $(\mathbf{u}_0,B_0,\Pi_0,\Omega_0)$ is axisymmetric. Then the axisymmetry of a classical solution to (21) is preserved and (ω,B,Π,Ω) is axisymmetric solution to (21). Now we go back to (9) with $\nu=\kappa=0$, and set

$$\begin{aligned}&u_0\in H^m\left(\mathbb{R}^3\right),\\ &\frac{\operatorname{curl}u_0}{r}\in H^{m-1}\left(\mathbb{R}^3\right),\\ &B_0\in H^m\left(\mathbb{R}^3\right),\\ &\frac{B_0}{r}\in H^m\left(\mathbb{R}^3\right),\\ &m>\frac{9}{2}.\end{aligned} \tag{77}$$

Then for

$$\omega_0 = \text{curl}\, u_0,$$
$$\Omega_0 = \frac{\omega_0}{r}, \qquad (78)$$
$$\Pi_0 = \frac{B_0}{r},$$

we know that there exists a unique solution $(\omega, B, \Omega) \in C(0, T_0; H^{m-1}) \cap \text{Lip}(0, T_0; H^{m-2})$, $\Pi \in C(0, T_0; H^m) \cap \text{Lip}(0, T_0; H^{m-1})$. But if we replace Ω with ω/r and Π with B/r, then $\omega, B, \omega/r, B/r$ also satisfy (21) with the initial data. So by the uniqueness, $\Omega = \omega/r$ in $C(0, T_0; H^{m-1}) \cap \text{Lip}(0, T_0; H^{m-2})$ and $\Pi = B/r$ in $C(0, T_0; H^m) \cap \text{Lip}(0, T_0; H^{m-1})$. Next, we can show that $\mathbf{u} = \Phi * \text{curl}\omega \in C(0, T_0; H^m) \cap \text{Lip}(0, T_0; H^{m-1})$. By the Poincaré lemma, $\text{curl}\mathbf{u} = \omega$ satisfies the ω equation of (9). Then we can deduce that $\mathbf{u}$ and B satisfy (9) almost every time by finding the axisymmetric scalar pressure p. Then the energy inequality (4) implies that $u \in L^\infty(0, T_0; L^2)$ and, almost every time,

$$\begin{aligned}
&\|u(t_1) - u(t_2)\|_{L^2(\mathbb{R}^3)} \le |t_1 - t_2| \\
&\quad \cdot \int_{t_2}^{t_1} \|\partial_t u(t_1) - \partial_t u(t_2)\|_{L^2(\mathbb{R}^3)}(t)\, dt \le |t_1 - t_2| \\
&\quad \cdot \sup_{0\le t\le T_0} \|\partial_t u\|_{L^2} \le |t_1 - t_2| \sup_{0\le t\le T_0} \left(\|u\|_{L^\infty} \|\nabla u\|_{L^2}\right. \qquad (79)\\
&\quad \left. + \|(\nabla\times B)\times B\|_{L^2} + \|\nabla p\|_{L^2}\right) \le |t_1 - t_2| \\
&\quad \cdot \sup_{0\le t\le T_0} \left(\|\omega\|_1^2 + \|B\|_2^2\right) < C|t_1 - t_2|,
\end{aligned}$$

which implies $\mathbf{u} \in C(0, T_0; H^m) \cap \text{Lip}(0, T_0; H^{m-1})$. The uniqueness of $\mathbf{u}$ can be obtained from the standard techniques and we omit the details. Finally we can show that $B \in C(0, T_0; H^m) \cap \text{Lip}(0, T_0; H^{m-1})$. Almost every time, we can rewrite the B equation with

$$\partial_t B + (u^r\partial_r + u^z\partial_z) B = u^r\Pi + 2\Pi\partial_z B, \quad \left(\frac{B}{r} = \Pi\right). \qquad (80)$$

Then we can obtain the conclusion through the standard H^m estimate with Gronwall's inequality. □

3. Proof of Theorem 3: Blow-Up of Axisymmetric Hall Equations

The proof of Theorem 3 is split into two propositions: local-in-time existence of a regular solution to (14) and the finite time blow-up of the local-in-time solution.

Proposition 10. *The equation*

$$\partial_t\Pi = -2\Pi\partial_z\Pi$$
$$\Pi(x, 0) = \Pi_0(x) \in H^m, \quad m > \frac{5}{2} \qquad (81)$$

has a unique local-in-time solution.

Proof. First, we find the global solution to the following regularized equation of (14) without assuming the axisymmetry,

$$\partial_t\Pi_\epsilon = -2\mathcal{J}_\epsilon(\mathcal{J}_\epsilon\Pi_\epsilon\cdot\partial_z\mathcal{J}_\epsilon\Pi_\epsilon)$$
$$\Pi(x, 0) = \Pi_0(x), \quad \Pi_0 \in H^m, \; m > \frac{5}{2} \qquad (82)$$

Before proceeding further, we note that the divergence theorem can be applicable due to the mollifier. Let

$$F(\Pi_\epsilon) = -2\mathcal{J}_\epsilon(\mathcal{J}_\epsilon\Pi_\epsilon\cdot\partial_z\mathcal{J}_\epsilon\Pi_\epsilon). \qquad (83)$$

Hence, the image of the function F defined on H^m is included in H^m for $m > 3/2$.

To use the Picard theorem on H^m space ($m > 5/2$), we first obtain that F is Lipschitz continuous on H^m, i.e.,

$$\begin{aligned}
&\left\|F(\Pi_\epsilon^1) - F(\Pi_\epsilon^2)\right\|_m \\
&\le 2\left\|\mathcal{J}_\epsilon\left\{\mathcal{J}_\epsilon\Pi_\epsilon^1\left(\partial_z\mathcal{J}_\epsilon\Pi_\epsilon^1 - \partial_z\mathcal{J}_\epsilon\Pi_\epsilon^2\right)\right\}\right\|_m \\
&\quad + 2\left\|\mathcal{J}_\epsilon\left\{\left(\mathcal{J}_\epsilon\Pi_\epsilon^1 - \mathcal{J}_\epsilon\Pi_\epsilon^2\right)\partial_z\mathcal{J}_\epsilon\Pi_\epsilon^2\right\}\right\|_m \\
&\le 2\left\|\mathcal{J}_\epsilon\Pi_\epsilon^1\right\|_m\left\|\partial_z\mathcal{J}_\epsilon\left(\Pi_\epsilon^1 - \Pi_\epsilon^2\right)\right\|_m \qquad (84)\\
&\quad + 2\left\|\mathcal{J}_\epsilon\left(\Pi_\epsilon^1 - \Pi_\epsilon^2\right)\right\|_m\left\|\partial_z\mathcal{J}_\epsilon\Pi_\epsilon^2\right\|_m \\
&\le \frac{C}{\epsilon^{m+1}}\left(\left\|\Pi_\epsilon^1\right\|_{L^2} + \left\|\Pi_\epsilon^2\right\|_{L^2}\right)\left\|\Pi_\epsilon^1 - \Pi_\epsilon^2\right\|_m.
\end{aligned}$$

F is a Lipschitz continous function on a bounded open set O in H^m. Now we can apply Picard theorem. For each $\epsilon > 0$, there exist a unique solution Π_ϵ and a finite time T_ϵ, such that $\Pi_\epsilon \in C^1(0, T_\epsilon : H^m)$. Following the standard process of constructing local-in-time solution, we obtain an implicit form of the solution

$$\Pi_\epsilon(t) = \Pi_0 + \int_0^t F(\Pi_\epsilon(s))\, ds, \quad 0 \le t < T_\epsilon. \qquad (85)$$

Since $F(0) = 0$, we have

$$\begin{aligned}
\|\Pi_\epsilon\|_m(t) &\le \|\Pi_0\|_m + \int_0^t \|F(\Pi_\epsilon(s))\|_m\, ds \\
&\le \|\Pi_0\|_m + \int_0^t C_{m,\epsilon}\|\Pi_\epsilon\|_{L^2}(s)\|\Pi_\epsilon\|_m(s)\, ds, \qquad (86)\\
&\qquad 0 \le t < T_\epsilon.
\end{aligned}$$

Since the above regularized equation satisfies an energy estimate, we deduce that

$$\frac{d}{dt}\int_{\mathbb{R}^3}\Pi_\epsilon^2 dx = -\frac{2}{3}\int_{\mathbb{R}^3}\partial_z(\mathcal{J}_\epsilon\Pi_\epsilon)^3 dx = 0, \qquad (87)$$

and hence

$$\|\Pi_\epsilon\|_{L^2}(t) = \|\Pi_0\|_{L^2}. \qquad (88)$$

For the higher order norm, Gronwall's inequality implies

$$\|\Pi_\epsilon\|_m(t) \le \|\Pi_0\|_m e^{C_{m,\epsilon}\|\Pi_0\|_{L^2(\mathbb{R}^3)}t}, \quad 0 \le t < T_\epsilon. \qquad (89)$$

The above inequality justifies that each solution Π_ϵ is a global solution to regularized equation, and

$$\Pi_\epsilon \in C^1(0,\infty : H^m), \quad \text{for all } \Pi_0 \in H^m \tag{90}$$

Second, we show that, for some finite time T, the sequence $\{\Pi_\epsilon\}_{\epsilon>0}$ is a Cauchy sequence in $C(0,T;L^2)$. We note that, for $T < 1/C_m\|\Pi_0\|_m$,

$$\begin{aligned}\sup_{0\le t\le T} \|\Pi_\epsilon\|_m(t) &\le \frac{1}{1-C_mT\|\Pi_0\|_m}, \\ \sup_{0\le t\le T} \|\partial_t\Pi_\epsilon\|_{m-1}(t) &\le C\left(\frac{1}{1-C_mT\|\Pi_0\|_m}\right)^2.\end{aligned} \tag{91}$$

By the standard energy estimates, we have, for $0\le s\le m$,

$$\begin{aligned}\frac{1}{2}\frac{d}{dt}\|\nabla^s\Pi_\epsilon\|_{L^2}^2 &= -\int_{\mathbb{R}^3}\nabla^s\{\partial_z(\mathcal{J}_\epsilon\Pi_\epsilon)^2\}\nabla^s(\mathcal{J}_\epsilon\Pi_\epsilon)\,dx \\ &= -\int_{\mathbb{R}^3}\partial_z\{\nabla^j(\mathcal{J}_\epsilon\Pi_\epsilon)\nabla^{s-j}(\mathcal{J}_\epsilon\Pi_\epsilon)\}\nabla^s(\mathcal{J}_\epsilon\Pi_\epsilon)\,dx.\end{aligned} \tag{92}$$

If $j\ne 0$ or $j\ne s$, then we obtain easily that

$$\begin{aligned}\frac{1}{2}\frac{d}{dt}\|\nabla^s\Pi_\epsilon\|_{L^2}^2 &\le \int_{\mathbb{R}^3}\left|\nabla^{j+1}(\mathcal{J}_\epsilon\Pi_\epsilon)\right|\left|\nabla^{s-j}(\mathcal{J}_\epsilon\Pi_\epsilon)\right|\left|\nabla^s(\mathcal{J}_\epsilon\Pi_\epsilon)\right|dx \\ &\le C\|\nabla\Pi_\epsilon\|_{L^\infty}\|\Pi_\epsilon\|_m^2 \le C\|\Pi_\epsilon\|_m^3.\end{aligned} \tag{93}$$

If $j=0$ or $j=s$, then we obtain

$$\begin{aligned}\frac{1}{2}\frac{d}{dt}\|\nabla^s\Pi_\epsilon\|_{L^2}^2 &= \int_{\mathbb{R}^3}(\mathcal{J}_\epsilon\Pi_\epsilon)\cdot\partial_z\nabla^s(\mathcal{J}_\epsilon\Pi_\epsilon)\cdot\nabla^s(\mathcal{J}_\epsilon\Pi_\epsilon)\,dx \\ &= \int_{\mathbb{R}^3}(\mathcal{J}_\epsilon\Pi_\epsilon)\partial_z(\nabla^s(\mathcal{J}_\epsilon\Pi_\epsilon))^2\,dx \\ &= -\int_{\mathbb{R}^3}\partial_z(\mathcal{J}_\epsilon\Pi_\epsilon)\left|\nabla^s(\mathcal{J}_\epsilon\Pi_\epsilon)\right|^2dx \\ &\le \|\partial_z\Pi_\epsilon\|_{L^\infty(\mathbb{R}^3)}\|\Pi_\epsilon\|_m^2 \le C\|\Pi_\epsilon\|_m^3.\end{aligned} \tag{94}$$

Combining the above inequalities (93) and (94), we have

$$\frac{d}{dt}\|\Pi_\epsilon\|_m \le C_m\|\Pi_\epsilon\|_m^2. \tag{95}$$

The above inequality gives us

$$\begin{aligned}\|\Pi_\epsilon\|_m(t) &\le \frac{\|\Pi_0\|_m}{1-tC_m\|\Pi_0\|_m} \\ &\qquad \text{for } 0\le t\le T<\frac{1}{C_m\|\Pi_0\|_m}, \\ \sup_{0\le t\le T}\|\Pi_\epsilon\|_m(t) &\le \frac{1}{1-C_mT\|\Pi_0\|_m} \\ &\qquad \text{for } T<\frac{1}{C_m\|\Pi_0\|_m}.\end{aligned} \tag{96}$$

By applying H^{m-1} *norm* at the regularized equation, we deduce that

$$\begin{aligned}\|\partial_t\Pi_\epsilon\|_{m-1} &= \|F(\Pi_\epsilon)\|_{m-1} \le C'_m\|\Pi_\epsilon\|_{m-1}\|\partial_z\Pi_\epsilon\|_{m-1} \\ &\le C'_m\|\Pi_\epsilon\|_m^2(t) \\ &\le C'_m\left(\frac{1}{1-C_mT\|\Pi_0\|_m}\right)^2 \\ &\qquad \text{for } T<\frac{1}{C_m\|\Pi_0\|_m}.\end{aligned} \tag{97}$$

Now we are ready to show that $\{\Pi_\epsilon\}\subset C(0,T;L^2)$ is a Cauchy sequence (as a sequence for $\epsilon_n \longrightarrow 0$), where T is chosen as above.

$$\begin{aligned}\frac{1}{2}\frac{d}{dt}\|\Pi_\epsilon-\Pi_{\epsilon'}\|_{L^2(\mathbb{R}^3)}^2 &= -\int_{\mathbb{R}^3}\{\mathcal{J}_\epsilon\partial_z(\mathcal{J}_\epsilon\Pi_\epsilon)^2-\mathcal{J}_{\epsilon'}\partial_z(\mathcal{J}_{\epsilon'}\Pi_{\epsilon'})^2\} \\ &\quad\cdot(\Pi_\epsilon-\Pi_{\epsilon'})\,dx = -\int\mathcal{J}_\epsilon(\Pi_\epsilon-\Pi_{\epsilon'}) \\ &\quad\cdot\partial_z(\mathcal{J}_\epsilon\Pi\epsilon)^2+\int\mathcal{J}_{\epsilon'}(\Pi_\epsilon-\Pi_{\epsilon'})\partial_z(\mathcal{J}_{\epsilon'}\Pi_{\epsilon'})^2 \\ &= -2\int(\mathcal{J}_\epsilon-\mathcal{J}_{\epsilon'})(\Pi_\epsilon-\Pi_{\epsilon'})\mathcal{J}_\epsilon\Pi_\epsilon\partial_z\mathcal{J}_\epsilon\Pi_\epsilon \\ &\quad-2\int\mathcal{J}_{\epsilon'}(\Pi_\epsilon-\Pi_{\epsilon'})(\mathcal{J}_\epsilon\Pi_\epsilon-\mathcal{J}_{\epsilon'}\Pi_\epsilon)\partial_z(\mathcal{J}_\epsilon\Pi_\epsilon) \\ &\quad-2\int\{\mathcal{J}_{\epsilon'}(\Pi_\epsilon-\Pi_{\epsilon'})\}^2\partial_z(\mathcal{J}_\epsilon\Pi_\epsilon) \\ &\quad-2\int\mathcal{J}_{\epsilon'}(\Pi_\epsilon-\Pi_{\epsilon'})\mathcal{J}_{\epsilon'}\Pi_{\epsilon'}\partial_z(\mathcal{J}_\epsilon\Pi_\epsilon-\mathcal{J}_{\epsilon'}\Pi_\epsilon) \\ &\quad-\int\mathcal{J}_{\epsilon'}\Pi_{\epsilon'}\partial_z\{\mathcal{J}_{\epsilon'}(\Pi_\epsilon-\Pi_{\epsilon'})\}^2\end{aligned} \tag{98}$$

By the properties of regularizer $\mathcal{J}_\epsilon$, for $m > 5/2$, we have

$$\frac{1}{2}\frac{d}{dt}\|\Pi_\epsilon - \Pi_{\epsilon'}\|_{L^2(\mathbb{R}^3)}^2 \le C\|\Pi_\epsilon - \Pi_{\epsilon'}\|_{L^2}(\epsilon + \epsilon') \cdot (\|\partial_z\Pi_\epsilon\|_{L^\infty}\|\Pi_\epsilon\|_1 + \|\partial_z\Pi_\epsilon\|_1\|\Pi_\epsilon\|_{L^\infty}) + C(\epsilon + \epsilon')\|\Pi_\epsilon\|_1\|\Pi_\epsilon - \Pi_{\epsilon'}\|_{L^2}\|\partial_z\Pi_\epsilon\|_{L^\infty} + C\|\Pi_\epsilon - \Pi_{\epsilon'}\|_{L^2}^2\|\partial_z\Pi_\epsilon\|_{L^\infty} + C(\epsilon + \epsilon')\|\Pi_\epsilon\|_2\|\Pi_\epsilon - \Pi_{\epsilon'}\|_{L^2} \cdot \|\Pi_{\epsilon'}\|_{L^\infty} + C\|\Pi_\epsilon - \Pi_{\epsilon'}\|_{L^2}^2\|\partial_z\Pi_{\epsilon'}\|_{L^\infty} \le C(\epsilon + \epsilon')\|\Pi_\epsilon\|_m\|\Pi_{\epsilon'}\|_m\|\Pi_\epsilon - \Pi_{\epsilon'}\|_{L^2} + C(\|\Pi_\epsilon\|_m + \|\Pi_{\epsilon'}\|_m)\|\Pi_\epsilon - \Pi_{\epsilon'}\|_{L^2}^2 \le C\|\Pi_\epsilon - \Pi_{\epsilon'}\|_{L^2} \cdot \{(\epsilon + \epsilon')\|\Pi_\epsilon\|_m\|\Pi_{\epsilon'}\|_m + \|\Pi_\epsilon - \Pi_{\epsilon'}\|_{L^2}(\|\Pi_\epsilon\|_m + \|\Pi_{\epsilon'}\|_m)\} \tag{99}$$

In summary, we have

$$\frac{d}{dt}\|\Pi_\epsilon - \Pi_{\epsilon'}\|_{L^2(\mathbb{R}^3)} \le C(\epsilon + \epsilon')\|\Pi_\epsilon\|_m\|\Pi_{\epsilon'}\|_m + C\|\Pi_\epsilon - \Pi_{\epsilon'}\|_{L^2}(\|\Pi_\epsilon\|_m + \|\Pi_{\epsilon'}\|_m) \tag{100}$$

By Gronwall's inequality, we can conclude that $\{\Pi_\epsilon\}$ is a Cauchy sequence in the $C(0,T;L^2)$ space. And by H^m boundness, if we apply the interpolation inequality, then we can see that $\{\Pi_\epsilon\}$ is a cauchy sequence in $C(0,T;H^s)$, $\forall s < m$. So we have the limit function $\Pi \in C(0,T;H^s)$. And $\{\partial_t\Pi_\epsilon\}$ is also a cauchy sequence in the $C(0,T;L^2)$ space by the following estimates:

$$\begin{aligned}
&\|\partial_t\Pi_\epsilon - \partial_t\Pi_{\epsilon'}\|_{L^2} \\
&\le 2\|\mathcal{J}_\epsilon(\mathcal{J}_\epsilon\Pi_\epsilon\partial_z\mathcal{J}_\epsilon\Pi_\epsilon) - \mathcal{J}_{\epsilon'}(\mathcal{J}_{\epsilon'}\Pi_{\epsilon'}\partial_z\mathcal{J}_{\epsilon'}\Pi_{\epsilon'})\|_{L^2} \\
&\le 2\|(\mathcal{J}_\epsilon - \mathcal{J}_{\epsilon'})(\mathcal{J}_\epsilon\Pi_\epsilon\partial_z\mathcal{J}_\epsilon\Pi_\epsilon)\|_{L^2} \\
&\quad + 2\|(\mathcal{J}_\epsilon - \mathcal{J}_{\epsilon'})\Pi_\epsilon\partial_z\mathcal{J}_\epsilon\Pi_\epsilon\|_{L^2} \\
&\quad + 2\|\mathcal{J}_{\epsilon'}(\Pi_\epsilon - \Pi_{\epsilon'})\partial_z\mathcal{J}_\epsilon\Pi_\epsilon\|_{L^2} \\
&\quad + 2\|\mathcal{J}_{\epsilon'}\Pi_{\epsilon'}(\mathcal{J}_\epsilon - \mathcal{J}_{\epsilon'})\partial_z\Pi_\epsilon\|_{L^2} \\
&\quad + 2\|\mathcal{J}_{\epsilon'}\Pi_{\epsilon'}\mathcal{J}_{\epsilon'}\partial_z(\Pi_\epsilon - \Pi_{\epsilon'})\|_{L^2} \\
&\le 6(\epsilon + \epsilon')(\|\Pi_\epsilon\|_s^2 + \|\Pi_{\epsilon'}\|_s^2) \\
&\quad + 4(\|\Pi_\epsilon\|_s + \|\Pi_{\epsilon'}\|_s)\|\Pi_\epsilon - \Pi_{\epsilon'}\|_1
\end{aligned} \tag{101}$$

For $5/2 < s < m$, we have the limit function $\partial_t\Pi \in C(0,T;H^{s-1})$.

Finally, we can show that $\Pi \in C(0,T;H^m) \cap \mathrm{Lip}(0,T;H^{m-1})$. By the Banach Alaoglu theorem, we have $\Pi \in L^\infty(0,T_0;H^m)$ and $\partial_t\Pi \in L^\infty(0,T_0;H^{m-1})$, because we know that, for any $s < m$, $\Pi \in C(0,T;H^s)$. It implies $\Pi \in C_w(0,T;H^m)$ by the following estimate:

$$\begin{aligned}
[\varphi, \Pi_\epsilon - \Pi] &\le \|\varphi - \tilde{\varphi}\|_{H^{-m}}\|\Pi_\epsilon - \Pi\|_{H^m} \\
&\quad + [\tilde{\varphi}, \Pi_\epsilon - \Pi] \le \delta + [\tilde{\varphi}, \Pi_\epsilon - \Pi] \\
&\le 2\delta
\end{aligned} \tag{102}$$

for any given $\varphi \in H^{-m}$, for some $\tilde{\varphi} \in H^{-s}$, $s < m$. Now we show that

$$\liminf_{t\to 0+}\|\Pi(t)\|_m = \limsup_{t\to 0+}\|\Pi(t)\|_m = \|\Pi_0\|_m. \tag{103}$$

By the weak continuity, for any $\delta > 0$, there exists $r > 0$ such that if $0 < t < r$, then $-\delta < [\Pi(t) - \Pi_0, \varphi] < \delta$, for all $\varphi \in H^{-m}$. Choose $\varphi = \nabla^{2s}\Pi_0$ with $s \le m$. Then it gives us

$$\begin{aligned}
\|\nabla^s\Pi_0\|_{L^2}^2 - \delta &\le \|\nabla^s\Pi(t)\|_{L^2}\|\nabla^s\Pi_0\|_{L^2} \\
&\le \frac{\|\nabla^s\Pi(t)\|_{L^2}^2}{2} + \frac{\|\nabla^s\Pi_0\|_{L^2}^2}{2}.
\end{aligned} \tag{104}$$

Also we have

$$\|\Pi_0\|_m^2 \le \|\Pi(t)\|_m^2 + 2^{m+1}\delta, \tag{105}$$

which implies that

$$\liminf_{t\to 0+}\|\Pi(t)\|_m \ge \|\Pi_0\|_m \tag{106}$$

By the H^m boundness with weak convergence, it is deduced that

$$\|\Pi(t)\|_m \le \frac{\|\Pi_0\|_m}{1 - tC_m\|\Pi_0\|_m}, \quad 0 \le t \le T \tag{107}$$

which implies

$$\limsup_{t\to 0+}\|\Pi(t)\|_m \le \|\Pi_0\|_m. \tag{108}$$

Thus we have the time continuity of Π at 0. For any $t_0 \le T$ and initial value $\Pi(t_0)$, we can obtain a right continuity at t_0 by the time translation invariant property. By the fact that $\Pi(-x, t_0 - t)$ is also a solution to the euqation for $0 \le t \le t_0$, we have a left continuity at t_0. Of course by the above process, the left continuity at T also can be obtained. We have proved that $\Pi \in C(0,T;H^m)$. □

Proposition 11. *Let Π be an axisymmetric global classical solution to*

$$\partial_t\Pi + 2\Pi\partial_z\Pi = 0. \tag{109}$$

Then $\Pi \equiv 0$.

Proof. Define $\phi_{r_0}(t) : [0,\infty) \longrightarrow \mathbb{R}$ which satisfies the equation $\phi'_{r_0}(t) = \Pi(r_0, 2\phi_{r_0}(t), t)$. By our assumption, $\Pi \in C([0,\infty); H^m(\mathbb{R}^3)) \cap \mathrm{Lip}([0,\infty); H^{m-1}(\mathbb{R}^3))$ for $m > 5/2$,

$\phi''_{r_0}(t) = 0$ almost every time. So we can find the explicit form of it by

$$\phi_{r_0}(t) = t\Pi_0\left(r_0, 2\phi_{r_0}(0)\right) + \phi_{r_0}(0). \tag{110}$$

Now we choose initial values $\phi_{r_0}(0)$ and $\widetilde{\phi}_{r_0}(0)$ such that $\phi_{r_0}(0) \le \widetilde{\phi}_{r_0}(0)$. Then it satisfies

$$\Pi_0\left(r_0, 2\phi_{r_0}(0)\right) \le \Pi_0\left(r_0, 2\widetilde{\phi}_{r_0}(0)\right). \tag{111}$$

Because if we suppose that $\Pi_0(r_0, 2\phi_{r_0}(0)) > \Pi_0(r_0, 2\widetilde{\phi}_{r_0}(0))$, then by the explicit form of ϕ, for some $t_0 > 0$, we have $\phi_{r_0}(t_0) = \widetilde{\phi}_{r_0}(t_0)$ which makes a contradiction. Hence, $\Pi_0(r_0, z)$ is a nondecreasing funcion with respect to z. Since this process is independent of the choice of r_0, we can find that $\Pi_0 \equiv 0$ by the continuity and L^2 boundness. □

4. Proof of Theorem 5: Blow-Up Criterion

In this section, we provide the proof of Theorem 5 which is the blow-up criterion for the axisymmetric Hall-MHD equations with $\nu = 0$ and $\kappa = 1$:

$$\frac{\partial \Omega}{\partial t} + (u^r\partial_r + u^z\partial_z)\Omega + 2\Pi\partial_z\Pi = 0, \tag{112}$$

$$\begin{aligned}\frac{\partial \Pi}{\partial t} &+ (u^r\partial_r + u^z\partial_z)\Pi - 2\Pi\partial_z\Pi \\ &= \left(\partial_r^2 + \frac{3}{r}\partial_r + \partial_z^2\right)\Pi,\end{aligned} \tag{113}$$

where $\Omega = \omega^\theta/r$ and $\Pi = B^\theta/r$.

Known blow-up criterion for the partial viscous Hall-MHD equations (1) without symmetry ($\nu = 0$ and $\kappa = 1$) is as follows (see [13]).

Proposition 12. *Assume that $(u_0, B_0) \in H^m(\mathbb{R}^3)$, $m \ge 3$ with $\nabla\cdot u_0 = \nabla\cdot B_0 = 0$. Let (u, B) be a smooth solution to (1) ($\nu = 0$ and $\kappa = 1$) for $0 \le t < T$. If (u, B) satisfies*

$$\int_0^T \left(\|\nabla\times u(t)\|_{L^\infty} + \|\nabla\times B(t)\|_{L^\infty}^2\right)dt < \infty, \tag{114}$$

then the solution (u, B) can be extended beyond $t = T$.

With the axial symmetry, we can derive the following apriori estimates.

Proposition 13. *If (Ω, Π) is a solution to (112)–(113) satisfying $(\Omega, \Pi) \in C([0, T); H^m)$, with $m \ge 3$ then it holds that*

$$\begin{aligned}&\|\Pi\|_{L^\infty(0,T;L^p(\mathbb{R}^3))} \le C(\Pi_0), \\ &\|\nabla\Pi\|_{L^2(0,T;L^2(\mathbb{R}^3))}^2 \le \|\Pi_0\|_{L^2}^2, \\ &\|\Omega\|_{L^\infty(0,T;L^2)} \le C\left(\|\Omega_0\|_{L^2}, \|\Pi_0\|_{L^2}\right) < \infty.\end{aligned} \tag{115}$$

Proof. We first consider $p = 2n$ with $n \in \mathbb{N}$. Taking scalar product of (113) with Π^{p-1}, we deduce that

$$\begin{aligned}&\frac{1}{p}\frac{d}{dt}\|\Pi(t)\|_{L^p}^p \\ &\quad - 2\pi\int_{-\infty}^{\infty}\int_0^\infty\left(\left(\partial_r^2 + \frac{3}{r}\partial_r + \partial_z^2\right)\Pi\right)\cdot\Pi^{p-1}rdrdz \\ &= \frac{2}{p+1}\int_{\mathbb{R}^3}\partial_z(\Pi)^{p+1}dx \\ &\quad - \int_{\mathbb{R}^3}(u^r\partial_r + u^z\partial_z)\Pi\cdot\Pi^{p-1}dx.\end{aligned} \tag{116}$$

From the divergence free condition and the decay conditions like

$$\lim_{r\longrightarrow\infty}\Pi(r, z, t) = \lim_{|z|\longrightarrow\infty}\Pi(r, z, t) = 0, \tag{117}$$

(116) can be reduced to

$$\begin{aligned}\frac{1}{p}\frac{d}{dt}\|\Pi(t)\|_{L^p}^p &+ \frac{4(p-1)}{p^2}\int_{\mathbb{R}^3}\left|\partial_r\Pi^{p/2}\right|^2 \\ &+ \left|\partial_z\Pi^{p/2}\right|^2dx = 0.\end{aligned} \tag{118}$$

It implies that

$$\|\Pi(t)\|_{L^\infty(0,T;L^p)}^p \le \|\Pi_0\|_{L^p}^p. \tag{119}$$

For any $R > 0$, we have

$$\left(\int_{B_R}|\Pi_0(x)|^{2n}dx\right)^{1/2n} \le C\|\Pi_0\|_{L^\infty}R^{3/2n}. \tag{120}$$

As $n \longrightarrow \infty$, we have $\|\Pi\|_{L^\infty(0,T;L^\infty)} \le C\|\Pi_0\|_{L^\infty}$. If $p \in (2n, 2(n+1))$, then we have

$$\|\Pi\|_{L^\infty(0,T;L^p)}^p \le \|\Pi_0\|_{L^{2n}}^{n(2n+2-p)}\|\Pi_0\|_{L^{2(n+1)}}^{(n+1)(p-2n)}. \tag{121}$$

Taking L^2 scalar product of (112) with Ω, we have

$$\begin{aligned}\frac{1}{2}\frac{d}{dt}\|\Omega\|_{L^2}^2 &= -2\int_{\mathbb{R}^3}\Pi\partial_z\Pi\Omega dx \\ &\le 2\|\Pi\|_{L^\infty}\|\partial_z\Pi\|_{L^2}\|\Omega\|_{L^2}\end{aligned} \tag{122}$$

Then we have

$$\begin{aligned}\|\Omega\|_{L^\infty(0,T;L^2)} &\lesssim \|\Omega_0\|_{L^2} \\ &\quad + \int_0^T\|\Pi(t)\|_{L^\infty}^2 + \|\partial_z\Pi(t)\|_{L^2}^2dt\end{aligned} \tag{123}$$

This completes the proof. □

From the energy estimates of the velocity and magnetic fields, we have

$$\begin{aligned}&\|u^r\|_{L^\infty(0,T;L^2)}^2 + \|u^z\|_{L^\infty(0,T;L^2)}^2 + \|B^\theta\|_{L^\infty(0,T;L^2)}^2 \\ &\quad + \int_0^T\|\partial_rB^\theta\|_{L^2}^2 + \|\partial_zB^\theta\|_{L^2}^2 + \left\|\frac{B^\theta}{r}\right\|_{L^2}^2dt \\ &\le \|u_0^r\|_{L^2}^2 + \|u_0^z\|_{L^2}^2 + \|B_0^\theta\|_{L^2}^2.\end{aligned} \tag{124}$$

Proof of Theorem 5. First, we assume that assumption (19) holds. If we consider the equation of the vorticity ω^θ, then we have

$$\partial_t \omega^\theta + (u^r \partial_r + u^z \partial_z)\,\omega^\theta = \frac{u^r}{r}\omega^\theta - 2\left(\frac{B^\theta}{r}\right)\partial_z B^\theta. \tag{125}$$

Taking L^2 scalar product of (125) with ω^θ, we have

$$\begin{aligned}\frac{d}{dt}\left\|\omega^\theta\right\|_{L^2}^2 &\le \int_{\mathbb{R}^3} \frac{u_+^r}{r}\left(\omega^\theta\right)^2 dx \\ &\quad + 2\int_{\mathbb{R}^3} \left|\partial_z B^\theta\right| |\Pi| \left|\omega^\theta\right| dx \\ &\le \left\|\frac{u_+^r}{r}\right\|_{L^\infty(\mathbb{R}^3\setminus C_R)} \left\|\omega^\theta\right\|_{L^2(\mathbb{R}^3\setminus C_R)}^2 \\ &\quad + \left\|r u_+^r\right\|_{L^\infty(C_R)} \|\Omega\|_{L^2(C_R)}^2 \\ &\quad + 2\left\|\partial_z B^\theta\right\|_{L^2} \|\Pi\|_{L^\infty} \left\|\omega^\theta\right\|_{L^2}.\end{aligned} \tag{126}$$

Using Gronwall's inequality, we have

$$\begin{aligned}\left\|\omega^\theta\right\|_{L^\infty(0,T;L^2)} &\le \left(\left\|\omega_0^\theta\right\|_{L^2} + \int_0^T \left\|r u_+^r\right\|_{L^\infty(C_R)} dt + C\right) \\ &\quad \cdot \exp\left(\int_0^T \left\|\frac{u_+^r}{r}\right\|_{L^\infty(\mathbb{R}^3\setminus C_R)} dt + C\right).\end{aligned} \tag{127}$$

Hence we have $\omega^\theta \in L^\infty(0,T;L^2)$ if we assume (19).

If we consider the equations for $\widetilde{\nabla}\Pi$ ($\widetilde{\nabla} = (\partial_r, \partial_z)$), then we obtain

$$\begin{aligned}&\partial_t \partial_z \Pi + (u^r \partial_r + u^z \partial_z)\,\partial_z \Pi \\ &\quad + (\partial_z u^r \partial_r + \partial_z u^z \partial_z)\,\Pi = \left(\Delta + \frac{2}{r}\partial_r\right)\partial_z \Pi + \partial_z^2 \Pi^2 \\ &\partial_t \partial_r \Pi + (u^r \partial_r + u^z \partial_z)\,\partial_r \Pi + (\partial_r u^r \partial_r + \partial_r u^z \partial_z)\,\Pi \\ &\quad + \frac{2}{r^2}\partial_r \Pi = \left(\Delta + \frac{2}{r}\partial_r\right)\partial_r \Pi + \partial_r \partial_z \Pi^2\end{aligned} \tag{128}$$

Via an interpolation inequality and Young's inequality, we have

$$\begin{aligned}&\left\|\omega^\theta\right\|_{L^2(\mathbb{R}^3)} \left\|\widetilde{\nabla}\Pi\right\|_{L^6(\mathbb{R}^3)} \left\|\widetilde{\nabla}\Pi\right\|_{L^3(\mathbb{R}^3)} \\ &\quad \le \left\|\omega^\theta\right\|_{L^2(\mathbb{R}^3)} \left\|\widetilde{\nabla}^2\Pi\right\|_{L^2(\mathbb{R}^3)}^{3/2} \left\|\widetilde{\nabla}\Pi\right\|_{L^2(\mathbb{R}^3)}^{1/2} \\ &\quad \le C\left\|\omega^\theta\right\|_{L^2(\mathbb{R}^3)}^4 \left\|\widetilde{\nabla}\Pi\right\|_{L^2(\mathbb{R}^3)}^2 + \frac{1}{2}\left\|\widetilde{\nabla}^2\Pi\right\|_{L^2(\mathbb{R}^3)}^2.\end{aligned} \tag{129}$$

Taking scalar product of (128) with $\widetilde{\nabla}\Pi$, we deduce that

$$\begin{aligned}&\frac{1}{2}\frac{d}{dt}\int_{\mathbb{R}^3} \left|\widetilde{\nabla}\Pi\right|^2 dx + \int_{\mathbb{R}^3} \left|\widetilde{\nabla}^2\Pi\right|^2 dx \\ &\quad + \int_{\mathbb{R}^3} \frac{2}{r^2} |\partial_r \Pi|^2 dx \\ &\quad \le C\Bigg(\left\|\omega^\theta\right\|_{L^2} \left\|\widetilde{\nabla}\Pi\right\|_{L^6} \left\|\widetilde{\nabla}\Pi\right\|_{L^3} \\ &\quad + \int_{\mathbb{R}^3} \frac{1}{r}\partial_r\left(\left|\widetilde{\nabla}\Pi\right|^2\right) dx + \int_{\mathbb{R}^3} |\partial_z \Pi|^3 dx \\ &\quad + C\left\|\widetilde{\nabla}^2\Pi\right\|_{L^2} \left\|\widetilde{\nabla}\Pi\right\|_{L^2} \|\Pi\|_{L^\infty} \\ &\quad + \int_{\mathbb{R}^3} |\partial_r \Pi|^2 |\partial_z \Pi|\, dx\Bigg) \le C\left(\left\|\omega^\theta\right\|_{L^2}^4 + \left\|\widetilde{\nabla}\Pi\right\|_{L^2}^2 \right. \\ &\quad \left. + \|\Pi\|_{L^\infty}^2\right)\left\|\widetilde{\nabla}\Pi\right\|_{L^2}^2 + \frac{1}{2}\left\|\widetilde{\nabla}\Pi\right\|_{L^2}^2.\end{aligned} \tag{130}$$

In the above, we used the fact that $(1/r)\partial_r(|\partial_r \Pi|^2 + |\partial_z \Pi|^2) \in L^1(\mathbb{R}^3)$ when $t < T^*$. Gronwall's inequality again gives us

$$\begin{aligned}\widetilde{\nabla}\Pi &\in L^\infty\left(0,T;L^2\right), \\ \widetilde{\nabla}^2\Pi &\in L^2\left(0,T;L^2\right).\end{aligned} \tag{131}$$

Multiplying both sides of (112) with $|\Omega|^4\Omega$ and integrating over $\mathbb{R}^3$, we have

$$\begin{aligned}\frac{1}{6}\frac{d}{dt}\|\Omega\|_{L^6(\mathbb{R}^3)}^6 &\le 2\int_{\mathbb{R}^3} |\partial_z \Pi|\,|\Pi|\,|\Omega|^5 dx \\ &\le \left\|\widetilde{\nabla}\Pi\right\|_{L^6} \|\Pi\|_{L^\infty} \|\Omega\|_6^5.\end{aligned} \tag{132}$$

Then it is immediate that $\Omega \in L^\infty(0,T;L^6)$.

Following the ideas in [18], we introduce the angular stream function ψ^θ such that

$$-\left(\partial_r^2 + \frac{1}{r}\partial_r - \frac{1}{r^2} + \partial_z^2\right)\psi^\theta = \omega^\theta. \tag{133}$$

For all $1 < p < \infty$, we have

$$\int \left|\nabla^2 \frac{\psi^\theta}{r}\right|^p dx \le C\int \left|\frac{\omega^\theta}{r}\right|^p dx. \tag{134}$$

By the interpolation inequality $\|f\|_{L^\infty} \le C\|\nabla f\|_{L^2}^{1/2}\|\nabla f\|_{L^6}^{1/2}$, we have

$$\begin{aligned}&\int_0^T \left\|\frac{u^r}{r}\right\|_{L^\infty} dt \\ &\quad \le \int_0^T \left\|\nabla\partial_z\left(\frac{\psi^\theta}{r}\right)\right\|_{L^2}^{1/2} \left\|\nabla\partial_z\left(\frac{\psi^\theta}{r}\right)\right\|_{L^6}^{1/2} dt \\ &\quad \le \sup_t \|\Omega\|_{L^2}^{1/2} \int_0^T \|\Omega\|_{L^6}^{1/2} dt.\end{aligned} \tag{135}$$

Therefore, we have $u^r/r \in L^1(0,T;L^\infty)$. Also we can have $u^r/r \in L^p(0,T;L^\infty)$ for all $p < \infty$.

If we multiply $(B^\theta)^{2n-1}$ on the both sides of $(9)_4$ and integrate over $\mathbb{R}^3$, then we have

$$\begin{aligned}\frac{1}{2n}\frac{d}{dt}\left\|B^\theta\right\|_{L^{2n}}^{2n} &+ \frac{(2n-1)}{n^2}\left\|\tilde{\nabla}\left|B^\theta\right|^n\right\|_{L^2}^2 + \left\|\frac{\left(B^\theta\right)^n}{r}\right\|_{L^2}^2 \\ &= \int_{\mathbb{R}^3}\frac{u^r}{r}\left(B^\theta\right)^{2n}dx \\ &+ 2\pi\int_0^\infty\int_{-\infty}^\infty \partial_z\left(B^\theta\right)^2\left(B^\theta\right)^{2n-1}dzdr \\ &\leq \left\|\frac{u^r}{r}\right\|_{L^\infty}\left\|B^\theta\right\|_{L^{2n}}^{2n}.\end{aligned} \tag{136}$$

Hence we deduce that

$$\left\|B^\theta\right\|_{L^\infty(0,T;L^{2n})} \leq \left\|B_0^\theta\right\|_{L^{2n}}\exp\left(\int_0^T\left\|\frac{u^r}{r}\right\|_{L^\infty}dt\right). \tag{137}$$

Similarly, we have $\|B^\theta\|_{L^\infty(0,T;L^\infty)} \leq \|B_0^\theta\|_{L^{2n}}\exp(\int_0^T\|u^r/r\|_{L^\infty}dt)$.

Setting $\mathbf{B} = B^\theta e_\theta$, we have

$$\partial_t\mathbf{B} + (u\cdot\nabla)\mathbf{B} - \Delta\mathbf{B} = \frac{u^r}{r}\mathbf{B} + 2\partial_z\mathbf{B}\Pi. \tag{138}$$

By the maximal inequality, we have

$$\begin{aligned}\|\Delta\mathbf{B}\|_{L_{x,t}^{4,4}(\mathbb{R}^3\times(0.T))}^4 \leq C\Bigg(&\|\mathbf{B_0}\|_{W^{2,4}}^4 \\ &+ \|u\|_{L_{x,t}^{6,\infty}(\mathbb{R}^3\times(0.T))}^4\|\nabla\mathbf{B}\|_{L_{x,t}^{12,4}(\mathbb{R}^3\times(0.T))}^4 \\ &+ \left\|\frac{u^r}{r}\right\|_{L_{x,t}^{\infty,4}(\mathbb{R}^3\times(0.T))}^4\|\mathbf{B}\|_{L_{x,t}^{4,\infty}(\mathbb{R}^3\times(0.T))}^4 \\ &+ \|\Pi\|_{L_{x,t}^{\infty,\infty}(\mathbb{R}^3\times(0.T))}^4\|\nabla\mathbf{B}\|_{L_{x,t}^{4,4}(\mathbb{R}^3\times(0.T))}^4\Bigg).\end{aligned} \tag{139}$$

By the Gagliardo-Nirenberg inequality

$$\begin{aligned}\|\nabla\mathbf{B}\|_{L^{12}} &\leq C\|\mathbf{B}\|_{L^\infty}^{2/5}\|\Delta\mathbf{B}\|_{L^4}^{3/5}, \\ \|\nabla\mathbf{B}\|_{L^4} &\leq C\|\mathbf{B}\|_{L^\infty}^{4/5}\|\Delta\mathbf{B}\|_{L^4}^{1/5},\end{aligned} \tag{140}$$

(139) can be reduced to

$$\begin{aligned}\|\Delta\mathbf{B}\|_{L_{x,t}^{4,4}(\mathbb{R}^3\times(0.T))}^4 &\leq C\Big(\|\mathbf{B_0}\|_{W^{2,4}}^4 + \|\mathbf{B}\|_{L_{x,t}^{4,\infty}(\mathbb{R}^3\times(0.T))}^4 \\ &+ \|\mathbf{B}\|_{L_{x,t}^{\infty,4}(\mathbb{R}^3\times(0.T))}^{8/5}\|\Delta\mathbf{B}\|_{L_{x,t}^{4,4}(\mathbb{R}^3\times(0.T))}^{12/5} \\ &+ \|\mathbf{B}\|_{L_{x,t}^{\infty,4}(\mathbb{R}^3\times(0.T))}^{16/5}\|\Delta\mathbf{B}\|_{L_{x,t}^{4,4}(\mathbb{R}^3\times(0.T))}^{4/5}\Big) \leq C\Big(1 \\ &+ \|\mathbf{B}\|_{L_{x,t}^{4,\infty}(\mathbb{R}^3\times(0.T))}^4 + \|\mathbf{B}\|_{L_{x,t}^{\infty,4}(\mathbb{R}^3\times(0.T))}^4\Big) \\ &+ \frac{1}{2}\|\Delta\mathbf{B}\|_{L_{x,t}^{4,4}(\mathbb{R}^3\times(0.T))}^4.\end{aligned} \tag{141}$$

Since the last term in the above can be absorbed in the left hand side, we have

$$\begin{aligned}\Delta\mathbf{B} &\in L^4\left(0,T;L^4\right), \\ \nabla\mathbf{B} &\in L^4\left(0,T;L^\infty\right).\end{aligned} \tag{142}$$

Then, from (125), we have, for all $1 < p < \infty$,

$$\begin{aligned}\frac{1}{p}\frac{d}{dt}\left\|\omega^\theta\right\|_{L^p}^p \leq &\left\|\frac{u^r}{r}\right\|_{L^\infty}\left\|\omega^\theta\right\|_{L^p}^p \\ &+ 2\|\Pi\|_{L^\infty}\left\|\nabla B^\theta\right\|_{L^p}\left\|\omega^\theta\right\|_{L^p}^{p-1}.\end{aligned} \tag{143}$$

By Gronwall's inequality, we have

$$\begin{aligned}\left\|\omega^\theta\right\|_{L^\infty(0,T;L^p)} &\leq \left(\left\|\omega_0^\theta\right\|_{L^p} + 2\int_0^T\|\Pi\|_{L^\infty}\left\|\nabla B^\theta\right\|_{L^p}dt\right) \\ &\cdot\exp\left(\int_0^T\left\|\frac{u^r}{r}\right\|_{L^\infty}dt\right).\end{aligned} \tag{144}$$

If we let $p \longrightarrow \infty$, we obtain $\omega^\theta \in L^\infty(0,T;L^\infty)$. Hence, for any $T < \infty$, we obtain that $\omega^\theta, \nabla\mathbf{B} \in L^2(0,T;L^\infty)$ and conclude that there does not exist a finite time blow-up if we assume (19).

Next, we assume that condition (20) holds. If we apply (129) to (128), we obtain

$$\begin{aligned}\frac{1}{2}\frac{d}{dt}\int_{\mathbb{R}^3}\left|\tilde{\nabla}\Pi\right|^2dx &+ \int_{\mathbb{R}^3}\left|\tilde{\nabla}^2\Pi\right|^2dx \\ &+ \int_{\mathbb{R}^3}\frac{2}{r^2}\left|\partial_r\Pi\right|^2dx \leq C\Big(\left\|\omega^\theta\right\|_{L^2(\mathbb{R}^3)}^4\left\|\tilde{\nabla}\Pi\right\|_{L^2(\mathbb{R}^3)}^2 \\ &+ \frac{1}{2}\left\|\tilde{\nabla}^2\Pi\right\|_{L^2(\mathbb{R}^3)}^2 + \int_{\mathbb{R}^3}\frac{1}{r}\partial_r\left(\left|\partial_r\Pi\right|^2 + \left|\partial_z\Pi\right|^2\right)dx \\ &+ \int_{\mathbb{R}^3}\left|\partial_z\Pi\right|^3dx + \int_{\mathbb{R}^3}\left|\partial_z^2\Pi\right|\left|\partial_z\Pi\right|\left|\Pi\right|dx \\ &+ \int_{\mathbb{R}^3}\left|\partial_r\Pi\right|^2\left|\partial_z\Pi\right|dx + \int_{\mathbb{R}^3}\left|\tilde{\nabla}^2\Pi\right|\left|\partial_r\Pi\right|\left|\Pi\right|dx\Big)\end{aligned} \tag{145}$$

Using an inequality

$$\left\|\omega^\theta\right\|_{L^2(\mathbb{R}^3)}^4 \leq 2\left\|\omega^\theta\right\|_{L^2(\mathbb{R}^3\setminus C_R)}^4 + 2R^4\|\Omega\|_{L^2(C_R)}^4, \tag{146}$$

we have

$$\begin{aligned}
&\frac{1}{2}\frac{d}{dt}\int_{\mathbb{R}^3}\left|\widetilde{\nabla}\Pi\right|^2 dx+\int_{\mathbb{R}^3}\left|\widetilde{\nabla}^2\Pi\right|^2 dx \\
&\quad+\int_{\mathbb{R}^3}\frac{2}{r^2}\left|\partial_r\Pi\right|^2 dx \\
&\le C\left(\left\|\omega^\theta\right\|^4_{L^2(\mathbb{R}^3\setminus C_R)}+R^4\left\|\Omega\right\|^4_{L^2(C_R)}\right)\left\|\widetilde{\nabla}\Pi\right\|^2_{L^2(\mathbb{R}^3)} \\
&\quad+\frac{1}{2}\left\|\widetilde{\nabla}^2\Pi\right\|^2_{L^2(\mathbb{R}^3)} \\
&\quad+\int_{\mathbb{R}^3}\frac{1}{r}\partial_r\left(\left|\partial_r\Pi\right|^2+\left|\partial_z\Pi\right|^2\right)dx \\
&\quad+\int_{\mathbb{R}^3}\left|\partial_z\Pi\right|^3 dx+\int_{\mathbb{R}^3}\left|\partial_z^2\Pi\right|\left|\partial_z\Pi\right|\left|\Pi\right| dx \\
&\quad+\int_{\mathbb{R}^3}\left|\partial_r\Pi\right|^2\left|\partial_z\Pi\right| dx \\
&\quad+\int_{\mathbb{R}^3}\left|\widetilde{\nabla}^2\Pi\right|\left|\partial_r\Pi\right|\left|\Pi\right| dx.
\end{aligned} \tag{147}$$

By Gronwall's inequality, we conclude that

$$\begin{aligned}
\widetilde{\nabla}\Pi &\in L^\infty\left(0,T;L^2\right), \\
\widetilde{\nabla}^2\Pi &\in L^2\left(0,T;L^2\right).
\end{aligned} \tag{148}$$

The estimate of Ω, u^r/r, $\mathbf{B}$, and $\widetilde{\nabla}\Pi$ can be obtained similarly to the proof of the condition of (19). This completes the proof. □

Conflicts of Interest

The authors declare that they have no conflicts of interest.

Acknowledgments

E. Jeong was partially supported by Chung-Ang University Excellent Student Scholarship and J. Kim and J. Lee were partially supported by NRF Grant No. 2016R1A2B3011647. The authors thank Professor Dongho Chae for helpful remarks.

References

[1] T. G. Forbes, "Magnetic Reconnection in Solar Flares," *Geophysical & Astrophysical Fluid Dynamics*, vol. 62, no. 1-4, pp. 15–36, 1991.

[2] M. J. Lighthill, "Studies on magneto-hydrodynamic waves and other anisotropic wave motions," *Philosophical Transactions of the Royal Society of London. Series A. Mathematical and Physical Sciences*, vol. 252, pp. 397–430, 1960.

[3] P. A. Sturrock, *Plasma Physics*, Cambridge University Press, 1994.

[4] S. A. Balbus and C. Terquem, "Linear analysis of the hall effect in protostellar disks," *The Astrophysical Journal*, vol. 552, no. 1, pp. 235–247, 2001.

[5] H. Homann and R. Grauer, "Bifurcation analysis of magnetic reconnection in Hall-MHD-systems," *Physica D: Nonlinear Phenomena*, vol. 208, no. 1-2, pp. 59–72, 2005.

[6] P. D. Mininni, D. O. Gómez, and S. M. Mahajan, "Dynamo action in magnetohydrodynamics and Hall-magnetohydrodynamics," *The Astrophysical Journal*, vol. 587, no. 1 I, pp. 472–481, 2003.

[7] D. A. Shalybkov and V. A. Urpin, "The Hall effect and the decay of magnetic fields," *Astron. Astrophys*, vol. 32, pp. 685–690, 1997.

[8] M. Wardle, "Star formation and the hall effect," *Astrophysics and Space Science*, vol. 292, no. 1-4, pp. 317–323, 2004.

[9] M. Acheritogaray, P. Degond, A. Frouvelle, and J. Liu, "Kinetic formulation and global existence for the Hall-Magneto-hydrodynamics system," *Kinetic and Related Models*, vol. 4, no. 4, pp. 901–918, 2011.

[10] D. Chae, P. Degond, and J. Liu, "Well-posedness for Hall-magnetohydrodynamics," *Annales de l'Institut Henri Poincaré (C) Analyse Non Linéaire*, vol. 31, no. 3, pp. 555–565, 2014.

[11] D. Chae and J. Lee, "On the blow-up criterion and small data global existence for the Hall-magnetohydrodynamics," *Journal of Differential Equations*, vol. 256, no. 11, pp. 3835–3858, 2014.

[12] M. Dai, "Regularity criterion for the 3D Hall-magneto-hydrodynamics," *Journal of Differential Equations*, vol. 261, no. 1, pp. 573–591, 2016.

[13] Y.-Z. Wang and W. Zuo, "On the blow-up criterion of smooth solutions for Hall-magnetohydrodynamics system with partial viscosity," *Communications on Pure and Applied Analysis*, vol. 13, no. 3, pp. 1327–1336, 2014.

[14] D. Chae and M. Schonbek, "On the temporal decay for the Hall-magnetohydrodynamic equations," *Journal of Differential Equations*, vol. 255, no. 11, pp. 3971–3982, 2013.

[15] S. Weng, "On analyticity and temporal decay rates of solutions to the viscous resistive Hall-MHD system," *Journal of Differential Equations*, vol. 260, no. 8, pp. 6504–6524, 2016.

[16] O. A. Ladyzhenskaya, "Unique global solvability of the three-dimensional Cauchy problem for the Navier-Stokes equations in the presence of axial symmetry," *Zap. Nauchn. Sem. Leningrad. Otdel. Mati. Inst. Steklov. (LOMI)*, vol. 7, pp. 155–177, 1968.

[17] M. R. Ukhovskii and V. I. Iudovich, "Axially symmetric flows of ideal and viscous fluids filling the whole space," *Journal of Applied Mathematics and Mechanics*, vol. 32, pp. 52–61, 1968.

[18] Z. Lei, "On axially symmetric incompressible magnetohydrodynamics in three dimensions," *Journal of Differential Equations*, vol. 259, no. 7, pp. 3202–3215, 2015.

[19] J. Fan, S. Huang, and G. Nakamura, "Well-posedness for the axisymmetric incompressible viscous Hall-magnetohydrodynamic equations," *Applied Mathematics Letters*, vol. 26, no. 9, pp. 963–967, 2013.

[20] D. Chae and S. Weng, "Singularity formation for the incompressible Hall-MHD equations without resistivity," *Ann. Inst. Henri Poincare-Analyse Nonlineaire*, vol. 33, no. 4, pp. 1009–1022, 2016.

[21] D. Chae, R. Wan, and J. Wu, "Local well-posedness for the Hall-MHD equations with fractional magnetic diffusion," *Journal of Mathematical Fluid Mechanics*, vol. 17, no. 4, pp. 627–638, 2015.

[22] A. J. Majda and A. L. Bertozzi, *Vorticity and Incompressible Flow*, Cambridge University Press, Cambridge, UK, 2002.

4

A Study of the Transport of Marine Pollutants using Adjoint Method of Data Assimilation with Method of Characteristics

Xiaona Li,[1] Minjie Xu,[1] Xianqing Lv,[1] and Kai Fu[2]

[1]*Physical Oceanography Laboratory, Ocean University of China, Qingdao 266100, China*
[2]*School of Mathematical Sciences, Ocean University of China, Qingdao 266100, China*

Correspondence should be addressed to Kai Fu; kfu@ouc.edu.cn

Academic Editor: Xin Yu

An adjoint method of data assimilation with the characteristic finite difference (CFD) scheme is applied to marine pollutant transport problems and the temporal and spatial distribution of marine pollutants are simulated. Numerical tests of two-dimensional problems of pollutant transport with two different schemes indicate that the error of CFD is smaller than that of central difference scheme (CDS). Then the inversion experiments of the initial field and the source and sink terms of pollutants are carried out. Applying CFD in the adjoint method of data assimilation cannot only reduce simulation error to get a good inversion but can also enable larger time step size to decrease computation time and improve the calculation efficiency.

1. Introduction

With the rapid development of the coastal economy, offshore waters have suffered severe pollution damage and the ecological environment is gradually deteriorated, which is an important topic that attracted the attention all over the world, especially countries with long coastlines.

Many scholars have used mathematical models and methods to make numerical analysis in various areas. Gupta et al. [1] applied a two-dimensional model considering organized wastewater discharges to determine the waste water assimilative capacity of Tane creek; Harms et al. [2] applied a three-dimensional coupled ice-ocean-models of different horizontal resolution to simulate the dispersion of water from these rivers; Grell et al. [3] built the WRF/Chem model to simulate the distribution of atmospheric pollutants in the northeastern United States; Guo et al. [4] used the surface spline interpolation in the inversion of bottom friction coefficients in a two-dimensional tidal model to get a smoother surface; Liu et al. [5] presented a modified Cressman interpolation method for the simulation of routine monitoring data of total nitrogen in the Bohai Sea, which reduces interpolation errors by decreasing the influence radius and introducing background value.

The variational adjoint data assimilation can be applied to assimilate observations data into model by optimizing initial values and other parameters, which improves the model performance remarkably. Elbern et al. [6] used the adjoint method of data assimilation in the European air pollution dispersion model system and found the method allows them to analyze initial data even when sparse observations are available only; Peng and Xie [7] studied the inversion of the initial conditions of storm surge disaster and discovered that using the adjoint method of data assimilation can reduce error caused by uncertain initial condition; Zhang et al. [8] studied the space varying bottom friction coefficient using the adjoint method of data assimilation and get simulation result which is much better than that of traditional methods; Lv and Fan [9] applied the adjoint method of data assimilation in the inversion of spatially varying parameters of a marine ecosystem model and validated the efficiency of the method; Wang et al. [10] used the adjoint method of data assimilation to study the process of pollutant transport in Bohai Sea and studied the inversion of initial field using the assimilated routine monitoring data of pollutants. In the study of Fan and Lv [11], SeaWiFS chlorophyll-a data were assimilated into a NPZD (Nutrient-Phytoplankton-Zooplankton-Detritus) model by the adjoint method; Pan et al. [12] studied the open boundary condition of the M_2 tidal constituent using the adjoint method of data assimilation with spline interpolation; Zhang et al. [13] applied this method to study the similarities and the differences between the Ekman (linear) and the Quadratic (nonlinear) bottom friction parameters of a two-dimensional tidal model; and many other researches (Yu and O'Brien [14], Lawson et al. [15], Zhao et al. [16], Zhao and Lu [17], and Qi et al. [18]) have also proven the validity and rationality of the adjoint method.

Method of characteristics and the schemes it derives have been used to solve problems in several areas for its high accuracy and ability to use large time step size. Douglas Jr. and Russell [19] proposed characteristic method to solve convection-diffusion equations; Shen et al. [20] presented a characteristic finite difference method and its stability and convergence were analyzed; Fu and Liang [21] developed a conservative characteristic finite difference method to predict the distribution of atmospheric aerosols; Xu et al. [22] used the adjoint assimilation method with the characteristic finite difference scheme to solve aerosol transport problems.

In this paper, we construct an adjoint data assimilation model using the characteristic finite difference (CFD) scheme which has high accuracy and enables large time steps. Numerical experiments show that CFD can get more accurate results than central difference scheme (CDS) [10]. Ideal experiments of inverse problems for model variables are carried out. Applying CFD in the adjoint data assimilation model, simulation errors are reduced and time step sizes can be increased, which improves the calculation efficiency a lot.

The paper is structured as follows. Section 2 introduces the pollutant transport model, the adjoint model, and the CFD. In Section 3, numerical experiments are carried out and results are analyzed. Finally, conclusions are given in Section 4.

2. Model and Method

2.1. Three-Dimensional Marine Pollutant Transport Model. For the simulation of the pollutant transport in Bohai Sea, the initial field and the source and sink terms of pollutants have significant influences on the results. In this paper, we mainly consider the convection and diffusion processes, while other chemical and biological changes are included in the source and sink terms. The three-dimensional marine pollutant transport model [10] is given as below:

$$\frac{\partial C}{\partial t}+u\frac{\partial C}{\partial x}+v\frac{\partial C}{\partial y}+w\frac{\partial C}{\partial z} = \frac{\partial}{\partial x}\left(A_H\frac{\partial C}{\partial x}\right)+\frac{\partial}{\partial y}\left(A_H\frac{\partial C}{\partial y}\right)+\frac{\partial}{\partial z}\left(K_H\frac{\partial C}{\partial z}\right) + \theta, \quad (1)$$

meanings of symbols in (1) are presented in Table 1.

TABLE 1: The symbols in the pollutant transport model (1).

Symbol	Meaning
C	Pollutant concentration
u	Convection velocity in the east-west direction
v	Convection velocity in the north-south direction
w	Convection velocity in vertical direction
A_H	Horizontal diffusion coefficient
K_H	Vertical diffusion coefficient
θ	The source and sink terms of pollutants

The boundary conditions of the above model are set to

$$\frac{\partial C}{\partial t}=0,\quad V_n\le 0,\qquad \frac{\partial C}{\partial n}=0,\quad V_n>0, \quad (2)$$

where $\vec{n}$ is the outward normal to the boundary and V_n is the normal velocity of the boundary.

Assuming the pollutant concentrations at the grid points are known at $t = t^n$, in order to obtain the pollutant concentration at $t = t^{n+1}$, characteristic method is used here. Following the characteristic curve from a point (x_i, y_j, z_k) at $t = t^{n+1}$, the intersection point $(\overline{x}_i, \overline{y}_j, \overline{z}_k)$ of the curve with time level $t = t^n$ can be obtained. We approximate the point by $\overline{x}_i \approx x_i - u^n_{i,j,k}\cdot\Delta t$, $\overline{y}_j \approx y_j - v^n_{i,j,k}\cdot\Delta t$, and $\overline{z}_k \approx z_k - w^n_{i,j,k}\cdot\Delta t$.

The characteristic finite difference (CFD) scheme of the pollutant transport model (1) is given as

$$\frac{C1_{i,j,k}-\overline{C}^n_{i,j,k}}{\Delta t} = A_H\frac{C1_{i+1,j,k}-2C1_{i,j,k}+C1_{i-1,j,k}}{\left(\Delta x_j\right)^2}+\theta^n_{i,j,k}, \quad (3)$$

$$\frac{C2_{i,j,k}-\overline{C1}_{i,j,k}}{\Delta t} = A_H\frac{C2_{i,j+1,k}-2C2_{i,j,k}+C2_{i,j-1,k}}{(\Delta y)^2}, \quad (4)$$

$$\frac{C^{n+1}_{i,j,k}-\overline{C2}_{i,j,k}}{\Delta t} = K_H\left(\frac{C^{n+1}_{i,j,k+1}-C^{n+1}_{i,j,k}}{\Delta z_k\cdot\Delta z_{k+1/2}}-\frac{C^{n+1}_{i,j,k}-C^{n+1}_{i,j,k-1}}{\Delta z_k\cdot\Delta z_{k-1/2}}\right), \quad (5)$$

where $\overline{C}^n_{i,j,k}$, $\overline{C1}_{i,j,k}$, and $\overline{C2}_{i,j,k}$ are obtained from the following schemes:

$$\overline{C}^n_{i,j,k} = \frac{x_{r+1}-\overline{x}_i}{\Delta x_j}C^n_{r,j,k}+\frac{\overline{x}_i-x_r}{\Delta x_j}C^n_{r+1,j,k}, \quad (6)$$

$$\overline{C1}_{i,j,k} = \frac{y_{m+1}-\overline{y}_j}{\Delta y}C1_{i,m,k}+\frac{\overline{y}_j-y_m}{\Delta y}C1_{i,m+1,k}, \quad (7)$$

$$\overline{C2}_{i,j,k} = \frac{z_{l+1}-\overline{z}_k}{\Delta z_l}C2_{i,j,l}+\frac{\overline{z}_k-z_l}{\Delta z_{l+1}}C2_{i,j,l+1}. \quad (8)$$

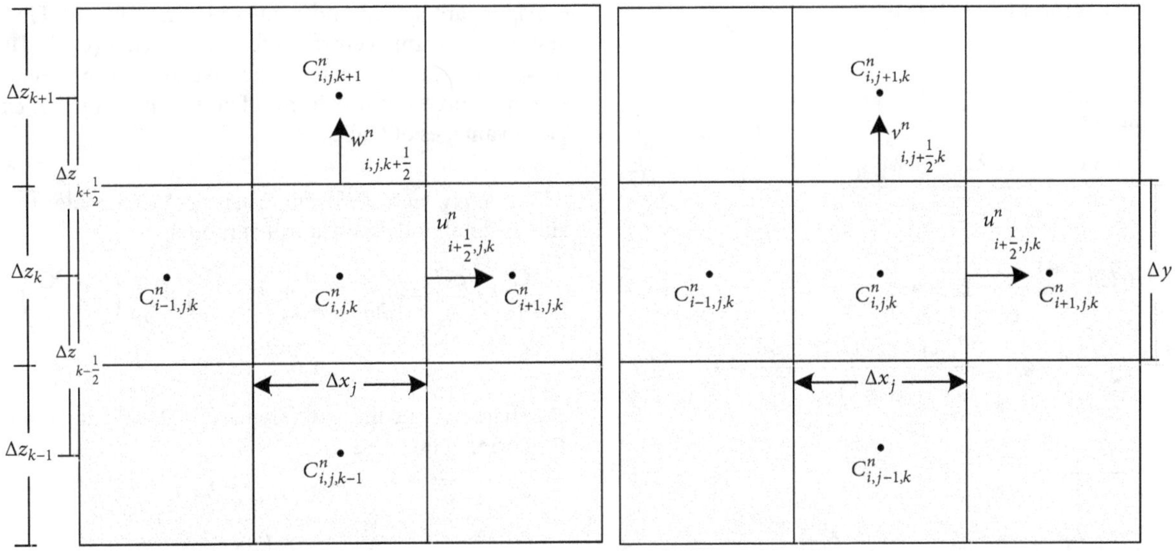

FIGURE 1: $C^n_{i,j,k}$ at grid center is the pollutant concentration. $u^n_{i+1/2,j,k}$, $v^n_{i,j+1/2,k}$, and $w^n_{i,j,k+1/2}$ are the convection velocities in the east-west, north-south, and vertical directions. Δx_j, Δy, Δz_k and $\Delta z_{k+1/2}$ indicate grid steps.

where the point $(\overline{x}_i, \overline{y}_j, \overline{z}_k)$ satisfies $x_r < \overline{x}_i < x_{r+1}$, $y_m < \overline{y}_j < y_{m+1}$, $z_l < \overline{z}_k < z_{l+1}$. And the meanings of $C^n_{i,j,k}$, $u^n_{i+1/2,j,k}$, $v^n_{i,j+1/2,k}$, $w^n_{i,j,k+1/2}$, Δx_j, Δy, Δz_k, and $\Delta z_{k+1/2}$ are shown in Figure 1.

2.2. The Adjoint Model. In order to get solution of the pollutant transport model, the adjoint model is applied here. We define the cost function J [23] that denotes the gap between the numerical solution and the observation data.

$$J = \frac{1}{2}\int_\Omega \left(C - C'\right)^T K\left(C - C'\right) d\Omega, \tag{9}$$

where C is the numerical solution of the pollutant transport model; C' is the observation data; T denotes matrix transposition; K is the weighting matrix of the observation data C' and is defined as

$$K = \begin{cases} 1, & \text{if the observations are available;} \\ 0, & \text{otherwise.} \end{cases} \tag{10}$$

Then we construct a Lagrange function:

$$\begin{aligned} L(\lambda, C, \theta) = \int_\Omega \lambda \cdot \Bigg[& \frac{\partial C}{\partial t} + u\frac{\partial C}{\partial x} + v\frac{\partial C}{\partial y} + w\frac{\partial C}{\partial z} \\ & - \frac{\partial}{\partial x}\left(A_H \frac{\partial C}{\partial x}\right) - \frac{\partial}{\partial y}\left(A_H \frac{\partial C}{\partial y}\right) - \frac{\partial}{\partial z}\left(K_H \frac{\partial C}{\partial z}\right) \\ & - \theta \Bigg] d\Omega + J, \end{aligned} \tag{11}$$

where λ is the Lagrange multiplier. According to Lagrange multiplier method, the first-order derivatives of the Lagrange function should equal zero when the minimum of the cost function is got,

$$\frac{\partial L}{\partial \lambda} = 0, \tag{12}$$

$$\frac{\partial L}{\partial C} = 0, \tag{13}$$

$$\frac{\partial L}{\partial \theta} = 0. \tag{14}$$

Equation (12) is the control equation of pollutant transport model (1) actually. And the adjoint (15) can be derived from (13).

$$\begin{aligned} -\frac{\partial \lambda}{\partial t} = & \frac{\partial (u\lambda)}{\partial x} + \frac{\partial (v\lambda)}{\partial y} + \frac{\partial (w\lambda)}{\partial z} - \frac{\partial}{\partial x}\left(A_H \frac{\partial \lambda}{\partial x}\right) \\ & - \frac{\partial}{\partial y}\left(A_H \frac{\partial \lambda}{\partial y}\right) - \frac{\partial}{\partial z}\left(K_H \frac{\partial \lambda}{\partial z}\right) \\ & - K\left(C - C'\right), \end{aligned} \tag{15}$$

with boundary conditions:

$$\begin{aligned} \frac{\partial \lambda}{\partial t} &= 0, \quad V_n \le 0, \\ \frac{\partial \lambda}{\partial n} &= 0, \quad V_n > 0. \end{aligned} \tag{16}$$

Assume the Lagrange multiplier λ at each grid point is known at $t = t^{n+1}$. In order to obtain the Lagrange multiplier λ at $t = t^n$, characteristic method is adopted. Following the characteristic curve from a point (x_i, y_j, z_k) at $t = t^n$, the intersection point $(\overline{x}_i, \overline{y}_j, \overline{z}_k)$of the curve with time level $t = t^{n+1}$ can be obtained. We approximate the point by $\overline{x}_i \approx x_i + u^{n+1}_{i,j,k} \cdot \Delta t$, $\overline{y}_j \approx y_j + v^{n+1}_{i,j,k} \cdot \Delta t$, and $\overline{z}_k \approx z_k + w^{n+1}_{i,j,k} \cdot \Delta t$.

The CFD of the adjoint (15) is

$$\frac{\lambda 1_{i,j,k}-\overline{\lambda}_{i,j,k}^{n+1}}{\Delta t} = A_H\frac{\lambda 1_{i+1,j,k}-2\lambda 1_{i,j,k}+\lambda 1_{i-1,j,k}}{\left(\Delta x_j\right)^2} + K\left(C_{i,j,k}^{n+1}-C'^{n+1}_{i,j,k}\right), \quad (17)$$

$$\frac{\lambda 2_{i,j,k}-\overline{\lambda 1}_{i,j,k}}{\Delta t} = A_H\frac{\lambda 2_{i,j+1,k}-2\lambda 2_{i,j,k}+\lambda 2_{i,j-1,k}}{\left(\Delta y\right)^2}, \quad (18)$$

$$\frac{\lambda_{i,j,k}^{n}-\overline{\lambda 2}_{i,j,k}}{\Delta t} = K_H\left(\frac{\lambda_{i,j,k+1}^{n}-\lambda_{i,j,k}^{n}}{\Delta z_k\cdot\Delta z_{k+1/2}}-\frac{\lambda_{i,j,k}^{n}-\lambda_{i,j,k-1}^{n}}{\Delta z_k\cdot\Delta z_{k-1/2}}\right), \quad (19)$$

where $\overline{\lambda}_{i,j,k}^{n+1}$, $\overline{\lambda 1}_{i,j,k}$, and $\overline{\lambda 2}_{i,j,k}$ are obtained from the following schemes:

$$\overline{\lambda}_{i,j,k}^{n+1} = \frac{x_{r+1}-\overline{x}_i}{\Delta x_j}\lambda_{r,j,k}^{n+1}+\frac{\overline{x}_i-x_r}{\Delta x_j}\lambda_{r+1,j,k}^{n+1}, \quad (20)$$

$$\overline{\lambda 1}_{i,j,k} = \frac{y_{m+1}-\overline{y}_j}{\Delta y}\lambda 1_{i,m,k}+\frac{\overline{y}_j-y_m}{\Delta y}\lambda 1_{i,m+1,k}, \quad (21)$$

$$\overline{\lambda 2}_{i,j,k} = \frac{z_{l+1}-\overline{z}_k}{\Delta z_l}\lambda 2_{i,j,l}+\frac{\overline{z}_k-z_l}{\Delta z_{l+1}}\lambda 2_{i,j,l+1}. \quad (22)$$

Based on (13), we can get the gradient of the cost function on the initial conditions of pollutant concentration $C_{i,j,k}^0$ [22]

$$\frac{\partial J}{\partial C^0} = \left[\frac{\partial\lambda}{\partial t}+u\frac{\partial\lambda}{\partial x}+v\frac{\partial\lambda}{\partial y}+w\frac{\partial\lambda}{\partial z}-\frac{\partial}{\partial x}\left(A_H\frac{\partial\lambda}{\partial x}\right) - \frac{\partial}{\partial x}\left(A_H\frac{\partial\lambda}{\partial y}\right)-\frac{\partial}{\partial x}\left(K_H\frac{\partial\lambda}{\partial z}\right)\right]^0. \quad (23)$$

Then the optimization of the initial condition can be obtained using the steepest descent method. The relationship between C^0 and the gradient is as follows:

$$C_{new}^0 = C_{old}^0-\alpha\frac{\partial J}{\partial C^0}, \quad (24)$$

where α is the step size of the steepest descent method.

With the initial condition obtained using the adjoint method, we can get more accurate simulation results of the pollutant transport model.

3. Numerical Experiments

In this section, we will first carry out numerical tests to observe the performance of the characteristic finite difference (CFD) scheme. Two-dimensional problems of pollutant transport are solved and results obtained from CFD and a first order in time central difference scheme (CDS) [10] are compared. Then we analyze the inversion of the initial field and the source and sink terms of pollutants to further explain the advantages of CFD.

3.1. Comparison of Different Schemes. We consider the two-dimensional pollutant transport model:

$$\frac{\partial C}{\partial t}+u\frac{\partial C}{\partial x}+v\frac{\partial C}{\partial y} = \frac{\partial}{\partial x}\left(A_H\frac{\partial C}{\partial x}\right)+\frac{\partial}{\partial y}\left(A_H\frac{\partial C}{\partial y}\right) + \theta. \quad (25)$$

The characteristic finite difference (CFD) scheme of the two-dimensional model is

$$\frac{C_{i,j}^{n+1}-\overline{C}_{i,j}^{n}}{\Delta t} = A_H\frac{C_{i+1,j}^{n+1}-2C_{i,j}^{n+1}+C_{i-1,j}^{n+1}}{\left(\Delta x_j\right)^2} + A_H\frac{C_{i,j+1}^{n+1}-2C_{i,j}^{n+1}+C_{i,j-1}^{n+1}}{\left(\Delta y\right)^2}+\theta_{i,j}^n, \quad (26)$$

where $\overline{C}_{i,j}^{n}$ is the pollutant concentration at the point $(\overline{x}_i,\overline{y}_j)$, which is obtained from the following scheme:

$$\overline{C}_{i,j}^{n} = a_{00}C_{r,m}^n+a_{01}C_{r,m+1}^n+a_{10}C_{r+1,m}^n+a_{11}C_{r+1,m+1}^n, \quad (27)$$

with

$$a_{00} = \frac{x_{r+1}-\overline{x}_i}{\Delta x}\cdot\frac{y_{m+1}-\overline{y}_j}{\Delta y}, \quad a_{01} = \frac{x_{r+1}-\overline{x}_i}{\Delta x}\cdot\frac{\overline{y}_j-y_m}{\Delta y}, \quad a_{10} = \frac{\overline{x}_i-x_r}{\Delta x}\cdot\frac{y_{m+1}-\overline{y}_j}{\Delta y}, \quad a_{11} = \frac{\overline{x}_i-x_r}{\Delta x}\cdot\frac{\overline{y}_j-y_m}{\Delta y}. \quad (28)$$

And the central difference scheme (CDS) [10] is

$$\frac{C_{i,j}^{n+1}-C_{i,j}^{n}}{\Delta t} = -u_{i,j}^n\frac{C_{i+1,j}^{n}-C_{i-1,j}^{n}}{2\Delta x_j}-v_{i,j}^n\frac{C_{i,j+1}^{n}-C_{i,j-1}^{n}}{2\Delta y} + A_H\frac{C_{i+1,j}^{n}-2C_{i,j}^{n}+C_{i-1,j}^{n}}{\left(\Delta x_j\right)^2} + A_H\frac{C_{i,j+1}^{n}-2C_{i,j}^{n}+C_{i,j-1}^{n}}{\left(\Delta y\right)^2}+\theta_{i,j}^n. \quad (29)$$

Consider model (25) with the initial condition:

$$C(x,y,0) = \exp\left(-\frac{(x-x_0)^2+(y-y_0)^2}{2\sigma_0{}^2}\right), \quad (x,y)\in\Omega, \quad (30)$$

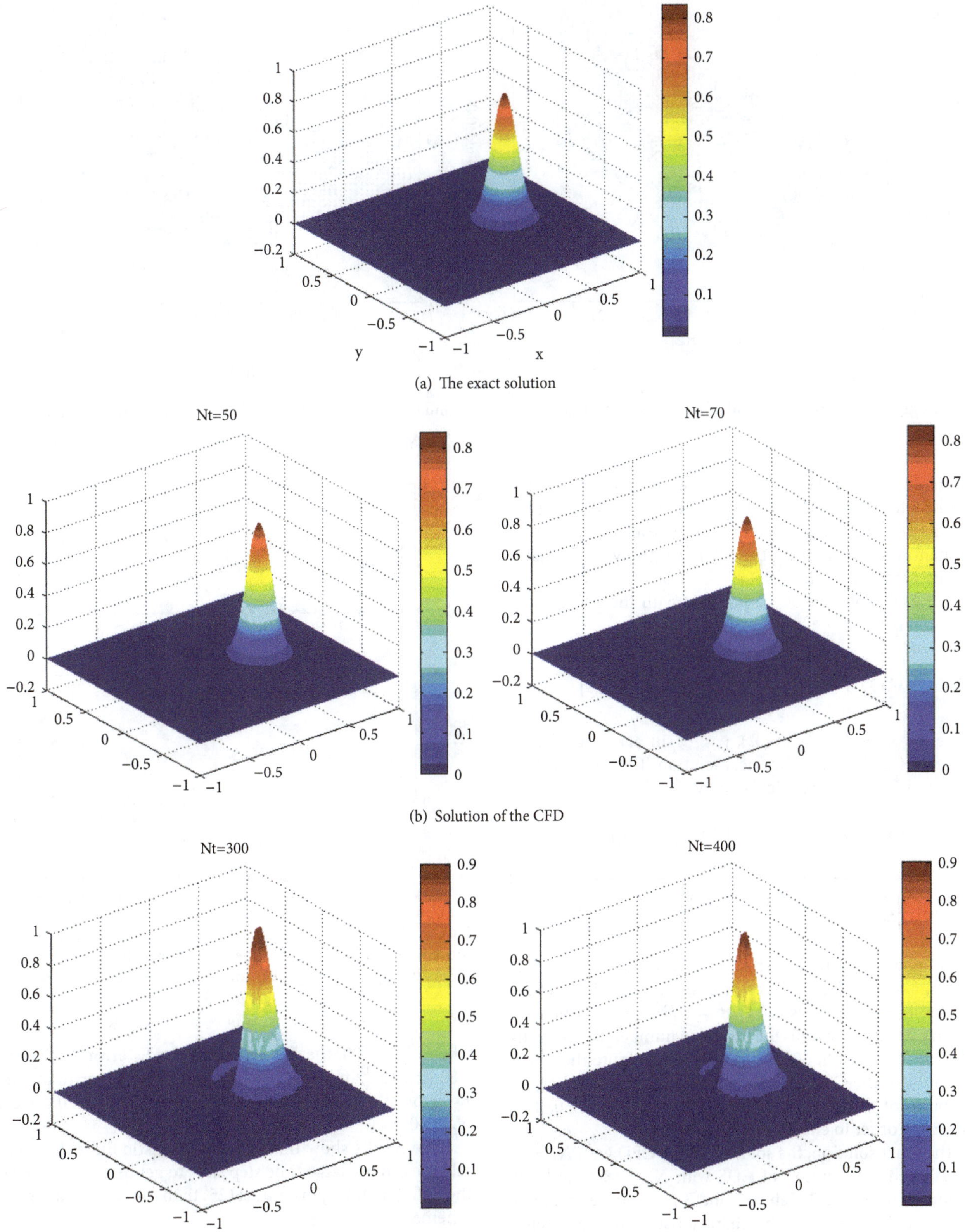

(a) The exact solution

(b) Solution of the CFD

(c) Solution of the CDS

FIGURE 2: The exact solution and approximate solutions of different schemes with different time steps.

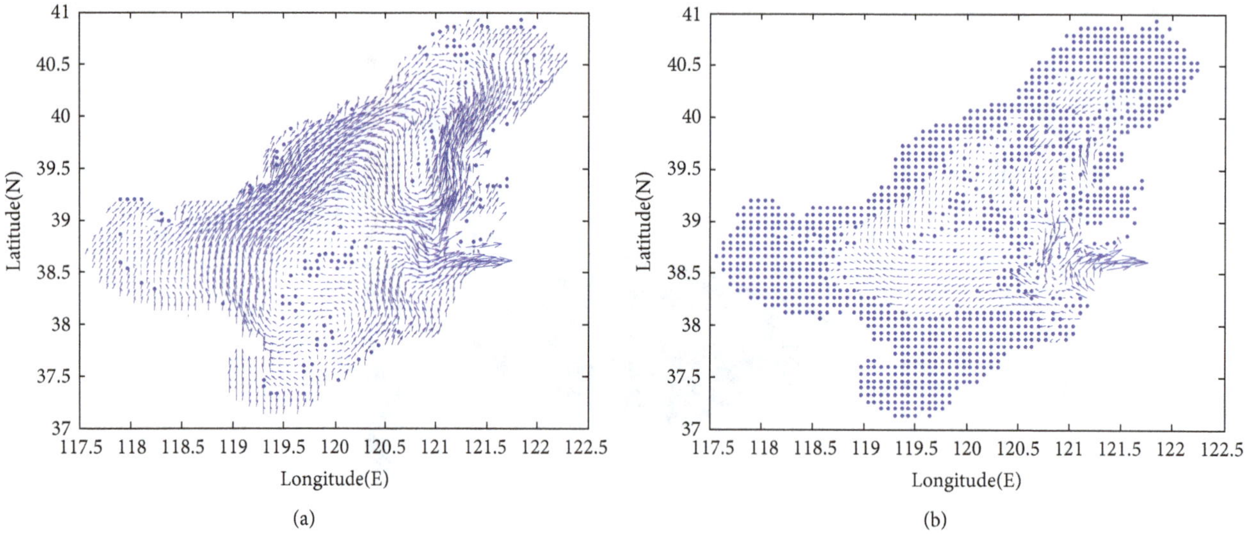

FIGURE 3: The average flow field of Bohai Sea in May 2009. Figures (a) and (b) are the average flow field of the first and third layers.

in domain $\Omega = [-1,1] \times [-1,1]$ and on $[0,T] = [0,\pi/4]$. The initial center is $(x_0, y_0) = (-0.4, 0)$ and $\sigma_0{}^2 = 0.01$. Besides, the diffusion coefficient A_H and the source and sink terms of pollutants θ are set to 0.0001 and 0, respectively. The variable velocity field is $u = -4y$, $v = 4x$.

The exact solution of the problem with the given initial condition is

$$C(x, y, t) = \exp\left(-\frac{(x - x_0)^2 + (y - y_0)^2}{2\sigma_0{}^2 + 4Kt}\right), \quad (31)$$

where $x = x\cos(4t) + y\sin(4t)$, $y = -x\sin(4t) + y\cos(4t)$.

Let $C^n(x, y)$ be the numerical solution. The errors in L_∞-norm and L_2-norm are calculated by

$$E_\infty = \max_i \max_j \left\{\left|C(x_i, y_j, t^n) - C^n(x_i, y_j)\right|\right\},$$
$$E_2 = \sqrt{\sum_i \sum_j \Delta x \Delta y \left(C(x_i, y_j, t^n) - C^n(x_i, y_j)\right)^2}. \quad (32)$$

We now compute the errors and ratios in time. In order to eliminate the effect of the error in space, a small space step size $\Delta x = \Delta y = 1/30$ is used. $\Delta t = T/N_t$ denotes the time step size, where N_t means the time step number. By choosing different N_t =40, 50, 60, 70, and 80, we compute the errors and ratios of the CFD in time, while we compute the errors and ratios of the CDS by choosing N_t =250, 300, 350, 400, and 450 in order to ensure stability.

The exact solution, the solution of the CFD with $N_t = 50$ and 70, and the solution of CDS with $N_t = 300$ and 400 are shown in Figure 2. Tables 2 and 3 present the results of the L_∞ and L_2 error of the characteristic finite difference (CFD) scheme and the central difference scheme (CDS). It is clearly shown that both two different schemes have first-order accuracy in time. However, the numerical simulation errors of CFD are smaller, even though the time step sizes of the CFD are much larger than those of the CDS. For example, when $N_t = 300$, the E_∞ error and the E_2 error of CDS are 3.8256×10^{-1} and 6.6713×10^{-2}, while the E_∞ error and the E_2 error of CFD are 2.4370×10^{-1} and 4.5245×10^{-2} when $N_t = 50$.

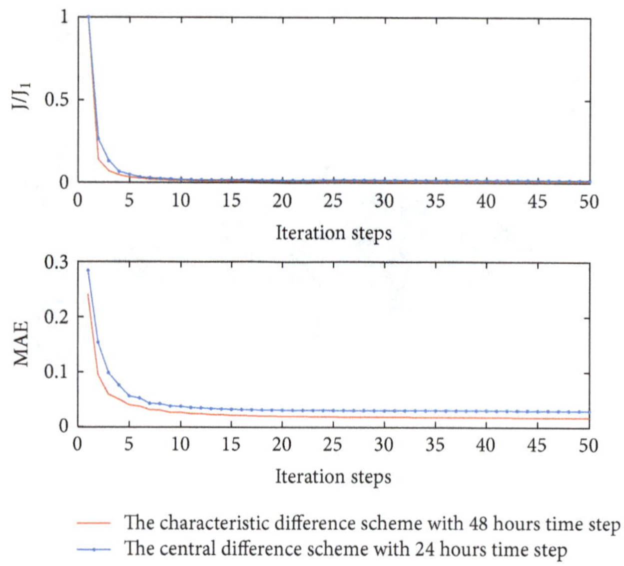

FIGURE 4: The descending curves of the relative magnitude of the cost function J/J_1 and the mean absolute error (MAE) of observation points using CFD and CDS with different time steps.

The results show that the characteristic finite difference scheme can use large time step sizes to get better solutions of the pollutant transport model (25) than the central difference scheme.

3.2. The Inversion of the Initial Field and the Source and Sink Terms. In this section, we study the inversion of the initial field and the source and sink terms of pollutants through ideal experiments; the "observation data" used in the

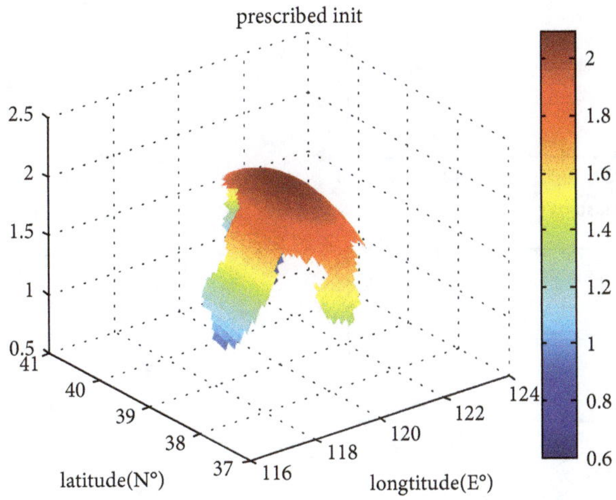

(a) The given initial field

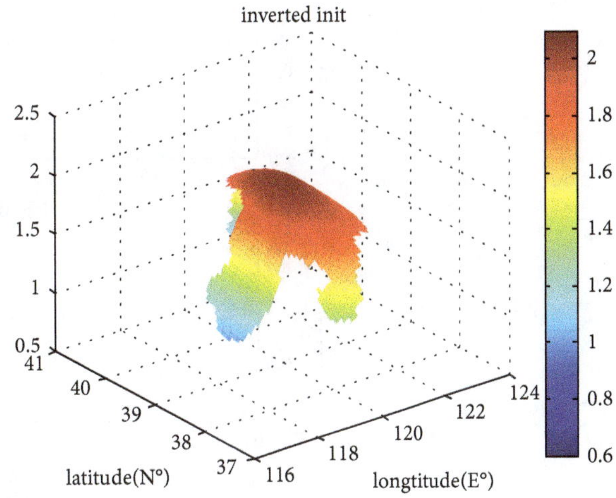

(b) The inversion result of the given initial field obtained by CFD with a 48 hours' time step size

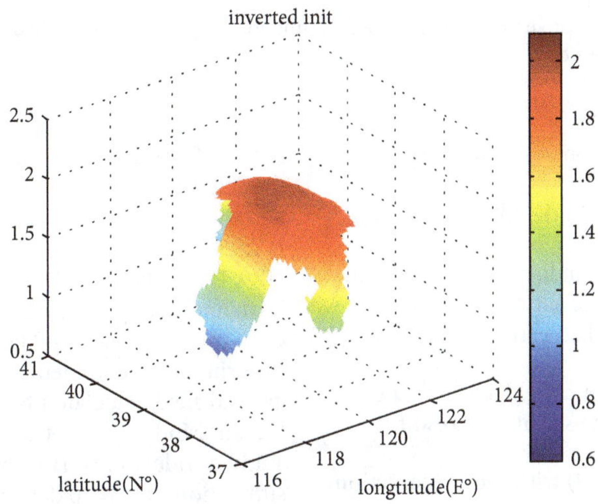

(c) The inversion result of the given initial field obtained by CDS with a 24 hours' time step size

FIGURE 5: Figures of the initial distribution inverted by the adjoint method of data assimilation with CFD and CDS.

TABLE 2: Errors and ratios in time of the characteristic finite difference scheme (CFD).

N_t	40	50	60	70	80
E_∞	3.0017E-01	2.4370E-01	2.0544E-01	1.7835E-01	1.5751E-01
Ratio	-	0.93394	0.93686	0.91736	0.93042
E_2	5.5471E-02	4.5245E-02	3.8185E-02	3.3043E-02	2.9150E-02
Ratio	-	0.91318	0.93054	0.93812	0.93893

TABLE 3: Errors and ratios in time of the central difference scheme (CDS).

N_t	250	300	350	400	450
E_∞	4.3826E-01	3.8256E-01	3.4791E-01	3.2436E-01	3.0875E-01
Ratio	-	0.74551	0.61569	0.52489	0.41870
E_2	7.8419E-02	6.6713E-02	5.9804E-02	5.5299E-02	5.2169E-02
Ratio	-	0.88673	0.70920	0.58651	0.49468

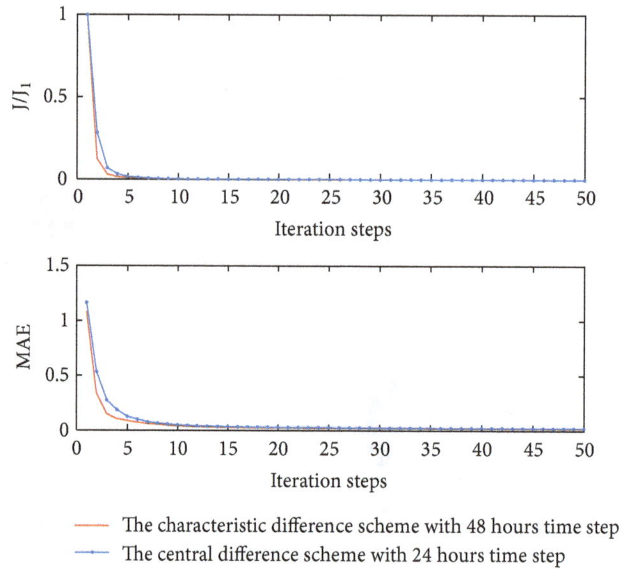

FIGURE 6: The descending curves of the relative magnitude of the cost function J/J_1 and the mean absolute error (MAE) of observation points using CFD and CDS with different time steps.

assimilation process is the simulation result obtained from pollutant transport model.

The experiments are carried out in the following steps.

Step 1. Give an initial field of pollutants and operate the forward model (1). The simulation result obtained from the forward model is regarded as the "observation data."

Step 2. Give a guess value of the initial field and operate the forward model again. We will get the simulation result.

Step 3. Compute the cost function (9) with the "observation data" of Step 1 and the simulation results of Step 2.

Step 4. Operate the adjoint model. Here we compute the gradient of the cost function on initial conditions and adjust the initial field of pollutants with the gradient. A new predicted value is got and then go to Step 2. The iterative stops when the cost function is decreased to a given small value or the iteration steps reaches to a given number.

3.2.1. Model Settings. The domain of pollutant transport model (1) is set to $37^\circ N \sim 41^\circ N$ and $117.5^\circ E \sim 122.5^\circ E$ and the horizontal resolution is $4' \times 4'$. The vertical direction is divided into 6 layers and the thickness of each layer is $10m$, $10m$, $10m$, $20m$, $25m$, and $25m$ from top to bottom. The horizontal diffusion coefficient and vertical diffusion coefficient are $100m^2/s$ and $0.00001m^2/s$, respectively. Numerical experiments are implemented with the hydrodynamic background field calculated by FVCOM (Finite Volume Coastal Ocean Model) [24], which has been widely used in the study of tide and storm surge in the Bohai Sea [25, 26]. The simulation time is 30 days and the average flow field of Bohai Sea in May 2009 is used here, the first and third layers of which are shown in Figure 3.

3.2.2. The Inversion of the Initial Field. In order to further explain the advantages of the characteristic finite difference (CFD) scheme, we set the time step size of the CFD to be 48 hours, while we set that of the central difference scheme (CDS) to be 24 hours in the inversion of the initial field.

We first consider an initial field which presents a downwardly directed opening and satisfies the following:

$$C(i,j,k) = \begin{cases} 0, & 1.75 * lon(i) - 3.5 * \dfrac{120.5}{2} + 37 - lat(j) > 0 \text{ or } gs(i,j,k) = 0, \\ -\dfrac{(lon(i)-120)^2 + (lat(j)-39)^2}{5} + 2.1, & otherwise, \end{cases} \quad (33)$$

where $lon(i)$ and $lat(j)$ denote the longitude and latitude at the grid point (i, j) in the simulation domain, and $gs(i, j, k)$ is the wet and dry condition at the point (i, j, k) and takes 0 for land and 1 for water.

Figure 4 presents the fact that the relative magnitude of the cost function J/J_1 and the mean absolute error (MAE) of observation points of the adjoint model using the characteristic finite difference (CFD) scheme decline more

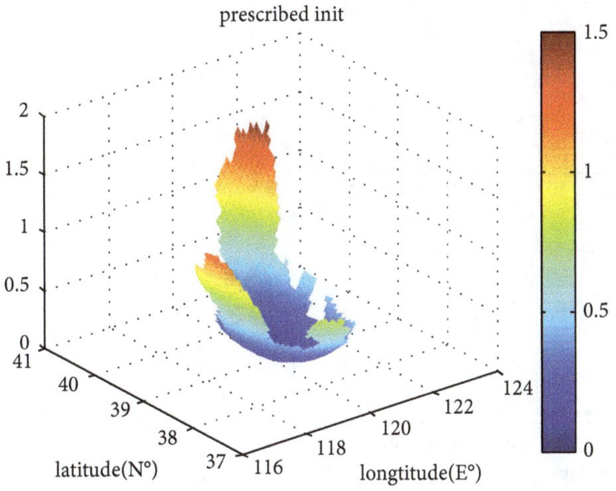

(a) The given initial field

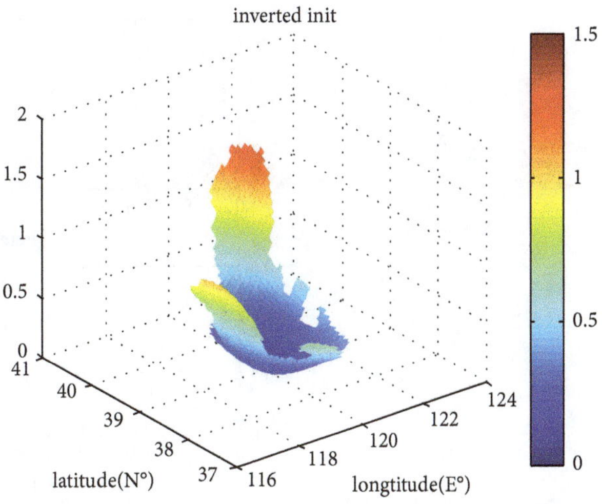

(b) The inversion result of the given initial field obtained by CFD with a 48 hours' time step size

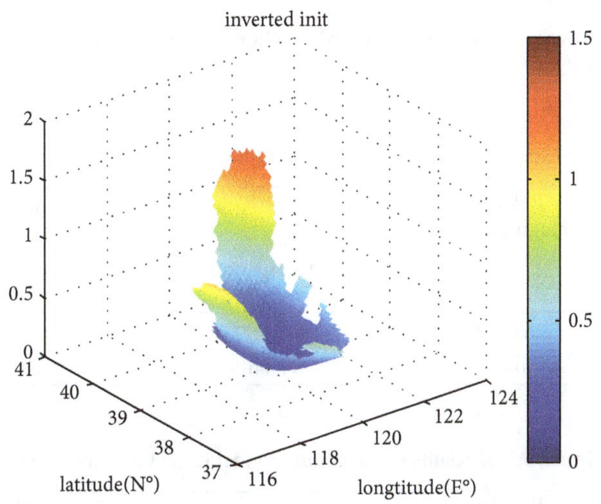

(c) The inversion result of the given initial field obtained by CDS with a 24 hours' time step size

FIGURE 7: Figures of the initial distribution inverted by the adjoint method of data assimilation with CFD and CDS.

quickly. Table 4 shows that when using CFD with a 48 hours' time step size, J/J_1 decreases by 3 orders of magnitude to 6.0650×10^{-3} and the MAE of observation points decreases by 92.87%, from 0.24121 mg/L to 0.01720 mg/L. However, when using the central difference scheme (CDS) with a 24 hours' time step size that is a half of the CFD's, J/J_1 is reduced to 1.3303×10^{-2} and the MAE of observation points decreases by 89.58% only.

Figure 5(a) is the given initial field and Figures 5(b) and 5(c) are the inversion results obtained by CFD with a 48 hours' time step size and CDS with a 24 hours' time step size, respectively. From Figure 5, we can see that the inversion result obtained by CFD with 48 hours is almost the same as the original distribution and better than the CDS with a 24 hours' time step size. That is to say, CFD gets better inversion of the initial distribution of pollutants with a larger time step size.

Then we consider an initial field which presents an upwardly directed opening and satisfies the following:

$$C(i,j,k) = \begin{cases} 0, & 1.75 * lon(i) - 3.5 * \dfrac{120.5}{2} + 37 - lat(j) > 0 \text{ or } gs(i,j,k) = 0, \\ \dfrac{(lon(i) - 120)^2 + (lat(j) - 39)^2}{5} + 0.05, & otherwise. \end{cases} \quad (34)$$

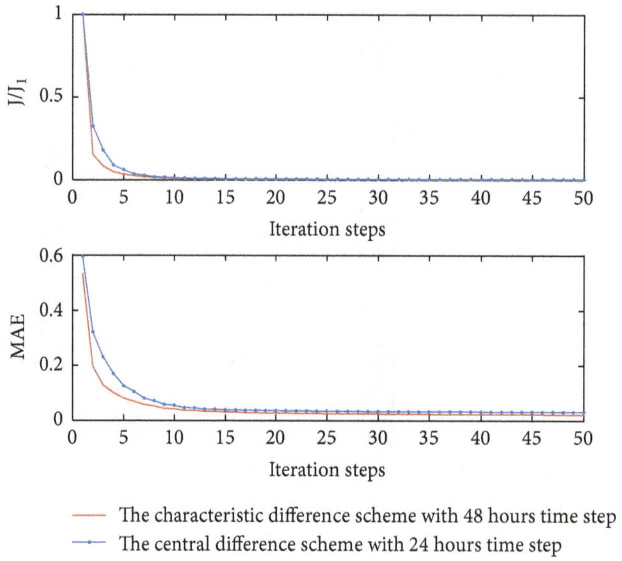

FIGURE 8: The descending curves of the relative magnitude of the cost function J/J_1 and the mean absolute error (MAE) of observation points using CFD and CDS with different time steps.

TABLE 4: The results of the adjoint model using CFD and CDS.

Method	Time step size (h)	Relative magnitude of the cost function	Mean absolute error of observation points		
			Before assimilation (mg/L)	After assimilation (mg/L)	Rate of decline
CFD	48	6.0650×10^{-3}	0.24121	0.01720	92.87%
CDS	24	1.3303×10^{-2}	0.28406	0.02959	89.58%

TABLE 5: The results of the adjoint model using CFD and CDS.

Method	Time step size (h)	Relative magnitude of the cost function	Mean absolute error of observation points		
			Before assimilation (mg/L)	After assimilation (mg/L)	Rate of decline
CFD	48	5.0118×10^{-4}	1.08499	0.01873	98.27%
CDS	24	6.1274×10^{-4}	1.16529	0.02102	98.20%

TABLE 6: The results of the adjoint model using CFD and CDS.

Method	Time step size (h)	Relative magnitude of the cost function	Mean absolute error of observation points		
			Before assimilation (mg/L)	After assimilation (mg/L)	Rate of decline
CFD	48	2.7084×10^{-3}	0.53359	0.02083	96.10%
CDS	24	4.4638×10^{-3}	0.59472	0.03060	94.86%

The relative magnitude of the cost function J/J_1 and the mean absolute error (MAE) of observation points of the adjoint model using CFD decline more quickly, which are shown in Figure 6. The given initial field and the inversion results obtained by CFD with a 48 hours' time step size and CDS with a 24 hours' time step size are given in Figure 7. Table 5 shows that when using CFD with a 48 hours' time step size, the relative magnitude of the cost function

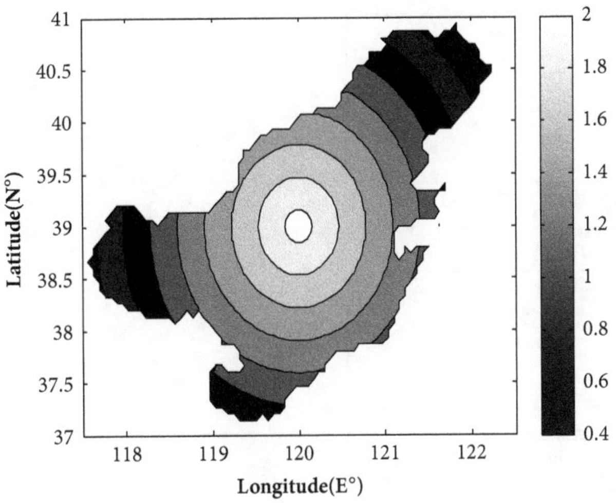

(a) The given source and sink terms

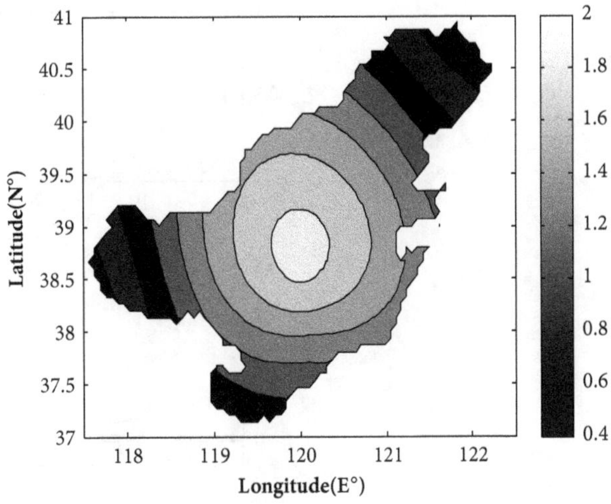

(b) The inversion result of the given source and sink terms obtained by CFD with a 48 hours' time step size

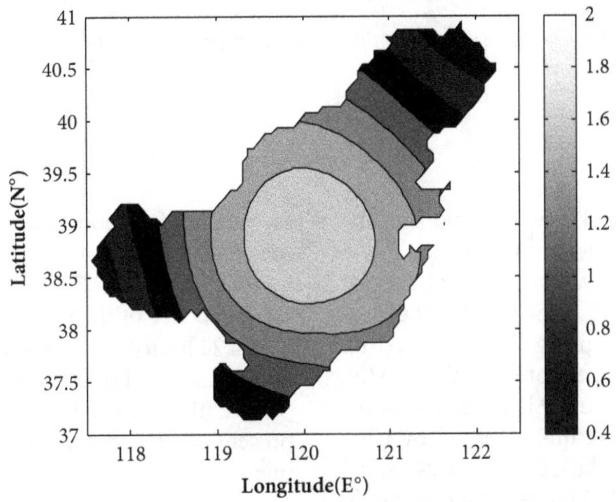

(c) The inversion result of the given source and sink terms obtained by CDS with a 24 hours' time step size

FIGURE 9: Figures of the source and sink terms inverted by the adjoint method of data assimilation with CFD and CDS.

J/J_1 decreases to 5.0118×10^{-4} and the mean absolute error (MAE) of observation points decreases by 98.27%, from 1.08499 mg/L to 0.01873 mg/L. When using CDS with a 24 hours' time step size, J/J_1 is reduced to 6.1274×10^{-4} and the MAE of observation points decreases by 98.20%. The inversion result of the adjoint model obtained by the CDS is inferior to that obtained by CFD with a large time step size.

Based on the inversion of the initial field, it is clearly shown that applying the characteristic finite difference scheme in the adjoint model can reduce the simulation error and enable using large time steps to improve the calculation efficiency.

3.2.3. The Inversion of the Source and Sink Terms. For the inversion of the source and sink terms, we also set the time step size of the CFD to be 48 hours, while the time step size of the CDS is set to be 24 hours.

We first consider the source and sink terms which present a downwardly directed opening and satisfy the following:

$$\theta(i,j) = \begin{cases} 0, & sur\_gs(i,j,1) = 0, \\ -\sqrt{\left[(lon(i)-120)^2 + (lat(j)-39)^2\right]} * 0.41 + 2.1, & otherwise, \end{cases} \tag{35}$$

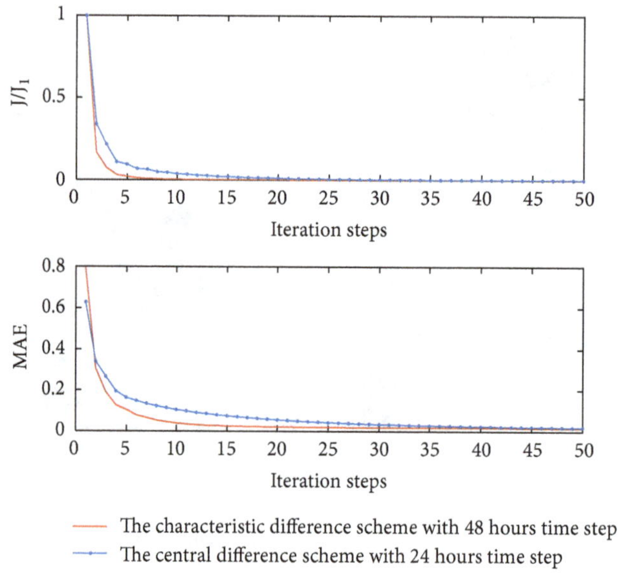

FIGURE 10: The descending curves of the relative magnitude of the cost function J/J_1 and the mean absolute error (MAE) of observation points using CFD and CDS with different time steps.

where $sur\_gs(i, j, 1)$ is the wet and dry condition at the surface point (i, j) and takes 0 for land and 1 for water.

Figure 8 also shows that the relative magnitude of the cost function J/J_1 and the mean absolute error (MAE) of observation points of the adjoint model using the characteristic finite difference (CFD) scheme decline more quickly than the central difference scheme (CDS). Figure 9 is the given source and sink terms and the inversion results obtained by CFD with a 48 hours' time step size and the CDS with a 24 hours' time step size. It is clear that in Table 6 the relative magnitude of the cost function J/J_1 decreases to 2.7084×10^{-3} and the mean absolute error (MAE) of observation points decreases by 96.10%, from 0.53359 mg/L to 0.02083 mg/L, when the time step size of the CFD is 48 hours. And when using CDS with a 24 hours' time step size, J/J_1 is reduced to 4.4638×10^{-3} and the MAE of observation points decreases by 94.86%.

We then consider the source and sink terms which present an upwardly directed opening and satisfy the following:

$$\theta(i,j) = \begin{cases} 0, & sur\_gs(i,j,1) = 0, \\ \sqrt{\left[(lon(i) - 120)^2 + (lat(j) - 39)^2\right]} * 0.41 + 0.05, & otherwise. \end{cases} \tag{36}$$

In Figure 10, the relative magnitude of the cost function J/J_1 and the mean absolute error (MAE) of observation points of the adjoint model using the characteristic finite difference (CFD) scheme are presented, which decline more quickly. The given source and sink terms and the inversion results obtained by CFD with a 48 hours' time step size and CDS with a 24 hours' time step size are shown in Figure 11. From Table 7, we can see when using the characteristic finite difference (CFD) scheme with a 48 hours' time step size that the relative magnitude of the cost function J/J_1 decreases to 7.5274×10^{-4} and the mean absolute error (MAE) of observation points decreases by 97.90%, from 0.79497 mg/L to 0.01667 mg/L. And J/J_1 is reduced to 1.7942×10^{-3} and the MAE of observation points decreases by 96.89% using the central difference scheme (CDS) with a 24 hours' time step size.

From simulation results of this part, we can see that, by applying CFD in the adjoint data assimilation model, simulation errors can be reduced when time step sizes are increased, which improves the calculation efficiency a lot.

4. Conclusion

In this paper, we adopt the adjoint method of data assimilation with the characteristic finite difference (CFD) scheme to solve the pollutant transport problem of Bohai Sea. Comparing the results obtained using the CFD and the central difference scheme (CDS) with different time step sizes, it can be seen that the simulation error of the CFD using large time step is smaller than that of the CDS using small time step. From the inversion of the initial field and the source and sink terms of pollutants, we come to the conclusion that the

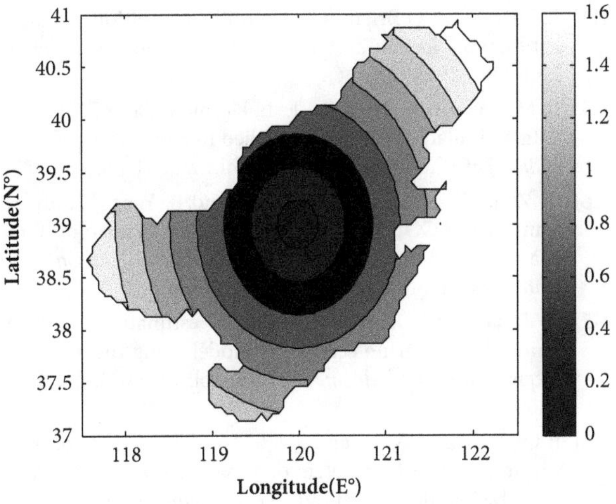

(a) The given source and sink terms

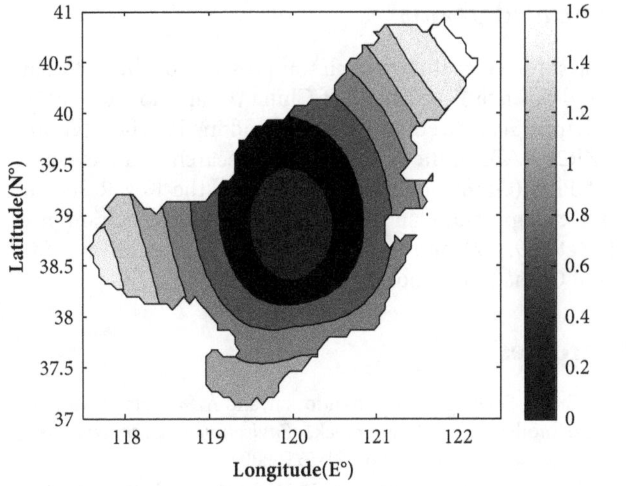

(b) The inversion result of the given source and sink terms obtained by CFD with a 48 hours' time step size

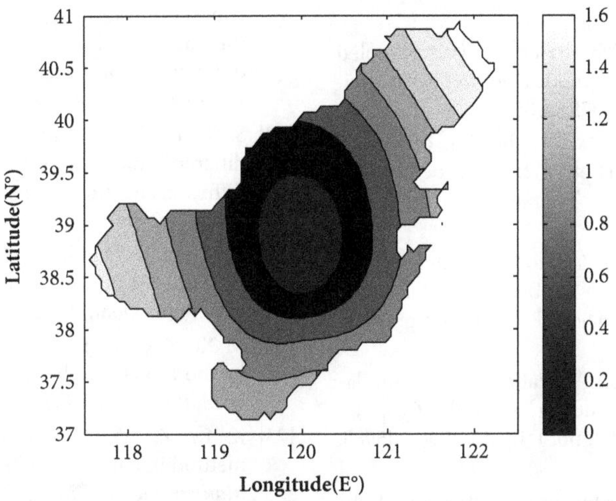

(c) The inversion result of the given source and sink terms obtained by CDS with a 24 hours' time step size

FIGURE 11: Figures of the source and sink terms inverted by the adjoint method of data assimilation with CFD and CDS.

TABLE 7: The results of the adjoint model using CFD and CDS.

Method	Time step size (h)	Relative magnitude of the cost function	Mean absolute error of observation points		
			Before assimilation (mg/L)	After assimilation (mg/L)	Rate of decline
CFD	48	7.5274×10^{-4}	0.79497	0.01667	97.90%
CDS	24	1.7942×10^{-3}	0.62949	0.01960	96.89%

adjoint model with CFD can reduce the simulation error and solve the problems effectively.

equations and parameters, researchers can use modeling simulations to replicate the tables and figures presented in the article.

Conflicts of Interest

The authors declare that there are no conflicts of interest regarding the publication of this paper.

Acknowledgments

Partial support for this research was provided by the National Natural Science Foundation of China (Grant no. 11601497), the Natural Science Foundation of Shandong Province (Grant no. ZR2016AB16), the National Key Research and Development Plan (Grant no. 2016YFC1402304), the Key Research and Development Plan of Shandong Province (Grant no. 2016ZDJS09A02), and the National Natural Science Foundation of China (Grant no. 41606006).

References

[1] I. Gupta, S. Dhage, A. A. Chandorkar, and A. Srivastav, "Numerical modeling for Thane creek," *Environmental Modeling and Software*, vol. 19, no. 6, pp. 571–579, 2004.

[2] I. H. Harms, M. J. Karcher, and D. Dethleff, "Modelling Siberian river runoff - Implications for contaminant transport in the Arctic Ocean," *Journal of Marine Systems*, vol. 27, no. 1, pp. 95–115, 2000.

[3] G. A. Grell, S. E. Peckham, R. Schmitz et al., "Fully coupled 'online' chemistry within the WRF model," *Atmospheric Environment*, vol. 39, no. 37, pp. 6957–6975, 2005.

[4] Z. Guo, H. Pan, W. Fan, and X. Lv, "Application of surface spline interpolation in inversion of bottom friction coefficients," *Journal of Atmospheric and Oceanic Technology*, vol. 34, no. 9, pp. 2021–2028, 2017.

[5] Y. Liu, J. Yu, Y. Shen, and X. Lv, "A modified interpolation method for surface total nitrogen in the Bohai Sea," *Journal of Atmospheric and Oceanic Technology*, vol. 33, no. 7, pp. 1509–1517, 2016.

[6] H. Elbern, H. Schmidt, and A. Ebel, "Variational data assimilation for tropospheric chemistry modeling," *Journal of Geophysical Research: Atmospheres*, vol. 102, no. D13, pp. 15967–15985, 1997.

[7] S.-Q. Peng and L. Xie, "Effect of determining initial conditions by four-dimensional variational data assimilation on storm surge forecasting," *Ocean Modelling*, vol. 14, no. 1-2, pp. 1–18, 2006.

[8] J.-C. Zhang and X.-Q. Lu, "Inversion of the bottom friction coefficient in 2D tidal model of the bohai, the yellow and east China seas," *Chinese Journal of Computational Mechanics*, vol. 24, no. 4, pp. 430–435, 2007.

[9] X.-Q. Lv and W. Fan, "Numerical study on spatially varying parameters of a marine ecosystem model with the adjoint method," *Periodical of Ocean University of China*, vol. 30, no. 1, pp. 7–14, 2009.

[10] C. Wang, X. Li, and X. Lv, "Numerical study on initial field of pollution in the Bohai sea with an adjoint method," *Mathematical Problems in Engineering*, pp. 389–405, 2013.

[11] W. Fan and X. Lv, "Data assimilation in a simple marine ecosystem model based on spatial biological parameterizations," *Ecological Modelling*, vol. 220, no. 17, pp. 1997–2008, 2009.

[12] H. Pan, Z. Guo, and X. Lv, "Inversion of tidal open boundary conditions of the M2 constituent in the bohai and yellow seas," *Journal of Atmospheric and Oceanic Technology*, vol. 34, no. 8, pp. 1661–1672, 2017.

[13] J. Zhang, X. Lu, P. Wang, and Y. P. Wang, "Study on linear and nonlinear bottom friction parameterizations for regional tidal models using data assimilation," *Continental Shelf Research*, vol. 31, no. 6, pp. 555–573, 2011.

[14] L. Yu and J. J. O'Brien, "On the Initial Condition in Parameter Estimation," *Journal of Physical Oceanography*, vol. 22, no. 11, pp. 1361–1364, 1992.

[15] L. M. Lawson, Y. H. Spitz, E. E. Hofmann, and R. B. Long, "A data assimilation technique applied to a predator-prey model," *Bulletin of Mathematical Biology*, vol. 57, no. 4, pp. 593–617, 1995.

[16] Q. Zhao, X. Hu, X. Lü, X. Xiong, and B. Yang, "Study on the transport of COD in the sea area around Maidao off Qingdao coast using data assimilation," *Journal of Ocean University of China*, vol. 6, no. 4, pp. 339–344, 2007.

[17] Q. Zhao and X. Lu, "Parameter estimation in a three-dimensional marine ecosystem model using the adjoint technique," *Journal of Marine Systems*, vol. 74, no. 1, pp. 443–452, 2008.

[18] P. Qi, C. Wang, X. Li, and X. Lv, "Numerical study on spatially varying control parameters of a marine ecosystem dynamical model with adjoint method," *Acta Oceanologica Sinica*, vol. 30, no. 1, pp. 7–14, 2011.

[19] J. Douglas Jr. and T. F. Russell, "Numerical methods for convection-dominated diffusion problems based on combining the method of characteristics with finite element or finite difference procedures," *SIAM Journal on Numerical Analysis*, vol. 19, no. 5, pp. 871–885, 1982.

[20] S. Shen, F. Liu, V. Anh, I. Turner, and J. Chen, "A characteristic difference method for the variable-order fractional advection-diffusion equation," *Applied Mathematics and Computation*, vol. 42, no. 1-2, pp. 371–386, 2013.

[21] K. Fu and D. Liang, "The conservative characteristic FD methods for atmospheric aerosol transport problems," *Journal of Computational Physics*, vol. 305, pp. 494–520, 2016.

[22] M. Xu, K. Fu, and X. Lv, "Application of adjoint data assimilation method to atmospheric aerosol transport problems," *Advances in Mathematical Physics*, vol. 2017, no. 4, pp. 1–14, 2017.

[23] X. Lv, Z. Wu, Y. Gu, and J. Tian, "Study on the adjoint method in data assimilation and the related problems," *Applied Mathematics and Mechanics -English Edition*, vol. 25, no. 6, pp. 636–646, 2004.

[24] C. Chen, H. Liu, and R. C. Beardsley, "An unstructured grid, finite-volume, three-dimensional, primitive equations ocean model: Application to coastal ocean and estuaries," *Journal of Atmospheric and Oceanic Technology*, vol. 20, no. 1, pp. 159–186, 2003.

[25] Y. Ding and H. Wei, "Research of the influential factors on the simulation of storm surge in the bohai sea," *Open Mechanical Engineering Journal*, vol. 8, no. 1, pp. 151–156, 2014.

[26] P. Zhao and W. Jiang, "A numerical study of storm surges caused by cold-air outbreaks in the Bohai Sea," *Natural Hazards*, vol. 59, no. 1, pp. 1–15, 2011.

5

Plasma Parameters Effects on Dust Acoustic Solitary Waves in Dusty Plasmas of Four Components

Abeer A. Mahmoud, Essam M. Abulwafa, Abd-alrahman F. Al-Araby, and Atalla M. Elhanbaly

Theoretical Physics Research Group, Physics Department, Faculty of Science, Mansoura University, Mansoura 35516, Egypt

Correspondence should be addressed to Essam M. Abulwafa; abulwafa@mans.edu.eg

Academic Editor: P. Areias

The presence and propagation of dust-acoustic solitary waves in dusty plasma contains four components such as negative and positive dust species beside ions and electrons are studied. Both the ions and electrons distributions are represented applying nonextensive formula. Employing the reductive perturbation method, an evolution equation is derived to describe the small-amplitude dust-acoustic solitons in the considered plasma system. The used reductive perturbation stretches lead to the nonlinear KdV and modified KdV equations with nonlinear and dispersion coefficients that depend on the parameters of the plasma. This study represents that the presence of compressive or/and rarefactive solitary waves depends mainly on the value of the first-order nonlinear coefficient. The structure of envelope wave is undefined for first-order nonlinear coefficient tends to vanish. The coexistence of the two types of solitary waves appears by increasing the strength of nonlinearity to the second order using the modified KdV equation.

1. Introduction

The dust acoustic solitary waves (DASWs) propagation (known as weak frequency-dusty plasma density vibrations) attracted the researchers since many years due to its eventual functions in different implementations. These implementations are such as lower ionosphere of the Earth, lower and upper mesosphere, asteroid zones, space environments, comet tails, interplanetary spaces and interstellar media [1–4]. In addition, it is instituting in different industrial process and laboratory devices like solar cells, semiconductor chips, plasma devices, fusion devices and so on [5–7]. Many researchers found that the spectra of the plasma waves are modified because of the existence of dust species. Ion waves in dusty plasma are studied by Angelis et al. [1] and they founded the equation, which described the perturbations due to the dust particles inhomogeneity distributions. Their results are employed to analyze the data for Halley's Comet collected from Vega and Giotto space probes. Rao et al. [8] showed theoretically that a new kind of dust acoustic waves (DAWs) propagates through dusty plasma with linear and nonlinear modes known as supersonic solitons. The DAW has been studied experimentally in laboratory under microgravity conditions by Piel and coworkers [7, 9]. Cassini spacecraft had taken high-resolution images for the Saturn's rings. These images investigated for proof of Mach cons DAW, which could supply information on the physical properties of the ring particles [10].

Since the existence of the dust charged particles was not neglected, because not only they modify the spectra of the plasma but also they introduce new eigenmodes such as dust ion acoustic waves (DIAWs), DAWs, dust lattice waves (DLWs), dust cyclotron acoustic, and dust drift mode, etc. [11–14]. Mamun et al. [15] studied two-component dusty plasma with Boltzmann distributed ions. They found that only negative solitary wave is admitted where the Kortweg-de Vries (KdV) equation was derived. In addition, they studied the same dusty plasma with vortex-like or nonthermal distribution for ions [16]. They concluded that the small amplitude dynamics of this system were described via the modified Kortweg-de Vries (mKdV) equation. Mamun [17] studied the ion temperature effect on the electrostatic solitons in nonthermal plasmas, which contain nonthermally distributed electrons and warm adiabatic ions using Sagdeev potential technique. It was noticed that ion temperature affects the Mach number for which the solitons exist and the nonthermal parameter. In addition, both the width and the amplitude of these solitary waves are affected.

There are broad mathematical ways to deal with waves in dusty plasma, but there are two major methods used. Studying the arbitrary amplitude of waves in plasmas using the Sagdeev pseudo-potential (SPP) approach is the most common way [18]. While for the small amplitude electrostatic waves, a reductive perturbation theory (RPT) can be employed to extract the evolution equation [19]. Mamun [20] applied the RPT to study the properties of the DAWs for unmagnetized three-component dusty plasma system composed of dusty grain carries negative charge beside Boltzmann distributed electrons and ions. Bedi et al. [21] hired the reductive perturbation method to deduce the KdV equation for four-component magnetized dusty plasma consisting of non-thermal electrons; Maxwellian distributed ions and two dust grains with negative and positive charges. Tasnim et al. [22] studied the propagation of electrostatic potential through adiabatic dusty plasma consisting of four components such as adiabatic electrons, adiabatic ions, adiabatic dust species of positive charge, and negatively charged warm dust grains. Akhter et al. [23] utilized the RPT to drive the KdV equation for magnetized dusty plasma composed of four components such ions and electrons with Maxwellian distributions alongside negatively and positively charged dust grains. They studied the effects of different dust plasma parameters on the dust acoustic solitary structures. Sebastian et al. [24] investigated the DA solitary waves in five-component dusty plasma system contain hot electrons and ions, kappa distributed photo-electrons, and two dust species of positive and negative charges.

Due to Boltzmann-Gibbs entropy generalization in the nonequilibrium states with the nonextensive entropy suggested by Tsallis [25], many researchers have paid more attention to employ the nonextensive distribution for the number density of particles in plasma [26–28].

Through this paper, a dusty plasma system composed of four components such as electrons and ions described by q-nonextensive distributions beside two species of dust having positive and negative charges is studied. The reductive perturbation method is applied to investigate the small amplitude electrostatic structures through this plasma system. The effects of some different plasma parameters (traveling wave velocity, nonextensive parameter, and dust temperature ratios) on the electrostatic solitary wave structure (shape, width, and height) are inspected.

The rest of the paper is ordered as follows: fluid equations that describe the considered plasma system are represented in Section 2. Through Sections 3 and 4, RPT is appointed to deduce the evolution equations that describe electrostatic waves propagate in the system. Section 5 is devoted for conclusion on the obtained results.

2. Fluid Equations for the Plasma System

A system of dusty plasma composed of ions and electrons with nonextensive distributions along with negative and positive dust sorts is treated in the following. The one-dimensional fluid equations that depict this kind of plasma are represented as [20, 29–31]:

$$\frac{\partial}{\partial t}N_j(x,t) + \frac{\partial}{\partial x}\left[N_j(x,t)U_j(x,t)\right] = 0, \tag{1a}$$

$$\begin{aligned}\beta_j\left[\frac{\partial}{\partial t}U_j(x,t) + U_j(x,t)\frac{\partial}{\partial x}U_j(x,t)\right] + \frac{\partial}{\partial x}\phi(x,t) \\ + \frac{\sigma_j}{N_j(x,t)}\frac{\partial}{\partial x}P_j(x,t) = 0,\end{aligned} \tag{1b}$$

$$\begin{aligned}\frac{\partial}{\partial t}P_j(x,t) + U_j(x,t)\frac{\partial}{\partial x}P_j(x,t) \\ + 3P_j(x,t)\frac{\partial}{\partial x}U_j(x,t) = 0.\end{aligned} \tag{1c}$$

where $j = p$ for the positive dusty fluid and $j = n$ for the negative one. $N_j(x,t)$ refers to the dust fluid density number normalized by its equilibrium value N_{j0}; $U_j(x,t)$ is the dust particle velocity normalized by the dust-acoustic velocity $C_s = \sqrt{Z_n k_B T_i/m_n}$. $\beta_p = Z_n m_p/(Z_p m_n) = \mu$ and $\beta_n = 1$. Z_j is the electronic charges number on the dust grain surface, m_j is the dust particle mass, T_i is the ions temperature, and k_B is the Boltzmann constant. $P_j(x,t)$ is the dust particles pressure normalized by $N_{j0}k_B T_j$ and the dust temperatures ratio is $\sigma_j = T_j/(Z_j T_i)$, where T_j is the jth dust fluid temperature. $\phi(x,t)$ is the electrostatic potential normalized by $k_B T_i/e$. x and t are the space and time variables, which are normalized with the positive dust Debye length $\lambda_{Dp} = \sqrt{k_B T_i/(4\pi N_{n0}e^2 Z_n)}$ and inverse frequency $\omega_p^{-1} = \sqrt{m_n/(4\pi N_{n0}e^2 Z_n)}$, respectively.

Equations ((1a), (1b), and (1c)) consist a set of six coupled nonlinear partial differential equations in seven unknown functions. To make the used plasma system equations ((1a), (1b), and (1c)) self-consistent, Poisson's equation is proceeded as

$$\begin{aligned}\frac{\partial^2}{\partial x^2}\phi(x,t) - \mu_e N_e(x,t) - N_n(x,t) + \mu_i N_i(x,t) \\ + \mu_p N_p(x,t) = 0,\end{aligned} \tag{2}$$

where $\mu_e = N_{e0}/(N_{n0}Z_n)$, $\mu_i = N_{i0}/(N_{n0}Z_n)$, and $\mu_p = N_{p0}Z_p/(N_{n0}Z_n)$ satisfy the neutrality condition $\mu_p + \mu_i - \mu_e = 1$, where N_{i0} and N_{e0} are the equilibrium density values of ions and electrons, respectively. $N_i(x,t)$ and $N_e(x,t)$ are the ions and electrons fluid densities, which are taken by q-nonextensive distribution forms as

$$N_l(x,t) = \left[1 + \sigma_l(q_l - 1)\phi(x,t)\right]^{(q_l+1)/[2(q_l-1)]}, \tag{3}$$

where the subscript l refers to i for ions and e for electrons, $\sigma_i = -1$, $\sigma_e = T_i/T_e$ and T_e is the electrons temperature.

Equations (1a)-(3) are complicated coupled nonlinear partial differential equations, so the reductive-perturbation method will be employed to discuss the acoustic waves of small-amplitude through our system of plasma.

3. Small-Amplitude Waves and KdV Equation

To examine the properties of dust acoustic solitons of small-amplitude, the reductive perturbation theory [22, 23] is utilized with the stretching variables:

$$\tau = \varepsilon^{3/2} t, \qquad \xi = \varepsilon^{1/2}(x - \lambda t), \tag{4}$$

where ε represents the amplitude of the perturbation and λ is the phase speed of the wave. All the plasma parameters $\Psi(x,t) = [N_j(x,t), U_j(x,t), P_j(x,t), \phi(x,t)]$ appearing in (1a)-(3) are expanding as a power series in ε about their equilibrium values as

$$\Psi(x,t) = \sum_{n=0}^{\infty} \varepsilon^n \psi_n(x,t), \tag{5}$$

with the conditions $N_p = N_n = P_p = P_n = 1$ and $U_p = U_n = \phi = 0$ as $|\xi| \longrightarrow \infty$ and with realization of the neutrality condition. Substituting (4) and (5) into (1a)-(3) and equating the coefficients of similar powers of ε, one has the following.

The coefficients of first order of ε give relations between the first perturbed orders of all physical quantities. The first perturbed order of Poisson's equation gives the following compatibility condition (linear dispersion relation (LDR)) as

$$LDR1 = \frac{\mu_p}{(\lambda^2\mu - 3\sigma_p)} + \frac{1}{(\lambda^2 - 3\sigma_n)} + \alpha_1 = 0, \tag{6a}$$

$$\alpha_1 = -\frac{[\mu_i(q_i+1) + \mu_e\sigma_e(q_e+1)]}{2}. \tag{6b}$$

This linear dispersion relation should be solved to derive the acoustic phase speed λ in terms of the plasma parameters as

$$\lambda_\pm = \left[\frac{\left\{3\alpha_1(\sigma_p + \mu\sigma_n) - (\mu + \mu_p) \pm \sqrt{[3\alpha_1(\sigma_p - \mu\sigma_n) - (\mu + \mu_p)]^2 + 12\alpha_1\mu(\sigma_p - \mu\sigma_n)}\right\}}{(2\alpha_1\mu)}\right]^{1/2}. \tag{6c}$$

The positive sign refers to fast acoustic speed while the negative one represents slow acoustic speed [32, 33]. In our numerical calculations, we considered the case of low acoustic speed whereas the fast acoustic speed is completely ignored. Also, it can be observed that the fast acoustic speed vanishes in the limiting case as $\sigma_p = \sigma_n = 0$.

The next ε-order handles a set of differential equations, by eliminating the second-order of the perturbed quantities $\psi_2(\xi,\tau)$; the following Kortweg de-Vreis equation is obtained as

$$\frac{\partial\phi(\xi,\tau)}{\partial\tau} + A\phi(\xi,\tau)\frac{\partial\phi(\xi,\tau)}{\partial\xi} + B\frac{\partial^3\phi(\xi,\tau)}{\partial\xi^3} = 0, \tag{7a}$$

where $\phi(\xi,\tau) = \phi_1(\xi,\tau)$ and A and B are the first-order nonlinear coefficient and the dispersion coefficient, respectively, which are specified by the following formulae:

$$A = \left[\frac{3\mu_p(\lambda^2\mu + \sigma_p)}{(\lambda^2\mu - 3\sigma_p)^2} + \frac{2\lambda}{(\lambda^2 - 3\sigma_n)^2} + 2\alpha_2\right]B, \tag{8a}$$

$$B = \frac{1}{\{2\lambda[\mu\mu_p/(\lambda^2\mu - 3\sigma_p)^2 + 1/(\lambda^2 - 3\sigma_n)^2]\}}, \tag{8b}$$

$$\alpha_2 = -\frac{[\mu_i(q_i+1)(q_i-3) + \mu_e\sigma_e^2(q_e+1)(q_e-3)]}{8}. \tag{8c}$$

The stationary solitary wave solution of (7a) is obtained using the following traveling wave transformation:

$$\zeta = \xi - v\tau, \qquad \Phi(\zeta) = \phi(\xi,\tau), \tag{9}$$

where v is the traveling wave velocity. The KdV equation (7a) becomes ordinary differential equation, which are known by the reduced KdV equation in the following form:

$$-v\frac{d\Phi(\zeta)}{d\zeta} + A\Phi(\zeta)\frac{d\Phi(\zeta)}{d\zeta} + B\frac{d^3\Phi(\zeta)}{d\zeta^3} = 0. \tag{10}$$

The well-known solution of (10) is represented by the following form:

$$\Phi(\zeta) = \Phi_m \operatorname{sech}^2\left(\frac{\zeta}{\Delta}\right), \tag{11a}$$

where

$$\Phi_m = \frac{3v}{A}, \qquad \Delta = \sqrt{\frac{4B}{v}}. \tag{11b}$$

Integrating (10) with boundary conditions $\Phi(\zeta) \longrightarrow 0$, $d\Phi(\zeta)/d\zeta \longrightarrow 0$ and $d^2\Phi(\zeta)/d\zeta^2 \longrightarrow 0$ as $\zeta \longrightarrow \pm\infty$, one gets

$$\frac{1}{2}\left[\frac{d\Phi(\zeta)}{d\zeta}\right]^2 + V_s(\Phi) = \mathrm{E}. \quad (12)$$

This is the form of a conservation law. The right-hand side E is the constant of integration, which is supposed as the total energy of a hypothetical body with unit mass, while the first term in the left-hand side is assumed as kinetic energy while the second term represents a potential energy that has the following form:

$$V_s(\Phi) = -\left[\frac{v}{(2B)}\right]\Phi(\zeta)^2 + \left[\frac{A}{(6B)}\right]\Phi(\zeta)^3. \quad (13)$$

This potential has the characteristics that its value and first derivative at $\Phi = 0$ vanish while the second derivative is less than zero where both v and B have positive values. This means that the potential has a root and maximum value at the origin of motion. The other root of the potential is given as $\Phi = 3v/A = \Phi_m$, which is the amplitude of the solitary wave. The amplitude value is negative or positive depending mainly on the value of A (the nonlinear coefficient), which depends on the plasma parameters. Therefore, plasma system provides two types of potential structure known as rarefactive (dark) solitons or compressive (bright). The dark (rarefactive) soliton exists if $A < 0$ while the bright (compressive) soliton exists if $A > 0$.

The change of the coefficient of nonlinear term A with different plasma parameters (non-extensive parameter q, positive dust temperatures ratio σ_p, and negative dust temperatures ratio σ_n) shows that the existence of the two sorts of solitons (compressive and rarefactive) depends mainly on the values of these parameters. To investigate the influences of plasma parameters (nonextensive parameter q and positive and negative dust temperatures ratios σ_p and σ_n), the other plasma parameters are used as $\mu = 2.0$, $\mu_p = 0.7$, $\mu_e = 0.2$, $\sigma_e = 0.5$ [34, 35]. Figure 1(a) shows that the soliton converted from compressive to rarefactive at nonextensive parameter value depending on the positive dust temperature ratio as shown in Figure 1(a) and on the negative dust temperature ratio as appeared in Figure 1(b). We have discussed the impacts of these different plasma parameters on the shape structure of the envelope soliton by sketching in two dimensions both the small amplitude potential $V_s(\Phi)$ versus electrostatic potential Φ and the first perturbed electrostatic potential $\Phi(\zeta)$ versus travelling wave parameter ζ as represented in Figures 2–5. It is seen that from Figure 2 one type of solitons only envelopes with varying the travelling wave velocity v, while the amplitude (width) of the wave increases (decreases) with increasing v without effect on changing from soliton type to another. It is obvious from Figures 3, 4, and 5 that the compressive soliton amplitude and width increase with q and σ_n while they decrease with σ_p. On the other hand, the amplitude and width of rarefactive soliton decrease with both q and σ_n but increase with σ_p. Figures 3–5 read that the soliton type transferred from compressive to rarefactive depending on the values of the plasma parameters q, σ_p and σ_n at which the nonlinear parameter A tends to zero. Therefore, nearly at $A \approx 0$, this system cannot be described and a higher order nonlinearity evolution equation must be considered to describe the system in this region. This evolution equation will be described in the following section.

4. Small Amplitude Waves and the Modified KdV Equation

To study the solitary waves at the critical region, the following stretching coordinates are as

$$\begin{aligned}\tau &= \varepsilon^3 t,\\ \xi &= \varepsilon(x - \lambda t).\end{aligned} \quad (14)$$

Substituting (5) and (14) into (1a)-(3) and comparing the coefficients of same powers of ε, the coefficients of second order of ε give relations between the first-order and the second-order perturbed physical quantities. The coefficients of the third order of ε give a set of equations in terms of the first-, second-, and third-order perturbed quantities. These relations lead to another linear dispersion relation of the following form:

$$\begin{aligned}LDR2 &= \frac{3\mu_p(\lambda^2\mu - \sigma_p)}{\left[2(\lambda^2\mu - 3\sigma_p)^3\right]} - \frac{3(\lambda^2 - \sigma_n)}{\left[2(\lambda^2 - 3\sigma_n)^3\right]} + \alpha_2\\ &= 0,\end{aligned} \quad (15)$$

Eliminating the third-order perturbations $\psi_3(\xi, \tau)$ from the fourth-order perturbation coefficients of ε, the following modified Kortweg de-Vreis equation is obtained in the following form:

$$\frac{\partial\phi(\xi,\tau)}{\partial\tau} + D\phi(\xi,\tau)^2\frac{\partial\phi(\xi,\tau)}{\partial\xi} + B\frac{\partial^3\phi(\xi,\tau)}{\partial\xi^3} = 0, \quad (16)$$

where D is the second-order nonlinear coefficient, which is given by the following formulae:

$$D = \left\{\frac{3\mu_p(5\lambda^4\mu^2 + 30\lambda^2\mu\sigma_p + 9\sigma_p^2)}{\left[2(\lambda^2\mu - 3\sigma_p)^5\right]} + \frac{3(5\lambda^4 + 30\lambda^2\sigma_n + 9\sigma_n^2)}{\left[2(\lambda^2 - 3\sigma_n)^5\right]} + 3\alpha_3\right\}B, \quad (17a)$$

$$\alpha_3 = -\frac{\left[\mu_i(q_i + 1)(q_i - 3)(3q_i - 5) + \mu_e\sigma_e^3(q_e + 1)(q_e - 3)(3q_e - 5)\right]}{48}. \quad (17b)$$

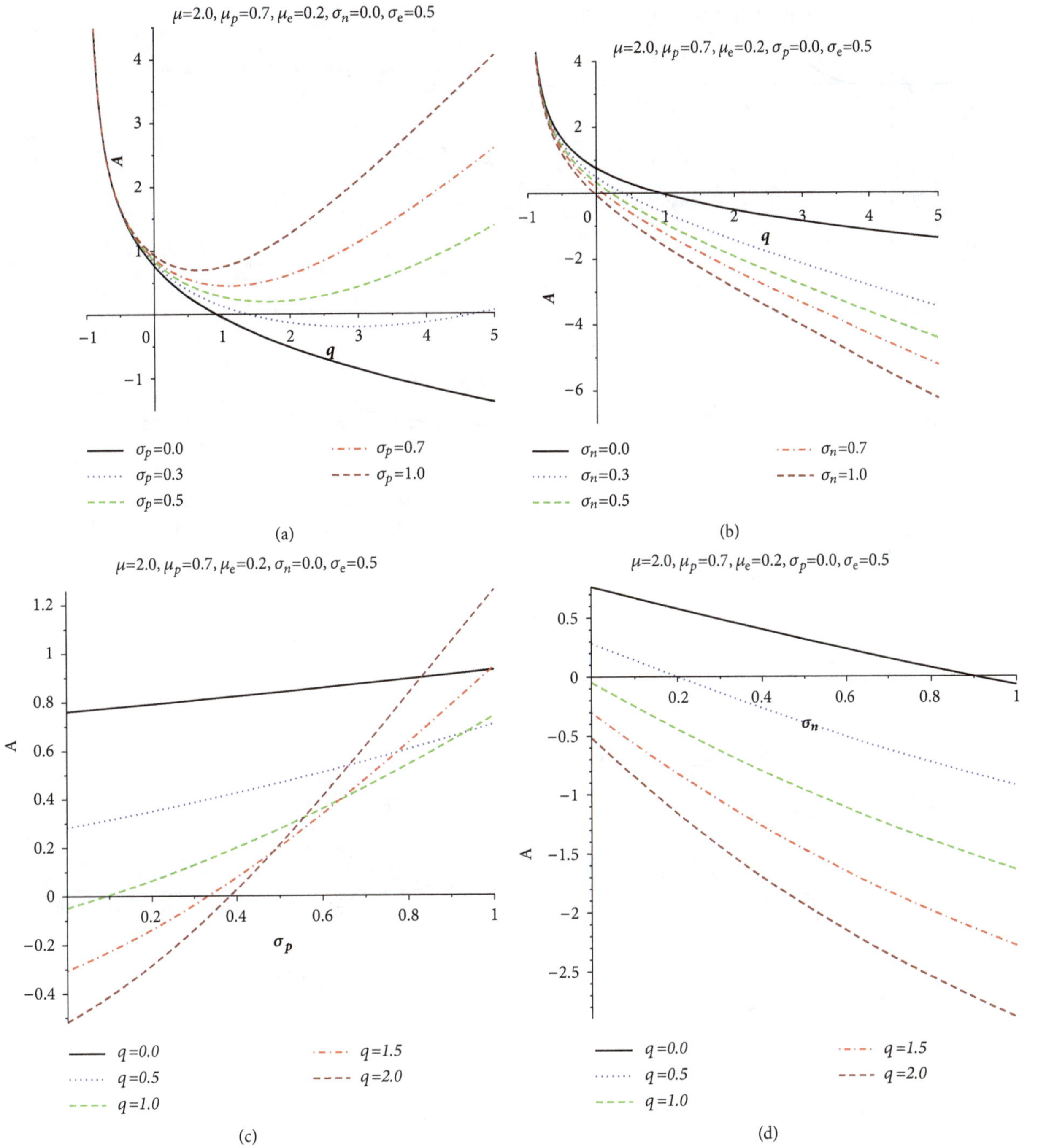

FIGURE 1

The stationary solitary wave solution of (16) can be deduced applying the traveling wave transformation (9). Therefore, the mKdV equation (16) is transformed into the following reduced mKdV equation:

$$-v\frac{d\Phi(\zeta)}{d\zeta} + D\Phi(\zeta)^2\frac{d\Phi(\zeta)}{d\zeta} + B\frac{d^3\Phi(\zeta)}{d\zeta^3} = 0, \quad (18)$$

where $\Phi(\zeta) = \phi(\xi, \tau)$ with two solitary solutions as

$$\Phi_1(\zeta) = \Phi_m \operatorname{sech}\left(\frac{\zeta}{\Delta}\right),$$

$$\Phi_2(\zeta) = -\Phi_m \operatorname{sech}\left(\frac{\zeta}{\Delta}\right), \quad (19a)$$

where Φ_m is the amplitude and Δ is the width of the wave envelop that are defined, respectively, as

$$\Phi_m = \sqrt{\frac{6M}{D}},$$
$$\Delta = \sqrt{\frac{4B}{M}}. \quad (19b)$$

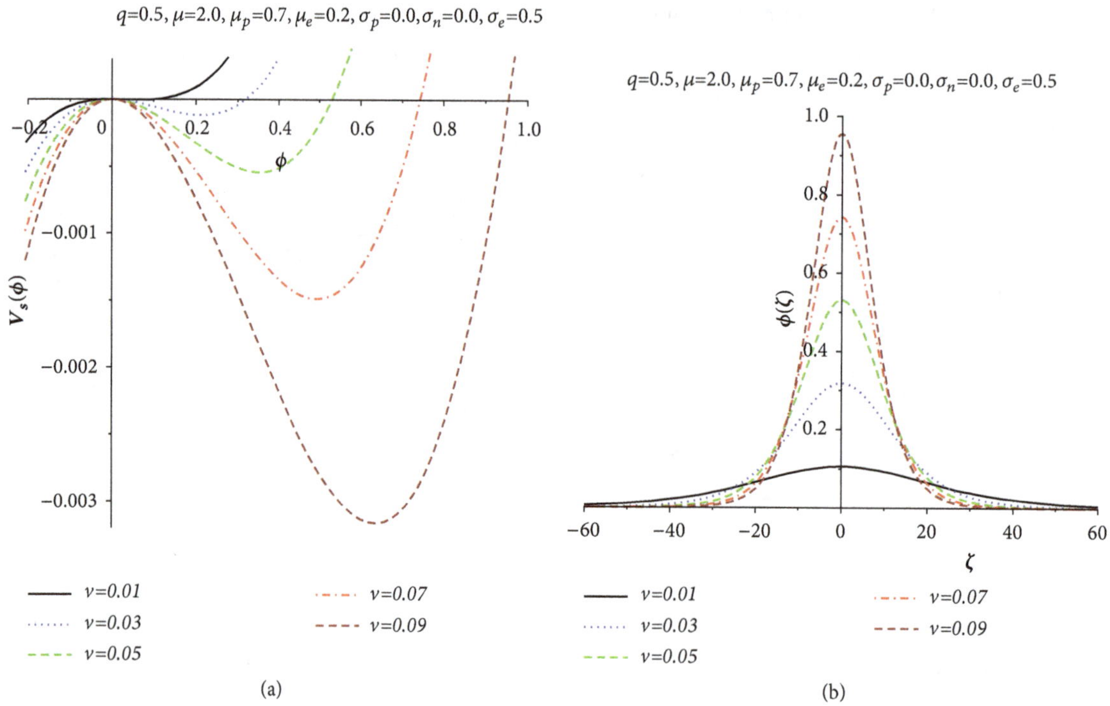

FIGURE 2

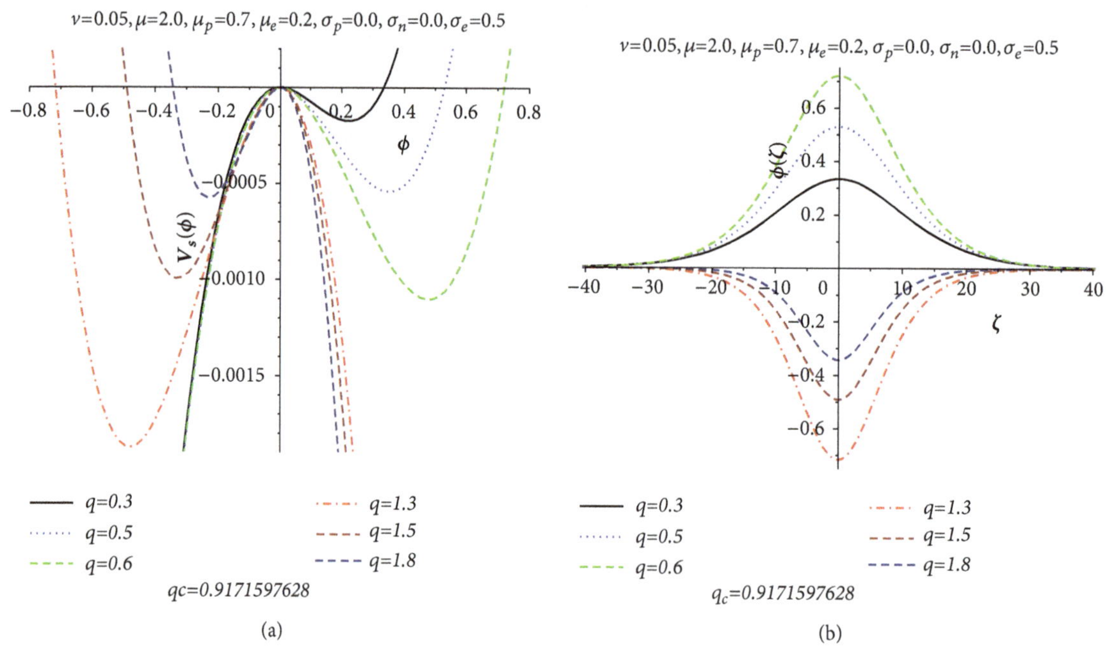

FIGURE 3

Here from the formulae of both amplitude and width of the envelop wave, we can say that, there is no envelop solitary wave for $D < 0$ where $B > 0$ from definition (8b).

Integrating (18) with boundary conditions $\Phi(\zeta) \longrightarrow 0$, $d\Phi(\zeta)/d\zeta \longrightarrow 0$ and $d^2\Phi(\zeta)/d\zeta^2 \longrightarrow 0$ as $\zeta \longrightarrow \pm\infty$, one gets the small amplitude potential energy as

$$V_s(\Phi) = -\left[\frac{v}{(2B)}\right]\Phi(\zeta)^2 + \left[\frac{D}{(12B)}\right]\Phi(\zeta)^4. \qquad (20)$$

This potential has the characteristics that its value and first derivative at $\Phi = 0$ vanish while the second derivative is negative. This means that there is a root and maximum value of $V_s(\Phi)$ at the origin. Both formulae ((19a), (19b)) and (20) appear the coexistence of the two types of solitons (compressive and rarefactive) in the region of vanishing the first-order nonlinear parameter A.

Here, we have treated the effects of non-extensive parameter q, positive dust temperature ratio σ_p and negative dust

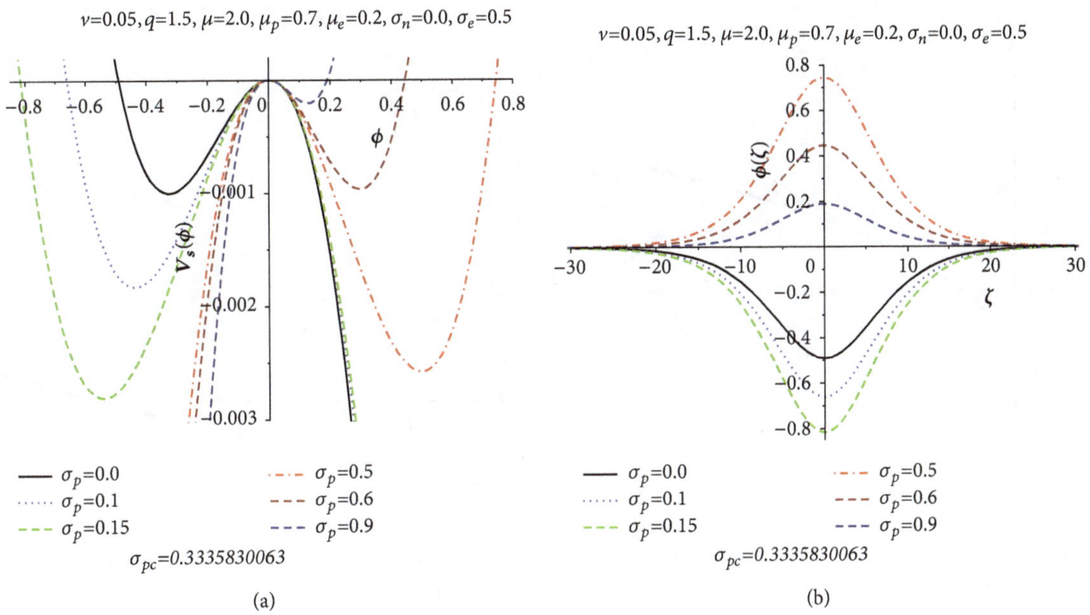

FIGURE 4

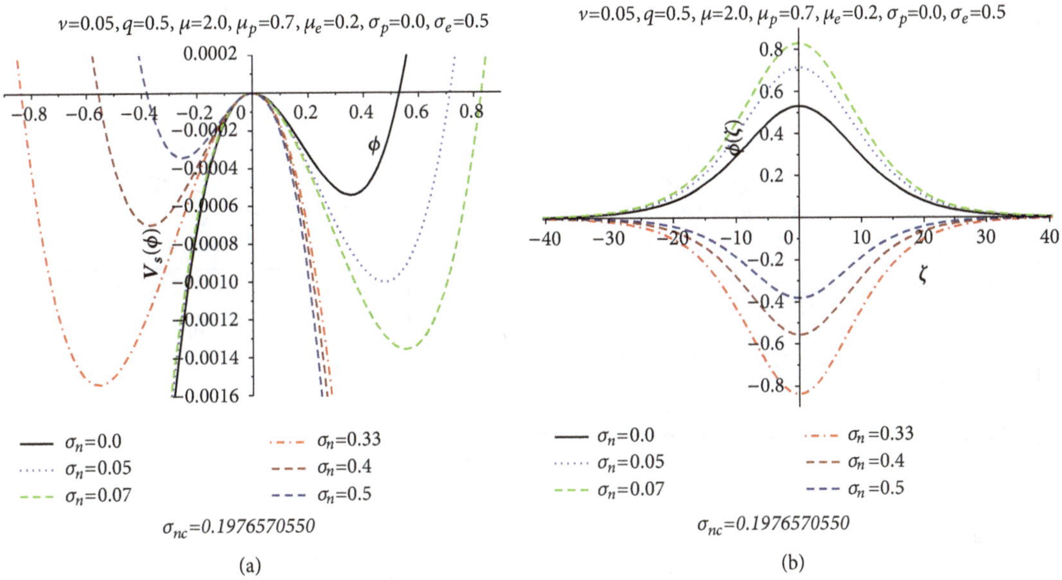

FIGURE 5

temperature ratio σ_n with other real plasma parameters used as $\mu = 2.0$, $\mu_p = 0.7$, $\mu_e = 0.2$ and $\sigma_e = 0.5$ [34, 35]. As shown in Figure 6, the nonlinear coefficient D increases with each of q, σ_p and σ_n, but as shown from (19b) that D must be greater than zero to envelope a soliton wave solution. The coexistence of both compressive and rarefactive solitons is indicated from the plotting of both small amplitude potential $V_s(\Phi)$ and the electrostatic potential $\Phi(\zeta)$ in Figures 7–10. It is evident that the amplitude (width) of the two types of solitons increases (decreases) with the increasing of the travelling wave velocity v as in Figure 7. Figures 8–10 represent that both the width and amplitude of soliton envelope decrease with increasing of q, σ_p and σ_n.

5. Conclusion

A plasma system consisting of four components such as negative and positive dust species beside ions and electrons is studied in this work. The hydrodynamics equations for the two dust species with pressure terms are employed to depict the dust particles motion, while the electrons and ions number densities are described in terms of nonextensive

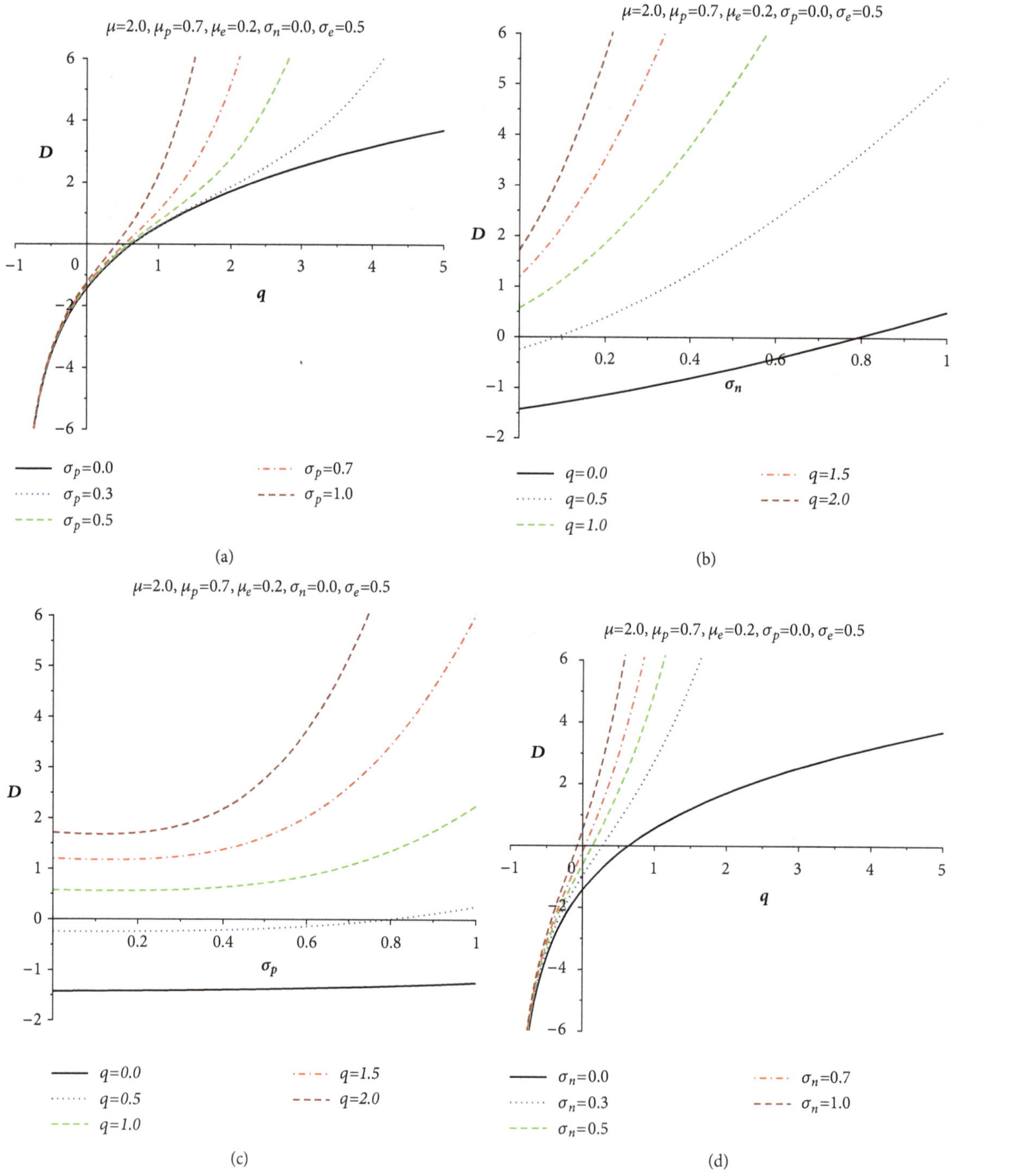

FIGURE 6

distributions. Poisson's equation is taking to make this system self-consistent.

The reductive perturbation method is utilized to deduce the KdV and mKdV equations that describe the small amplitude dust acoustic solitary waves propagated in our four-component plasma system. These equations are integrated to give the equations of motion of a hypothetical particle of unit mass and virtual velocity $d\Phi/d\zeta$ under the effect of a virtual potential $V_s(\Phi)$ described by (13) and (20), where $\Phi(\zeta)$ is the electrostatic potential. In our numerical calculations, we considered the case of low acoustic speed whereas the fast acoustic speed completely vanishes in the limiting case as $\sigma_p = \sigma_n = 0$.

The variation of the coefficients of nonlinear terms with different plasma parameters (the travelling wave velocity v, nonextensive parameter q, positive dust temperatures ratio σ_p, and negative dust temperatures ratio σ_n) shows the existence of two types of solitons (compressive and rarefactive). The soliton is compressive or rarefactive predicating mainly on the value of A (first-order nonlinear coefficient), where

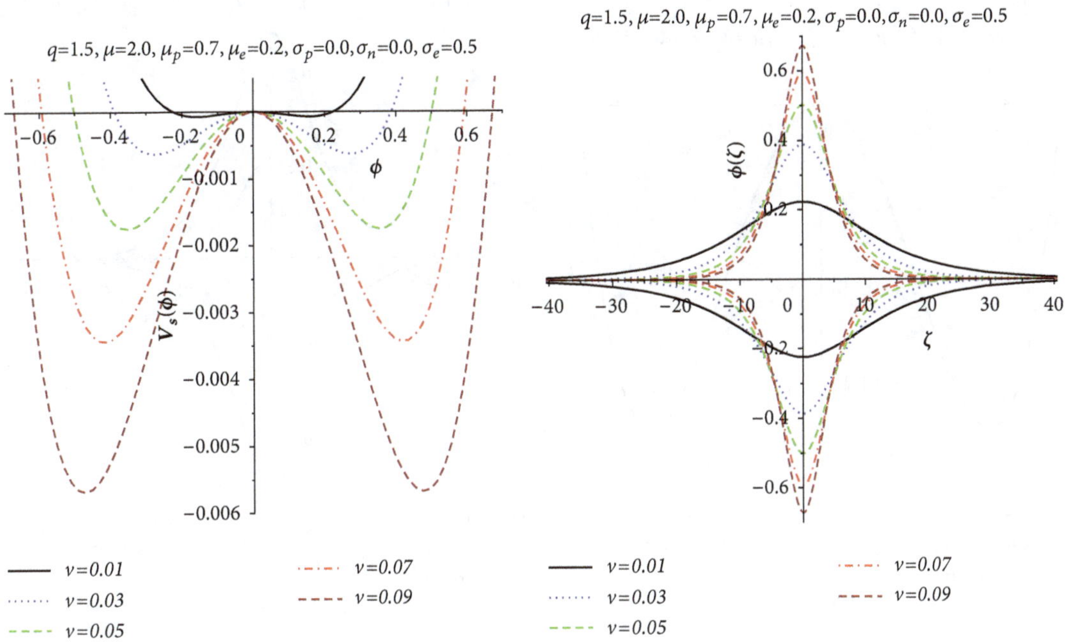

(a) (b)

FIGURE 7

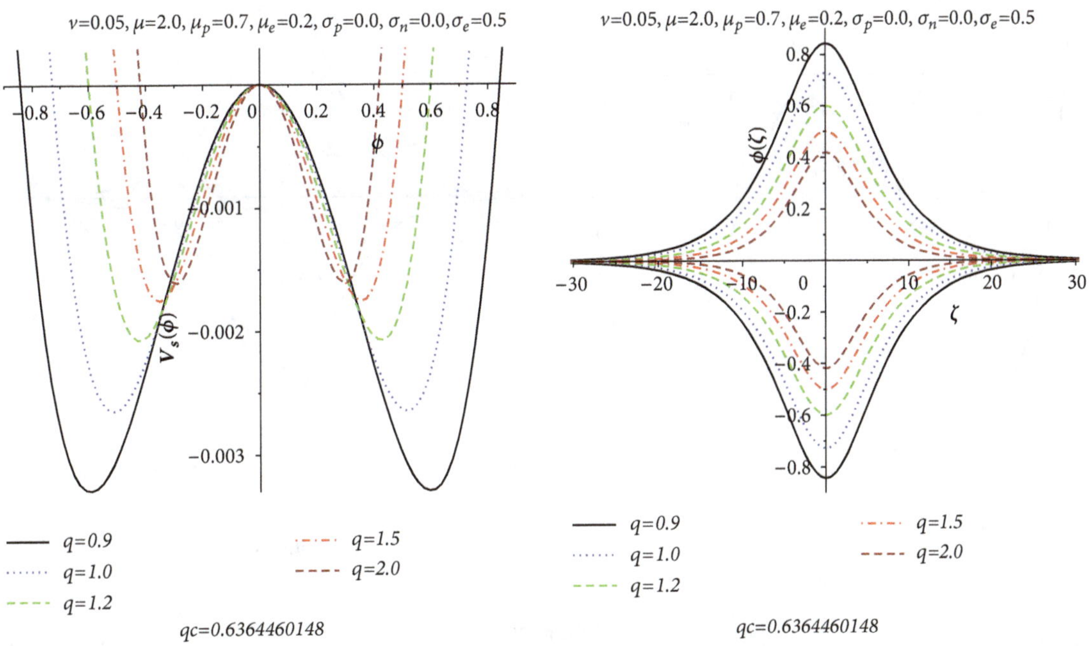

(a) (b)

FIGURE 8

rarefactive soliton exists if $A < 0$ while compressive soliton exists if $A > 0$, but at $A \approx 0$, this system cannot be described. Therefore, the mKdV equation is considered to describe the system in this region. The solution of the mKdV equation represents the coexistence of both compressive and rarefactive solitons, which appears from the plotting of both the small amplitude potential $V_s(\Phi)$ and the electrostatic potential $\Phi(\zeta)$ for different values of the nonextensive

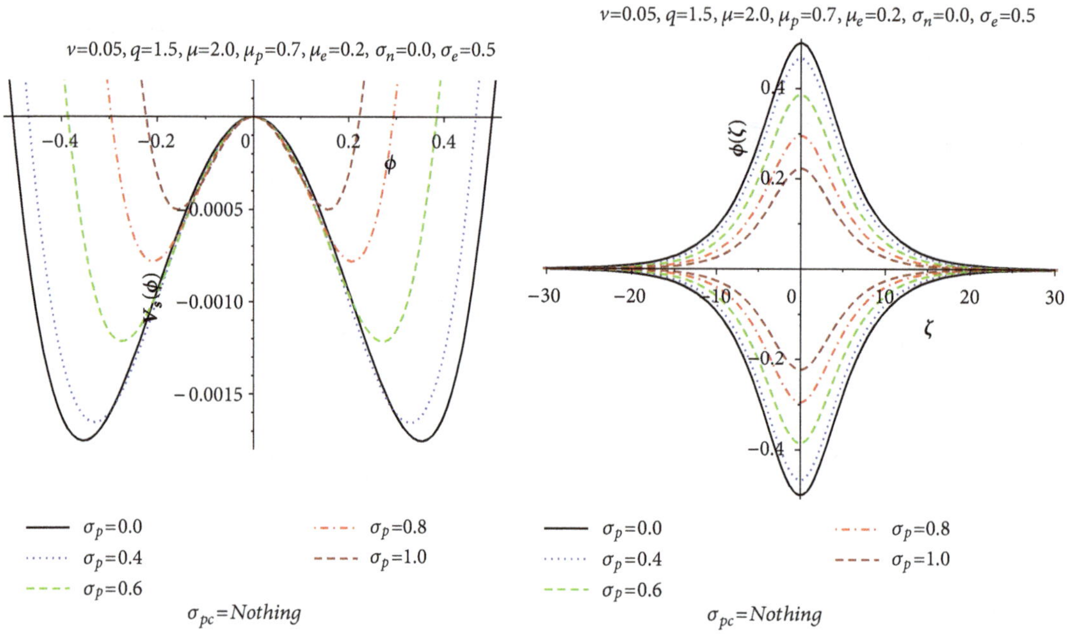

(a) (b)

FIGURE 9

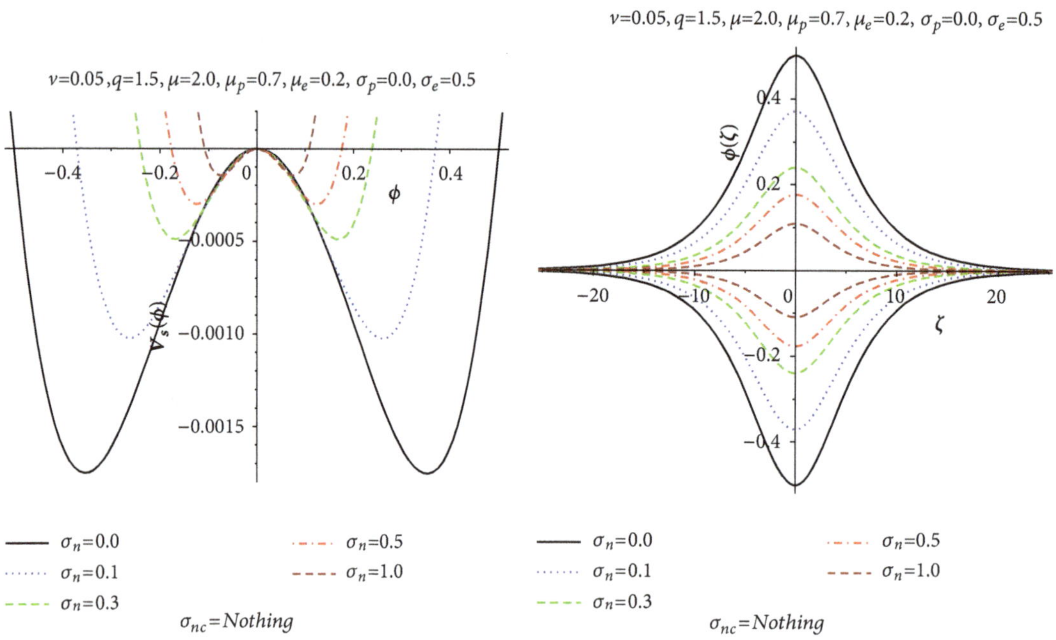

(a) (b)

FIGURE 10

parameter (q), and positive and negative dust temperature ratios (σ_p and σ_n).

Conflicts of Interest

The authors declare that they have no conflicts of interest.

Acknowledgments

The authors express their great acknowledgement to Prof. S. A. El-Wakil for his useful comments, continuous encouragement, and guidance.

References

[1] U. de Angelis, V. Formisano, and M. Giordano, "Ion Plasma Waves in Dusty Plasmas: Halley's Comet," *Journal of Plasma Physics*, vol. 40, no. 3, pp. 399–406, 1988.

[2] P. K. Shukla, "Low-frequency modes in dusty plasmas," *Physica Scripta*, vol. 45, no. 5, pp. 504-507, 1992.

[3] F. Melandsø, T. K. Aslaksen, and O. Havnes, "A new damping effect for the dust-acoustic wave," *Planetary and Space Science*, vol. 41, no. 4, pp. 321–325, 1993.

[4] P. K. Kaw and A. Sen, "Low frequency modes in strongly coupled dusty plasmas," *Physics of Plasmas*, vol. 5, no. 10, pp. 3552–3559, 1998.

[5] A. Barkan, N. D'Angelo, and R. L. Merlino, "Experiments on ion-acoustic waves in dusty plasmas," *Planetary and Space Science*, vol. 44, no. 3, pp. 239–242, 1996.

[6] R. L. Merlino, A. Barkan, C. Thompson, and N. D'Angelo, "Laboratory studies of waves and instabilities in dusty plasmas," *Physics of Plasmas*, vol. 5, no. 5, pp. 1607–1614, 1998.

[7] T. Reichstein, I. Pilch, and A. Piel, "Toroidal dust motion in magnetized plasmas," *Physics of Plasmas*, vol. 17, Article ID 093701, 2010.

[8] N. N. Rao, P. K. Shukla, and M. Y. Yu, "Dust-acoustic waves in dusty plasmas," *Planetary & Space Science*, vol. 38, no. 4, pp. 543–546, 1990.

[9] K. O. Menzel, O. Arp, and A. Piel, "Spatial frequency clustering in nonlinear dust-density waves," *Physical Review Letters*, vol. 104, no. 23, Article ID 235002, 2010.

[10] O. Havnes, T. Aslaksen, T. W. Hartquist et al., "Probing the properties of planetary ring dust by the observation of Mach cones," *Journal of Geophysical Research: Space Physics*, vol. 100, no. A2, pp. 1731–1734, 1995.

[11] F. Melandso, "Lattice waves in dust plasma crystals," *Physics of Plasmas*, vol. 3, no. 11, pp. 3890–3901, 1996.

[12] A. Homann, A. Melzer, S. Peters, R. Madani, and A. Piel, "Laser-excited dust lattice waves in plasma crystals," *Physics Letters A*, vol. 242, no. 3, pp. 173–180, 1998.

[13] P. K. Shukla and H. U. Rahman, "Instability of electrostatic dust-cyclotron waves and associated wake potential," *Planetary and Space Science*, vol. 46, no. 5, pp. 541–543, 1998.

[14] P. K. Shukla, M. Y. Yu, and R. Bharuthram, "Linear and nonlinear dust drift waves," *Journal of Geophysical Research: Space Physics*, vol. 96, no. A12, pp. 21343–21346, 1991.

[15] A. A. Mamun, R. A. Cairns, and P. K. Shukla, "Solitary potentials in dusty plasmas," *Physics of Plasmas*, vol. 3, no. 2, pp. 702–704, 1996.

[16] A. A. Mamun, R. A. Cairns, and P. K. Shukla, "Effects of vortex-like and non-thermal ion distributions on non-linear dust-acoustic waves," *Physics of Plasmas*, vol. 3, no. 7, pp. 2610–2614, 1996.

[17] A. A. Mamun, "Effects of ion temperature on electrostatic solitary structures in nonthermal plasmas," *Physical Review E: Statistical Physics, Plasmas, Fluids, and Related Interdisciplinary Topics*, vol. 55, no. 2, pp. 1852–1857, 1997.

[18] R. Z. Sagdeev, "Cooperative phenomena and shock waves in collisionless plasmas," *Reviews of Plasma Physics*, vol. 4, pp. 23–91, 1966.

[19] H. Washimi and T. Taniuti, "Propagation of ion-acoustic solitary waves of small amplitude," *Physical Review Letters*, vol. 17, no. 19, pp. 996–998, 1966.

[20] A. A. Mamun and P. K. Shukla, "Solitary potentials in cometary dusty plasmas," *Geophysical Research Letters*, vol. 29, no. 18, p. 1870, 2002.

[21] C. Bedi, T. S. Gill, and A. S. Bains, "Four component magnetized dusty plasma containing non-thermal electrons," *Journal of Physics: Conference Series*, vol. 208, no. 1, Article ID 012037, 2010.

[22] S. Tasnim, A. Mannan, and A. A. Mamun, "Electrostatic solitary structures in a four-component adiabatic dusty plasma," *Astrophysics and Space Science*, vol. 337, no. 1, pp. 261–267, 2012.

[23] T. Akhter, M. M. Hossain, and A. A. Mamun, "Obliquely propagating solitary waves in a four-component magnetized dusty plasma," *African Review of Physics*, vol. 8, no. 0012, pp. 75–81, 2013.

[24] S. Sebastian, G. Sreekala, M. Michael et al., "Solitary waves in a five-component dusty plasma with kappa described electrons and ions," *International Journal of Science and Research*, vol. 3, no. 8, pp. 1301–1306, 2014.

[25] C. Tsallis, "Possible generalization of Boltzmann-Gibbs statistics," *Journal of Statistical Physics*, vol. 52, no. 1-2, pp. 479–487, 1988.

[26] A. M. El-Hanbaly, E. K. El-Shewy, M. Sallah, and H. F. Darweesh, "Nonlinear dust acoustic waves in dissipative space dusty plasmas with superthermal electrons and nonextensive ions," *Communications in Theoretical Physics*, vol. 65, no. 5, pp. 606–612, 2016.

[27] M. Tribeche and P. K. Shukla, "Charging of a dust particle in a plasma with a non extensive electron distribution function," *Physics of Plasmas*, vol. 18, no. 10, p. 103702, 2011.

[28] R. Amour and M. Tribeche, "Variable charge dust acoustic solitary waves in a dusty plasma with a q -nonextensive electron velocity distribution," *Physics of Plasmas*, vol. 17, no. 6, 2010.

[29] S. A. El-Wakil, M. A. Zahran, E. K. El-Shewy, and H. G. Abdelwahed, "Solitary waves in space dusty plasma with dust of opposite polarity," in *Proceedings of the 7th Conference on Nuclear and Particle Physics*, pp. 323–333, Sharm El-Sheikh, Egypt, 11-15 Nov. 2009.

[30] K. Roy, P. Chatterjee, and S. Kundu, "Dust acoustic dressed solitons in a four component dusty plasma with nonthermal electron," *Advances in Space Research*, vol. 50, no. 9, pp. 1288–1293, 2012.

[31] W. F. El-Taibany, "Nonlinear dust acoustic waves in inhomogeneous four-component dusty plasma with opposite charge polarity dust grains," *Physics of Plasmas*, vol. 20, no. 9, 2013.

[32] A. E. Dubinov, "On a widespread inaccuracy in defining the Mach number of solitons in a plasma," *Plasma Physics Reports*, vol. 35, no. 11, pp. 991–993, 2009.

[33] V. V. Prudskikh, "Ion-acoustic solitons in Bi-Ion dusty plasma," *Plasma Physics Reports*, vol. 34, no. 11, pp. 955–962, 2008.

[34] A. E. Dubinov and D. Y. Kolotkov, "Ion-acoustic super solitary waves in dusty multispecies plasmas," *IEEE Transactions on Plasma Sciences*, vol. 40, no. 5, pp. 1429–1433, 2012.

[35] A. E. Dubinov, D. Y. Kolotkov, and M. A. Sazonkin, "Supernonlinear waves in plasma," *Plasma Physics Reports*, vol. 38, no. 10, pp. 833–844, 2012.

6

Loss, Gain, and Singular Points in Open Quantum Systems

Hichem Eleuch [1,2] **and Ingrid Rotter** [3]

[1] *Institute for Quantum Science and Engineering, Texas A&M University, College Station, Texas 77843, USA*
[2] *Department of Applied Science and Mathematics, College of Arts and Sciences, Abu Dhabi University, Abu Dhabi, UAE*
[3] *Max Planck Institute for the Physics of Complex Systems, 01187 Dresden, Germany*

Correspondence should be addressed to Ingrid Rotter; rotter@pks.mpg.de

Academic Editor: Soheil Salahshour

Non-Hermitian quantum physics is used successfully for the description of different puzzling experimental results, which are observed in open quantum systems. Mostly, the influence of exceptional points on the dynamical properties of the system is studied. At these points, two complex eigenvalues $\mathscr{E}_i \equiv E_i + i\Gamma_i/2$ of the non-Hermitian Hamiltonian $\mathscr{H}$ coalesce (where E_i is the energy and Γ_i is the inverse lifetime of the state i). We show that also the eigenfunctions Φ_i of the two states play an important role, sometimes even the dominant one. Besides exceptional points, other critical points exist in non-Hermitian quantum physics. At these points $a = a^{\rm cr}$ in the parameter space, the biorthogonal eigenfunctions of $\mathscr{H}$ become orthogonal. For illustration, we show characteristic numerical results.

1. Introduction

The properties of open quantum systems are described usually by averaging over the spectroscopic features of the individual states of the system. Open quantum systems are characterized by, e.g., their Markovian and non-Markovian behavior or by equilibrium and nonequilibrium properties. The present-day high-resolution experimental studies provide however more information, namely, concrete information on the spectroscopic properties of the individual states. It is a challenge for the theory to describe these results.

In many current theoretical studies, a non-Hermitian formalism is used for the description of individual states of an open quantum system. In these papers, the part of the system, which is localized in a certain finite space region, is considered to be embedded into an infinitely extended environment. Mathematically, the total function space consists of two parts: the localized part of the system and the extended environment. Mostly, the environment is assumed to be the continuum of scattering wavefunctions[1].

This formalism is, from the point of view of mathematics, much more complicated than the familiar Hermitian formalism. For example, the eigenfunctions of a non-Hermitian operator $\mathscr{H}$ are biorthogonal with nontrivial consequences for their normalization. Most important is however the existence of singularities.

In mathematics, the so-called exceptional points (EPs) are known for a long time; see the book [1] on the perturbation theory for linear operators. At the EPs, two eigenvalues of $\mathscr{H}$ coalesce. This coalescence is, on principle, different from the well-known degeneration of two eigenstates of a Hermitian operator H. In the last case, the eigenfunctions of the two states are different and orthogonal while those of two coalescing eigenstates of $\mathscr{H}$ are biorthogonal and differ only by a phase from one another at the EP. The related interesting topological features are studied also experimentally in detail [2–4].

Due to the important role of EPs in non-Hermitian systems, in most theoretical studies of non-Hermitian quantum physics only the eigenvalues of the non-Hermitian Hamiltonian $\mathscr{H}$ are considered in detail. The eigenfunctions are often assumed to be of minor interest. Examples are the method of complex scaling [5] and the review [6] on superradiance. The meaning of the eigenvalues as well as of the eigenfunctions of $\mathscr{H}$ is discussed theoretically in the review [7] and compared to experimental results in the review [8]. The eigenfunctions around EPs play an important role for the explanation of some experimental results that are puzzling from the point of view of Hermitian quantum physics. Examples are the so-called phase lapses [9] observed experimentally in mesoscopic systems and the high efficiency of the photosynthesis [10]. The role of EPs in quantum physics is discussed in many other papers [11–16].

The meaning of the eigenvalues and that of the eigenfunctions of $\mathscr{H}$ are less different from one another than it seems to be. On the one hand, the eigenfunctions of $\mathscr{H}$ play a role also in those cases in which they are usually not considered [7, 8]. Among others, they are responsible for the fact that the influence of the EPs occurs in a larger parameter range around the EP than expected from the behavior of the eigenvalues. Moreover, higher-order EPs are shielded from observation in physical systems [17] because of the mixing of the eigenfunctions of $\mathscr{H}$ via the environment. On the other hand, the eigenvalues of $\mathscr{H}$ determine the position of the EPs and influence therefore also the behavior of the eigenfunctions of $\mathscr{H}$.

Further studies in non-Hermitian quantum physics have shown the following nontrivial results. The coupling between system and environment may occur via an exchange of particles or of information (mostly in terms of excitons). We have the following in detail.

(i) Every state i of the system is related to different decay channels c according to the values of the so-called partial widths of the state i relative to the channels c. The coupling of the whole system to the environment is characterized by the sum of partial widths. This case is realized in many-body systems. In standard Hermitian quantum physics, the exchange of particles between system and environment is described as tunneling of particles.

(ii) Every state of the system is coupled exclusively to only one channel with, respectively, some gain from the environment and some loss to the environment. This case is realized, e.g., in the photosynthesis in which visible light is captured in the light-harvesting complex [10]. The exciton transfer occurs without excitation of any eigenstates of $\mathscr{H}$. It cannot be described in standard theory.

Another problem of non-Hermitian quantum physics is the following. In mathematics, the EPs are defined in relation to one environment (mostly called channel). Physical systems are related however generally to more than one channel. For example, transmission of particles through a system needs at least two channels, entrance and exit channel. In this case, the different channels are parts of the total environment and are orthogonal to one another. Another example is the above-mentioned processes with gain and loss each of which occurs relative to another channel. The role played by EPs in physical systems is therefore not at all clear.

Consideration of the eigenfunctions of the non-Hermitian Hamiltonian $\mathscr{H}$ in detail has shown further nontrivial properties of non-Hermitian quantum physics. For example, the EPs are not the only singular points in non-Hermitian quantum systems. Other singularities occur at maximum width bifurcation or level repulsion where the eigenfunctions of $\mathscr{H}$ are orthogonal (instead of biorthogonal). They are related to obviously nonlinear irreversible processes [17].

In the present paper, we will illustrate the relation between the eigenvalues $\mathscr{E}_i \equiv E_i + i/2\Gamma_i$ [2] and eigenfunctions Φ_i of the non-Hermitian Hamiltonian $\mathscr{H}$, on the one hand, as well as the appearance of gain and loss and different types of singularities in non-Hermitian quantum physics, on the other hand.

For this aim, we start from a system with two states $i = 1, 2$ the complex energies of which are $\varepsilon_i \equiv e_i + i/2\gamma_i$ [2]. We then consider the coupling of the system to two channels under the assumption that each state is coupled, above all, to only one channel: $\gamma_1^{(1)} \gg \gamma_1^{(2)}; \gamma_2^{(2)} \gg \gamma_2^{(1)}$. We vary parametrically the energies e_i of the states and determine the eigenvalues $\mathscr{E}_i \equiv E_i + i/2\Gamma_i$ of the non-Hermitian Hamiltonian $\mathscr{H}$. In such a case, width bifurcation causes automatically a transition from loss (determined by $\gamma_1^{(2)} < 0$ and $\gamma_2^{(1)} < 0$) to gain (determined by $\gamma_1^{(2)} > 0$ and $\gamma_2^{(1)} > 0$).

The paper is organized in the following manner. In Section 2, we sketch the formalism which is used in performing the calculations. Some numerical results are shown in Section 3. First we show a few eigenvalue trajectories together with the corresponding eigenfunction trajectories in Section 3.1. The resonance structure of the transmission and the transmission probability are given in Section 3.2. In the last Section 4 we summarize and discuss the results and draw some conclusions on non-Hermitian quantum physics of open quantum systems.

2. Non-Hermitian Hamiltonian $\mathscr{H}$ of an Open Quantum System

2.1. Eigenvalues and Eigenfunctions of $\mathscr{H}^{(2,1)}$. To begin with, we sketch the features typical for an open quantum system embedded in one common continuum. Details can be found in [7] and, above all, in [17]. They can be discussed by means of the genuine 2×2 non-Hermitian matrix:

$$\mathscr{H}^{(2,1)} = \begin{pmatrix} \varepsilon_1^{(1)} \equiv e_1^{(1)} + \frac{i}{2}\gamma_1^{(1)} & \omega^{(1)} \\ \omega^{(1)} & \varepsilon_2^{(1)} \equiv e_2^{(1)} + \frac{i}{2}\gamma_2^{(1)} \end{pmatrix}. \quad (1)$$

Here, $\varepsilon_i^{(1)}$ are the complex energies of the two states $i = 1, 2$ which are coupled to the environment 1 (called also channel 1)[1]. $e_i^{(1)}$ are the energies of the states i and $\gamma_i^{(1)}$ are their widths. $\omega^{(1)}$ stand for the coupling matrix elements of the two states via the common environment 1. They are complex where $\mathrm{Re}(\omega^{(1)})$ arises from the principal value integral and $\mathrm{Im}(\omega^{(1)})$ from the residuum [7]. The complex eigenvalues $\mathscr{E}_i^{(1)} \equiv E_i^{(1)} + (1/2)\Gamma_i^{(1)}$ of $\mathscr{H}^{(2,1)}$ give energies $E_i^{(1)}$ and widths $\Gamma_i^{(1)}$ of the states of the localized part of the system[1].

In (1), a singular point may appear, the so-called EP, at which two eigenvalues of $\mathscr{H}^{(2,1)}$ coalesce [1]. At these points, the two corresponding eigenfunctions are not orthogonal. Instead

$$\Phi_1^{\mathrm{cr}} \longrightarrow \pm i\Phi_2^{\mathrm{cr}}; \qquad \Phi_2^{\mathrm{cr}} \longrightarrow \mp i\Phi_1^{\mathrm{cr}} \quad (2)$$

according to analytical and numerical results [18–20]. An EP is, according to its definition, related to the common environment in which the system is embedded. In other words, it is well defined under the condition that the system is embedded in only one continuum.

The eigenfunctions of a non-Hermitian Hamilton operator are biorthogonal

$$\mathscr{H}\left|\Phi_i\right\rangle = \mathscr{E}_i\left|\Phi_i\right\rangle \qquad \left\langle\Psi_i\right|\mathscr{H} = \mathscr{E}_i\left\langle\Psi_i\right|. \tag{3}$$

In the case of the symmetric 2×2 Hamiltonian (1), it is

$$\Psi_i = \Phi_i^* \tag{4}$$

and the eigenfunctions should be normalized according to

$$\left\langle\Phi_i^* \mid \Phi_j\right\rangle = \delta_{ij} \tag{5}$$

in order to smoothly describe the transition from a closed system with discrete states to a weakly open one with narrow resonance states. As a consequence of (5), the values of the standard expressions are changed:

$$\left\langle\Phi_i \mid \Phi_i\right\rangle = \operatorname{Re}\left(\left\langle\Phi_i \mid \Phi_i\right\rangle\right); \qquad A_i \equiv \left\langle\Phi_i \mid \Phi_i\right\rangle \geq 1 \tag{6}$$

$$\left\langle\Phi_i \mid \Phi_{j\neq i}\right\rangle = i\ \operatorname{Im}\left(\left\langle\Phi_i \mid \Phi_{j\neq i}\right\rangle\right) = -\left\langle\Phi_{j\neq i} \mid \Phi_i\right\rangle \tag{7}$$

$$\left|B_i^j\right| \equiv \left|\left\langle\Phi_i \mid \Phi_{j\neq i}\right\rangle\right| \geq 0. \tag{8}$$

Furthermore, the phase rigidity, which is a quantitative measure for the biorthogonality of the eigenfunctions,

$$r_k \equiv \frac{\left\langle\Phi_k^* \mid \Phi_k\right\rangle}{\left\langle\Phi_k \mid \Phi_k\right\rangle} = A_k^{-1}, \tag{9}$$

is smaller than 1. Far from an EP, $r_k \approx 1$ while it approaches the value $r_k = 0$ when an EP is approached.

Additionally to the Hamiltonian (1), we will consider the non-Hermitian matrix

$$\mathscr{H}_0^{(2,1)} = \begin{pmatrix} \varepsilon_1^{(1)} \equiv e_1^{(1)} + \frac{i}{2}\gamma_1^{(1)} & 0 \\ 0 & \varepsilon_2^{(1)} \equiv e_2^{(1)} + \frac{i}{2}\gamma_2^{(1)} \end{pmatrix} \tag{10}$$

which describes the system without any mixing of its states via the environment. In other words, $\omega = 0$ corresponds to vanishing EM of the eigenstates. The eigenfunctions Φ_i of $\mathscr{H}^{(2,1)}$ can be represented in the set of eigenfunctions $\{\Phi_i^0\}$ of $\mathscr{H}_0^{(2,1)}$,

$$\Phi_i = \sum b_{ij}\Phi_j^0; \qquad b_{ij} = \left\langle\Phi_j^{0*} \mid \Phi_i\right\rangle \tag{11}$$

under the condition that b_{ij} are normalized by $\sum_j (b_{ij})^2 = 1$. The coefficients $|b_{ij}|^2$ differ from the $(b_{ij})^2$. They contain the information on the strength of EM.

The main features characteristic of open quantum systems are described well by the eigenvalues and eigenfunctions of (1). Typical values related to the eigenfunctions are the phase rigidity (9) and the contribution of EM (11) to their purity. All these values contain the influence of the environment. They are proven experimentally; for details see [7, 8, 17].

2.2. Schrödinger Equation with $\mathscr{H}^{(2,1)}$. The Schrödinger equation $(\mathscr{H}^{(2,1)} - \mathscr{E}_i^{(1)}|\Phi_i^{(1)}\rangle = 0$ may be rewritten into a Schrödinger equation with source term [7, 17],

$$\left(\mathscr{H}_0^{(2,1)} - \mathscr{E}_i^{(1)}\right)\left|\Phi_i^{(1)}\right\rangle = -\begin{pmatrix} 0 & \omega \\ \omega & 0 \end{pmatrix}\left|\Phi_i^{(1)}\right\rangle. \tag{12}$$

In this representation, the coupling ω of the states i and $j \neq i$ of the localized system via the common environment of scattering wavefunctions (EM) is contained in the source term of the Schrödinger equation; for details see [7].

Far from EPs, the coupling of the localized system to the environment influences the spectroscopic properties of the system only marginally [7, 17]. The influence of the environment is however nonvanishing also in this case; see, e.g., the experimental results [21]. That means, the eigenstates i of $\mathscr{H}$ will never be the same as those of a Hermitian operator.

In the neighborhood of EPs, the coupling between system and environment cause, according to mathematical studies, nonlinear effects in the Schrödinger equation (12). Among others, these effects lead to a conservation of the resonance structure of the cross section in the one-channel case which is finally unaffected by EM and by the existence of EPs [17]. Thus, the one-channel case cannot be used in order to test the results of the non-Hermitian formalism.

2.3. Two-Channel Case with $\mathscr{H}^{(2,2)}$. Let us consider now the genuine 4×4 non-Hermitian matrix

$$\mathscr{H}^{(2,2)} = \begin{pmatrix} \varepsilon_1^{(1)} & \omega^{(1)} & 0 & 0 \\ \omega^{(1)} & \varepsilon_2^{(1)} & 0 & 0 \\ 0 & 0 & \varepsilon_1^{(2)} & \omega^{(2)} \\ 0 & 0 & \omega^{(2)} & \varepsilon_2^{(2)} \end{pmatrix}. \tag{13}$$

Here, $\varepsilon_i^{(1)} \equiv e_i^{(1)} + (i/2)\gamma_i^{(1)}$ and $\varepsilon_i^{(2)} \equiv e_i^{(2)} + (i/2)\gamma_i^{(2)}$ are the complex energies of the two states $i = 1, 2$ relative to the two different channels $c = 1, 2$ [1]. Usually $\varepsilon_i^{(1)} \neq \varepsilon_i^{(2)}$. $\omega^{(1)}$ and $\omega^{(2)}$ stand for the coupling matrix elements of the two states via environments 1 and 2, respectively. The Hamiltonian (13) includes the fact that the complex energy $\varepsilon_i^{(c)}$ of the localized state i may be different relative to the different channels c.

Also the eigenvalues $\mathscr{E}_i^{(c)} \equiv E_i^{(c)} + (i/2)\Gamma_i^{(c)}$ and eigenfunctions $\Phi_i^{(c)}$ of (13) are characterized by two numbers: the number i of the state ($i = 1, 2$) of the localized part of the system and the number c of the channel ($c = 1, 2$), called environment, in which the system is embedded. Usually, $E_i^{(1)} \neq E_i^{(2)}$ and $\Gamma_i^{(1)} \neq \Gamma_i^{(2)}$. Further, the wave functions $\Phi_i^{(1)}$ and $\Phi_i^{(2)}$ differ from one another due to the EM of the eigenstates via environments 1 and 2, respectively. That means the system has four states, from a mathematical point of view.

The two environments are different from and orthogonal to one another. Further, the two eigenstates with equal i and different c arise from the same state i of the localized part of the system. The zeros in the matrix (13) express the corresponding fact that the two eigenstates i relative to the two channels 1 and 2 of $\mathcal{H}^{(2,2)}$ cannot interact with one another.

We mention here that the Hamiltonian (13) is formally the same as the Hamiltonian (24) or rather (1) in [10]. There is however a fundamental difference: in (1) in [10], the transition of excitons is considered while (13) describes primarily the transition of particles. Furthermore, in (1) in [10] the whole system is fully embedded into both environments ($c = 1$) and ($c = 2$) which are both of completely different nature and exist independently of one another. In contrast to this, the Hamiltonian (13) describes a system, the states of which are primarily embedded partially in each of the two different environments. These two different environments are also independent of one another. They are, however, nothing but parts of the total environment.

Without an EP in the considered parameter range in relation to both channels, we have $E_i^{(1)} \approx E_i^{(2)}$, $\Gamma_i^{(1)} \approx \Gamma_i^{(2)}$, and $\Phi_i^{(1)} \approx \Phi_i^{(2)}$. Under the influence of an EP relative to $c = 1$ and/or relative to $c = 2$, the eigenvalues and eigenfunctions will be, however, different from one another; $E_i^{(1)} \neq E_i^{(2)}$, $\Gamma_i^{(1)} \neq \Gamma_i^{(2)}$, and $\Phi_i^{(1)} \neq \Phi_i^{(2)}$ in the corresponding parameter range.

In analogy to (10), we will consider also the non-Hermitian Hamiltonian:

$$\mathcal{H}_0^{(2,2)} = \begin{pmatrix} \varepsilon_1^{(1)} & 0 & 0 & 0 \\ 0 & \varepsilon_2^{(1)} & 0 & 0 \\ 0 & 0 & \varepsilon_1^{(2)} & 0 \\ 0 & 0 & 0 & \varepsilon_2^{(2)} \end{pmatrix} \quad (14)$$

which describes the system without any mixing of its states via any environment. In other words, $\omega^{(1)} = \omega^{(2)} = 0$ corresponds to vanishing EM of the eigenstates via an environment. The mixing of the eigenstates of (13) can be represented in a set of eigenfunctions of (14) in complete analogy to the relation (11) for two states coupled to one common environment.

2.4. Schrödinger Equation with $\mathcal{H}^{(2,2)}$. Using (14), we can write down the Schrödinger equation with source term for the two-channel case in analogy to (12) for the one-channel case. The corresponding equation reads

$$\left(\mathcal{H}_0^{(2,2)} - \mathcal{E}_i^{(c)}\right) \left|\Phi_i^{(c)}\right\rangle = - \begin{pmatrix} 0 & \omega^{(1)} & 0 & 0 \\ \omega^{(1)} & 0 & 0 & 0 \\ 0 & 0 & 0 & \omega^{(2)} \\ 0 & 0 & \omega^{(2)} & 0 \end{pmatrix} \left|\Phi_i^{(c)}\right\rangle . \quad (15)$$

The source depends on the coupling of the system to both channels, i.e., on $\omega^{(1)}$ and on $\omega^{(2)}$. It does not depend on the energies $\varepsilon_i^{(c)}$.

We repeat here that, according to their definition [1], EPs occur only in the one-channel case, i.e., only in the two submatrices related to channel 1 and channel 2, respectively. They are not defined in the 4×4 matrix (13). However, the EPs of the two submatrices in (13) may influence the dynamics of the open two-channel system.

2.5. Non-Hermitian Hamiltonian and Resonance Structure of the S Matrix. The resonance structure of the S matrix containing two resonance states can be obtained from the expression

$$S = \frac{\left(E - E_1 - (i/2)\,\Gamma_1\right)\left(E - E_2 - (i/2)\,\Gamma_2\right)}{\left(E - E_1 + (i/2)\,\Gamma_1\right)\left(E - E_2 + (i/2)\,\Gamma_2\right)} \quad (16)$$

which is derived in [22] by rewriting the Breit-Wigner expression for one or more isolated resonances that are coupled to a common continuum of scattering wavefunctions; see also [7, 17]. The expression (16) is unitary. According to (16), the resonance structure of the cross section is determined exclusively by the spectroscopic values of the localized part of the system, i.e., by the eigenvalues $\mathcal{E}_i = E_i + i/2\Gamma_i$ of the non-Hermitian Hamiltonian $\mathcal{H}$. The expression (16) allows us therefore to receive reliable results also in the two-channel case and, moreover, when the phase rigidity of the eigenfunctions of $\mathcal{H}$ is reduced, $r_k < 1$ and when the eigenfunctions of $\mathcal{H}$ contain EM, i.e., when they are mixed in the set of eigenfunctions $\{\Phi_i^0\}$ of $\mathcal{H}_0$ according to (11).

According to the results obtained in [17], the evolution of the system near to an EP is driven exclusively by the nonlinear source term of the Schrödinger equation (12) which describes the coupling of the localized part of the system to the common environment and is characteristic of the open quantum system embedded in one environment. The calculations in [17] are performed for fixed ω; i.e., ω can not be responsible for the width bifurcation occurring in these calculations under the influence of an EP. The only conclusion is therefore that the nonlinear source term is the driving force. It is able, in the one-channel case, to largely conserve the resonance structure of the cross section.

The one-channel case does, therefore, not allow us to prove the existence of the nonlinear effects and of EM, since the resonance structure of the cross section calculated with and without EM is the same in this case [17]. This result agrees, on the one hand, with the experience received from many different numerical studies in realistic cases which are performed without taking into account EM. On the other hand, it is not in contradiction with the conclusions received from the study of non-Hermitian physics of open quantum systems.

The conservation of the resonance structure of the cross section, which is possible in the one-channel case, is expected to be impossible, generally, in the two-channel (or more-channel) case.

3. Numerical Results

3.1. Eigenvalues and Eigenfunctions of $\mathcal{H}^{(2,2)}$. The aim of our numerical studies is first to show the influence of a singularity onto the eigenvalues and eigenfunctions of the non-Hermitian Hamilton operator $\mathscr{H}^{(2,2)}$. Knowing these results, we study then, in the following Section 3.2, the influence of the singularity onto observable values such as resonance structure and contour plot of the transmission probability.

In Figure 1, we provide numerical results for the parameter dependence of eigenvalues $\mathscr{E}_i^{(1,2)} \equiv E_i^{(1,2)} + (i/2)\Gamma_i^{(1,2)}$ and eigenfunctions $\Phi_i^{(1,2)}$ of the Hamiltonian $\mathscr{H}^{(2,2)}$. We consider the case that one of the states is coupled to a special channel much stronger than the other one, $\gamma_i^{(c)} \ll \gamma_{j\neq i}^{(c)}$. Without loss of generality, we choose the extreme case $\gamma_1^{(1)} \neq 0$; $\gamma_2^{(2)} \neq 0$; $\gamma_1^{(2)} = \gamma_2^{(1)} = 0$.

We are interested, above all, in the appearance of a critical parameter value $a = a^{\text{cr}}$ at which the spectroscopic values of the localized part of the system change. All our calculations show that a^{cr} is well defined: it is the value of maximum width bifurcation. At a^{cr}, the phase rigidity approaches the value 1, meaning that the two eigenstates of the non-Hermitian Hamiltonian $\mathscr{H}$ become orthogonal at this parameter value. However, the EM of the states via the continuum cannot be neglected at a^{cr}.

In Figure 1 we show a few typical numerical results. The results of the left column are obtained with $\omega^{(1)} = \omega^{(2)}$ and those of the right column with $\omega^{(1)} \neq \omega^{(2)}$. The other parameters are given in the figure caption.

The results are the following. The energies E_i cross at $a = a^{\text{cr}}$ (Figures 1(a) and 1(f)), while the widths Γ_i bifurcate (Figures 1(b) and 1(g)). Notably, the state with vanishing width gets a positive width due to the width bifurcation; i.e., it does not decay around $a = a^{\text{cr}}$.

Around the critical parameter value, the phase rigidity r_i decreases slightly (Figures 1(c), 1(d), 1(h), and 1(i)). It approaches the value 1 when $a \longrightarrow a^{\text{cr}}$. That means, the eigenfunctions of the non-Hermitian operator $\mathscr{H}$ become orthogonal when $a \longrightarrow a^{\text{cr}}$. They are mixed via the environment (Figures 1(e) and 1(j)).

When we start from parameter independent energies and parameter dependent widths in contrast to the case considered in Figure 1, then the biorthogonal eigenfunctions of $\mathscr{H}$ become orthogonal at maximum level repulsion. In any case, there is a critical parameter value at which the biorthogonal eigenfunctions of $\mathscr{H}$ become orthogonal.

We remark that not only do eigenstates with positive widths appear in the present study on systems embedded in two environments but also they are well known from different studies on one-channel systems.

3.2. Resonance Structure and Contour Plot of the Transmission Probability. Observable information on the spectroscopic properties of the localized part of the system is contained in the resonance structure of the cross section. Using (16), we have calculated the resonance structure of the cross section at the critical parameter value $a = a^{\text{cr}}$ and at the two values $a_1 = -3$ and $a_2 = 3$ of a far from a^{cr}.

In all cases we see the double-hump structure of the transmission which is characteristic of the resonance structure of a two-level system coupled to one channel [7, 17, 23]. An example is shown in Figure 2. That means the cross section in the two-channel case is very much like that in the one-channel case when the system is coupled with the same strength to both channels and the EM of the states is the same. The results obtained for other values of the coupling strength (e.g., for $\omega = 0.01i$; $0.1i$ or $0.5i$) are very similar.

Additionally, we have performed some calculations with different values of the original widths ($\gamma_i^{(1)} \neq \gamma_i^{(2)}$) under the condition $\omega^{(1)} = \omega^{(2)} \equiv \omega$. As a result, differences in the values of the original widths $\gamma_i^{(c)}$ almost do not influence the resonance structure of the cross section in the critical parameter range. This corresponds to the fact that $\gamma_i^{(c)}$ are not involved in the source term of the Schrödinger equation (15) and therefore do not contribute to the EM of the states via the environment.

When the EM of the eigenstates via the continuum of scattering wavefunctions is different for the two channels and different from zero, $\omega^{(1)} \neq \omega^{(2)}$, then the minimum of the cross section at a^{cr} will fill up. An example is shown in Figure 3. Similar results (not shown) are received with $\omega^{(1)} = 0.4i$ and $\gamma_1/2 = 0.4$. In both cases, several different values are used for $|\omega^{(2)}| < |\omega^{(1)}|$. The filling-up of the minimum causes a broad maximum of the cross section (after averaging) around $a = a^{\text{cr}}$ instead of the typical resonance peak. This can be seen well when $|\omega^{(1)}| \approx \gamma_1/2$.

Although the double-hump structure of the cross section, appearing under the condition $\omega^{(1)} = \omega^{(2)} \equiv \omega \neq 0$, is well pronounced, it differs, even for small ω, from that obtained for vanishing coupling strength $\omega = 0$ between system and environment (see Figure 4). This result differs from those obtained in the one-channel case; see [17]. That means, EM can generally not be neglected in the two-channel case.

Further information is contained in the contour plots of the cross section some of which are shown in the lower parts of Figures 2, 3, and 4. Most interesting is the padding of the minimum of the double-hump structure of the cross section in the critical parameter range around a^{cr} when the EM of the two states via the environment is different from one another. It causes an enhancement of the cross section around the critical parameter value a^{cr}.

In all cases, the two eigenstates of the non-Hermitian operator $\mathscr{H}^{(2,2)}$ become orthogonal in approaching the critical value $a = a^{\text{cr}}$. Here the eigenfunctions of $\mathscr{H}^{(2,2)}$ are mixed in the set of eigenfunctions of (14).

4. Discussion of the Results and Conclusions

The results of our calculations show very clearly that not only the eigenvalues $\mathscr{E}_i = E_i + i\Gamma_i/2$ of $\mathscr{H}$ play an important role in the non-Hermitian quantum physics of open quantum systems but also the eigenfunctions of $\mathscr{H}$ are of similar importance. It depends on the physical problem that is considered, whether the eigenvalues are more significant than the eigenfunctions or vice versa. In any case, the eigenvalues *and* the eigenfunctions must be considered in describing and explaining a physical problem.

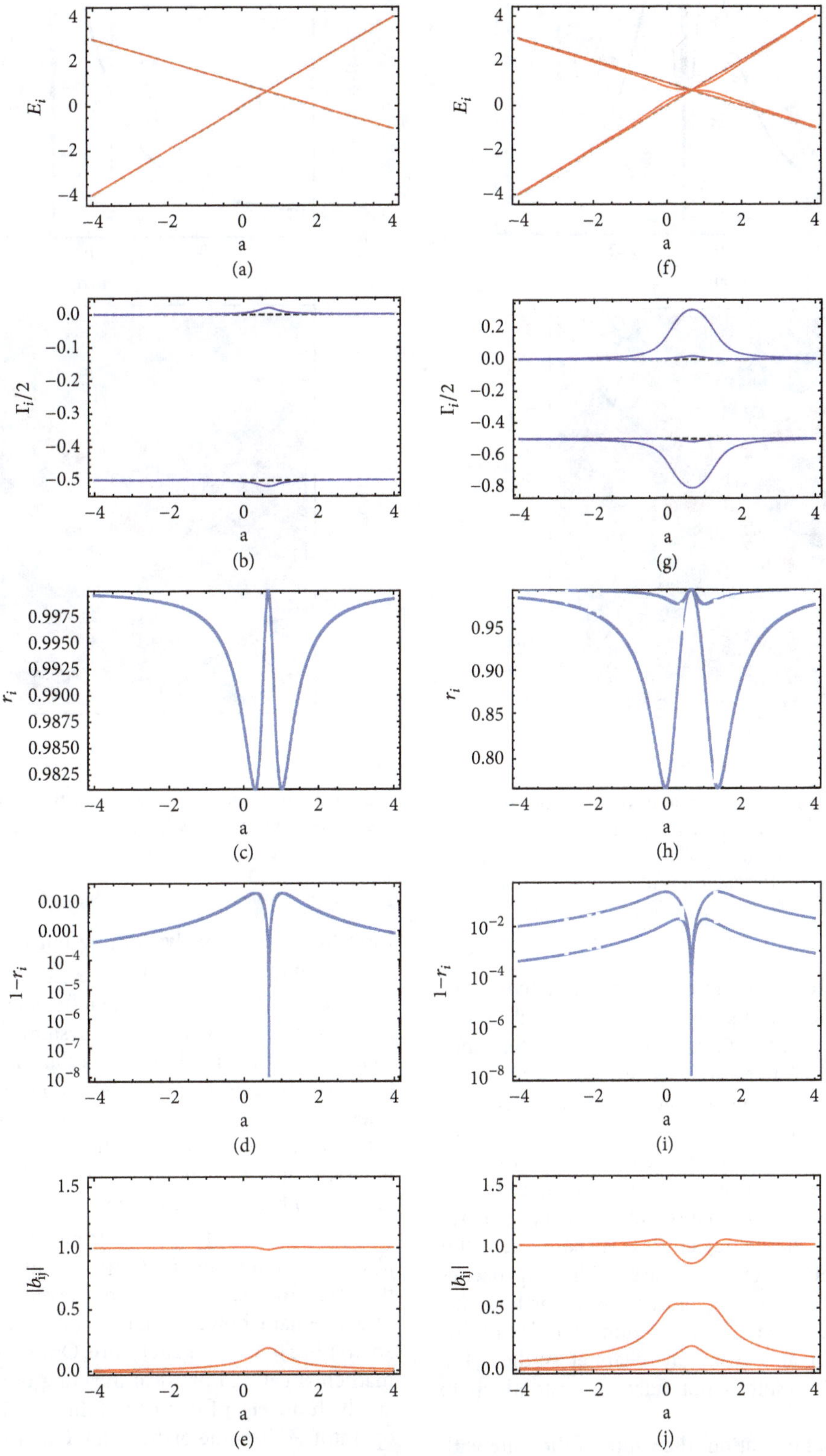

FIGURE 1: Eigenvalues $\mathcal{E}_i^{(1,2)} \equiv E_i^{(1,2)} + (i/2)\Gamma_i^{(1,2)}$ and eigenfunctions $\Phi_i^{(1,2)}$ of the Hamiltonian $\mathcal{H}^{(2,2)}$ as a function of a. $\omega^{(1)} = \omega^{(2)} = 0.1i$ (left), $\omega^{(1)} = 0.5i$; $\omega^{(2)} = 0.1i$ (right). Parameters: $e_1 = 1 - a/2$; $e_2 = a$; $\gamma_1^{(1)}/2 = \gamma_2^{(2)}/2 = -0.5$; $\gamma_1^{(2)}/2 = \gamma_2^{(1)}/2 = 0$ (dashed lines in a, b, f, and g). At the critical parameter value $a = a^{\text{cr}} = 0.6494$, the phase rigidity r_i approaches value 1.

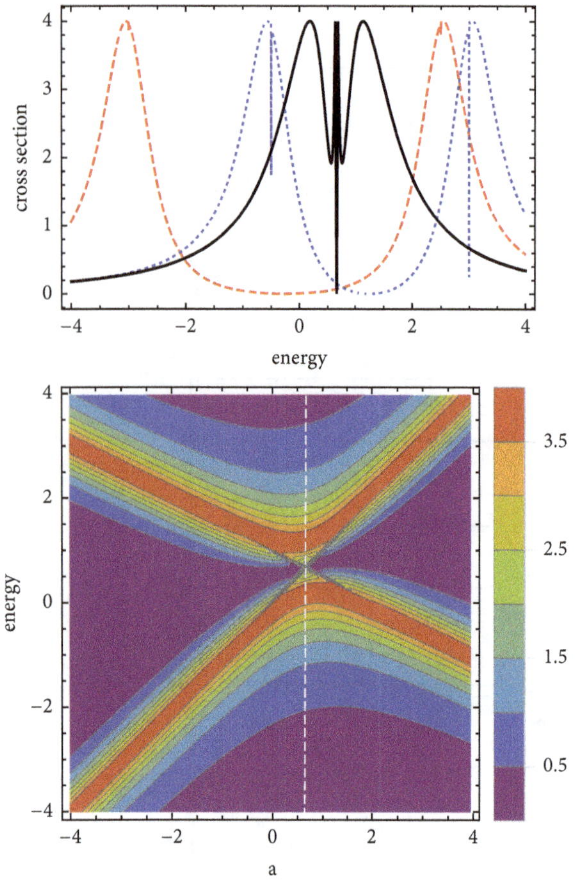

FIGURE 2: Resonance structure (above) and contour plot (below) of the transmission with $\omega^{(1)} = \omega^{(2)} \neq 0$. The parameters are the same as in Figure 1, left. Above: black: $a = a^{\mathrm{cr}}$; red: $a_1 = -3$; blue: $a_2 = 3$. Below: the white line marks the value $a = a^{\mathrm{cr}}$.

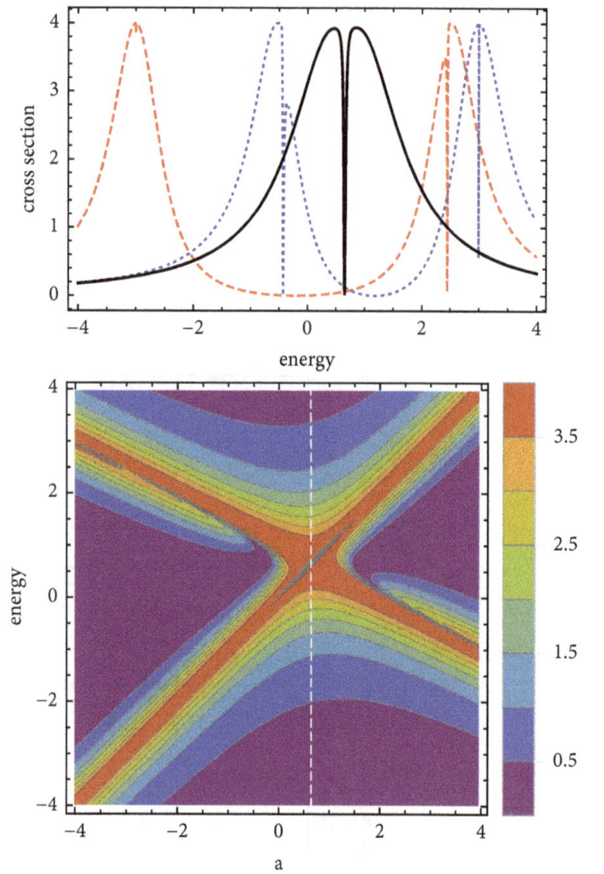

FIGURE 3: The same as Figure 2 but with $\omega^{(1)} \neq \omega^{(2)}$. The parameters are the same as in Figure 1, right.

While the influence of the eigenvalues is restricted, above all, to a small parameter range around some critical points, the eigenfunctions influence a much larger parameter range around these points. An example is the EM (which is a second-order effect). Its influence cannot be neglected over a comparably large parameter range. This behavior is known from the one-channel case [17] and holds true also in the two-channel case as can be seen from the results in Figure 1.

Instead of an EP, we see in Figure 1 the critical parameter value a^{cr} at which the energy trajectories cross and the width bifurcation is maximum. Here, the phase rigidity approaches the value 1, meaning that the eigenfunctions are orthogonal. They are strongly mixed via the continuum. Similar results are very well known also from one-channel studies [17]. The meaning of this result is not fully understood up to now.

Due to the width bifurcation, the width of the state with originally vanishing coupling strength to one of the channels becomes automatically positive. That means that this state gains something from the environment. Thus some gain from the environment in an open quantum system is not at all an exotic process. The different characteristic features of non-Hermitian quantum physics, which are considered in the Introduction, are really nothing but different sides of non-Hermitian quantum physics.

In conclusion we state the following. According to long-time experience, open quantum systems are described usually well by standard methods. There are however some exceptions which occur in a parameter range around singularities. These may be the well-known exceptional points (EPs). Others are related to the fact that the biorthogonal eigenfunctions of $\mathscr{H}$ become orthogonal when width bifurcation (or level repulsion) becomes maximum.

In any case, the non-Hermitian formalism is a very powerful method and is able to explain different phenomena that are puzzling in standard Hermitian quantum physics. There remain however many open questions to which an answer has to be found in future. One of these questions is the mathematical and physical meaning of the critical point a^{cr} at which the eigenfunctions of the non-Hermitian Hamilton operator $\mathscr{H}$ become orthogonal. Other questions arise from the meaning of EPs in many-channel systems.

Conflicts of Interest

The authors declare that there are no conflicts of interest regarding the publication of this paper.

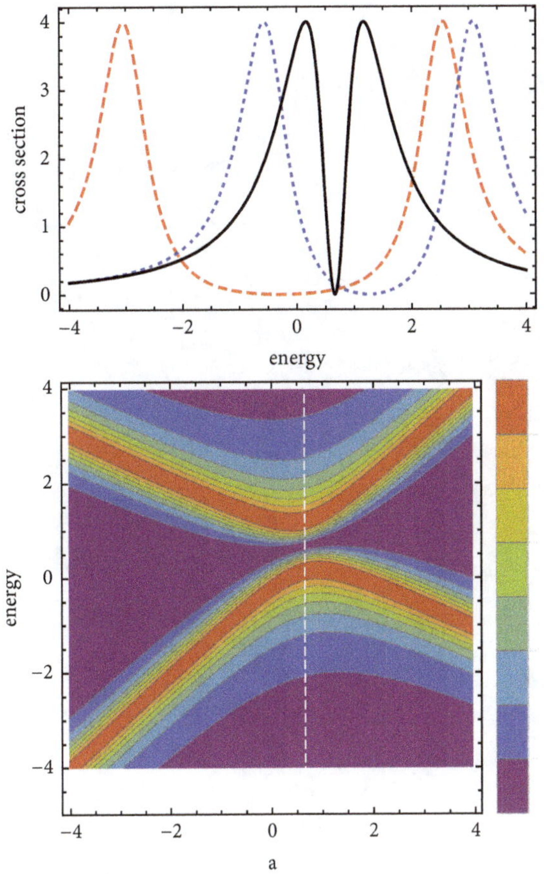

FIGURE 4: The same as Figure 2 but with vanishing external mixing, $\omega^{(1)} = \omega^{(2)} = 0$.

Acknowledgments

The authors thank Jon Bird for valuable discussions.

Endnotes

1. We consider open quantum systems described by a non-Hermitian Hamilton operator. This should not be confused with the consideration of PT-symmetric systems which are neither open nor closed, but nonisolated according to the definition in, e.g., C.M. Bender, Journal of Physics: Conference Series 631, 012002 (2015).

2. In contrast to the definition that is used in, for example, nuclear physics, we define the complex energies before and after diagonalization of $\mathscr{H}$ by $\varepsilon_k = e_k + (i/2)\gamma_k$ and $\mathscr{E}_k = E_k + (i/2)\Gamma_k$, respectively, with $\gamma_k \le 0$ and $\Gamma_k \le 0$ for decaying states. This definition will be useful when discussing systems with gain (positive widths) and loss (negative widths).

References

[1] T. Kato, *Perturbation Theory for Linear Operators*, vol. 132 of *Grundlehren der Mathematischen Wissenschaften*, Springer, New York, NY, USA, 1966.

[2] C. Dembowski, H. Gräf, H. L. Harney et al., "Experimental Observation of the Topological Structure of Exceptional Points," *Physical Review Letters*, vol. 86, no. 5, pp. 787–790, 2001.

[3] C. Dembowski, B. Dietz, H. Gräf et al., "Observation of a Chiral State in a Microwave Cavity," *Physical Review Letters*, vol. 90, no. 3, 2003.

[4] T. Gao, E. Estrecho, K. Y. Bliokh et al., "Observation of non-Hermitian degeneracies in a chaotic exciton-polariton billiard," *Nature*, vol. 526, no. 7574, pp. 554–558, 2015.

[5] N. Moiseyev, *Non-Hermitian Quantum Mechanics*, Cambridge University Press, Cambridge, UK, 2011.

[6] N. Auerbach and V. Zelevinsky, "Super-radiant dynamics, doorways and resonances in nuclei and other open mesoscopic systems," *Reports on Progress in Physics*, vol. 74, Article ID 106301, 2011.

[7] I. Rotter, "A non-Hermitian HAMilton operator and the physics of open quantum systems," *Journal of Physics A: Mathematical and General*, vol. 42, no. 15, 153001, 51 pages, 2009.

[8] I. Rotter and J. P. Bird, "A review of progress in the physics of open quantum systems: theory and experiment," *Reports on Progress in Physics*, vol. 78, Article ID 114001, 2015.

[9] M. Müller and I. Rotter, "Phase lapses in open quantum systems and the non-Hermitian Hamilton operator," *Physical Review A*, vol. 80, Article ID 042705, 2009.

[10] H. Eleuch and I. Rotter, "Gain and loss in open quantum systems," *Physical Review E: Statistical, Nonlinear, and Soft Matter Physics*, vol. 95, no. 6, Article ID 062109, 11 pages, 2017.

[11] W. D. Heiss, M. Müller, and I. Rotter, "Collectivity, phase transitions, and exceptional points in open quantum systems," *Physical Review E: Statistical, Nonlinear, and Soft Matter Physics*, vol. 58, no. 3, pp. 2894–2901, 1998.

[12] O. Atabek, R. Lefebvre, M. Lepers, A. Jaouadi, O. Dulieu, and V. Kokoouline, " Proposal for a Laser Control of Vibrational Cooling in ," *Physical Review Letters*, vol. 106, no. 17, 2011.

[13] A. Jaouadi, M. Desouter-Lecomte, R. Lefebvre, and O. Atabek, "Signatures of exceptional points in the laser control of non-adiabatic vibrational transfer," *Journal of Physics B*, vol. 46, Article ID 145402, 2013.

[14] I. Gilary and N. Moiseyev, "Asymmetric effect of slowly varying chirped laser pulses on the adiabatic state exchange of a molecule," *Journal of Physics B*, vol. 45, Article ID 051002, 2012.

[15] G. A. Álvarez, E. P. Danieli, P. R. Levstein, and H. M. Pastawski, "Environmentally induced quantum dynamical phase transition in the spin swapping operation," *The Journal of Chemical Physics*, vol. 124, Article ID 194507, 2006.

[16] A. Ruderman, A. D. Dente, E. Santos, and H. M. Pastawski, "Molecular dissociation in presence of a catalyst: II. The bond breaking role of the transition from virtual to localized states," *Materials Research Express*, vol. 3, no. 8, Article ID 085017, 2016.

[17] H. Eleuch and I. Rotter, "Resonances in open quantum systems," *Physical Review A*, vol. 95, no. 6, Article ID 022117, 2017.

[18] I. Rotter, "Dynamics of quantum systems," *Physical Review E*, vol. 64, no. 3, Article ID 036213, 2001.

[19] B. Wahlstrand, I. I. Yakimenko, and K. Berggren, "Wave transport and statistical properties of an open non-Hermitian quantum dot with parity-time symmetry," *Physical Review E: Statistical, Nonlinear, and Soft Matter Physics*, vol. 89, no. 6, Article ID 062910, 2014.

[20] F. Tellander and K. Berggren, "Spectra, current flow, and wave-function morphology in a model," *Physical Review A: Atomic, Molecular and Optical Physics*, vol. 95, no. 4, Article ID 042115, 2017.

[21] J. Gros, U. Kuhl, O. Legrand, F. Mortessagne, E. Richalot, and D. Savin, "Experimental Width Shift Distribution: A Test of Nonorthogonality for Local and Global Perturbations," *Physical Review Letters*, vol. 113, no. 22, Article ID 224101, 2014.

[22] I. Rotter, "Effective Hamiltonian and unitarity of the S matrix," *Physical Review E: Statistical, Nonlinear, and Soft Matter Physics*, vol. 68, no. 1, 016211, 13 pages, 2003.

[23] M. Müller, F. Dittes, W. Iskra, and I. Rotter, "Level repulsion in the complex plane," *Physical Review E: Statistical, Nonlinear, and Soft Matter Physics*, vol. 52, no. 6, pp. 5961–5973, 1995.

7

Magnetoacoustic Heating in Nonisentropic Plasma caused by Different Kinds of Heating-Cooling Function

Anna Perelomova

Gdansk University of Technology, Faculty of Applied Physics and Mathematics, Ul. Narutowicza 11/12, 80-233 Gdansk, Poland

Correspondence should be addressed to Anna Perelomova; anna.perelomova@pg.edu.pl

Academic Editor: Alexander Iomin

The nonlinear phenomena which associate with magnetoacoustic waves in a plasma are analytically studied. A plasma is an open system with external inflow of energy and radiation losses. A plasma's flow may be isentropically stable or unstable. The nonlinear phenomena occur differently in dependence on stability or instability of a plasma's flow. The nonlinear instantaneous equation which describes dynamics of nonwave entropy mode in the field of intense magnetoacoustic perturbations is the result of special projecting of the conservation equations in the differential form. It is analyzed in some physically meaningful cases; those are periodic magnetoacoustic perturbations and particular cases of heating-cooling function. A plasma is situated in the straight magnetic field with constant equilibrium magnetic strength which form constant angle with the direction of wave propagation. A plasma is initially uniform and equilibrium. The conclusions concern nonlinear effects of fast and slow magnetoacoustic perturbations and may be useful in direct and inverse problems.

1. Introduction

The study of propagation of magnetohydrodynamic (MHD) perturbations has a key role in the astrophysical applications. It is of crucial importance in solar hydrodynamics, in particular in formation of coronal loops, fluid flows in thermonuclear reactors, interstellar gases, and planetary nebulae. Gaseous plasma is an open system with disturbed adiabaticity of wave perturbations. This may be modeled by some heating-cooling function which reflects the physical conditions of energy inflow and radiation losses [1–3]. Linear dynamics of the MHD waves, that is, waves of infinitely small magnitude in an open plasma which may be isentropically unstable, has been well studied in the last decade. The main conclusions are common in hydrodynamics of open systems [4, 5]. Flows of other fluids in thermodynamical nonequilibrium reveal analogous features which are described by the similar equations [6, 7]. Apart from complexity of the MHD system of equations which imposes coexistence of slow, fast sound modes, and Alfvén modes, along with the nonwave modes, other linear effects go into play, such as mechanical damping and thermal conduction of a plasma.

The general nonlinear dynamics of perturbations in plasma is not far well understood. As for the peculiarities of nonlinear evolution of wave MHD perturbations without accounting for their interaction with other modes, they are well understood starting from some particular but important cases which concern one-dimensional flows along and across the magnetic field [8–10] and ending by various numerical investigations. Nakariakov and coauthors in [11] advanced in the analysis of one-dimensional propagation of the MHD wave perturbations at arbitrary angle between the straight magnetic field and direction of wave's propagation. They have derived the generic weakly nonlinear evolutionary equation which describes slow and fast perturbations in active plasma. Authors of [12] concluded about possibility of self-organization of MHD waves and described theoretically slow magnetoacoustic shock autowaves with magnitude independent on the initial or boundary conditions. They discovered that the extension of a heating-cooling function in the Taylor series accounting for quadratic nonlinear terms "introduces new physics such as existence of solitary waves". The conclusions concern nonlinear dynamics of individual slow and fast wave perturbations. The formation of the asymmetric autowave pulse in the similar geometry of a flow was described in [13].

While perturbations of infinitely small magnitude develop independently, nonlinearity makes different types of a fluid's motion to interact in physical reality which deals with finite-magnitude perturbations. The wave modes may excite nonwave modes which form inhomogeneities in turn impacting on wave propagation; waves may be scattered by other waves, and so on [14–16]. Apart from nonlinearity, the reasons for interaction are irreversible losses in energy and momentum or/and the disturbance of adiabaticity. In description of nonlinear interaction of modes, we face with mathematical difficulties much serious as compared to the case of nonlinear dynamics of individual mode. The nonlinear excitation of the entropy mode in the field of slow and fast magnetosonic perturbations for different kinds of external heating-cooling function is the subject of this study. Following Nakariakov and coauthors, we consider weakly nonlinear flow of completely ionized gas affected by some generic heating-cooling function. Weak nonlinearity means the smallness of the Mach number in a flow: $M \ll 1$. Also, the impact of nonlinearity may be obtained as corrections to the linear results. This concerns the links between specific perturbations inside every mode and dynamic equations. The method which has been applied by the author in studies of nonlinear interaction of wave and nonwave modes in a number of different fluid flows is useful in investigation of magnetic flows as well. It allows subdividing nonlinear dynamic equations for perturbations in a specific mode from the system of conservation equations in a differential form. The procedure is in fact some linear combination of equations which eliminates all liner terms containing nonspecific perturbations. The dynamic coupling equations include the first-order derivatives with respect to time. The quadratic nonlinear terms which are products of nonspecific perturbations and their derivatives form the "foreign forces". In the context of weak nonlinearity, the quadratic terms including magnetosonic perturbations are of importance. Nonlinear corrections in the dynamic equations for fast and slow magnetosonic perturbations may be also considered as a result of self-interaction of corresponding wave mode. Nonlinearity, even weak, accumulates in time and is responsible for formation of shock fronts, enrichment of perturbations' spectrum, and nonlinear transfer of energy and momentum. In regard to magnetoacoustic heating, nonlinear transfer of the wave energy into the energy of the entropy mode leads to forming of the thermal lenses which in turn scatter sound. The effects of plasma's boundaries are not considered. Nonlinear dynamic equations are valid with the accepted accuracy up to quadratic nonlinear terms, that is, up to terms proportional to the squared Mach number, M^2. That concerns also terms originating from the heating-cooling function. The perturbations which specify the entropy mode enhance with time. In the frames of a model, their magnitudes should not exceed magnitudes of corresponding wave perturbations. When these magnitudes equalize, the entropy mode may have impact on the wave process. Hence, the equations are valid over some temporal and spatial domains, where wave perturbations remain dominant.

One may expect that not only the wave processes, but the magnetoacoustic heating reveals anomalous behavior in the isentropically unstable active plasma, as it happens to all acoustically active media independently on the physical reason of acoustical activity [6, 7]. The subsequent analysis confirms this conclusion. We do not consider mechanical and thermal losses in a plasma and its finite electrical conductivity which are well studied. They introduce additional attenuation and dispersion. The magnetoacoustic heating due to propagation of the MHD perturbations across the straight magnetic field has been studied by the author in [17, 18]. In this study, we analyze the instantaneous dynamic equation of excitation of the entropy mode by fast and slow MHD sound modes in the field of planar waves which form an arbitrary angle with the magnetic strength and discuss it in some physically meaningful cases of wave perturbations and the heating-cooling function. That concerns also the case of nonadiabatic instability. The first results in a special case of heating-cooling function which depends exclusively on temperature may be found in the recent author's paper [19]. In this study, we make conclusions about entropy perturbations due to magnetosonic heating in general case.

2. Perturbations of Infinitely Small Magnitude in the MHD Flow

The system of PDE equations consists of continuity equation, momentum equation for ionized gas, energy balance equation, and electrodynamic equations:

$$
\begin{gathered}
\frac{\partial \rho}{\partial t} + \vec{\nabla} \cdot \left(\rho \vec{v}\right) = 0, \\
\rho \frac{d\vec{v}}{dt} = -\vec{\nabla} p + \mu_0 \left(\vec{\nabla} \times \vec{B}\right) \times \vec{B}, \\
\frac{dp}{dt} - \gamma \frac{p}{\rho} \frac{d\rho}{dt} = (\gamma - 1) L(p, \rho), \\
\frac{\partial \vec{B}}{\partial t} = \vec{\nabla} \times \left(\vec{v} \times \vec{B}\right), \\
\vec{\nabla} \cdot \vec{B} = 0,
\end{gathered}
\tag{1}
$$

where $\vec{v}$, $\vec{B}$, p, and ρ are the velocity of a plasma, the magnetic field strength, plasma's pressure, and density, respectively, and μ_0 is the magnetic permeability of free space. L is the heating-cooling function which disturbs adiabaticity of wave perturbations in a plasma. The choice of the heating-cooling function which depends on pressure and density makes using these basic thermodynamic variables and expressing all other thermodynamic quantities such as internal energy as functions of pressure and density convenient. Following [11], it is assumed to be dependent on pressure and density. The third equation in the set (1) relates to an ideal gas with the caloric equation of state

$$
e = \frac{p}{(\gamma - 1)\rho}, \tag{2}
$$

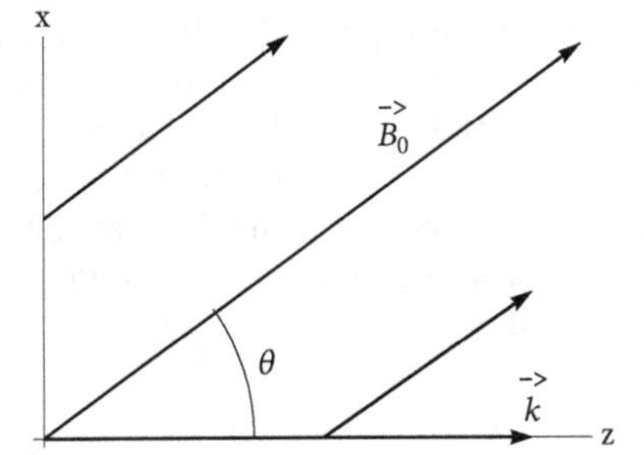

FIGURE 1

where e denotes the internal energy of a gas and $\gamma = C_P/C_V$ is the ratio of specific heats under constant pressure and constant density. We consider the same geometry of a flow as in [11]. The equilibrium magnetic strength $\vec{B}_0$ lies in the (x, z) plane. Its equilibrium magnitude B_0 is constant, and it forms constant angle θ with the positive direction of axis z, so

$$B_{0,x} = B_0 \sin(\theta), \qquad B_{0,z} = B_0 \cos(\theta) \tag{3}$$

denote projections of $\vec{B}_0$ onto axis x and z. Axis z points to the direction of wave propagation. Figure 1 recalls the geometry of a planar flow.

All field quantities are expanded in the vicinity of the equilibrium state as $f(z,t) = f_0 + f'(z,t)$. We consider initially homogeneous stationary plasma without bulk flows. That imposes $v_{0,x} = v_{0,y} = v_{0,z} = 0$. The starting point represents the leading-order equations which follow from (1):

$$\begin{aligned}
&\frac{\partial \rho'}{\partial t} + \rho_0 \frac{\partial v_z}{\partial z} = -\rho' \frac{\partial v_z}{\partial z} - v \frac{\partial \rho'}{\partial z}, \\
&\frac{\partial v_x}{\partial t} - \frac{B_{0,z}}{\rho_0 \mu_0} \frac{\partial B'_x}{\partial z} = -\frac{B_{0,z}}{\rho_0 \mu_0} \rho' \frac{\partial B'_x}{\partial z} - v_z \frac{\partial v_x}{\partial z}, \\
&\frac{\partial v_y}{\partial t} - \frac{B_{0,z}}{\rho_0 \mu_0} \frac{\partial B'_y}{\partial z} = -\frac{B_{0,z}}{\rho_0 \mu_0} \rho' \frac{\partial B'_y}{\partial z} - v_z \frac{\partial v_y}{\partial z}, \\
&\frac{\partial v_z}{\partial t} + \frac{1}{\rho_0} \frac{\partial p'}{\partial z} + \frac{B_{0,x}}{\rho_0 \mu_0} \frac{\partial B'_x}{\partial z} \\
&\quad = -\frac{B_{0,z}}{\rho_0 \mu_0} \rho' \frac{\partial B'_x}{\partial z} - \frac{\rho'}{\rho_0} \frac{\partial p'}{\partial z} - v_z \frac{\partial v_z}{\partial z} \\
&\quad\quad - \frac{1}{2\rho_0 \mu_0} \frac{\partial}{\partial z} \left(B'^2_x + B'^2_y \right), \\
&\frac{\partial p'}{\partial t} + c_0^2 \rho_0 \frac{\partial v_z}{\partial z} - (\gamma - 1) \left(L_p p' + L_\rho \rho' \right) \\
&\quad = -v_z \frac{\partial p'}{\partial z} - \gamma p' \frac{\partial v_z}{\partial z} \\
&\quad\quad + (\gamma - 1) \left(0.5 L_{pp} p'^2 + 0.5 L_{\rho\rho} \rho'^2 + L_{p\rho} p' \rho' \right), \\
&\frac{\partial B'_x}{\partial t} + \frac{\partial}{\partial z} \left(B_{0,x} v_z - B_{0,z} v_x \right) = -v_z \frac{\partial B'_x}{\partial z} - B'_x \frac{\partial v_z}{\partial z}, \\
&\frac{\partial B'_y}{\partial t} - B_{0,z} \frac{\partial v_y}{\partial z} = -v_z \frac{\partial B'_y}{\partial z} - B'_y \frac{\partial v_z}{\partial z},
\end{aligned} \tag{4}$$

where

$$L_p = \frac{\partial L}{\partial p}, \quad L_\rho = \frac{\partial L}{\partial \rho}, \quad L_{pp} = \frac{\partial^2 L}{\partial p^2}, \quad L_{\rho\rho} = \frac{\partial^2 L}{\partial \rho^2}, \quad L_{p\rho} = \frac{\partial^2 L}{\partial p \partial \rho} \tag{5}$$

are partial derivatives of the heating-cooling function $L(p, \rho)$ evaluated at equilibrium state (p_0, ρ_0) and

$$c_0 = \sqrt{\frac{\gamma p_0}{\rho_0}} \tag{6}$$

is the acoustic speed in unmagnetized gas in equilibrium.

As usual, establishment of the dispersion relations is the primary procedure in the linear flows of a fluid. All perturbations are thought as a sum of planar waves proportional to $\exp(i\omega(k_z)t - ik_z z)$, where k_z designates the wave number of any individual planar wave which propagates along axis z:

$$f'(z,t) = \int_{-\infty}^{\infty} \tilde{f}(k_z) \exp(i\omega(k_z)t - ik_z z)\, dk_z, \tag{7}$$

and $(\tilde{f}(k_z) \exp(i\omega(k_z)t) = \tilde{f}(k_z, t)$ denotes the Fourier transform of $f'(z,t)$, so $\tilde{f}(k_z, t) = (1/2\pi) \int_{-\infty}^{\infty} f(z,t) e^{ik_z z} dz)$. The dispersion relations in a planar flow of a magnetic fluid reflect the solvability of (4). They take the form of seven solutions to the algebraic equation which demands zero main determinant of the system of linear equations:

$$\begin{aligned}
\omega_{1,2} &= \pm \frac{B_{0,z} k_z}{\sqrt{\mu_0 \rho_0}} = \pm C_{A,z} k_z, \\
\omega_{3,4,5,6} &= C k_z + i \frac{S}{c_0^2} \left(c_0^2 L_p + L_\rho \right), \\
\omega_7 &= \frac{i(\gamma - 1) L_\rho}{c_0^2},
\end{aligned} \tag{8}$$

where

$$S = -\frac{\left(C^2 - C_A^2\right)(\gamma - 1)}{2\left(2C^2 - c_0^2 - C_A^2\right)},$$
$$C_A = \frac{B_0}{\sqrt{\mu_0 \rho_0}}, \tag{9}$$
$$C_{A,z} = C_A \cos(\theta).$$

C_A designates the Alfvén speed, $C_{A,z}$ is its z-component, and C is one from four roots of the equation

$$B_{0,x}^2 C^4 + \left(C^2 - c_0^2\right)\left(B_{0,z}^2 - \mu_0 \rho_0 C^2\right) = 0. \tag{10}$$

All magnetosonic speeds are evaluated at the equilibrium state (p_0, ρ_0). Four roots C are real values for any θ. With accounting for equalities

$$\frac{B_{0,x}}{B_0} = \frac{C_{A,x}}{C_A} = \sin(\theta),$$
$$\frac{B_{0,z}}{B_0} = \frac{C_{A,z}}{C_A} = \cos(\theta), \tag{11}$$

(10) which determines acoustic speeds takes the form

$$C^4 - C^2\left(c_0^2 + C_A^2\right) + c_0^2 C_{A,z}^2 = 0. \tag{12}$$

The first two roots ω_1, ω_2 relate to the magnetic Alfvén waves, the next four roots specify magnetosonic waves of different directions of propagation (the so-called slow and fast MHD waves of different direction of propagation), and the last root ω_7 corresponds to the entropy mode. The dispersion relations in (8) are calculated with accuracy up to terms proportional to the first powers of L_p, L_ρ. They have been established by Nakariakov and coauthors [11, 12] and were in fact the starting point for analysis of individual weakly nonlinear wave perturbations in a plasma. The condition of acoustic (isentropic) instability is

$$c_0^2 L_p + L_\rho > 0, \tag{13}$$

since S is real negative value for any C and θ, apart form the particular case of $\theta = 0$, when C^2 equals C_A^2 and $S = 0$. This condition has been discovered in the earlier studies of nonequilibrium processes in fluids [4, 5]. In the case of isentropic instability, the MHD wave perturbations of infinitely small magnitude enhance in the course of propagation. The condition of thermal isobaric instability sounds as

$$L_\rho < 0. \tag{14}$$

Note that the heating-cooling function is supposed to depend on pressure and density. The choice of the reference thermodynamic variables evidently impacts on the form of instability conditions. These inequalities are common for open systems and do not certainly overlap.

The linearized system (4) has an equivalent form

$$\frac{\partial \widetilde{\psi}}{\partial t} + \widetilde{L}\widetilde{\psi} = \left(i\omega + \widetilde{L}\right)\widetilde{\psi} = 0, \tag{15}$$

where

$$\widetilde{\psi} = \begin{pmatrix} \widetilde{\rho} \\ \widetilde{v}_x \\ \widetilde{v}_y \\ \widetilde{v}_z \\ \widetilde{p} \\ \widetilde{B}_x \\ \widetilde{B}_y \end{pmatrix},$$

$$\widetilde{L} = \begin{pmatrix} 0 & 0 & 0 & -i\rho_0 k_z & 0 & 0 & 0 \\ 0 & 0 & 0 & 0 & 0 & \frac{ik_z B_{0,z}}{\rho_0 \mu_0} & 0 \\ 0 & 0 & 0 & 0 & 0 & 0 & \frac{ik_z B_{0,z}}{\rho_0 \mu_0} \\ 0 & 0 & 0 & 0 & -\frac{ik_z}{\rho_0} & -\frac{ik_z B_{0,x}}{\rho_0 \mu_0} & 0 \\ -(\gamma - 1) L_\rho & 0 & 0 & -ik_z c_0^2 \rho_0 & -(\gamma - 1) L_p & 0 & 0 \\ 0 & ik_z B_{0,z} & 0 & -ik_z B_{0,x} & 0 & 0 & 0 \\ 0 & 0 & ik_z B_{0,z} & 0 & 0 & 0 & 0 \end{pmatrix} \tag{16}$$

denote vector of the Fourier transforms of perturbations and matrix operating in the space of Fourier transforms. In fact, seven roots of dispersion equation (8) are the eigenvalues of matrix $i\widetilde{L}$. The total perturbation is represented by a sum of specific disturbances which are eigenvectors of $i\widetilde{L}$ corresponding to seven eigenvalues. Magnetoacoustic modes are represented by four eigenvectors which connect the Fourier transforms of specific perturbations:

$$\widetilde{\psi}_{3,4,5,6} = \begin{pmatrix} \widetilde{\rho} \\ \widetilde{v}_x \\ \widetilde{v}_y \\ \widetilde{v}_z \\ \widetilde{p} \\ \widetilde{B}'_x \\ \widetilde{B}'_y \end{pmatrix}_{3,4,5,6} = \begin{pmatrix} 1 \\ -\dfrac{B_{0,z}\left(Cc_0^2\left(C^2-c_0^2\right)k_z + \left(ic_0^2(\gamma-1) + iS\left(C^2+c_0^2\right)\right)\left(c_0^2L_p + L_\rho\right)\right)}{B_{0,x}C^2c_0^2k_z\rho_0} \\ 0 \\ \dfrac{Cc_0^2k_z + iS\left(c_0^2L_p + L_\rho\right)}{k_zc_0^2\rho_0} \\ \dfrac{Cc_0^2k_z - i(\gamma-1)\left(c_0^2L_p + L_\rho\right)}{Ck_z} \\ \dfrac{\left(Cc_0^2\left(C^2-c_0^2\right)k_z + i\left(c_0^2(\gamma-1) + 2SC^2\right)\left(c_0^2L_p + L_\rho\right)\right)\mu_0}{CB_{0,x}k_z} \\ 0 \end{pmatrix} \widetilde{\rho}_{3,4,5,6}, \quad (17)$$

where the ordering number by C is omitted. The relations between perturbations may be readily written on in z,t variables, reminding that the factor $-ik_z$ corresponds to the operator $\partial/\partial z$ and $(-ik_z)^{-1}$ corresponds to $\int dz$. The upper limit of integration is z, and the lower one depends on the physical context of a flow. In the case $B_{0,x} = 0$, the magnetoacoustic x-component of velocity and perturbation in magnetic strength in all four modes are zero: $\widetilde{v}_x = 0$, $\widetilde{B}_x = 0$, so $v_x = 0$, $B'_x = 0$. The specific perturbations representing the Alfvén modes are linked in the following way:

$$\psi_1 = \begin{pmatrix} \rho' \\ v_x \\ v_y \\ v_z \\ p' \\ B'_x \\ B'_y \end{pmatrix}_1 = \begin{pmatrix} 0 \\ 0 \\ 1 \\ 0 \\ 0 \\ -\dfrac{B_0}{C_A} \\ 0 \end{pmatrix} v_{y,1},$$

$$\psi_2 = \begin{pmatrix} 0 \\ 0 \\ 1 \\ 0 \\ 0 \\ \dfrac{B_0}{C_A} \\ 0 \end{pmatrix} v_{y,2}. \quad (18)$$

The entropy mode is established by the relations

$$\psi_7 = \begin{pmatrix} \rho' \\ v_x \\ v_y \\ v_z \\ p' \\ B'_x \\ B'_y \end{pmatrix}_7 = \begin{pmatrix} 1 \\ \dfrac{(\gamma-1)C_{A,x}L_\rho}{C_{A,z}c_0^2\rho_0}\displaystyle\int dz \\ 0 \\ \dfrac{(\gamma-1)L_\rho}{c_0^2\rho_0}\displaystyle\int dz \\ 0 \\ 0 \\ 0 \end{pmatrix} \rho_7. \quad (19)$$

All eigenvectors are derived with accuracy up to terms including the first powers of L_p and L_ρ. The case $C_{A,z} = 0$ corresponds to zero v_x in the entropy mode. Any total perturbation is a sum of specific disturbances:

$$v_x = \sum_{i=1}^{7} v_{x,i},$$

$$v_y = \sum_{i=1}^{7} v_{y,i},$$

$$v_z = \sum_{i=1}^{7} v_{z,i},$$

$$B'_x = \sum_{i=1}^{7} B'_{x,i},$$

$$B'_y = \sum_{i=1}^{7} B'_{y,i},$$

$$p' = \sum_{i=1}^{7} p_i,$$
$$\rho' = \sum_{i=1}^{7} \rho_i. \tag{20}$$

Index in summation i denotes the ordering number of specific mode. The row which distinguishes individual excess density corresponding to the entropy mode is determined by the system of seven algebraic equations:

$$P_7 \begin{pmatrix} \rho' \\ v_x \\ v_y \\ v_z \\ p' \\ B'_x \\ B'_y \end{pmatrix} = \rho_7. \tag{21}$$

It takes the form [19]

$$P_7 = \left(1 \quad -\frac{(\gamma-1) C_{A,x}\rho_0}{C_{A,z}c_0^4} \left(c_0^2 L_p + L_\rho\right) \int dz \quad 0 \quad -\frac{(\gamma-1)\rho_0}{c_0^4} \left(c_0^2 L_p + L_\rho \int dz\right) \quad -\frac{1}{c_0^2} \quad 0 \quad 0 \right). \tag{22}$$

Analogously, the projecting rows which subdivide velocity specifying the Alfvén modes are determined by the systems

$$P_1 \begin{pmatrix} \rho' \\ v_x \\ v_y \\ v_z \\ p' \\ B'_x \\ B'_y \end{pmatrix} = v_{y,1},$$
$$P_2 \begin{pmatrix} \rho' \\ v_x \\ v_y \\ v_z \\ p' \\ B'_x \\ B'_y \end{pmatrix} = v_{y,2} \tag{23}$$

and take the forms

$$P_1 = \left(0 \quad 0 \quad \frac{1}{2} \quad 0 \quad 0 \quad 0 \quad -\frac{C_A}{2B_0} \right),$$
$$P_2 = \left(0 \quad 0 \quad \frac{1}{2} \quad 0 \quad 0 \quad 0 \quad \frac{C_A}{2B_0} \right). \tag{24}$$

Projectors P_1,P_2 are also evaluated with accuracy up to terms proportional to the first powers of L_p and L_ρ. When P_1, P_2, P_7 apply at the linearized equation (4), they reduce all therms containing perturbations which do not belong to the corresponding mode. Application of projectors yield the linear dynamic equations which govern $v_{y,1}$, $v_{y,2}$, and ρ_7, respectively. In the following section, we consider the nonlinear excitation of these modes in the field of intense magnetoacoustic waves.

3. Nonlinear Effects Which Associate with the Intense Magnetoacoustic Wave

Among variety of nonlinear phenomena, we should distinguish nonlinear effects associated with the individual magnetoacoustic wave without accounting for its interaction with other modes and these ones which reflect interaction between different modes. In the context of nonlinear acoustics, distortions and nonlinear phenomena of intense wave are of the major importance. That means that magnetosonic perturbations are larger than that of other modes, at least in some temporal and special domains under consideration. We suppose that only one magnetosonic wave is dominant with the linear speed C which is arbitrary of four roots of (10). The dynamic equation which describes the nonlinear distortion of magnetoacoustic wave and its nonlinear phenomena follow by means of application of corresponding projector at system (4). Since (4) include only quadratic nonlinear terms, the resulting dynamical equations are valid with accuracy up to quadratic nonlinear terms, also up to these one which stand by L_p,L_ρ, L_{pp},$L_{p\rho}$, and $L_{\rho\rho}$.

The correct description of nonlinear effects associating with quadratic nonlinear terms imposes that the linear relations of perturbations in a wave should be corrected by means of involving terms which make them isentropic in the leading order [18]. The corrected links in terms of the longitudinal velocity v_z are as follows:

$$\psi_{3,4,5,6}^{cor} = \begin{pmatrix} \rho' \\ v_x \\ v_y \\ v_z \\ p' \\ B'_x \\ B'_y \end{pmatrix}_{3,4,5,6}$$

$$= \begin{pmatrix} \frac{\rho_0}{C} - \frac{\rho_0 S\left(c_0^2 L_p + L_\rho\right)}{C^2 c_0^2} \int dz \\ \frac{c_0^2 C_{A,z}}{C^2 C_{A,x}} - \frac{C_{A,z}}{C_{A,x}} - \frac{C_{A,z}\left(c_0^2 L_p + L_\rho\right)(\gamma - 1 + 2S)}{C^3 C_{A,x}} \int dz \\ 0 \\ 1 \\ \frac{c_0^2 \rho_0}{C} - \frac{\left(c_0^2 L_p + L_\rho\right)(\gamma - 1 + S)}{C^2} \int dz \\ \frac{\left(C^2 - c_0^2\right) B_0}{C C_A C_{A,x}} + B_0 \frac{\left(c_0^2 L_p + L_\rho\right)\left(c_0^2 (\gamma - 1) + S\left(C^2 + c_0^2\right)\right)}{C^2 c_0^2 C_A C_{A,x}} \int dz \\ 0 \end{pmatrix} v_{z,3,4,5,6} + \begin{pmatrix} A_1 \\ A_2 \\ 0 \\ 0 \\ A_3 \\ A_4 \\ 0 \end{pmatrix} v_{z,3,4,5,6}^2, \tag{25}$$

where

$$\begin{aligned} A_1 &= \frac{\left(c_0^2 + C^4 (\gamma - 4) - C^2 C_A^2 (\gamma - 3)\right) \rho_0}{4C^4 \left(c_0^2 + C_A^2 - 2C^2\right)}, \\ A_2 &= \frac{c_0^2 \left(C^2 - c_0^2\right)\left(c_0^4 - C^2 \left(2c_0^2 + (\gamma - 1) C_A^2\right) + \gamma C^4\right)}{2C^5 \left(C^2 - C_A^2\right)\left(c_0^2 + C_A^2 - 2C^2\right)} \frac{C_{A,z}}{C_{A,x}}, \\ A_3 &= \frac{c_0^2 \left(c_0^4 - 3\gamma C^4 + C^2 \left(2c_0^2 (\gamma - 1) + C_A^2 (\gamma + 1)\right)\right) \rho_0}{4C^4 \left(c_0^2 + C_A^2 - 2C^2\right)}, \\ A_4 &= \frac{B_0 \left(C^2 - c_0^2\right)\left(C^6 - C^2 c_0^4 - C^4 \left(3C_{A,z}^2 + c_0^2 (\gamma - 3)\right) + c_0^2 C_{A,z}^2 \left(C_{A,z}^2 (\gamma + 1) - c_0^2\right)\right)}{4C^2 \left(C^4 - c_0^2 C_{A,z}^2\right) C_A C_{A,x}^3}. \end{aligned} \tag{26}$$

Relations (25) may be obtained from the set of algebraic equation. Substituting into (4), they must yield the equivalent dynamic quadratic nonlinear equations for the MHD velocity $v_z(z,t)$. In fact, these four algebraic equations determine four unknowns $A_1, \ldots, A_4$. The weakly nonlinear equation which governs the component of velocity v_z in all MHD wave modes takes the form

$$\frac{\partial v_z}{\partial t} + C\frac{\partial v_z}{\partial z} + CDv_z + \varepsilon v_z \frac{\partial v_z}{\partial z} = 0, \tag{27}$$

where

$$\begin{aligned} D &= \frac{S}{C c_0^2}\left(c_0^2 L_p + L_\rho\right), \\ \varepsilon &= \frac{3c_0^2 + C_A^2 (\gamma + 1) - C^2 (\gamma + 4)}{2\left(c_0^2 - 2C^2 + C_A^2\right)}. \end{aligned} \tag{28}$$

It may be readily obtained by application of one from four magnetosonic projectors, $P_{3,4,5,6}$, and it was firstly derived by other method in [11, 12]. Equation (27) describes individual dynamics of fast and slow magnetosonic perturbations. It does not reflect the nonlinear interaction between different modes. In the absence of magnetic field, these amendments coincide with the well-known terms which make the progressive Riemann's wave isentropic [20]. This is the case $B_0 = 0$, $C = c_0$, $C_A = C_{A,x} = C_{A,z} = 0$, and

$$\begin{aligned} A_1 &= -\frac{(\gamma - 3)\rho_0}{4c_0^2}, \\ A_2 &= 0, \\ A_3 &= \frac{\gamma + 1}{4}\rho_0, \\ A_4 &= 0. \end{aligned} \tag{29}$$

Equation (27) describes the nonlinear propagation of any MHD wave taken alone. It recalls dynamic equations for perturbations in other media which may be acoustically active [6, 7]. Equation (27) without accounting for heating-cooling function has been firstly derived and analyzed in the context of propagation of a saw-tooth impulse in [10]. Equation (27) may be readily rearranged into the leading-order pure nonlinear equation, if $D \neq 0$:

$$\frac{\partial V}{\partial Z} - \frac{\varepsilon}{C^2} V \frac{\partial V}{\partial \tau} = 0, \tag{30}$$

by means of new variables

$$V = v_z \exp(Dz),$$
$$Z = \frac{1 - e^{-Dz}}{D}, \quad (31)$$
$$\tau = t - \frac{z}{C}.$$

Equation (30) may be solved by the method of characteristics. Note that Z is always positive for nonzero D. It has been established that discontinuity in the waveform always forms in acoustically active media (that is the case of $D < 0$) and may not arise otherwise due to attenuation [6].

As for the Alfvén modes, they are not excited by the magnetosonic modes in the leading order. This has been discovered by Nakariakov and coauthors in [11]. The reason is that the Alfvén modes are determined exclusively by v_y and B'_y, while the magnetoacoustic mode is determined by all other field perturbations. Magnetoacoustic and Alfvén modes only weakly couple in the nonlinear flows. That means that there are only mixed quadratic nonlinear terms which may form "the driving forces" in interactions between modes and hence have impact on the nonlinear dynamics. The projecting also confirms conclusions about weak coupling of magnetosonic and Alfvén modes.

Application of P_7 at the system (4) leads to weakly nonlinear evolutionary equation with properly distributed quadratic nonlinear terms. By means of projecting equation (1) into the different subspaces which are designated by dispersion relations, one readily obtains the equation describing dynamics of excess density which specifies the entropy mode, ρ_7. The row P_7 projects the system onto relative dynamic equation, eliminating all foreign terms in the linear part of the final equation. Among all variety of quadratic nonlinear terms, only MHD terms belonging to one from four wave modes are kept. They form the magnetoacoustic forces of magnetosonic heating or cooling. As a result of application of P_7, one arrives at the equation which governs an excess density in the entropy mode:

$$\begin{aligned}
\frac{\partial \rho_7}{\partial t} + \frac{(\gamma+1) L_\rho}{c_0^2}\rho_7 &= \frac{(\gamma-1)\left(C_{A,x}C^4 + C_{A,z}\left(C^2 - c_0^2\right)^2\right)\left(3c_0^2 + (\gamma+1)C_A^2 - (\gamma+4)C^2\right)}{4C_{A,x}C^4 c_0^2\left(c_0^2 + C_A^2 - 2C^2\right)}\left(c_0^2 L_p + L_\rho\right)\rho_0 v_z^2 \\
&- \frac{\rho_0}{4C^2 c_0^2\left(C^4 - c_0^2 C_{A,z}^2\right)}\Big((\gamma-1)\left(C^2 c_0^2 (4-\gamma)\left(c_0^2 L_p + L_\rho\right)\right. \\
&+ C^4\left(L_\rho + 2\rho_0 c_0^2 L_{pp} + 2\rho_0 L_{\rho\rho} + c_0^2\left((2\gamma-1)L_p + 4\rho_0 L_{p\rho}\right)\right) \\
&- c_0^2 C_{A,z}^2\left((5-\gamma)L_\rho + 2c_0^4\rho_0 L_{pp} + 2\rho_0 L_{\rho\rho} + c_0^2\left((\gamma+3)L_p + 4\rho_0 L_{p\rho}\right)\right)\Big)v_z^2 + 2\left(2C^4\gamma(\gamma-1) - C^2 c_0^2(\gamma-1)^2\right. \\
&\left.- C_{A,z}^2 c_0^2\left(\gamma^2-1\right)\right)\left(c_0^2 L_p + L_\rho\right)\frac{\partial v_z}{\partial z}\int v_z\left(z', t\right)dz'\Big).
\end{aligned} \quad (32)$$

Excitation of an excess density in the entropy mode is described by (32), where C is one of four solutions to (27), that is, the speed of slow or fast MHD wave. Equation (32) includes only instantaneous perturbations and it is valid in description of heating excited by any type of magnetoacoustic force which is represented by its right-hand side. Equation (32) is written on in terms of excess density which specifies the entropy mode. The entropy mode is isobaric in accordance with (19). An excess temperature in the isobaric processes readily follows from the Clapeyron equation,

$$\frac{\rho_7}{\rho_0} = -\frac{T_7}{T_0}. \quad (33)$$

Isobaric cooling of a media is always followed by enlargement in its excess density, and isobaric heating corresponds to negative excess density.

3.1. Magnetoacoustic Heating for Different Cases of Heating-Cooling Function. Equation (32) is much more simple for analysis in the case of periodic or quasiperiodic magnetoacoustic perturbations. In the leading order,

$$\overline{v_z^2} = -\overline{\frac{\partial v_z}{\partial z}\int v_z dz}, \quad (34)$$

where top line denotes the temporal average over period of the magnetoacoustic wave. We reproduce averaged over the sound period form of (32) in some physically meaningful cases. The first example refers to the case when the heating-cooling function depends exclusively on temperature, $L = L(T)$. In this particular case, one may readily rearrange (32) into the dynamic equation

$$\begin{aligned}
&\overline{\frac{\partial \rho_7}{\partial t}} - \frac{(\gamma+1)}{C_V(\gamma-1)\gamma\rho_0}L_T\overline{\rho_7} \\
&= \frac{(\gamma-1)\left(3C_A^4 + 2(\gamma+1)c_0^4 + (3\gamma+7)c_0^2 C_A^2\right)}{4\gamma c_0^2\left(c_0^2 + C_A^2\right)^2 C_V} \\
&\cdot L_T\overline{v_z^2},
\end{aligned} \quad (35)$$

if $C_{A,z} = 0$, that is, in the case of magnetic strength perpendicular to the wave vector, when speeds of fast modes equal $C = \pm\sqrt{c_0^2 + C_A^2}$, and

$$\frac{\partial\overline{\rho_7}}{\partial t} - \frac{(\gamma+1)L_T}{C_V(\gamma-1)\gamma\rho_0}\overline{\rho_7} = \frac{\gamma^2-1}{2\gamma c_0^2 C_V}L_T\overline{v_z^2}, \tag{36}$$

in the case of parallel propagation of sound (this corresponds to $C_{A,z} = C_A$ and fast modes with speeds $C = \pm c_0$). They are readily integrated for zero initial condition with the result

$$\overline{\rho_7} = \exp\left(\frac{(\gamma+1)L_T}{C_V(\gamma-1)\gamma\rho_0}t\right) \cdot \int_0^t \exp\left(-\frac{(\gamma+1)L_T}{C_V(\gamma-1)\gamma\rho_0}t'\right)I(z,t')\,dt', \tag{37}$$

where $I(z,t)$ is the averaged over magnetosonic period right-hand force. This solution corresponds to zero initial ρ_7. In turn, $v_z(z,t)$ is a solution to (27). It may be approximately let sinusoidal function with constant I proportional to $\overline{v_z^2} = 0.5V_0^2$, where V_0 is the amplitude of v_z. In particular, (36) takes the form

$$\overline{\rho_7} = \left(\exp\left(\frac{(\gamma+1)L_T}{C_V(\gamma-1)\gamma\rho_0}t\right) - 1\right)\frac{V_0^2(\gamma-1)^2\rho_0}{4c_0^2}. \tag{38}$$

The case $L_T < 0$ reflects the enlargement in the background temperature which achieves some maximum with time which is independent on L_T:

$$\overline{T_7} = \frac{V_0^2(\gamma-1)^2 T_0}{4c_0^2}. \tag{39}$$

If $L_T > 0$, a gas is acoustically active. Linear conditions of isentropic and thermal instability in this particular case overlap. Sound enhances in a medium taking energy from the background. This is followed by isobaric cooling of a gas which is associated with enlargement of excess density specifying the entropy mode in time, $\rho_7 > 0$, and $T_7 < 0$. Magnetoacoustic forces in (35), (36) are positive, and theoretically they make excess entropy density to increase with time infinitely. This growth in fact is suppressed by the nonlinear transfer of energy between different modes in dependence on the ratio of their intensities and nonlinear attenuation of the wave modes. In (35), (36), the second-order derivative L_{TT} is omitted. That is valid if

$$|L_{TT}| \ll \frac{(\gamma-1)\gamma C_V}{c_0^2}|L_T| = \frac{\gamma^2}{T_0}|L_T|. \tag{40}$$

The dynamic equation becomes simple in the case of constant heating per unit mass ($L_p = 0$). In the cold molecular gas, the external heating occurs mainly due to grain photoelectrons and cosmic rays [3]. This is the case of $L_p = 0$. Dropping terms standing by the second derivatives L_{pp}, $L_{\rho\rho}$, and $L_{p\rho}$, we rearrange (27) as

$$\frac{\partial\overline{\rho_7}}{\partial t} + \frac{(\gamma+1)L_\rho}{c_0^2}\overline{\rho_7} = \frac{(\gamma-1)\left(3C_A^4 + 2(2\gamma-1)c_0^4 + (5\gamma+3)c_0^2C_A^2\right)\rho_0}{4c_0^4\left(c_0^2+C_A^2\right)^2} \cdot L_\rho\overline{v_z^2} \tag{41}$$

in the case of magnetic strength perpendicular to the wave vector and the force induced by the fast modes and

$$\frac{\partial\overline{\rho_7}}{\partial t} + \frac{(\gamma+1)L_\rho}{c_0^2}\overline{\rho_7} = \frac{(\gamma-1)(2\gamma-1)\rho_0}{2c_0^4}L_\rho\overline{v_z^2} \tag{42}$$

in the case of magnetic strength parallel to the wave vector and the force generated by the fast modes. Equations (41) and (42) are readily integrated for zero initial condition with the result

$$\overline{\rho_7} = \exp\left(-\frac{(\gamma+1)L_\rho}{c_0^2}t\right) \cdot \int_0^t \exp\left(\frac{(\gamma+1)L_\rho}{c_0^2}t'\right)I(z,t')\,dt', \tag{43}$$

with I standing in place of magnetoacoustic force of heating. The conditions of acoustic and thermal instabilities sound as $L_\rho > 0$ and $L_\rho < 0$.

The example of L which depends exclusively on pressure is also of importance. In particular, it specifies heating by coronal current dissipation. In this case, $L_\rho = 0$ and (32) may be rearranged in the dynamic equation

$$\frac{\partial\overline{\rho_7}}{\partial t} = \frac{(\gamma-1)\left(3C_A^4 + 2\gamma c_0^4 + (3\gamma+5)c_0^2C_A^2\right)\rho_0}{4c_0^2\left(c_0^2+C_A^2\right)^2}L_p\overline{v_z^2} \tag{44}$$

in the case of perpendicular direction of magnetic strength and

$$\frac{\partial\overline{\rho_7}}{\partial t} = \frac{\gamma(\gamma-1)\rho_0}{2c_0^2}L_p\overline{v_z^2} \tag{45}$$

in the case of parallel direction of magnetic strength. The magnetic gas is thermally neutral, but it is acoustically active if $L_p > 0$. That ensures positive magnetoacoustic forces in the right-hand sides of (44) and (45) and enlargement in time of entropy excess density which follows an isobaric cooling of a gas. Equations (44) and (45) are readily integrated with respect to time.

4. Concluding Remarks

The main result of this study is analysis of the nonlinear instantaneous dynamic equation (32), which describes excitation of the excess density in the entropy mode in the field of intense magnetosonic waves. This is the nonlinear phenomenon which takes place in flows with irreversible internal losses or/and due to inflow of energy into a system. We consider nonadiabaticity due to external heating-cooling balance exclusively. For proper description of the magnetoacoustic heating (or cooling), the dynamic equation which governs perturbations in the entropy mode is extracted from the general system of equations by means of projecting. The projecting in fact eliminates perturbations of foreign modes in the linear part of equation but results to coupling terms which form "the driving force" responsible for the nonlinear excitation. The procedure is instantaneous and leads to the instantaneous dynamic equation (32). Its right-hand side represents the magnetoacoustic force of heating (cooling), manifesting two origins of the phenomenon, nonlinearity and nonadiabaticity. Equation (32) is valid for periodic and aperiodic sound. The only limitations are

(1) the weak nonlinearity of a flow, so that only quadratic nonlinear terms are considered;

(2) intense exciting wave perturbations as compared to that of other modes;

(3) weak attenuation (or enhancement) of wave mode over its characteristic period in the course of propagation.

The first condition means smallness of the Mach number, and the last condition ensures the wave process and specifies in fact the spectrum of magnetoacoustic disturbances. There is no restriction concerning strength of the magnetic field in this study.

Perturbation in density which specifies the entropy mode is followed by the isobaric perturbation in temperature which in turn forms the new background of the magnetoacoustic waves. As it always happens to the open systems, a gas may be acoustically active in dependence on the heating-cooling function. That means that a magnetic gas may be cooled by MHD wave perturbations. If the condition of acoustical activity is not valid, the temperature of a gas enlarges as it occurs in fluids due to irreversible losses, that is, shear, bulk viscosity, and thermal conductivity, though the reasons for damping are different and a flow is described by the different equations. The condition of thermal stability prevents unlimited theoretical growth of absolute excess temperature and density in the entropy mode. In fact, this enlargement is suppressed for other reasons. Not to mention attenuation, the dynamic equations should be rearranged in view of new ratio of magnitudes of specific perturbations. The flow is determined by a number of other factors, apart from the heating-cooling function. There are the angle between the magnetic strength and the wave vector, the equilibrium magnetic strength, and the equilibrium thermodynamic state of a gas. As for the magnetoacoustic waves in the accepted geometry of a flow, their individual nonlinear dynamics without accounting for interaction with other modes is analyzed analytically in detail in the number of papers; see [12] and the reference therein. The recent studies consider also impact of the second partial derivatives of L on dynamics of MHD waves. The longitudinal velocity in MHD wave which contributes in the magnetosonic force must be solution to the corresponding dynamic equation (27).

Equations(35), (36), (41), (42), (44), and (45) are averaged over the period of magnetosonic perturbations for different kinds of L. They are valid for any angle between the equilibrium magnetic strength and wave number by means of linear magnetosonic speed C which is one of roots of (10) and may correspond to slow or fast wave. For example, we have considered fast periodic magnetosonic perturbations and particular cases of the heating-cooling function:

(1) L depends exclusively on temperature. In this case, the magnetoacoustic force of heating for perpendicular magnetic strength and wave vector is larger than in the case of parallel orientation, if $c_0 > \sqrt{(2\gamma-1)/(3-\gamma)}C_A$, that is, if the plasma β is larger than $2(2\gamma-1)/(\gamma(3-\gamma))$ ($\beta = (2/\gamma)c_0^2/C_A^2$ is the ratio of hydrostatic and magnetic pressures). The ratio of magnetoacoustic forces tends to 1 when c_0/C_A tends to infinity and achieves maximum $(4{,}5\gamma^3 + 18\gamma^2 + 30{,}5\gamma + 25)/4(\gamma+1)(\gamma+2)^2$. The next two cases refer to the heating-cooling function which depends on pressure and density, $L(p,\rho)$.

(2) $L_p = 0$; $L_\rho \neq 0$. In the case, the ratio of "perpendicular" and "parallel" magnetoacoustic forces tends to 1 when β tends to infinity achieving maximum value $(25\gamma^2 - 18\gamma + 33)/8(2\gamma-1)(\gamma+2)$. The ratio is greater than unity if $\beta > 2(4\gamma-5)/\gamma(7-3\gamma)$.

(3) $L_\rho = 0$; $L_p \neq 0$. The ratio of "perpendicular" and "parallel" magnetoacoustic forces also tends to 1 when c_0/C_A tends to infinity and achieves maximum value $(9\gamma^2 + 6\gamma + 25)/8\gamma(\gamma+2)$. It is greater than unity if $\beta > 2(2\gamma-3)/\gamma(5-\gamma)$. All estimations concern equal magnetoacoustic powers of "perpendicular" and "parallel" MHD perturbations (proportional to $\overline{v_z^2}$). All second derivatives of the heating-cooling function are dropped.

Magnetoacoustic heating in general leads to the nonuniformity of the equilibrium temperature of a plasma, that is, to the thermal lenses which in turn have impact on the speed of magnetosonic perturbations and dynamics of their wave fronts. These inhomogeneities may be of especial and substantive interest in the plasma's applications. The results of this study may be addressed to a hot atomic plasma with temperature greater than 10^4K and a cold molecular gas with temperature less than 10^3K and to different kinds of the function $L(p,\rho)$ which describes effects of nonadiabaticity. The radiation function also contributes in L. In particular, the optically thin hot plasma radiates proportionally to $\rho^{2-\eta}p^{\eta}$ [1]. Various models of coronal radiative losses and coronal heating may be found in [21] and references therein. The results may be useful in the inverse problems. The slow variations of equilibrium thermodynamic state may be measured remotely, which would provide data for analytical establishment of the kind of heating-cooling function in a plasma and magnetosonic perturbations. This may be of importance in many applications, also these relating to pulsed perturbations.

Theoretical predictions are of especial importance in view of the new studies which revealed existence of isentropically unstable regions in the solar corona with nonlinear enhancement of magnetoacoustic waves. The astrophysical media with adiabatic instability have been recently discussed in [22–28]. Papers [24, 26, 28–33] were devoted to the features of the nonlinear magnetosound waves distortions and nonlinear interaction of waves in a magnetic fluid flow. The acoustical activity and nonlinear phenomena in acoustically active media have counterparts in many flows with other kind of anomalous thermodynamic relaxation corresponding dispersion: in flows of gases with excited vibrational degrees of molecules and chemically reacting gases. That is why the analytical methods and results may find application in the similar problems of fluid flows, both well-known as these less studied. The plasma applications are much more difficult in view of involving of magnetic equations and strong dependence of fluid dynamics on the angle between particles velocity and the magnetic strength. In particular, thermal self-action and self-refraction, which concern propagation of divergent beams with varying direction of velocity can not be considered by the analytical methods successful in the unmagnetized flows [34, 35].

Conflicts of Interest

The author declares that they have no conflicts of interest.

References

[1] J. F. Vesecky, S. K. Antiochos, and J. H. Underwood, "Numerical modeling of quasi-static coronal loops. I - Uniform energy input," *The Astrophysical Journal*, vol. 233, no. 3, pp. 987–997, 1979.

[2] R. B. Dahlburg and J. T. Mariska, "Influence of heating rate on the condensational instability," *Solar Physics*, vol. 117, no. 1, pp. 51–56, 1988.

[3] H. I. S. Miguel and A. Parravano, "On the thermal structure and stability of configurations with heat diffusion and a gain-loss function. 3: Molecular gas," *The Astrophysical Journal*, vol. 424, no. 2, pp. 763–771, 1994.

[4] G. B. Field, "Thermal instability," *The Astrophysical Journal*, vol. 142, pp. 531–567, 1965.

[5] E. N. Parker, "Instability of thermal fields," *The Astrophysical Journal*, vol. 117, pp. 431–436, 1953.

[6] A. I. Osipov and A. V. Uvarov, "Kinetic and gasdynamic processes in nonequilibrium molecular physics," *Soviet Physics Uspekhi*, vol. 35, no. 11, pp. 903–923, 1992.

[7] N. E. Molevich, "Sound amplification in inhomogeneous flows of nonequilibrium gas," *Acoustical Physics*, vol. 47, no. 1, pp. 102–105, 2001.

[8] N. S. Anderson, "Longitudinal magneto-hydrodynamic waves," *The Journal of the Acoustical Society of America*, vol. 25, no. 3, pp. 529–532, 1953.

[9] E. A. Ponomarev, "On the propagation of low-frequency oscillations along the magnetic field in a viscous compressible plasma," *Soviet Astronomy*, vol. 5, pp. 673–677, 1961.

[10] V. D. Sharma, L. P. Singh, and R. Ram, "The progressive wave approach analyzing the decay of a sawtooth profile in magnetogasdynamics," *Physics of Fluids*, vol. 30, no. 5, pp. 1572–1574, 1987.

[11] V. M. Nakariakov, C. A. Mendoza-Briceño, and M. H. Ibáñez S., "Magnetoacoustic waves of small amplitude in optically thin quasi-isentropic plasmas," *The Astrophysical Journal*, vol. 528, no. 2, pp. 767–775, 2000.

[12] R. Chin, E. Verwichte, G. Rowlands, and V. M. Nakariakov, "Self-organization of magnetoacoustic waves in a thermally unstable environment," *Physics of Plasmas*, vol. 17, no. 32, pp. 107–118, 2010.

[13] D. I. Zavershinsky, N. E. Molevich, and D. S. Ryashchikov, "Structure of acoustic perturbations in heat-releasing medium," *Procedia Engineering*, vol. 106, pp. 363–367, 2015.

[14] G. Brodin, L. Stenflo, and P. K. Shukla, "Nonlinear interactions between kinetic alfvén and ion-sound waves," *Solar Physics*, vol. 236, no. 2, pp. 285–291, 2006.

[15] D. I. Zavershinsky and N. E. Molevich, "Alfvén wave amplification as a result of nonlinear interaction with a magnetoacoustic wave in an acoustically active conducting medium," *Technical Physics Letters*, vol. 40, no. 8, pp. 701–703, 2014.

[16] O. Lyubchyk and Y. Voitenko, "Nonlocal nonlinear coupling of kinetic sound waves," *Annales Geophysicae*, vol. 32, no. 11, pp. 1407–1413, 2014.

[17] A. Perelomova, "On the nonlinear effects of magnetoacoustic perturbations in a perfectly conducting viscous and thermo-conducting gas," *Acta Physica Polonica A*, vol. 130, no. 3, pp. 727–733, 2016.

[18] A. Perelomova, "On the nonlinear distortions of sound and its coupling with other modes in a gaseous plasma with finite electric conductivity in a magnetic field," *Archives of Acoustics*, vol. 41, no. 4, pp. 691–699, 2016.

[19] A. Perelomova, "Magnetoacoustic heating in a quasi-isentropic magnetic gas," *Physics of Plasmas*, vol. 25, no. 4, p. 042116, 2018.

[20] O. V. Rudenko and S. I. Soluyan, *Theoretical Foundations of Nonlinear Acoustics*, Plenum, New York, NY, USA, 1977.

[21] R. Rosner, W. H. Tucker, and G. S. Vaiana, "Dynamics of the quiescent solar corona," *The Astrophysical Journal*, vol. 220, pp. 643–665, 1978.

[22] N. E. Molevich, D. I. Zavershinsky, R. N. Galimov, and V. G. Makaryan, "Traveling self-sustained structures in interstellar clouds with the isentropic instability," *Astrophysics and Space Science*, vol. 334, no. 1, pp. 35–44, 2011.

[23] R. Soler, J. L. Ballester, and S. Parenti, "Stability of thermal modes in cool prominence plasmas," *Astronomy & Astrophysics*, vol. 540, A7, 2012.

[24] D. I. Zavershinskiy and N. E. Molevich, "Parametrical amplification of Alfvén waves in heat-releasing ionized media with magnetoacoustic instability," *Astrophysics and Space Science*, vol. 358, no. 1, p. 22, 2015.

[25] N. E. Molevich, D. S. Ryashchikov, and D. I. Zavershinskiy, "Influence of thermal conduction on MHD waves properties in thermally unstable plasma," *Magnetohydrodynamics*, vol. 52, no. 1, pp. 199–208, 2016.

[26] K. V. Krasnobaev, R. R. Tagirova, S. I. Arafailov, and G. Y. Kotova, "Evolution and saturation of Autowaves in photodissociation regions," *Astronomy Letters*, vol. 42, no. 7, pp. 460–473, 2016.

[27] D. S. Ryashchikov, N. E. Molevich, and D. I. Zavershinskii, "Characteristic times of acoustic and condensation instability in heat-releasing gas media," *Procedia Engineering*, vol. 176, pp. 416–422, 2017.

[28] K. V. Krasnobaev and R. R. Tagirova, "Isentropic thermal instability in atomic surface layers of photodissociation regions,"

Monthly Notices of the Royal Astronomical Society, vol. 469, no. 2, pp. 1403–1413, 2017.

[29] S. A. Belov, N. E. Molevich, and D. I. Zavershinskii, "Amplification of alfvén waves due to nonlinear interaction with a fast magnetoacoustic wave in acoustically active conductive media," *Technical Physics Letters*, vol. 44, no. 3, pp. 199–202, 2018.

[30] V. G. Makaryan and N. E. Molevich, "Stationary shock waves in nonequilibrium media," *Plasma Sources Science and Technology*, vol. 16, no. 1, pp. 124–131, 2007.

[31] N. E. Molevich, D. I. Zavershinskiy, and D. S. Ryashchikov, "Investigation of the MHD wave dynamics in thermally unstable plasma," *Magnetohydrodynamics* , vol. 52, no. 1, pp. 191–198, 2016.

[32] D. I. Zavershinsky and N. E. Molevich, "A magnetoacoustic autowave pulse in a heat-releasing ionized gaseous medium," *Technical Physics Letters*, vol. 39, no. 8, pp. 676–679, 2013.

[33] S. Kumar, V. M. Nakariakov, and Y.-J. Moon, "Effect of a radiation cooling and heating function on standing longitudinal oscillations in coronal loops," *The Astrophysical Journal*, vol. 824, no. 8, p. 8, 2016.

[34] A. Perelomova, "Thermal self-action effects of acoustic beam in a gas with reversible or irreversible chemical reaction," *Acta Acustica united with Acustica*, vol. 99, no. 3, pp. 352–358, 2013.

[35] A. Perelomova, "Self-refraction of acoustic pulses with shock fronts in some nonequilibrium media," *Canadian Journal of Physics*, vol. 92, no. 5, pp. 401–405, 2014.

8

Lump Solutions to the (3+1)-Dimensional Generalized B-Type Kadomtsev-Petviashvili Equation

Xifang Cao

School of Mathematical Sciences, Yangzhou University, China

Correspondence should be addressed to Xifang Cao; xfcao@yzu.edu.cn

Academic Editor: Antonio Scarfone

This paper is devoted to the study of lump solutions to the (3+1)-dimensional generalized B-type Kadomtsev-Petviashvili equation. First we use a direct method to construct a class of exact solutions which contain six arbitrary real constants. Then we use these solutions to generate lump solutions with four real parameters. We also determine the amplitude and velocity of these lumps.

1. Introduction

Lump is a type of localized rational solutions. Lump solutions for the (2+1)-dimensional Kadomtsev-Petviashvili I equation

$$\left(u_t + 6uu_x + u_{xxx}\right)_x - u_{yy} = 0 \tag{1}$$

were first obtained by Manakov et al. [1]. Then Satsuma and Ablowitz [2] constructed more lump solutions by taking a long wave limit of the corresponding soliton solutions. More recently, by use of the bilinear form, Ma [3] generalized the results of [1, 2] and got a larger class of lump solutions. Over the past few decades, lump has been an active area in the study of nonlinear evolution equations. They can be used to describe nonlinear patterns in plasma [4], in optical media [5], in the Bose-Einstein condensate [6], and so forth. Lump solutions have been obtained for some other equations, such as the Ishimori equation [7], the Jimbo-Miwa equation [8], and the Sawada-Kotera equation [9].

Recently much literature is devoted to the study of (3+1)-dimensional B-type Kadomtsev-Petviashvili (BKP) equation

$$u_{yt} + 3u_{xz} - 3u_x u_{xy} - 3u_{xx}u_y - u_{xxxy} = 0, \tag{2}$$

which can be used to model fluid dynamics, plasma physics, and weakly dispersive media [10, 11]. Various methods have been applied to (2) to construct its soliton and multiple wave solutions [12–14]. In this paper, we develop another method to construct its lump solutions. We first use a direct method to obtain a class of exact solutions which contain six arbitrary real constants (see Theorem 1). Then we show that the limits of these solutions can generate lump solutions (see Theorem 4). We also determine the amplitude and velocity of these lumps.

2. Exact Solutions

In this section, we seek exact solutions to the (3+1)-dimensional BKP equation (2) in the following form:

$$u(x, y, z, t) = 2(\ln f)_x, \tag{3}$$

where

$$f(x, y, z, t) = \exp(-\eta) + m\cos\xi + n\exp\eta, \tag{4}$$

$\eta = p(x+qy+rz)$, $\xi = p(ax+by+cz+dt)$, and $m, n, p, q, r, a, b, c, d$ are real constants to be determined. Substituting (3) into the left hand side of (2) and then factoring we get the following three determining equations:

$$\begin{aligned} &6ac - 3r + bd + 3a^2r + p^2q + 4a^3bp^2 - 6a^2p^2q \\ &\quad + a^4p^2q + adq - 4abp^2 = 0, \end{aligned} \tag{5}$$

$$\begin{aligned} &3c + 6ar + dq - 3a^2c - bp^2 - 4ap^2q + 6a^2bp^2 - a^4bp^2 \\ &\quad + 4a^3p^2q - abd = 0, \end{aligned} \tag{6}$$

$$m^2\left(3r+18ac+3bd+9a^2r-p^2q+24a^3bp^2 - 6a^2p^2q+3a^4p^2q+3adq\right)+16n\left(4p^2q-3r\right) = 0. \quad (7)$$

Solving for r, c from (5) and (6) gives

$$r = \frac{p^2q+bd-a^4p^2q-2a^3bp^2-2abp^2-adq}{3(a^2+1)}, \quad (8)$$

and

$$c = \frac{bp^2+2a^3p^2q+2ap^2q-a^4bp^2-abd-dq}{3(a^2+1)}. \quad (9)$$

Substituting (8) and (9) into (7) yields

$$n = -\frac{3ba^5p^2+2qa^4p^2+4ba^3p^2+2qa^2p^2+bap^2-dqa+bd}{4(qa^4p^2+2ba^3p^2+4qa^2p^2+2bap^2+3qp^2+dqa-bd)} \cdot m^2. \quad (10)$$

Therefore, we have the following.

Theorem 1. *The (3+1)-dimensional BKP equation (2) admits the following exact solutions:*

$$u(x,y,z,t) = 2(\ln f)_x, \qquad f(x,y,z,t) = \exp(-\eta)+m\cos\xi+n\exp\eta, \quad (11)$$

where $\eta = p(x+qy+rz)$, $\xi = p(ax+by+cz+dt)$, m, p, q, a, b, d *are arbitrary real constants, and* r, c, n *are given by (8), (9), and (10), respectively.*

3. Lump Solutions

In this section, we use (11) to construct lump solutions to the (3+1)-dimensional BKP equation (2). We first give some examples and then give a general result.

Example 1. Let $m=-2$, $q=1$, $a=0$, $b=1$, $d=1$. Then (11) becomes

$$u(x,y,z,t) = 2(\ln f)_x, \qquad f(x,y,z,t) = \exp(-\eta)-2\cos\xi+\frac{1}{1-3p^2}\exp\eta, \quad (12)$$

where $\eta = p(x+y+((1+p^2)/3)z)$, $\xi = p(y+((p^2-1)/3)z+t)$. As $p \longrightarrow 0$, the limit of (12) gives a lump solution to (2):

$$U(x,y,z,t) = \lim_{p\to 0} u(x,y,z,t) = \frac{4X}{3+X^2+Y^2}, \quad (13)$$

where

$$X = x+y+\frac{z}{3}, \qquad Y = y-\frac{z}{3}+t. \quad (14)$$

The lump (13) attains its maximum value $2/\sqrt{3}$ at $(X,Y) = (\sqrt{3},0)$ and minimum value $-2/\sqrt{3}$ at $(X,Y) = (-\sqrt{3},0)$. So its amplitude is $2/\sqrt{3}$. The profiles of the lump (13) in the (X,Y)-space are given in Figure 1.

From the point of view of the (x,y,z)-space, at each time t, the lump (13) attains its extreme values at the lines

$$x+y+\frac{z}{3} = \pm\sqrt{3}, \qquad -3y+z = 3t. \quad (15)$$

Note that the planes $x+y+z/3 = \pm\sqrt{3}$ are fixed (do not depend on t), while the plane $-3y+z=3t$ moves along its normal vector $\mathbf{N} = (1/\sqrt{10})(0,-3,1)$ with the speed $3/\sqrt{10}$. Therefore the velocity of the moving lines (15) is

$$\mathbf{V} = \left(V_x, V_y, V_z\right) = \frac{63}{5\sqrt{266}}(8,-11,9). \quad (16)$$

Example 2. Let $m=-2$, $q=0$, $a=1$, $b=1$, $d=1$. Then (11) becomes

$$u(x,y,z,t) = 2(\ln f)_x, \qquad f(x,y,z,t) = \exp(-\eta)-2\cos\xi+\frac{1+8p^2}{1-4p^2}\exp\eta, \quad (17)$$

where $\eta = p(x+((1-4p^2)/6)z)$, $\xi = p(x+y-(1/6)z+t)$. As $p \longrightarrow 0$, the limit of (17) gives a lump solution to (2):

$$U(x,y,z,t) = \lim_{p\to 0} u(x,y,z,t) = \frac{4(X+Y)}{12+X^2+Y^2}, \quad (18)$$

where

$$X = x+\frac{z}{6}, \qquad Y = x+y-\frac{z}{6}+t. \quad (19)$$

The lump (18) attains its maximum value $\sqrt{6}/3$ at $(X,Y) = (\sqrt{6},\sqrt{6})$ and minimum value $-\sqrt{6}/3$ at $(X,Y) = (-\sqrt{6},-\sqrt{6})$. So the amplitude of the lump (18) is $\sqrt{6}/3$. The profiles of the lump (18) in the (X,Y)-space are given in Figure 2.

From the point of view of the (x,y,z)-space, at each time t, the lump (18) attains its extreme values at the lines

$$x+\frac{z}{6} = \pm\sqrt{6}, \qquad x+y-\frac{z}{6}+t = \pm\sqrt{6}, \quad (20)$$

which move with the velocity

$$\mathbf{V} = \left(V_x, V_y, V_z\right) = \frac{1476}{73\sqrt{1517}}(-2,-37,12). \quad (21)$$

Example 3. Let $m=-2$, $q=1$, $a=-1$, $b=0$, $d=1$. Then (11) becomes

$$u(x,y,z,t) = 2(\ln f)_x, \qquad f(x,y,z,t) = \exp(-\eta)-2\cos\xi+\frac{1+4p^2}{1-8p^2}\exp\eta, \quad (22)$$

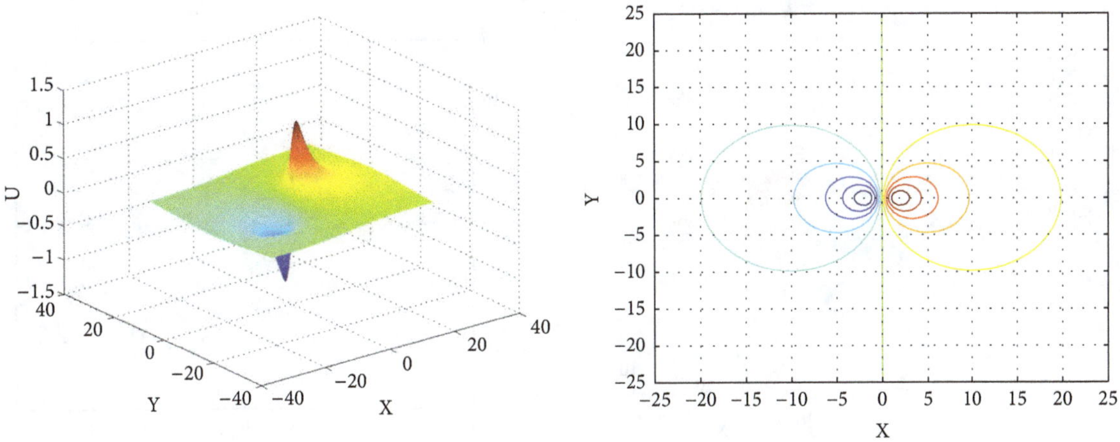

FIGURE 1: The 3D plot of lump (13) and the corresponding contour plot.

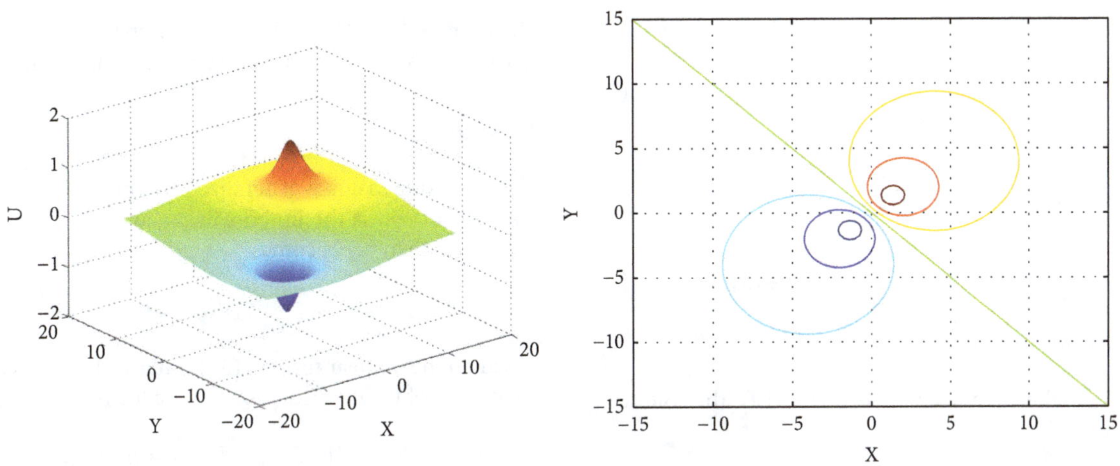

FIGURE 2: The 3D plot of lump (18) and the corresponding contour plot.

where $\eta = p(x+y+(1/6)z), \xi = p(-x+((4p^2-1)/6)z+t)$. As $p \longrightarrow 0$, the limit of (22) gives a lump solution to (2):

$$U(x,y,z,t) = \lim_{p\longrightarrow 0} u(x,y,z,t) = \frac{4(X-Y)}{12+X^2+Y^2}, \quad (23)$$

where

$$X = x+y+\frac{z}{6}, \qquad Y = -x-\frac{z}{6}+t. \quad (24)$$

The lump (23) attains its maximum value $\sqrt{6}/3$ at $(X,Y) = (\sqrt{6}, -\sqrt{6})$ and minimum value $-\sqrt{6}/3$ at $(X,Y) = (-\sqrt{6}, \sqrt{6})$. So the amplitude of the lump (23) is $\sqrt{6}/3$. The profiles of the lump (23) in the (X,Y)-space are given in Figure 3.

From the point of view of the (x,y,z)-space, at each time t, the lump (23) attains its extreme values at the lines

$$x+y+\frac{z}{6} = \pm\sqrt{6}, \qquad -x-\frac{z}{6}+t = \mp\sqrt{6}, \quad (25)$$

which move with the velocity

$$\mathbf{V} = (V_x, V_y, V_z) = \frac{1332}{37\sqrt{2701}}(36,-37,6). \quad (26)$$

In general, for arbitrary real constants q, a, b, d, let $m = -2$. Then as $p \longrightarrow 0$, the limit of (11) yields the following result.

Theorem 4. *The (3+1)-dimensional BKP equation (2) admits the following rational solutions:*

$$U(x,y,t) = 4\frac{X+aY}{A+X^2+Y^2}, \quad (27)$$

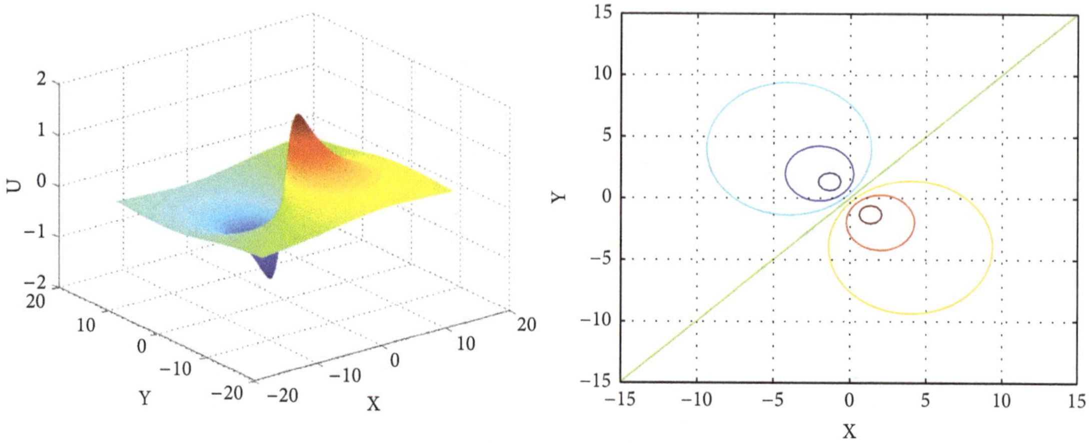

FIGURE 3: The 3D plot of lump (23) and the corresponding contour plot.

where

$$A = \frac{3(a^2+1)^2(q+ab)}{d(b-aq)}, \quad (28)$$

$$X = x + qy - \frac{d(aq-b)}{3(a^2+1)}z, \quad (29)$$

$$Y = ax + by - \frac{d(q+ab)}{3(a^2+1)}z + dt, \quad (30)$$

and q, a, b, d are arbitrary real constants.

Remark 5. If the parameters q, a, b, d satisfy the following condition,

$$\frac{q+ab}{d(b-aq)} > 0, \quad (31)$$

then (27) are lump solutions.

For given parameters q, a, b, d satisfying condition (31), the lump (27) attains its extreme values $\pm 2\sqrt{d(b-aq)/3(a^2+1)(q+ab)}$ at the lines

$$x + qy - \frac{d(aq-b)}{3(a^2+1)}z = \pm\sqrt{\frac{3(a^2+1)(q+ab)}{d(b-aq)}},$$
$$ax + by - \frac{d(q+ab)}{3(a^2+1)}z + dt = \pm a\sqrt{\frac{3(a^2+1)(q+ab)}{d(b-aq)}} \quad (32)$$

in the (x, y, z)-space. Therefore the amplitude of the lump (27) is

$$U_{\text{amplitude}} = 2\sqrt{\frac{d(b-aq)}{3(a^2+1)(q+ab)}}. \quad (33)$$

The velocity $\mathbf{V} = (V_x, V_y, V_z)$ of the lump (27) in the (x, y, z)-space can also be determined. Here we omit the details.

4. Discussion

The (3+1)-dimensional generalized B-type Kadomtsev-Petviashvili equation is the Jimbo-Miwa equation. If we replace t with $-t/2$, then (2) becomes

$$u_{xxxy} + 3u_y u_{xx} + 3u_x u_{xy} + 2u_{yt} - 3u_{xz} = 0. \quad (34)$$

Based on the bilinear form of (34), Yang and Ma [8] construct ten classes of its lump solutions in the form

$$u = 2(\ln f)_x, \quad f = g^2 + h^2 + a_{11}, \quad (35)$$

where $g = a_1x + a_2y + a_3z + a_4t + a_5$, $h = a_6x + a_7y + a_8z + a_9t + a_{10}$, and a_i $(i = 1, \ldots, 11)$ are some constants.

In this section, we point out that up to the invariance after a translation and a Galilean transformation, all the solutions obtained in [8] can be simplified and be rewritten in a unified form of (27). This is because, firstly, the parameters a_5 and a_{10} can be set zero by the translation $(x, y, z, t) \longrightarrow (x', y', z', t')$ defined by

$$\begin{aligned} x' &= x + x_0, \\ y' &= y + y_0, \\ z' &= z + z_0, \\ t' &= t + t_0, \end{aligned} \quad (36)$$

where x_0, y_0, z_0, t_0 are some constants; secondly, since the parameters a_1 and a_6 can not be zero simultaneously, without loss of generality, we may assume $a_1 \neq 0$, and then from the form of (35), we can set $a_1 = 1$; thirdly, since (34) is invariant under the Galilean transformation $(x, y, z, t) \longrightarrow (x', y', z', t')$ given by

$$\begin{aligned} x' &= x + 3\lambda t, \\ y' &= y + 2\lambda z, \\ z' &= z, \\ t' &= t, \end{aligned} \tag{37}$$

where λ is an arbitrary real number, we can set $a_4 = 0$. Then under the condition

$$\begin{aligned} a_1 &= 1, \\ a_4 &= 0, \\ a_5 &= 0, \\ a_{10} &= 0, \end{aligned} \tag{38}$$

g and h are of the forms of (29) and (30), and a direct calculation shows that each class of solutions in [8] can be rewritten in the unified form of (27).

From the mathematical point of view, the lump solutions (27) are limits of the exact solutions (11) as the parameter p approaches zero. It is worth studying the physical meaning of this limit.

Conflicts of Interest

The author declares that there are no conflicts of interest.

Acknowledgments

This work is supported by the Natural Science Foundation of Jiangsu Province through BK20151304.

References

[1] S. V. Manakov, V. E. Zakharov, L. A. Bordag, A. R. Its, and V. B. Matveev, "Two-dimensional solitons of the Kadomtsev-Petviashvili equation and their interaction," *Physics Letters A*, vol. 63, no. 3, pp. 205-206, 1977.

[2] J. Satsuma and M. J. Ablowitz, "Two-dimensional lumps in nonlinear dispersive systems," *Journal of Mathematical Physics*, vol. 20, no. 7, pp. 1496–1503, 1979.

[3] W. X. Ma, "Lump solutions to the Kadomtsev-Petviashvili equation," *Physics Letters A*, vol. 379, no. 36, pp. 1975–1978, 2015.

[4] M. J. Ablowitz and H. Segur, *Solitons and the Inverse Scattering Transform*, SIAM, Philadelphia, PA, USA, 1981.

[5] D. E. Pelinovsky, Y. A. Stepanyants, and Y. S. Kivshar, "Self-focusing of plane dark solitons in nonlinear defocusing media," *Physical Review E: Statistical, Nonlinear, and Soft Matter Physics*, vol. 51, no. 5, pp. 5016–5026, 1995.

[6] V. A. Mironov, A. I. Smirnov, and L. A. Smirnov, "Structure of vortex shedding past potential barriers moving in a Bose-Einstein condensate," *Journal of Experimental and Theoretical Physics*, vol. 110, no. 5, pp. 877–889, 2010.

[7] K. Imai, "Dromion and lump solutions of the Ishimori-I equation," *Progress of Theoretical Physics*, vol. 98, no. 5, pp. 1013–1023, 1997.

[8] J. Y. Yang and W. X. Ma, "Abundant lump-type solutions of the Jimbo-Miwa equation in (3+1)-dimensions," *Computers and Mathematics with Applications*, vol. 73, no. 2, pp. 220–225, 2017.

[9] L. L. Huang and Y. Chen, "Lump solutions and interaction phenomenon for (2 + 1)-dimensional Sawada-Kotera equation," *Communications in Theoretical Physics*, vol. 67, no. 5, pp. 473–478, 2017.

[10] E. Date, M. Jimbo, M. Kashiwara, and T. Miwa, "Transformation groups for soliton equations VI-KP hierarchies of orthogonal and symplectic type," *Journal of the Physical Society of Japan*, vol. 50, no. 11, pp. 3813–3818, 1981.

[11] E. Date, M. Jimbo, M. Kashiwara, and T. Miwa, "Transformation groups for soliton equations: IV. A new hierarchy of soliton equations of KP-type," *Physica D: Nonlinear Phenomena*, vol. 4, no. 3, pp. 343–365, 1982.

[12] W. X. Ma and E. G. Fan, "Linear superposition principle applying to Hirota bilinear equations," *Computers and Mathematics with Applications*, vol. 61, pp. 950–959, 2011.

[13] W. X. Ma and Z. Zhu, "Solving the (3 + 1)-dimensional generalized KP and BKP equations by the multiple exp-function algorithm," *Applied Mathematics and Computation*, vol. 218, no. 24, pp. 11871–11879, 2012.

[14] S. Liu, Y. Ding, and J. G. Liu, "A class of exact solutions of (3+1)-dimensional generalized B-type Kadomtsev–Petviashvili equation," *International Journal of Nonlinear Sciences and Numerical Simulation*, vol. 18, no. 2, pp. 137–144, 2017.

9

Vortex Structures inside Spherical Mesoscopic Superconductor Plus Magnetic Dipole

A. Ludu

Department of Mathematics, Embry-Riddle Aeronautical University, Daytona Beach, FL 32114, USA

Correspondence should be addressed to A. Ludu; ludua@erau.edu

Academic Editor: Alexander Iomin

We investigate the existence of multivortex states in a superconducting mesoscopic sphere with a magnetic dipole placed at the center. We obtain analytic solutions for the order parameter $\Psi(\vec{r})$ inside the sphere through the linearized Ginzburg-Landau (GL) model, coupled with mixed boundary conditions, and under regularity conditions and decoupling coordinates approximation. The solutions of the linear GL equation are obtained in terms of Heun double confluent functions, in dipole coordinates symmetry. The analyticity of the solutions and the associated eigenproblem are discussed thoroughly. We minimize the free energy for the fully nonlinear GL system by using linear combinations of linear analytic solutions, and we provide the conditions of occurring multivortex states. The results are not restricted to the particular spherical geometry, since the present formalism can be extended for large samples, up to infinite superconducting space plus magnetic dipole.

1. Introduction

The rapidly growing field of quantum computation requires nanoscale miniaturization of electronic circuits, way beyond the silicon era type of devices [1]. Therefore, the mesoscopic superconductors, having the size comparable to the coherence length or the magnetic penetration depth [2], are the prime candidates for construction of nanodevices among all other superconducting systems. Mesoscopic physics revealed a number of open fundamental problems like quantum confinement, quantum vortices and loops, spintronics, quantum dots, etc. [3] whose solutions can bring significant knowledge in fields like nanotechnology, synthesis of new materials, novel sensors, modern lithography, or molecular biology [4–6].

The most important feature of a mesoscopic superconductor is that its shape and size have an important effect on the interplay of the magnetic field and superconducting condensate. The properties of mesoscopic superconductors are very different compared to those of bulk superconductors. While in bulk superconductors penetrating vortices form a lattice due to the vortex-vortex repulsion, in mesoscopic superconductors there is a competition between the vortex lattice and the boundary which tries to impose its geometry on the vortex lattice. It is observed experimentally that flux quantum configurations have the same symmetry as the symmetry of the shape of the sample in a homogeneous magnetic field [7]. Such systems are studied via the Ginzburg-Landau (GL) model. The GL equations arise from the Euler-Lagrange equations for the free Gibbs energy for a mesoscopic superconductor sample in magnetic field. These equations must be solved under specific boundary conditions: the normal component of the superconducting current is equal to zero [8]. Near and below the transition temperature, theoretical calculations show that anti-vortex and giant-vortex can appear in order to maintain the symmetric vortex configuration.

The response of mesoscopic superconductor samples of different shapes (thin discs, spheres, cones, and rings) to an external magnetic field, as well as the effect of the geometry, has been theoretically [9–18] and experimentally [7] studied. In all these cases a constant external magnetic field is applied along the the revolution axis. The small volume to surface ratio of these mesoscopic structures brings new features not found in the bulk: there are two kinds of superconducting states, depending on the strength of the magnetic field, the sample geometry (external surface), and its size: giant and multivortex states. The giant vortex state has cylindrical symmetry and is the only kind stable in small size superconductors due to the confinement effect [13, 14]. If the size of the sample increases, such giant vortex states can break up into multivortices through saddle-point transitions [17, 18]. For three-dimensional objects (sphere or cone) the vortex lines need to intersect the surface perpendicularly in order to cancel the outward supercurrent component [9–12]. Consequently, the shape of the lines is strongly affected by the sample surface. For example, in the case of a mesoscopic sphere placed in uniform external magnetic field, the vortex lines are curved inside, packing denser in the equatorial plane, and spreading out towards the poles [10, 11].

In this paper we consider a different situation where the magnetic field is not anymore externally generated but is generated from inside the sample, *e.g.*, an infinitesimal magnetic dipole placed at the center of a mesoscopic sphere. Such a configuration can generate confined vortex loops. The topological transition between open and closed vortex loops is controlled by the geometry, *i.e.*, R, and the central magnetic dipole strength.

The goal of this paper is to demonstrate the occurrence of multivortex structures in the superconducting sphere plus magnetic dipole configuration, especially below and around the transition temperature. Our approach is based on the GL model of free energy for a finite volume V. Outside of this volume the Cooper pair density $\Psi(\vec{r})$ describing the superconducting phase, called the order parameter, is zero [5, 6, 9–18]. The free energy functional is given in the GL model by [10, 11]

$$F = \frac{1}{V}\int_V \left[\frac{\left|\vec{P}(\Psi)\right|^2}{2m} + \alpha|\Psi|^2 + \frac{\beta}{2}|\Psi|^4 + \frac{\left|\vec{h}\right|^2}{8\pi}\right] dv, \quad (1)$$

where m is the Cooper pairs mass, $\vec{P}$ is the quantum momentum operator in the presence of magnetic field, and $\vec{h}$ is the intensity of the magnetic field. The temperature dependent coefficient function $\alpha(T) < 0$ and the nonlinear term coupling constant $\beta > 0$ are typical Landau second-order phase transition parameters [9–12]. For mesoscopic samples, one can neglect the term responsible of the expulsion of magnetic flux from the superconductor, that is last term in Eq. (1) [9–12].

The traditional procedure for finding Ψ by minimizing the free energy functional Eq. (1) for constant volume V consists in expanding the order parameter Ψ in a basis of eigenfunctions of the corresponding linearized GL problem [10, 11, 16–18] and numerically evaluating the expansion coefficients which minimize the full nonlinear functional Eq. (1).

In this paper we solve the GL linear problem analytically and investigate the properties of the eigenfunctions and spectrum. The contribution of the infinitesimal magnetic dipole will be approached in the dipole system of coordinates. The linearized GL equation factorizes in two ordinary differential equations, for the two orthogonal dipole coordinates, respectively. From the physical point of view such a factorization seems natural because far enough from the sphere surface, the abstract surfaces containing vortex lines follow the magnetic field stream lines, but these are exactly the dipole magnetic field lines. Along these surfaces the order parameter has slow variation or is practically constant. This gives sufficient physical reason for neglecting higher order terms in the dipole variable going along the magnetic field lines. This approximation allows integrating the two ordinary differential equations. The linear solution consists in a product of angular momentum eigenfunctions in the azymuthal coordinate, exponential function for one of the dipole coordinates, and a double confluent Heun function in the third coordinate. The final step is to come back to the fully nonlinear GL problem, and write Ψ as a linear combination of eigenfunctions of the linear GL problem, with arbitrary coefficients, and then minimize the free energy with respect to these coefficients.

We dedicate a large part of the present calculations to solve exactly the linearized GL equation and to ensure the completeness and orthogonality of the linear basis because near and below the transition temperature, even the linear GL equation is sufficient to describe multivortex states. The order parameter Ψ is very small in this range, and higher order terms of Ψ can be neglected. Nevertheless, at lower temperatures, the vortex configuration does not have to match the symmetry of the system, and higher order terms of GL equation cannot be negligible [19]. It is the contribution of these nonlinear terms which generates the multivortex states at lower temperatures.

The paper is organized as follows. In Section 2 we formulate the nonlinear and auxiliary linear GL problem and write the partial differential equation associated with the GL problem. In Section 3 we discuss the importance of the infinitesimal central magnetic dipole from a potential aspect, introduce the dipole coordinates, and obtain general form of the dipole equation in azimuthal symmetry plus dipole coordinates. In Section 4 we reduce the general dipole equation to a double confluent Heun equation (DCHE) by the help of a geometric approximation and a decoupling of the dipole orthogonal modes. We obtain analytic solutions for the DCHE as Heun series around the point at infinity, and we present some examples. In Section 5 we show how the dipole equation plus the physical boundary conditions can be brought to a Sturm-Liouville problem, and we solve the associated GL linear eigenproblem and find the eigenvalues spectrum. Examples of spectra for different configurations are presented. In Section 6 we describe how one can use the exact solutions and spectra of the linear GL problem to build approximate solutions for full nonlinear GL problem, by minimizing the free energy. We describe the procedure to identify multivortex states, give the sufficient criteria, and provide an example of equipotential surfaces with vortex structure inside the sphere.

2. Ginzburg-Landau Equation for Dipole inside Sphere

In this section we obtain the governing ODEs for the GL problem given in Eq. (1) for a mesoscopic superconducting sphere in a local magnetic field $\vec{h} = \vec{\nabla} \times \vec{A}$ produced by a infinitesimal dipole placed at its center $\vec{r} = 0$. The Euler-Lagrange equations for the free energy functional Eq. (1), written in the absence of the last term in $\vec{h}^2$ as explained above, are known as the GL equation [9–12]

$$\frac{1}{2m}\vec{P}^2\Psi + \alpha(T)\Psi + \beta|\Psi|^2\Psi = 0, \tag{2}$$

$$\nabla \times \vec{h} = \frac{4\pi\vec{J}}{c}, \tag{3}$$

$$\vec{J} = \frac{q}{2m}\left[\Psi^*\vec{P}\Psi + \left(\vec{P}\Psi\right)^*\Psi\right], \tag{4}$$

where the electromagnetic momentum $\vec{P}$ is given by

$$\vec{P} = -i\hbar\vec{\nabla} - \frac{q}{c}\vec{A}, \tag{5}$$

where $\vec{A}$ is the vector potential, q is the charge of the Cooper pair, and $\vec{J}$ is the current density. The main approach for the nonlinear problem is to minimize the free energy, Eq. (1), with test functions chosen from a basis of eigenfunctions of the associated linear problem. The linearized GL equation (LGL) is obtained by neglecting the last term of Eq. (2)

$$\frac{1}{2m}\left(-i\hbar\vec{\nabla} - \frac{q}{c}\vec{A}\right)^2\Psi = \alpha(T)\Psi. \tag{6}$$

We rewrite Eq. (6) in the London electromagnetic gauge, $\nabla \cdot \vec{A} = 0$, and we rescale the variables $\vec{r} \longrightarrow \vec{r}\sqrt{-2m\alpha(T)}/\hbar$, $\vec{A} \longrightarrow \vec{A}q/(c\sqrt{-2m\alpha(T)})$ and $\Psi \longrightarrow \Psi\sqrt{-\beta/\alpha(T)}$. We follow the GL phenomenological model from [9–18] and use the temperature dependence

$$\alpha = \alpha_0\left(1 - \frac{T}{T_{c0}}\right), \qquad \xi(T) = \frac{\hbar}{\sqrt{-2m\alpha(T)}}, \tag{7}$$

where T_{c0} is the critical temperature in absence of magnetic field and $\xi(T)$ is the coherence length. In the new variables the LGL Eq. (6) can be written in the form of a linear operator equation

$$\widehat{\mathcal{O}}\Psi = \Delta\Psi - 2i\vec{A}\cdot\nabla\Psi - A^2\Psi + \Psi = 0. \tag{8}$$

Since the dipole magnetic field has axial symmetry we can use any axially symmetric orthogonal coordinates in azimuth angle φ in the form $\vec{r} \longrightarrow (\varphi, a, b)$. In this case the φ dependence in Eq. (8) can be separated, and $\widehat{\mathcal{O}}$ has $\exp(iL\varphi)$ as eigenfunctions labeled by the angular momentum L. This decomposes the space of solutions $\Psi(\vec{r})$ into subspaces parameterized by L

$$\mathscr{L}^2\left(R^3\right) = \bigoplus_{L\in N} V_L, \quad V_L = \left\{\frac{\Psi}{\Psi(\vec{r})} = \Phi(a,b)\,e^{iL\varphi}\right\}, \tag{9}$$

with $\Phi \in \mathscr{L}^2(R^2)$ square integrable function. The decomposition in Eq. (9) reduces Eq. (6) to a two-dimensional PDE. On each of these L subspaces the operator $\widehat{\mathcal{O}}$ has a particular form depending on L denoted by $\widehat{\mathcal{O}}_L$, and for each such restricted operator we can attach an eigenvalue problem

$$\widehat{\mathcal{O}}_L\Phi_{E,\kappa} = E\Phi_{E,\kappa}, \tag{10}$$

where κ is a provisional degeneracy label, which later on becomes the radial quantum number. Of course, E and $\Phi_{E,\lambda}$ carry the L dependence.

The vector potential associated with the infinitesimal (point-like) magnetic dipole can be expressed in a simple form in spherical coordinates

$$\vec{A} = \frac{\mu\sin\theta}{r^2}\vec{e}_\varphi, \tag{11}$$

where $\vec{e}_\varphi = (-\sin\varphi, \cos\varphi, 0)$ is the unit vector in the φ direction. The boundary conditions associated with this problem request the order parameter wave function to cancel at the origin and the flow at the mesoscopic volume surface Σ to be zero (the Saint-James-de Gennes conditions) [9–12, 15–18]

$$\Psi|_{r=0} = 0, \qquad \left.\frac{\partial\Psi}{\partial n}\right|_\Sigma = 0, \tag{12}$$

where $\vec{n}$ represents the outward normal to the spherical surface Σ defined by $|\vec{r}| = R$.

We can make coordinate free general statements concerning the spectrum of the eigenfunction problem in Eq. (10). Following the Leinfelder-Simader theorem [20] and taking into account that the integral over $\mathbf{R}^3 \setminus \{0\}$ of the forth power of the magnitude of the dipole vector potential is finite, we find out that the operator in Eq. (6) is essentially self-adjoint if applied on analytic functions. Moreover, from the Miller-Simon theorem [21], we know the eigenvalue problem of such type of magnetic multipole potential vector operators is given by a positive, unbounded from above, continuous spectrum. However, given the two-point boundary conditions, Eqs. (12), we expect the reduction of this spectrum to a discrete and bounded spectrum for energy.

3. The Generalized Dipole Equation

In this section we present analytic solutions for the system Eqs. (6), (8), and (10) with boundary conditions given by Eq. (12). The presence of the magnetic dipole renders the spherical or the cylindrical coordinates unsuitable to factorize the LGL Eqs. (8) and (10) because of the terms mixing inside the square of the electromagnetic momentum. Attempts to solve similar equations in spherical coordinates in the presence of an electric dipole have been made. In [22], for example, the Schrödinger equation is separable, and the authors find exact wave functions in terms of Bessel and Mathieu functions for $m = 0$ spherical modes in the presence of a dipolar external electric field. A similar generalized

Coulomb problem for a class of general Natanzon confluent potentials is exactly solved in [23] by reducing the corresponding system to confluent hypergeometric differential equations. More recently, in [24], the authors succeeded to solve the eigenvalue wave equation for an electron in the field of a molecule with an electric dipole moment by expanding the solutions of a second-order Fuchsian differential equation with regular singularities in $\cos\theta = 1, -1, 0$ in a series of Jacobi polynomials with "dipole polynomial" coefficients.

In all these situations separation and integration of the Schrödinger equations are possible because in spherical (cylindrical) coordinates the resulting second-order ordinary linear differential equations are of Fuchsian type, with maximum three regular singularities. Such equations can be mapped into several types of hypergeometric differential equations.

However, the presence of the dipole magnetic field makes the situation more complicated because the order of singularity growths above the hypergeometric one, and consequently the dynamics is governed by differential equations of Heun, Hill, or Riemann type [25–27]. A similar situation occurs in the case of a charged particle moving on a sphere under a radial magnetic field and Coulomb force. The corresponding Schrödinger equation is transformed into a Heun equation in canonical form, and exact solutions are obtained in terms of series of hypergeometric functions. The separation of variables is still possible since the vector potential depends only on one spherical variable, θ in this case [28]. Another example of the same type of difficulty is the two-dimensional case of interaction of three particles with a perpendicular magnetic field [29]. Here the Schrödinger equations map into biconfluent Heun's equations, too, through the higher order singularities induced by the Coulombian interaction between particles. Similar problems related to magnetic field (finite-gap potentials, Fokker-Planck, central two-point connection, generalized central potentials up to order $1/r^6$, Hawking radiation, etc.) were approached in the literature and usually the resulting leading differential equation for the wave function reduces to one of Heun's differential equations [30–34]. An interesting review and study on the use of the Heun's type of differential equations as generalizations of the hypergeometric ones, in relation to supersymmetry, are given in [35], where a two-Coulomb-center problem is solved based on a self-adjoint separation of coordinates in prolate spheroidal coordinates.

Before moving to the dipole coordinates we perform a qualitative analysis of Eq. (6) in spherical coordinates. In spherical coordinates Eqs. (6) and (10) reduce to a linear Schrödinger equation in the form

$$\left[\Delta_r + \frac{1}{r^2}\Delta_\Omega + \frac{1}{r^2}\left(-\frac{2i\chi}{r}\frac{\partial}{\partial\varphi} - \frac{\chi^2\sin^2\theta}{r^2}\right) + E\right]\Psi = 0, \tag{13}$$

where we denoted $\chi = 2\pi q\mu/(\hbar c)$, and the first two terms in the LHS are the radial and angular parts of the Laplace operator in spherical coordinates, respectively. We look for solutions of Eq. (13) in the subspaces described by

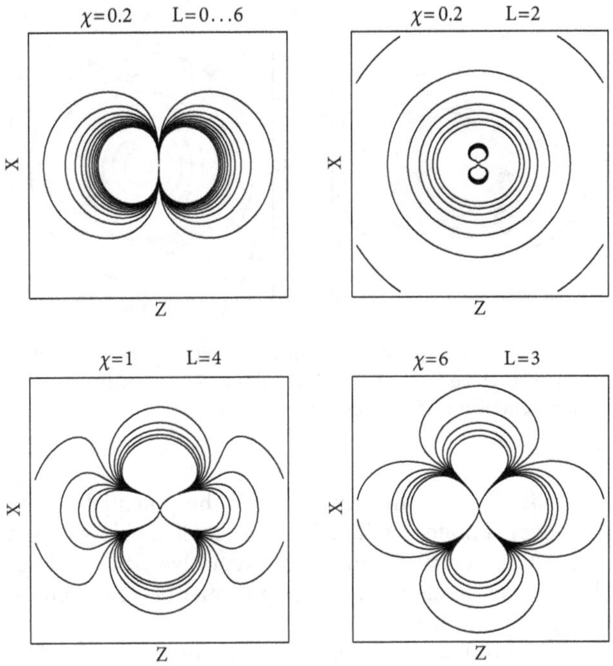

FIGURE 1: Equipotential contours of the effective potential $V(x,z)$ at $y = 0$, as given by Eq. (15). The dipole is along z-axis. Several values for dipole strength χ and L. For $L = 0$ the potential geometry is the same for any dipole strength, but for higher magnetic field the equipotential shapes become complicated and self-intersecting.

Eq. (9), meaning that the wave functions have good quantum numbers for the angular momentum L. We can write these wave functions in the form

$$\Psi(r,\theta,\varphi) = \Phi(r,\theta)\, e^{iL\varphi}, \tag{14}$$

with L being an arbitrary positive integer. Eq. (13) reduces to a 2-dimensional Schrödinger equation with an effective potential in the form

$$V(r,\theta) = -\frac{L^2}{\sin^2\theta} + \frac{2\chi L}{r^3} - \frac{\chi^2\sin^2\theta}{r^4}. \tag{15}$$

In Figures 1 and 2 we present some equipotential contours of the effective potential $V(x,z)$ at $y = 0$ and $y = 0.5$. This potential has a repulsive tail that drops to zero towards infinity like r^{-4}. Depending on θ the potential generates a finite barrier in a neighborhood of the origin. The resulting effective potential valley produces a finite spectrum of energies for the linearized equation, hence a finite space of eigenfunctions available for the nonlinear GL equation. The parameter χ controls the barrier height. For $\theta \sim \pi/4$ the barrier reduces to zero, and actually the potential is almost everywhere attractive. This is normal, since without magnetic dipole the spectrum is continuous, and the only possible state is spherical isotropic without any vortex. For $\theta \sim 0, \pi$ the barrier height increases and allows the accumulation of more eigenstates in the linear spectrum. This denser spectrum generates a higher probability of formation of open vortices that spring towards the sphere surface around the poles of

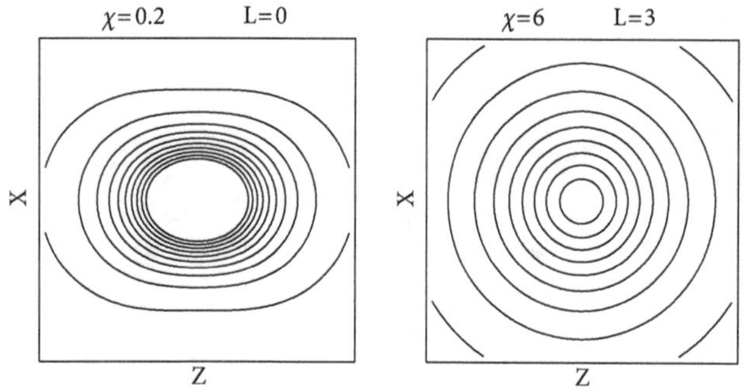

FIGURE 2: Equipotential contours of the effective potential $V(x,z)$ at $y = 0.5$, as given by Eq. (15). Dipole along z-axis. Far from center the isopotential surfaces are deformed spheres.

the z-axis. Next to the origin $r = 0$ the potential has an (attractive) infinite depth valley. For small dipole strength values the potential becomes pure repulsive, and for large values of the dipole strength the potential becomes almost attractive everywhere.

The Schrödinger equation with the potential in Eq. (15) cannot be resolved by traditional expansion in spherical harmonics with coefficients given by the monomials r^l. Since this equation contains nonhomogenous terms as powers of r, even if we use Laurent polynomials instead of these monomials, the coefficients of Y_{lL} will always contain irretrievable powers in r. This situation enforces the coefficients of spherical harmonics to include all powers of r. However, even if we use arbitrary (linear independent) functions $f_l(r)$ instead of constant coefficients for each Y_{lL}, the differential equations resulting for each f_l will contain infinite many terms. This happens because the term $\sin^2\theta$ multiplied with spherical harmonics of different orders decomposes (through the Wigner $3j$-symbols) in a sum of spherical harmonics of orders smaller than l. Consequently, spherical coordinates do not map Eq. (6) into an integrable system.

Given the symmetry of the magnetic dipole we introduce a suitable type of orthogonal curvilinear coordinates, namely, *dipole* coordinates denoted by (a,b,φ). These dipole coordinates are related to the spherical ones by the relations

$$\begin{aligned} a &= \frac{r}{\sin^2\theta}, \\ b &= \frac{r^2}{\cos\theta}, \\ \varphi &= \varphi. \end{aligned} \tag{16}$$

Dipole coordinates can be introduced in a variety of ways [36], and they are useful in a couple of physical systems controlled by magnetic dipolar terms. That is where the lines of force have strong anisotropy, like terrestrial ionosphere, solar corona, magnetostars, toroidal magnetic moments in atomic physics, etc. [37, 38]. In the following, we use the dipole coordinates in order to provide exact analytic solutions for the order parameter $\Psi(\vec{r})$ in the configuration magnetic dipole plus superconducting sphere. The dipole coordinates were used in [39] to solve the LGL equation for a magnetic dipole placed in a superconducting space. In that study we demonstrate that analytic solutions of the LGL equation in dipole coordinates can be obtained by transforming it in a double confluent Heun's differential equation. Combination of such linear solutions can numerically minimize the nonlinear GL free energy functional for multivortex states as confined vortex loops.

In the present paper we develop this method even further and, in addition, we apply Dirichlet type of boundary conditions on the surface of a finite superconducting sphere. This procedure maps the dipole equation plus boundary conditions into a regular Sturm-Liouville problem, for which we obtain and discuss in detail the resulting spectra and eigenfunctions. This analytic method was previously used and tested in comparison with numerical calculations and experimental results for the simpler geometry of a finite superconducting cylinder in exterior magnetic field [40, 41].

The dipole coordinates, their coordinate surfaces, and other properties are described in more detail in Appendix A. From there we notice that the field lines of the magnetic infinitesimal dipole,

$$\vec{h} = \frac{(\vec{\mu}\cdot\hat{r})\hat{r} - \vec{\mu}}{r^3}, \tag{17}$$

are given by the family of curves a constant. Surfaces defined by constant a follow the field lines of the magnetic dipole and are orthogonal to the surfaces of constant b. Consequently, the order parameter has a stronger dependence on a than on b, such that the surfaces $|\Psi|$ =const. are very close to the dipole coordinates surfaces a =constant. Along this reasonable approximation we can neglect the b dependence of the order parameter solutions.

In the dipole coordinates the electromagnetic momentum has a simpler expression

$$\vec{P} = \frac{\hbar}{i}\vec{\nabla}_{a,b,c} - \frac{q\mu}{ca}\vec{e}_{\varphi}. \tag{18}$$

By introducing Eqs. (16) and (18) in Eq. (6), under the hypothesis Eq. (14), we obtain the following partial differential equation in dipole variables (a,b) for $\Phi(r,\theta) \longrightarrow \Phi(a,b)$:

$$\begin{aligned} &\left(\frac{a}{r}\right)^3\left(1+3\frac{r^4}{b^2}\right)\left(\frac{d^2\Phi}{da^2}+\frac{4}{1+(3r^4/b^2)}\frac{1}{a}\frac{d\Phi}{da}\right) \\ &\quad+\left(1+3\frac{r^4}{b^2}\right)\frac{b^2}{r^6}\frac{d}{db}\left(b^2\frac{d\Phi}{db}\right)-\frac{a}{r^3}\left(L-\frac{\chi}{a}\right)^2\Phi \\ &\quad+E\Phi=0, \end{aligned} \tag{19}$$

where $L \in \mathbf{N}$, E are free parameters. We notice that it is not possible to fully eliminate the coordinate r from Eq. (19), because the inverse transformation $(a,b) \longrightarrow (r,\theta)$ involves an algebraic equation of order 4 whose solutions would introduce prohibitive long expressions in Eq. (19); see Eq. (A.3) in the Appendix A.

In order to obtain a separation of the dipole variables (a,b) in Eq. (19) we investigate solutions in the approximation $a \sim r$ (r being the radial spherical coordinate), which implies

$r^2/b << 1$ from Eq. (A.3). This approximation describes the order parameter in neighborhoods of $\theta = 0$, and $\theta = \pi$, that is around the poles of the bispherical surfaces a =constant; see Figure 11. Actually, this approximation is valid within a very large part of the superconducting sphere. If we perform the inverse transformation from dipole to spherical coordinates we can express r as the product between a and a power series in the parameter $\epsilon = a^2/b$, according to Eq. (A.7) in Appendix A. For example, if we just choose $\epsilon = 0.2$ we have a relative error of only 4% or less for $a \sim r$ within a range of the azimuth angle $\theta \in [0, 80^\circ] \cup [100^\circ, 180^\circ]$, which represents that 79% of the total volume of the sphere provides very good agreement with this approximation. From the geometrical perspective, this approximation limits our calculations to surfaces a =const. that are very close to the spherical surface r =const, Figure 11.

In the approximation $a \sim r$ Eq. (19) reduces to the form

$$\frac{\partial^2 \Phi}{\partial a^2} + \frac{4}{a}\frac{\partial \Phi}{\partial a} + \frac{b^2}{a^6}\frac{\partial}{\partial b}\left(b^2 \frac{\partial \Phi}{\partial b}\right) - \frac{1}{a^2}\left(L - \frac{\chi}{a}\right)^2 \Phi + E\Phi = 0, \tag{20}$$

and we can factorize further the b-dependence from the a-dependence by the substitution

$$\Phi(a,b) = Q(a)\left(e^{-c/b} + C_0 e^{c/b}\right), \tag{21}$$

where c, C_0 are arbitrary coupling constants between the equations in a and b. Consequently, we can obtain an equation for $Q(a)$ in the same way it was obtained in Eq. 910) in [39]

$$a^6 \frac{d^2 Q}{da^2} + 4a^5 \frac{dQ}{da} + \left(c^2 - \chi^2 a^2 + 2L\chi a^3 - L^2 a^4 + Ea^6\right) Q = 0, \tag{22}$$

This equation is associated with the boundary conditions Eq. (12)

$$\lim_{a \to 0} Q(a) = 0, \qquad \left.\frac{dQ}{da}\right|_{a_\infty} = 0, \tag{23}$$

namely, the order parameter needs to cancel at the center of the mesoscopic volume, and the normal component of the superconducting current has to cancel at the external boundary of this volume. In agreement with the approximation $r \sim a$ the external boundary condition can be written as $a = a_\infty = R$. At this point we describe the mesoscopic volume taken for the present study. The equation $r = R\sin^2\theta$ is a special case of the so-called rose curve, firstly described by the seventeenth century Italian mathematician, Guido Grandi. Therefore the mesoscopic volume has an "apple" shape, such that at the north ($\theta \approx 0^\circ$) and south ($\theta \approx 180^\circ$) poles the volume is depleted ($r \approx 0$) but not at the equator region ($r = R$). This choice of volume is key to achieve separation of variables in these coordinates. However this choice does not compromise our major results, namely, to prove the existence of confined loops in any mesoscopic volume. The boundary conditions Eq. (23) are called "separated" since each involves only one of the ends of the domain of definition. Eq. (22) is an ordinary differential equation of order two with meromorphic coefficients. It has two irregular singularities of rank 2 at $a = 0, \infty$, [42], and consequently there are no convergent solutions in terms of Frobenius series. We need to solve this equation in the interval $a \in (0, a_\infty]$, where $a_\infty = R$ is the radius of the superconducting volume.

The approach used to solve the homogenous two-point boundary value problem Eqs. (22) and (23) consists in performing a series of transformations that map Eq. (22) into a symmetric canonical form of a double confluent Heun's equation [27, 43–45].

4. Solutions of the Dipole Equation

4.1. Qualitative Analysis of the Dipole Equation. In order to perform a qualitative analysis of Eq. (22) we can transform it into other types of differential equations (e.g., $Q(a) \longrightarrow \widetilde{Q}(a) = a^p Q(a)$ for some real values of p) with some known physical significance. For example, by the substitution $\widetilde{Q}(a) = a^2 Q(a)$ we have

$$-\frac{d^2\widetilde{Q}}{da^2} + V_{1d}(a)\widetilde{Q} + E\widetilde{Q} = 0, \tag{24}$$

which is a 1-dimensional Schrödinger equation with an effective potential

$$V_{1d} = -\frac{c^2}{a^6} + \frac{\chi^2}{a^4} - \frac{2L\chi}{a^3} + \frac{L^2+2}{a^2}. \tag{25}$$

Here the last term represents the traditional centrifugal term plus an additional constant 2. This potential is repulsive for large values of a, and it becomes attractive in a short range close to the origin because of the coupling with the $\exp(-c/b) + C_0 \exp(c/b)$ part of the wave function, through c.

Another possible approach is provided by the substitution $\widetilde{A}(a) = aA(a)$, and we obtain the radial part of a Schrödinger equation in spherical coordinates

$$-\frac{d^2\widetilde{Q}}{da^2} - \frac{2}{a}\frac{d\widetilde{Q}}{da} + \left(-\frac{c^2}{a^6} + \frac{\chi^2}{a^4} - \frac{2L\chi}{a^3} + \frac{L^2+2}{a^2} - E\right)\widetilde{Q} = 0. \tag{26}$$

The effective potential for this case (the first four terms in the parenthesis of Eq. (26)) is presented in Figure 3. It has a repulsive asymptotic behavior for $V_{sph}(+\infty) = 0$, and $V_{sph}(0) \longrightarrow \pm\infty$ depending on the value of c. In the simpler case $c = 0$, for example, if $L \geq 4$ this potential has a valley centered at

$$a_1 = \chi \frac{3L - \sqrt{L^2 - 16}}{2(L^2+2)}. \tag{27}$$

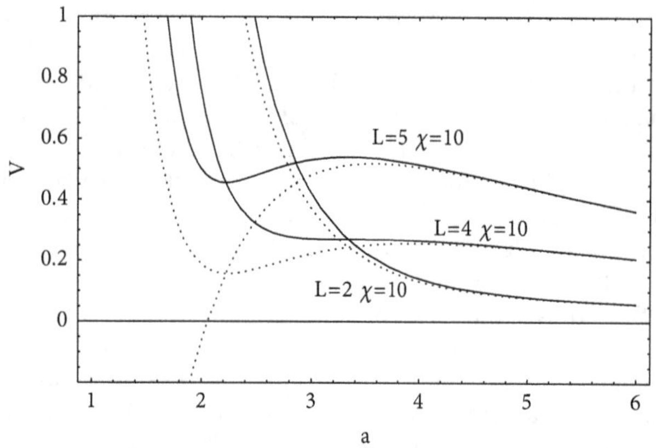

FIGURE 3: The effective potential $V_{sph}(a)$ in the radial part of the Schrödinger equation in spherical coordinates, Eq. (26). Solid lines represent $c = 0$ potentials, and dotted lines represent $c = 5$ potentials, for the same L, χ.

This valley creates a barrier of height

$$V_{bar} = \frac{8\left(L^2+2\right)^3\left(-4+L^2+L\sqrt{L^2-16}\right)}{\chi L\left(3L+\sqrt{L^2-16}\right)^4}, \quad (28)$$

and relative depth with respect to the top of the valley

$$V_0 = \frac{L\left(L^2-16\right)^{3/2}}{16\chi^2}. \quad (29)$$

From the aspect of this potential it results that for weak $a-b$ coupling ($c \sim 0$) the expected spectrum, in the absence of the boundary conditions, is continuous and positive. This is the case of very large superconducting volumes. By its richness of eigenstates, this configuration increases the chances for vortex creation. Indeed, for large superconducting samples it is known that more magnetic flux is expelled in the Meissner phase; hence superconductivity is suppressed at the boundaries. Consequently, more weak points are created that facilitate the entry of many vortices [9]. At higher values of angular quantum number and dipole intensity some resonant states may occur, like in the potential valley shown in Figure 3 for $L = 5, \chi = 10$.

For the substitution $\widetilde{Q} = a^{3/2}Q(a)$ we obtain the radial part of the Schrödinger equation in cylindrical coordinates

$$-\frac{d^2\widetilde{Q}}{da^2} - \frac{1}{a}\frac{d\widetilde{Q}}{da} + \left(-\frac{c^2}{a^6} + \frac{\chi^2}{a^4} - \frac{2L\chi}{a^3} + \frac{L^2+9/4}{a^2} - E\right)\widetilde{Q} = 0. \quad (30)$$

Eq. (30) is a Fuchsian differential equation with singularities at the point at infinity, and order six singularity at $a = 0$, presenting some similarities with the symmetries of the Bessel equation. Since the existence and uniqueness of this type of equation will be investigated elsewhere and since later on we will justify that the value of the decoupling parameter c is very small, we will further reduce this equation to a fourth-order singularity Fuchsian equation.

As we mentioned in Section 3, Eqs. (21), (22), and (30), we introduce the second hypothesis (in addition to $r \backsim a$) by considering a weak coupling between the a and the b coordinates in the order parameter dependence. This constraint is equivalent to $c \sim 0$, and it arises somehow naturally since the order parameter $\Psi(a, b, \varphi)$ has its dominant dependence on the a coordinate. This happens because the $|\Psi|^2$ iso-surfaces are basically laying along the dipole surface coordinates a =constant. Under this hypothesis, Eq. (22) becomes

$$\frac{d^2Q}{da^2} + \frac{4}{a}\frac{dQ}{da} + \left(-\frac{\chi^2}{a^4} + \frac{2L\chi}{a^3} - \frac{L^2}{a^2} + E\right)Q = 0, \quad (31)$$

and we will refer further to Eq. (31) as the *dipole* equation. The asymptotic solutions in zero and at infinity for Eq. (31) are described in [39] in terms of Bessel, Kummer, and Tricomi special functions, but here we are not interested in the asymptotical behavior of the solution at zero or infinity. An interesting limiting case is obtained if we keep only the even power terms $1/a^4, 1/a^2, E$, in the potential energy. The resulting equation is mapped into the Mathieu equation for $A(z)$ through the transformation $a = (-\chi^2/E)^{1/4}\exp(iz)$. It is interesting that even holding all the terms in the powers of $1/a$ we still can map the dipole equation into Hill's differential equation [46].

The existence of situations when one can map the dipole equation in traditional equations of mathematical physics (Bessel, confluent hypergeometric, Mathieu, etc.) by neglecting higher or lower powers in $1/a$ has a deep meaning. Although not obvious, Eq. (31) has an intrinsic symmetry that allows mapping it into a very symmetrical form, called the *symmetrical-canonical double confluent Heun equation* [27, 43, 44]. We will discuss in Section 4.2 multiple advantages of expressing the dipole equation in this form, especially related to the existence of simple analytic solutions. When we neglect some of the terms in the dipole equation, certain conditions of nondegeneracy fail, and mapping the dipole equation into the symmetrical-canonical DCHE becomes impossible. Hence, under such approximations, Eq. (31) degenerates into those classical equations mentioned above.

4.2. Analytic Solutions for the Dipole Equation. In the following we construct analytic solutions for the dipole equation, in the form of Eq. (22), or Eq. (31), in the limit of weak coupling with the orthogonal b coordinate, that is, for $c = 0$. Also in this section we will not take into account yet the boundary conditions and just obtain the most general solution sets.

The first step is to reduce Eq. (22) to a dimensionless form by the transformation $a \longrightarrow z = a/\chi$. The resulting equation is a general double confluent Heun equation (gDCHE [27]) in $Q(z)$ of the form

$$D^2Q + p(z)DQ + q(z)Q = 0,$$

$$p(z) = \sum_{i=-1}^{1} p_i z^i,$$

$$q(z) = \sum_{i=-2}^{2} q_i z^i, \tag{32}$$

where the operator $D = z(d/dz)$ and p_i, q_i are arbitrary complex coefficients. The properties and solutions of different confluent types of Heun's equation have been studied extensively in the last decade [31, 47–52]. There is also an extensive volume of applications of Heun functions and equations in physics; see, for example, [28]. In our case for the dipole equation we have $p_{-1} = p_1 = q_1 = 0, p_0 = 3, q_{-2} = -1, q_{-1} = 2L, q_0 = -L^2$ and $q_2 = \lambda$, where $\lambda = E/\chi^2$. By using a transformation of the dependent variable of the form

$$Q(z) = z^{\xi_0} e^{\xi_1 z - \xi_{-1}/z} y(z) \tag{33}$$

any gDCHE (for example, Eq. (32)) can be transformed into the so-called (nondegenerate) canonical form of the DCHE [27, 43]. For the dipole equation there are two possible transformations that map it into a DCHE canonical form, namely, $\xi_0 = -3/2, \xi_1 = \pm\sqrt{\lambda}, \xi_{-1} = \pm 1$. It is worth mentioning that one obtains a nondegenerate (a "good") canonical DCHE form if one has the condition

$$\left(q_2 - \frac{p_1^2}{2}\right)\left(q_{-2} - \frac{p_{-1}^2}{2}\right) \neq 0, \tag{34}$$

fulfilled (see Eq. (32)). This is actually the case for the dipole equation, but if we neglect some of its terms this condition fails, and we have a degenerated canonical DCHE that reduces to the traditional differential equations mentioned in Section 4.1.

The last step towards integrating the equation is to apply one of the similarity transformations in the independent variable: $z = (\sqrt{i}/\lambda^{1/4})\tilde{z}$, where the multiplicity comes from the square roots of the imaginary/negative parameters. After this transformation, the dipole equation has the form of the symmetric canonical DCHE for the function $y(\tilde{z})$

$$\begin{aligned} &\tilde{D}^2 y + \alpha\left(\tilde{z} + \frac{1}{\tilde{z}}\right)\tilde{D} y \\ &\quad + \left[\left(\beta_1 + \frac{1}{2}\right)\alpha\tilde{z} + \left(\frac{\alpha^2}{2} - \gamma\right) + \left(\beta_{-1} - \frac{1}{2}\right)\frac{\alpha}{\tilde{z}}\right] y \\ &= 0, \end{aligned} \tag{35}$$

where $\tilde{D} = \tilde{z}(d/d\tilde{z})$, and

$$\begin{aligned} \alpha &= \pm 2i\sqrt{i}\lambda^{1/4}, \\ \beta_1 &= 0, \\ \beta_{-1} &= \mp L, \\ \gamma &= L^2 + \frac{9}{4}, \end{aligned} \tag{36}$$

are the singular parameters of the symmetric canonical DCHE associated with the dipole equation. As one can note, there are several (more than two, but less than eight) representations of the dipole equation into the symmetrical form Eq. (35).

Using this symmetrical canonical form has several advantages [27]. First, it generates a recursion relation for the series coefficients of the analytic solutions in only three terms. This is a major advantage over all other representations. Second, it depends on less parameters, just four instead of eight, like in the case of original gDCHE. These four parameters control the leading terms of the asymptotical expansion of the solutions. The singular parameters $\alpha, \beta_{\pm 1}$ determine the leading coefficients of the asymptotical series of the solutions at ∞ and 0, respectively. However, in the case of the dipole equation we have $\beta_1 = 0$, and only the reduced energy (λ) controls the asymptotical behavior of the solution at infinity, through α. This behavior is in agreement with the fact that the boundary condition at a_∞ is the only responsible constraint for energy quantization in prescribed levels. The singular parameters α, β_{-1} determine the leading coefficients of the asymptotic solutions around zero, which means that in the origin the solution depends on both E and L. The γ "accessory" parameter just influences the monodromy behavior of the solutions, and it is also responsible for the eigenvalue problem related to some appropriate boundary conditions. Last advantage of the symmetric canonical form is that the central connection problem, i.e., the relations between asymptotical solutions at different singularities, becomes trivial. In that, changing the asymptotic series from the singularity 0 to ∞ just reduces to permuting the β parameters, modulo some gauge of multiplication with a constant phase [44].

The symmetrical canonical form of the DCHE emerging from the dipole equation, Eq. (35), also has two irregular singularities of rank 1, namely, $\tilde{z} = 0, \infty$. That means that, according to the general theory of complex ordinary differential equations with meromorphic coefficients, all solutions can be continued analytically along any path within $\mathbf{C}^* = \mathbf{C}\backslash\{0\}$. Hence,we can obtain fundamental analytic solutions as series around the singularities with a certain prescribed asymptotic behavior. This asymptotic behavior will also provide the Schmidt-Wolf uniqueness of the fundamental set. Either set of fundamental solutions can be continued analytically over $\mathbf{C}^*$, though because $\mathbf{C}^*$ is not simply connected the Poincaré lemma conditions are not fulfilled, and these resulting global solutions will not be single valued. This problem can be fixed by specifying the asymptotic behavior within special sectors of $\mathbf{C}^*$, bounded by the corresponding Stokes rays, depending on the argument of $\alpha\tilde{z}$ [45].

Since the physical boundary conditions at a_∞, Eq. (12), are important for the eigenvalue problem for energy we will construct the solutions of Eq. (35) based on the asymptotic behavior for $\tilde{z} \longrightarrow \infty$. We choose the specific asymptotic behavior in such a way that will match the asymptotic behavior of the dipole equation in the large a approximation and the corresponding Bessel solutions. This asymptotic behavior secures both uniqueness of the series solution through its Laplace transform and the application of Watson's lemma [27] and the possibility of applying the von Neumann boundary condition at a_∞. Then, this solution can be continued analytically towards the solution around $\tilde{z} = 0$. The continuation is performed through the central connection relations between asymptotic solutions, such that the behavior at zero of the fundamental solution fulfills the boundary condition in zero, too, Eq. (12).

The symmetric canonical DCHE form for the dipole equation is also useful because it has a simple group of transformations which maps any solution into another solution. In this way we can generate from any certain fundamental solution, constructed a series, another linear independent solution, however not necessarily linear independent. The group is infinite and it is generated by the transformations

$$T_1 \left[y\left(\alpha, \beta_1, \beta_{-1}, \gamma; z\right) \right] = \exp\left(-\alpha z\right) y\left(i\alpha, \beta_{-1}, -\beta_1, \gamma; iz\right), \tag{37}$$

$$T_{-1} \left[y \right] = \exp\left(\frac{\alpha}{z}\right) y\left(i\alpha, -\beta_{-1}, \beta_1, \gamma; -iz\right),$$

and

$$T_0 \left[y \right] = \exp\left(-\alpha z + \frac{\alpha}{z}\right) y\left(\alpha, \beta_1, \beta_{-1}, \gamma; \frac{1}{z}\right). \tag{38}$$

The two fundamental solutions for DCHE are built by using the asymptotic construction of solutions of Eq. (35), following the receipt in [27], through the procedure of expanding around the point at infinity described in detail in [39]:

$$y_1\left(\tilde{z}\right) = H_l\left(\alpha, \beta_1, \gamma; \tilde{z}\right), \tag{39a}$$

$$y_2\left(\tilde{z}\right) = ie^{-i\pi\beta_1} e^{-\alpha\tilde{z}} H_l\left(e^{i\pi/2}\alpha, \beta_{-1} - \beta_1, \gamma; e^{i\pi/2}\tilde{z}\right), \tag{39b}$$

where H_l is the holomorphic local Heun function as a solution of the DCHE (35) given by the series

$$H_l\left(\alpha, \beta, \gamma; \tilde{z}\right) = \left(\alpha\tilde{z}\right)^{-\beta-1/2} \sum_{n=0}^{\infty} \frac{C_n}{\left(\alpha\tilde{z}\right)^n}, \tag{40}$$

in the range on $\arg(\alpha\tilde{z}) \in (-\pi/2, 5\pi/2)$, with the coefficients $C_n = C_n(\alpha, \beta_1, \beta_{-1}, \gamma)$ uniquely given by the three-term recursion relation

$$C_n^{\pm} = \frac{1}{n} \left[\left(\frac{\alpha^2}{2} - \gamma + \left(n + \beta_1 - \frac{1}{2}\right)^2 + \right) C_{n-1}^{\pm} - \alpha^2 \left(n + \beta_1 - \beta_{-1} - 1\right) C_{n-2}^{\pm} \right], \tag{41}$$

with $n \in \mathbf{N}$, $C_{-1}^{\pm} = 0, C_0^{\pm} = 1$. The signs $\pm$ represent the choice for one of the two solutions for the reduced parameters α, β_{-1} in Eq. (36). We mention that the second solution y_2 is generated from the first one by application of the group operator T_1 on y_1, and it is linear independent from this one. Finally, one can perform the variable substitutions in reverse order and obtain the solutions Eqs. (39a) and (39b) in terms of the physical parameters a, χ, E, L. The Heun functions in Eqs. (39a) and (39b) represent the fundamental set of exact solutions of the linearized dipole equation (31).

An interesting feature of these solutions is that in the range of negative energies $E < 0$ the solutions $y_1(\tilde{z})$, that is, $Q_1(a)$, are bounded and have no oscillations, yet they can be truncated to quasipolynomials. Such special truncations, very useful for analytic and numeric calculation, are possible only for the series coefficients that have $\beta_{-1} = -L$ and only for odd values for L. For each such odd L value there is one unique value of the energy which provides the truncation of the series into a quasipolynomial (the series reduces to a polynomial, but the exponential and power in front of the polynomial do not vanish). This limitation occurs because of the special structure of the dipole ODE, namely, because of the linear term $2\chi L/a^3$. For Q_2, however, the majority of solutions have $E > 0$, so they cannot be reduced to quasipolynomials.

In the following, we present two examples of quasitruncated solutions $Q_1(a)$ obtained from the abstract solutions $y_1(\tilde{z})$:

$$Q_1\left(L = 1, a\right) = \frac{\chi^2}{\sqrt{3}a^2} e^{-3a/2\chi - \chi/a}, \tag{42}$$

with $E = -9\chi^2/4$, and

$$Q_1\left(L = 3, a\right) = \frac{\chi^2}{\sqrt{w_1}a^2} e^{-w_1 a/2\chi - \chi/a} \left[1 + C_1^- \frac{\chi}{w_1 a} + C_2^- \left(\frac{\chi}{w_1 a}\right)^2 \right], \tag{43}$$

with $E = -\chi^2 w_1^2/4$. In order to have a three-dimensional visual on the behavior of these solutions we plot in Figure 4 few surfaces of equation $|\Psi(a, \varphi)| = |\Psi(r/\sin^2\theta, \varphi)| = Q_1(r/\sin^2\theta)e^{iL\varphi}$const. The three surfaces presented in Figure 4 follow the shape of the surfaces of coordinates $a =$const., proving the hypothesis according to which the b dependence of the order parameter can be neglected.

5. Eigenvalue Problem for the Dipole Equation

Typical study of the quantization of the energy spectrum of Heun's equation involves the γ accessory parameter. Such an analysis is performed, for example, in [45] where the authors study the eigenvalues of a double confluent Heun equation with possible applications in gravitational theory. The boundary conditions used in this paper differ from our case, since they ask the solution to be bounded in origin, and to approach either a finite value, or infinity when $z \longrightarrow \infty$, but not faster than an exponential. Consequently, the [45] boundary conditions quantize the accessory parameter γ, and the recursion relation for the coefficients of the series becomes a four-term Poincaré-Perron type, which is more difficult to handle than our three-term recursion relations.

In order to solve our eigenvalue problem for $Q(a)$ and E from Eqs. (22) and (23) we substitute $a = a_\infty x$, $y(x) = a_\infty x Q(a_\infty x)$, where a_∞ is the upper bound of the a variable. The resulting equation is a regular Sturm-Liouville problem. The self-adjoint form of the corresponding linearized dipole equation (22) is

$$L\left[y\right] = Er\left(x\right) y, \tag{44}$$

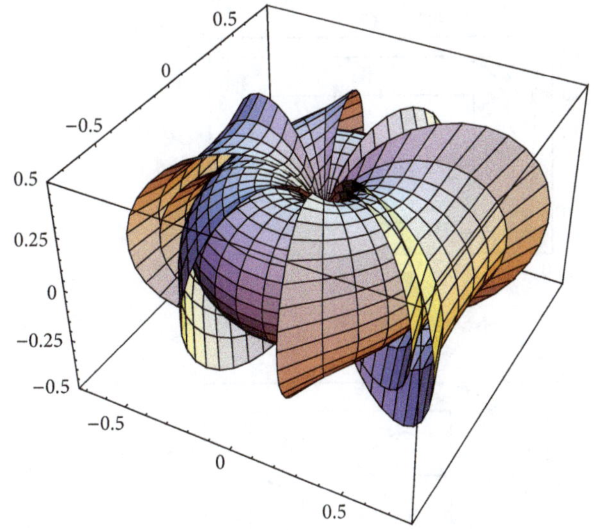

FIGURE 4: The linear solutions $|\Psi|^2 = 0.5, 0.6, 0.75$, and 0.9 isoplot surfaces for $\chi = 0.5$, $a_\infty = 10$, $L = 1$, calculated with the quasi-polynomial solutions Eq. (42) for $E = -5.3$. The surfaces are tori-like, as expected from the behavior of a =const. surface coordinates, Figure 11. Some of the exterior surfaces are plotted only partially, in order to make visible the torus shape of the inner surface.

with

$$L[y] = -\frac{d}{dx}\left(p(x)\frac{dy}{dx}\right) + \left[1 + \frac{\chi^2}{2a_\infty^2 x^2} - \frac{L\chi}{a_\infty x} + \frac{L^2}{2}\right]y, \tag{45a}$$

$$r(x) = \frac{a_\infty^2 x^2}{2}, \quad p(x) = \frac{x^2}{2}. \tag{45b}$$

In this $y(x)$ representation we have $x \in (0,1]$, and $p(x) > 0$, $r(x) > 0$ in this range. In this case [21], the general theory of Sturm-Liouville two-point boundary value problem provides the following results. All eigenvalues E of Eq. (44) are real and form an infinite sequence $E_1 < E_2 < \cdots < E_n < \ldots$ with $\lim_{n\to\infty} E_n = \infty$. According to the physical restriction $E < 1$, it results that the spectrum of interest of the linear dipole boundary problem is also finite. All eigenvalues E_n are not degenerate, and to every two distinct eigenvalues there correspond two linear independent eigenfunctions, orthogonal with respect to the weighted scalar product. That is, for $E_k \neq E_n$ we have

$$\langle y_k, y_n \rangle = \frac{a_\infty^2}{2}\int_0^1 x^2 y_k(x)\, y_n(x)\, dx = 0. \tag{46}$$

Normalized with respect to the weight $r(x)$, the eigenfunctions form a complete orthonormal basis of piecewise continuous functions, with piecewise square integrable continuous derivative in $a \in [0, a_\infty]$, $b \geq 0$.

In order to solve explicitly the eigenvalue problem for E we apply the boundary condition at a_∞, Eq. (23). Since the physical range of interest consists in large values for a_∞ (at least three times, to ten times larger than the coherence length, in order to identify multivortex structures) we use the asymptotic solution for the order parameter from [39], as we mentioned previously. In terms of Bessel functions, we can write the boundary conditions Eq. (23) in the form

$$\begin{aligned}\left.\frac{dQ^\infty}{da}\right|_{a=a_\infty} &= N_1\left[a_\infty^2 E^{1/2}\left(J_{\tilde{L}-1}(w) - J_{\tilde{L}+1}(w)\right) - \chi^2 J_{\tilde{L}}(w)\left(3a_\infty - 2\chi\right)\right] \\ &+ N_2\left[a_\infty^2 E^{1/2}\left(Y_{\tilde{L}-1}(w) - Y_{\tilde{L}+1}(w)\right) - \chi^2 Y_{\tilde{L}}(w)\left(3a_\infty - 2\chi\right)\right] = 0,\end{aligned} \tag{47}$$

where we denoted

$$\tilde{L} = \sqrt{\frac{9}{4} + L^2 + \frac{2iE^{1/2}}{\chi}}, \quad w = \frac{E^{1/2}a}{\chi^2}. \tag{48}$$

The boundary condition Eq. (47) contains two unknowns, $N_{1,2}$. One of them will be eliminated from the condition of smooth match between the Heun analytic solutions with the Bessel asymptotic approximation. Since this procedure will provide a proportionality between $N_{1,2}$ they will be actually eliminated from the homogenous boundary condition. Consequently the only thing provided by the boundary condition equation Eq. (47) will be the value of the energy.

On the one hand we use the Bessel asymptotic expansion [53] which gives

$$Q^{Bessel} \backsim \sqrt{\frac{2\chi}{\pi}}\cos\left(\frac{E^{1/2}a}{\chi^2}\right) \cdot \left[\cos\phi_{L,E}(N_1 + N_2) - \sin\phi_{L,E}(N_2 - N_1)\right], \tag{49}$$

where we denote

$$\phi_{L,E} = \frac{\pi}{2}\left(\sqrt{\frac{9}{4} + L^2 + \frac{2iE^{1/2}}{\chi}} + \frac{1}{2}\right). \tag{50}$$

On the other hand, we calculate the asymptotic behavior towards infinity of the two linear independent analytic Heun solutions $Q_{1,2}(a)$ from $y_{1,2}(\tilde{z})$, Eqs. (39a) and (39b). It results in an asymptotic term of the form

$$2\cos\left(\frac{E^{1/2}a}{\chi^2}\right)\left(\frac{2iE^{1/2}}{\chi^2}\right)^{-1/2}, \tag{51}$$

to be matched with Eq. (49). The solution of this matching condition results in

$$C_1 = \frac{\cos\phi_{L,E} + \sin\phi_{L,E}}{2}\sqrt{\frac{\pi\chi}{iE^{1/2}}}, \tag{52a}$$

$$C_2 = \frac{\cos\phi_{L,E} - \sin\phi_{L,E}}{2}\sqrt{\frac{\pi\chi}{iE^{1/2}}}, \tag{52b}$$

where $\phi_{L,E}$ has the same meaning as above, Eq. (50).

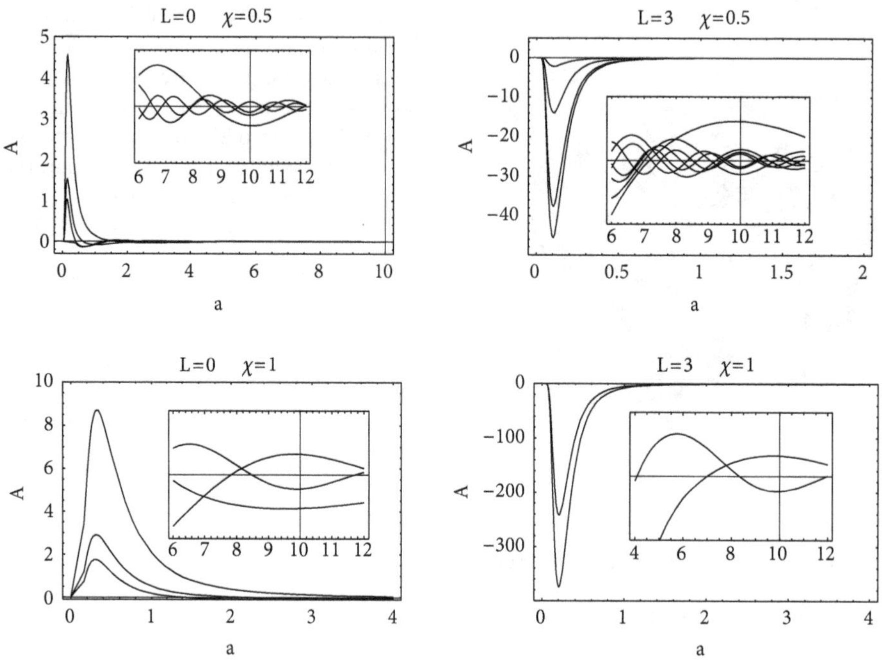

FIGURE 5: Examples of the application of the boundary conditions at $a_\infty = 10$, for the dipole equation analytic solution. The positive part of the spectrum $0 < E < 1$, for $L = 0, 3$, and $\chi = 0.5, 1$. In the main window we present the whole wave function, and in the smaller window we zoom in to the zero derivative point at $a = a_\infty$. The number of eigenstates (oscillations) increases with L, and decreases with χ, because for smaller dipole strength the solution is closer to the Bessel solutions. The eigenvalues of energy are $E(L,\chi) = E(0,0.5) = 0.06, 0.32, 0.78, 0.92$; $E(3,0.5) = 0.02, 0.16, 0.23, 0.41, 0.52, 0.64, 0.77$; $E(0,1) = 0.085, 0.44, 0.985$, and $E(3,1) = 0.32, 0.88$. Higher energy states have more oscillations.

With the coefficients $N_{1,2}$ obtained in Eqs. (52a) and (52b), with a choice for a_∞, L, χ as parameters, we solve the boundary condition relation Eq. (47) and find the corresponding spectrum for each such set of parameters. The solution is not unique, and we can write the resulting spectrum as $E = E_k(a_{\infty,L,\chi})$, where k is the "radial" quantum number.

The corresponding eigenfunctions are $Q_1(a; a_\infty, L, \chi, E_k(a_\infty, L, \chi))$. The resulting solutions of the linearized GL equation are

$$\Psi_{k,L}(a,\varphi) = Q_1\left(a; a_\infty, L, \chi, E_k\left(a_\infty, L, \chi\right)\right) e^{iL\varphi}. \quad (53)$$

We present some examples of such solutions in Figure 5 for $a_\infty = 10$ and two values of χ. For each situation we obtain a discrete spectrum for E, as predicted at the beginning of this section. In the investigated range we obtained between two and seven positive eigenvalues for E. For higher dipole strength we have less positive eigenvalues. This is because the oscillatory character of the free solution (Bessel asymptotic) is reduced by the real exponential in the solution. Also, the spectrum has more eigenvalues when L increases.

The final form of this equation is

$$\tan\left[\pi\sqrt{\frac{9}{4} + L^2 + \frac{2iE^{1/2}}{\chi^2}}\right] - \frac{\chi^2\left(3a_\infty + 2\chi\right)}{2a_\infty^2 E^{1/2}} = 0. \quad (54)$$

For any set of values for a_∞, χ and L the solutions of Eq. (54) provide the E spectrum. For every L value the spectrum is discrete and bounded, according to the general Sturm-Liouville results mentioned in the beginning of this section, and the physical constraint $E < 1$. We also mention that the finiteness of the spectrum is in agreement with the finite depth of the potential valley noticed in Figure 3.

Examples of energy spectra for different values of the parameters are given in Figures 6 and 7. In Figure 8 we present the eigenfunctions and spectra for small values of energy in the range $0 < E < 1$. We note that the spectrum becomes richer with the increase of the volume radius. Also, for large enough dipole strength, the energy of a certain level decreases with the increasing of the angular momentum. The increasing of the dipole strength also produces a dilation of the spectral levels. From Eq. (54) we note that the energy spectrum for the linear problem has three distinct domains. The positive energy domain is rich in states and is the most important source of potential multivortex structures. We define it by $0 < E < 1$. The maximum number of spectral levels k_{max} depends on L and it can be calculated from the condition $|\Psi|^2 < 1$. The next spectral domain is within the negative energies

$$-\frac{\chi}{2}\left(\frac{9}{4} + L^2\right) < E < 0, \quad (55)$$

and it also depends on all the parameters. In the normal range of values ($L = 1, \ldots, 6$, $\chi = 0.2 - 10$) this condition offers a wide range of energies (it can go down to $E \sim -75$) so this is the main range for negative energies. In this domain, because the argument of the tan function in Eq. (54)

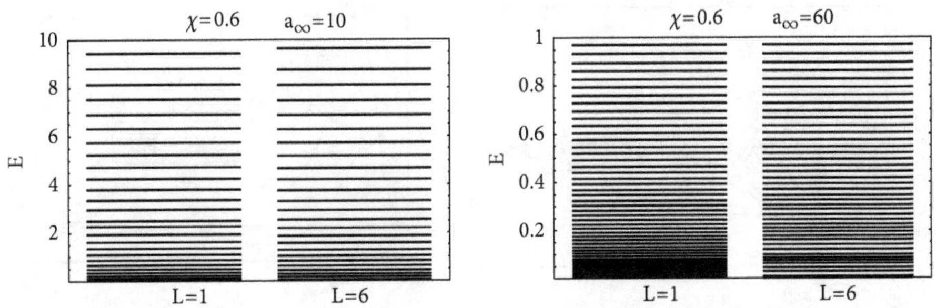

FIGURE 6: Energy spectrum E for the linear dipole problem with boundary conditions on the volume from Eq. (47), or Eq. (54), for small dipole strength, $\chi = 0.6$. We present two cases of volume radii, $R = a_\infty = 10, 60$. The coupling of energy with angular momentum is reduced, so we notice a weak dependence of L. For larger volumes the spectrum is richer, note the difference in scale of left and right frames.

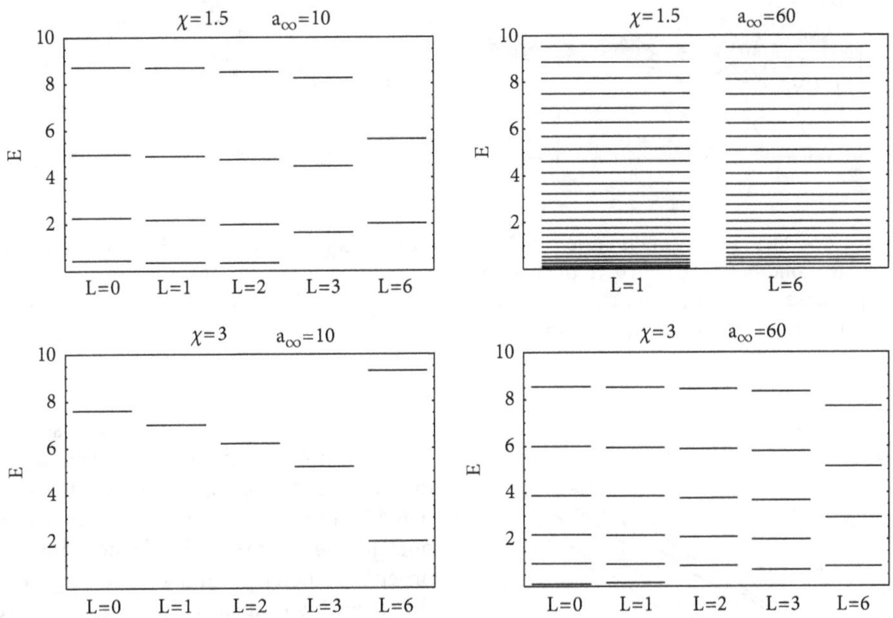

FIGURE 7: Energy spectrum E for the linear dipole problem with boundary conditions on the volume from Eq. (47) for two dipole strength, $\chi = 1.5, 3$ and two volume radii, $R = a_\infty = 10, 60$.

becomes imaginary, this tangent transforms into hyperbolic tangent. Consequently, for each value of L the spectrum contains just one negative energy level. The last domain is for energy less than the limit written above, which represents unstable physical situations, and we will ignore it in the following.

For weak dipole strength (basis of frames) the spectrum is rich and represents all the asymptotic oscillations of the Bessel solutions. There is always a range of the distance to the centrum after which the spectrum becomes sparser. At this linear level, it appears that the energy spectrum is not strongly dependent on the values of the angular momentum. In Figure 9 we present the action of the boundary condition on the analytic solutions, for positive energies, for example. We choose a fixed size of the sample ($a_\infty = 10$) and an average value of the relative dipole strength $\chi = 0.5$. The behavior of the eigenfunctions around the boundary limit is presented in the left frames, and the general aspect of the wave function all over its range is presented in the right frames, together within the spectrum.

For small values of the magnitude of the energy, and also for large enough χ, we can approximate the solutions of the spectral equation Eq. (54) with the analytical expression

$$E_k^{1/2} \sim \frac{\chi^2 \pi}{a_\infty} \left(\sqrt{\frac{9}{4} + L^2} + k \right), \tag{56}$$

with integer values of k such that $E < 1$. The number of eigenvalues for positive E in this approximation can be estimated

$$k_{max} = N_{eigenvalues}(L) = \frac{a_\infty}{\chi^2 \pi} - \sqrt{\frac{9}{4} + L^2}, \tag{57}$$

and we have such bounded states only if the following criterion is fulfilled:

$$a_\infty > \chi^2 \pi \sqrt{\frac{9}{4} + L^2}. \tag{58}$$

This lower bound for a_∞ (or equivalently, an upper bound for L) gives a criterium for the critical size, or dipole strength that

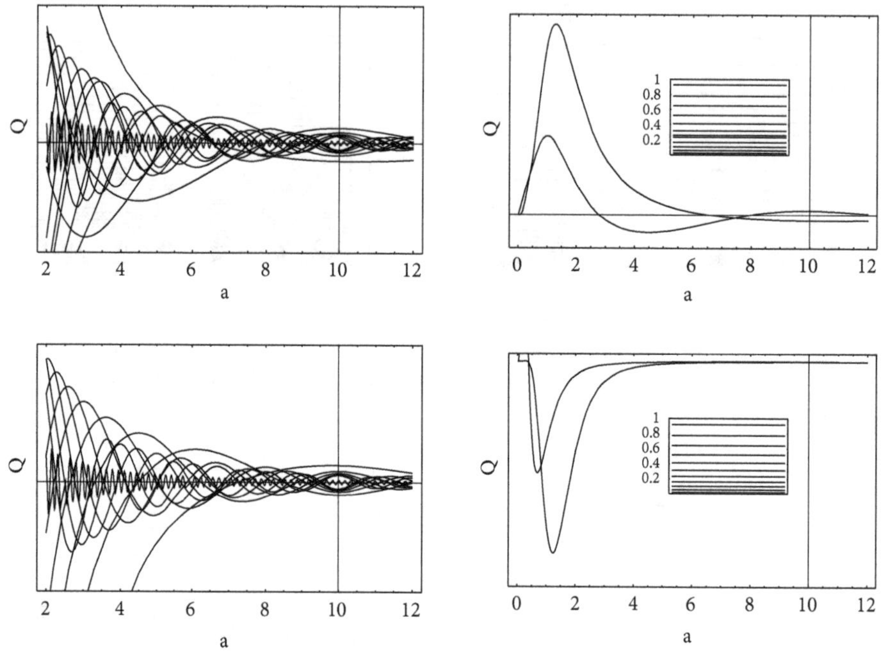

FIGURE 8: Eigenvalues ($0 < E < 1$) and eigenfunctions for the linearized dipole equation for $\chi = 0.5$ and $L = 0$ (upper frames) and $L = 3$ (lower frames) with boundary condition at $a_\infty = 10$. In the left frames we zoom in all the eigenfunctions for a given L around the exterior boundary. In the right frames, we plot the wave functions for the first two lowest energy values. The little windows inside the right frames show the whole spectrum of energies for these a_∞, χ, L values.

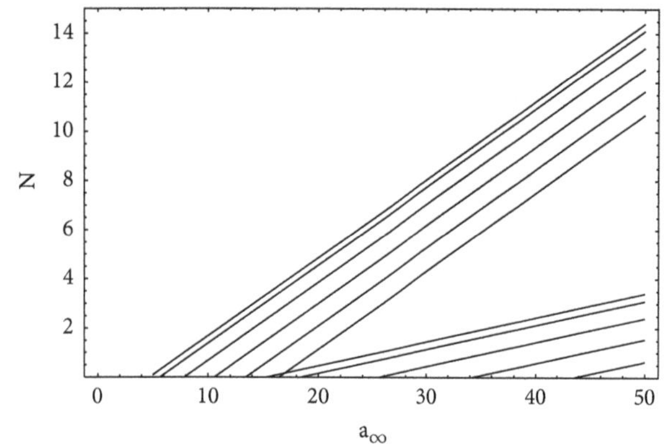

FIGURE 9: The number of eigenvalues E in the positive spectrum versus a_∞ for $L = 0, \ldots, 5$. The number increases with L. The upper band of lines is for $\chi = 1$, and the lower one for $\chi = 5$. Obviously the number of eigenvalues increases with the radius of the volume since larger samples allow more oscillations inside.

can trigger the generation of multivortex structures creation. For example, if we choose $a_\infty = 20$ we have 5 eigenstates for $L = 0, 1$ and 4 eigenstates for $L = 2, 3, 4$.

For larger radii of the volume the energy spectrum obtained as solution of Eq. (54) can be approximated with

$$E_{a_\infty} \sim -\frac{\chi^4}{4}\left(k^2 - L^2 - \frac{9}{4}\right)^2, \tag{59}$$

with k integer ("radial" quantum number) labeling the energy spectrum for fixed L.

The energy spectra presented so far depend only on two quantum numbers: L, k. Even if this is a full three-dimensional problem, a third quantum number was eliminated by the supposition that $\nabla\Psi$ is mainly directed orthogonal to the magnetic dipole field lines. Consequently, the order parameter is almost constant along these lines, and the coordinate surfaces a =const. describe the vortices surfaces. By this hypothesis we can neglect the b coordinate dependence in the order parameter, which eliminates the third quantum numbers, basically like taking into account just ground states with respect to b variable excitations.

6. Nonlinear Vortex Patterns

The goal of this paper is to obtain the nonlinear order parameter as the solution $\Psi(\vec{r})$ of the full nonlinear Ghinzburg-Landau problem Eq. (2), which minimizes the free energy functional F for the mesoscopic sphere with magnetic dipole, Eq. (1). This nonlinear solution, denoted Ψ^{NL}, is constructed in the following as a linear combination of exact analytic solutions of the associated linear GL problem, Eq. (6), or presented in a more rigorous form by Eqs. (8) and (10). These linear solutions form a complete orthogonal system of eigenfunctions, Eq. (53), over the space of dipole coordinates (a, b, φ), under boundary conditions, Eq. (12). The corresponding eigenvalues spectrum is described in Section 5.

The spherical surface of our mesoscopic sample is expressed in dipole coordinates as a volume $a_\infty \sim R$. In the approximation of weak coupling with the b orthogonal variable (i.e., $B(b)$ =constant) described above we introduce the following notation for the linear basis:

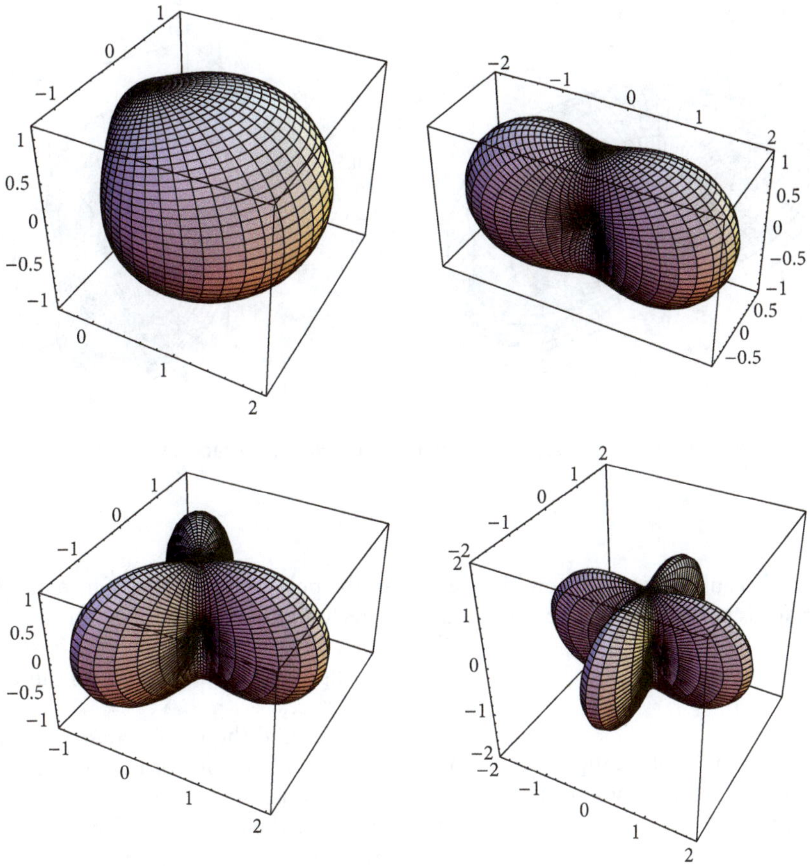

FIGURE 10: Isoplot surfaces $|\Psi^{NL}| = 0.9$ for $L = 1, 2, 3, 4$ from left upper frame, clockwise. We choose $a_\infty = 100$, $\chi = 1$, $L_1 = 0$, $k_1 = 1$ and the minimal energy was obtained for $k_2 = 3$.

$$\Psi_{L,E}(a, b, \varphi) \longrightarrow \Psi_{L,k}(a, \varphi) = Q_1(L, k; a)\, e^{iL\varphi}, \quad (60)$$

with the connection between the physical parameter E and the spectrum label k being provided by Eq. (54).

We search nonlinear solutions of the form

$$\Psi^{NL}(a, \varphi) = \sum_{L=0}^{L_{max}} \sum_{k=1}^{k_{max}(L)} C_{L,k} \Psi_{L,k}(a, \varphi), \quad (61)$$

with $C_{L,k}$ parameters that have to be determined. In order to generate physical solutions for the full nonlinear GL problem, the expression Eq. (61) is plugged in the Gibbs free energy expression Eq. (1) and this integral is minimized in the space of parameters $C_{L,k}$[7, 9–18]. In order to identify the vortex structures inside the superconducting sphere, in this model, one needs to find that it exists at least a value $a_v \in (0, a_\infty)$ of the dipole coordinate such that the equation $|\Psi^{NL}(a_v, \varphi)| = 0$ has a number of distinct solutions for $\varphi \in [0, 2\pi)$. If we denote by φ_j, $j = 1, 2, \ldots, n$ these solutions, by analytic prolongation theorem we can always find a neighborhood of a_v on which $|\Psi^{NL}|$ is arbitrary small for each of the roots φ_j. The surfaces described by the values of a in these neighborhoods are enveloping surfaces of the multivortex states. In this case the value of L_{max} is called the vorticity of the state. The center-lines of the vortices are the curves obtained by the intersections of the $a = a_v$ and $\varphi = \varphi_j$ surfaces.

In literature [15, 16] there are examples of such multivortex states obtained with as little as $k_{max}(L_{max}) = 2$ linear wave functions in Eq. (61). In [39], for example, we obtained multivortex states from the linear combination of two linear wavefunctions. We can apply here the same procedure, except we have to replace the solutions in the full space with the present solutions bounded by the spherical surface. Thus, we obtain for the angular positions of the vortices central lines ($k_{max} = 2, L \in \{L_1, L_2\}$) the expression

$$\varphi_j = \left[\arccos \frac{1 + \left|C_2 Q_{1,L_2,k_2}/C_1 Q_{1,L_1,k_1}\right|^2}{2\left|C_2 Q_{1,L_2,k_2}/C_1 Q_{1,L_1,k_1}\right|} + \arg\left(C_1 Q_{1,L_1,k_1}\right) - \arg\left(C_1 Q_{1,L_2,k_2}\right) \pm j\pi\right] \cdot (L_1 - L_2)^{-1} \quad (62)$$

with $j = 1, 2, \ldots, |L_1 - L_2|$ and all functions Q are evaluated at a_v. To visualize an example of multivortex state, we choose $a_\infty = 100$, and $L_1 = 0, k_1 = 0$, $L_2 = 0, \ldots, 4$. We plot in Figure 10 four isoplot surfaces with $|\Psi^{NL}| = 0.9$ for different vorticity numbers. Nevertheless, the topology of the multivortex surfaces is controlled mainly by the linear

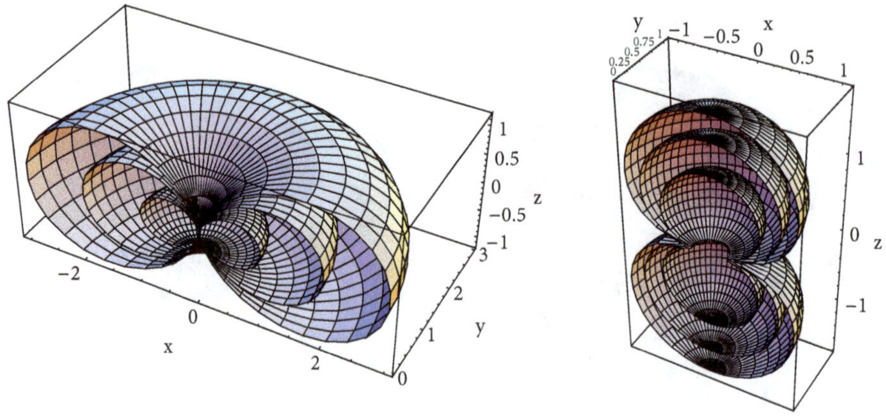

FIGURE 11: Left: coordinate surfaces of constant $a = 1, 2, 3$ (tori-like); Right: coordinate surfaces of constant $b = 1, 2, 3$ (double-volumes). The dipole is directed along z axis.

GL equation. This happens because close to the multivortex surface the order parameter is very small, so the nonlinear becomes negligible.

7. Conclusions

We investigate the existence of multivortex superconducting states in a mesoscopic sphere where the magnetic field is generated by an infinitesimal magnetic dipole placed at the center. We use the Ginzburg-Landau model in the approximation of weak magnetic field. The Euler-Lagrange equations emerging from the GL free energy functional reduce to a magnetic Schrödinger equation with mixed boundary conditions at the center and the surface of the superconducting sample. The goal of this paper is to obtain the nonlinear order parameter, i.e., the solution $\Psi(\overrightarrow{r})$ of the full nonlinear Ghinzburg-Landau problem Eq. (2), which minimizes the free energy functional F for the mesoscopic sphere with magnetic dipole, Eq. (1). The nonlinear solution is constructed as a linear combination of exact analytic solutions of the associated linear GL problem, Eqs. (6), (8), and (10). These linear solutions form a complete orthogonal Sturm-Liouville system of eigenfunctions, Eq. (53), over the space of dipole coordinates (a, b, φ), under the boundary conditions Eq. (12). As a first step, we use cylindrical symmetric coordinates. The linear solution can be separated in $\Psi = \Phi(a, b)e^{iL\varphi}$, Eq. (9), and it maps into the complete or general dipole equation, Eq. (19). Further on, by using a justified geometric approximation, we can separate further the dipole coordinates a and b, and the solution can be written as $\Phi(a, b) = Q(a)(e^{-c/b} + C_0 e^{c/b})$, Eqs. (20), (21), (22), and (30). We reduce further the resulting equation by the mode decoupling condition $c = 0$, and we obtain the dipole equation (31) for Q which is a double confluent Heun equation, Eq. (35). By performing a substitution, Eq. (33), in the form $z = a/\chi$ and $Q(a) = z^{\xi_0} e^{\xi_1 z - \xi_{-1}/z} y(z)$ we can finally solve the dipole equation and write its analytic solutions, Eqs. (39a) and (39b).

The two approximations introduced here are possible because we confine our calculations in the neighborhoods of the poles of the sphere. Since the gradient of Ψ is mainly directed radial and orthogonal to the magnetic dipole field lines, it results that Ψ is almost constant along these lines, and the dipole coordinates surfaces can describe the multivortex isosurfaces.

We build the nonlinear solution for the GL problem as linear combinations of the exact solutions of the linearized dipole equation and tune the combination's coefficients that minimize the free energy functional. Further on, we demonstrate that multivortex states are possible in this sphere plus dipole configuration, even without the presence of an external magnetic field. Our results are in very good agreement with the numerical calculation of the same model [10, 11].

Appendix

A. Dipole Coordinate System

The dipole coordinates defined in Eq. (16) have as domain of definition $a \in [0, \infty), b \in (-\infty, \infty), b \neq 0$ for $r \in (0, \infty), \theta \in (0, \pi), \theta \neq \pi/2$. The Jacobian of the transformation from spherical coordinates and the element of volume are given by

$$J = r^2 \frac{4 + \tan^2\theta}{\sin^3\theta},$$
$$dV = \frac{a^4}{b^2} da\, db\, d\varphi \tag{A.1}$$

The dipole coordinates are actually different from other apparently similar coordinates like bipolar, bispherical, or toroidal coordinates [54–56]. In the notation used above $(x^i) = (x, y, z) \longrightarrow (a, b, \varphi)$, these traditional orthogonal curvilinear systems of coordinates have symmetrical forms for the three Euclidean coordinates of the general form

$$x^i = \frac{f_1^i(a)\, f_2^i(b)\, f_3^i(\varphi)}{f_4(a) - f_5(b)}, \quad i = 1, 2, \tag{A.2}$$

where the functions f are trigonometric or hyperbolic functions.

Consequently, the two parameters a, b are in general an angle and a distance. In the case of dipole coordinates, the parameters a, b have different dimensions, but both are related to distances, $[a] = m, [b] = m^2$. The symmetry mentioned above for these traditional coordinates induces a certain symmetry in their coordinates curves and surfaces. All these curves represent circles either secant or not intersecting. On the contrary, in the case of dipole coordinates the coordinate curves are not circles, but either glued deformed circles (for $a = const.$) or tangent closed curves with a common singular point (for $b = const.$). The coordinate surfaces, Figure 11, are orthogonal curvilinear coordinates. In order to obtain the inverse transformation, back to spherical coordinates, we have to solve the equation

$$\frac{r}{a} + \frac{r^4}{b^2} = 1. \quad \text{(A.3)}$$

This equation is a particular case of a depressed quartic equation and it can be solved by the Ferrari method, hence reducing it to a depressed cubic equation, and then use Cardano's formulas. Since all coefficients of Eq. (A.3) are positive, this equation always has two complex conjugate solutions, a real negative solution and a nonnegative solution which represents the correct geometric one, as being the value of the magnitude of the position vector r. This solution reads

$$r(\epsilon, b) = \frac{\sqrt{b}}{2 \cdot 6^{1/3}} \left(-\sqrt{-\frac{8 \cdot 3^{1/3}\epsilon}{f(\epsilon)} + \frac{2^{1/3} f(\epsilon)}{\epsilon}} + \sqrt{\frac{8 \cdot 3^{1/3}\epsilon}{f(\epsilon)} - \frac{2^{1/3} f(\epsilon)}{\epsilon} + \frac{12}{\sqrt{-8 \cdot 3^{1/3}\epsilon^2 / f(\epsilon) + 2^{1/3} f(\epsilon)}}} \right), \quad \text{(A.4)}$$

where we define

$$f(\epsilon) = \left(9\epsilon^2 + \sqrt{768\epsilon^6 + 81\epsilon^4}\right)^{1/3}, \quad \text{(A.5)}$$

and

$$\epsilon = \frac{a^2}{b}. \quad \text{(A.6)}$$

We expand the above expression $r(\epsilon, b)$ in Taylor series with respect to ϵ around zero, and we obtain

$$r = \sqrt{b\epsilon}\left(1 - \epsilon^2 + 4\epsilon^4 - 22\epsilon^6 + 140\epsilon^8 + O\left(\epsilon^{10}\right)\right), \quad \text{(A.7)}$$

where obviously the dominant term is $r \sim \sqrt{b\epsilon} = a$. This approximation is valid around the North and South poles of the glued, vertically aligned, deformed circles of equation $a = const.$, Figure 11. The solutions of a similar type of quartic equation in θ,

$$a^2 \cos^4\theta - 2a^2 \cos^2\theta - b \cos\theta + a^2 = 0, \quad \text{(A.8)}$$

can provide the inverse $\theta = \theta(a, b)$.

A more compact expression can be obtained if we solve the inverse transformation Eq. (A.3) in cartesian coordinates, namely, $x = r \sin\phi, z = r\cos\phi$ as functions of (a, b), where $\phi = \pi/2 - \theta$. Beginning with relation Eq. (A.3) and the above definitions for x, z, with $r^2 = x^2 + z^2$ we obtain

$$r^4 = \left[1 - \left(\frac{x}{a}\right)^{2/3}\right] b^2, \qquad z = a\left(\frac{x}{a}\right)^{2/3} \sqrt{1 - \left(\frac{x}{a}\right)^{2/3}}. \quad \text{(A.9)}$$

After some elementary algebra we obtain the inverse in the form

$$x = \frac{a^{1/2}}{b^{3/4}} \left[\frac{\sqrt{4 + 1/a^{2/3}b}}{2} - \frac{1}{2a^{1/3}b^{1/2}}\right]^{3/4}, \qquad z = \frac{a^2}{b^{5/2}} \left[\frac{\sqrt{4 + 1/a^{2/3}b}}{2} - \frac{1}{2a^{1/3}b^{1/2}}\right]^{3/2}, \quad \text{(A.10)}$$

which can be compressed even more into

$$x = \left(\frac{W}{b}\right)^{3/2} f^{3/4}(W), \qquad z = \left(\frac{W}{b^{11/12}}\right)^{6} f^{3/2}(W), \quad \text{(A.11)}$$

where

$$W = a^{1/3} b^{1/2}. \quad \text{(A.12)}$$

The equations above prove the fact that dipole coordinates have the dimension unbalance between coordinates, as opposed to the other bipolar, bispherical, or toroidal coordinates.

The Lammè coefficients of the dipole coordinate system are

$$h_a = \left(\frac{r}{a}\right)^{3/2} \frac{1}{\left(1 + 3\left(r^4/b^2\right)\right)^{1/2}}, \qquad h_b = \frac{r^3}{b^2} \frac{1}{\left(1 + 3\left(r^4/b^2\right)\right)^{1/2}}, \qquad h_c = \frac{r^{3/2}}{a^{1/2}}. \quad \text{(A.13)}$$

Consequently, the Laplacian in (a, b) coordinates has the form

$$\Delta_{a,b,\varphi} = \frac{1 + 3\cos^2\theta}{\sin^6\theta} \frac{\partial^2}{\partial a^2} + \frac{4}{r \sin^4\theta} \frac{\partial}{\partial a} + \frac{2\left(1 + 3\cos^2\theta\right)}{\cos^3\theta} \frac{\partial}{\partial b} + \frac{r^2\left(1 + 3\cos^2\theta\right)}{\cos^4\theta} \frac{\partial^2}{\partial b^2} + \frac{1}{r^2 \sin^2\theta} \frac{\partial^2}{\partial \varphi^2}. \quad \text{(A.14)}$$

B. Solutions of the Dipole Equation as Expansion in Confluent Hypergeometric Functions

The Heun functions solutions represented as series are analytic in the complex plane within some sectors, which brings some constraints on the solutions. Also, being power series they are not efficiently fast convergent. An alternative to this problem is to expand these solutions in terms of confluent hypergeometric functions [27]. The advantage is that the new series provide a full analytic behavior. They are convergent absolutely and locally uniformly all over the complex plane. Also, the solutions built with these hypergeometric functions are of Floquet type; that is, they represent a series expansion around any point in the complex plane, not only about the singularities at 0 and ∞. We can write these series in the form

$$\begin{aligned} Q_(a)^{j,l,\tilde{l}} &= -2\pi\left(\frac{\chi}{a}\right)^{3/2} \\ &\cdot \exp\left[(-1)^{l+\tilde{l}}\frac{iE^{1/2}a}{\chi^2}+(-1)^{\tilde{l}+1}\frac{\chi}{a}\right]\times\sum_{\mu\in\nu+Z}\eta_\mu^{j,l} \\ &\cdot\frac{e^{(j-1)(\pi/2)+(j-2)\pi\mu}\left[(-1)^{l+1}\left(iE^{1/2}a/\chi^2\right)\right]^\mu}{\left(1-e^{-2\pi i\mu}\right)\Gamma(1/2-\mu)} \\ &\cdot\Phi\left(\mu+\frac{1}{2};1+2\mu;(-1)^l\frac{iE^{1/2}a}{\chi^2}\right), \end{aligned} \tag{B.1}$$

where $\Phi(a;b;z)$ is the Kummer function, the indices $j,l,\tilde{l}\in\{1,2\}$ independently (which generates 8 solutions), and μ is the summation label which is an integer plus an arbitrary complex parameter ν called the Floquet exponent of the DCHE. This exponent (is unique modulo $\mathbf{Z}$) should be conveniently chosen according to the point about which the series expands. Finally, the series coefficients $\eta_\mu^{j,l}$ are obtained from the three-term recursion relation

$$\begin{aligned} &(-1)^j\left(\frac{\alpha}{2}\right)^2\frac{\left(\mu+1/2-(-1)^j\beta_1\right)\left(\mu+1/2+\beta_{-1}\right)}{(\mu+1/2)(\mu+1)} \\ &\cdot\eta_{\mu+1}^{j,l}+\left[\left(\mu^2-\gamma\right)+\left(\frac{\alpha}{2}\right)^2\frac{2\beta_1\beta_{-1}}{(\mu-1/2)(\mu+1)}\right] \\ &\cdot\eta_\mu^{j,l}+(-1)^j\left(\frac{\alpha}{2}\right)^2 \\ &\cdot\frac{\left(\mu-1/2+(-1)^j\beta_1\right)\left(\mu-1/2-\beta_{-1}\right)}{(\mu-1/2)(\mu-1)}\eta_{\mu-1}^{j,l}=0, \end{aligned} \tag{B.2}$$

where α, $\beta_\pm$ and γ are the reduced parameters Eq. (36) for the DCHE associated with the linearized dipole equation (35), and $j,l\in\{1,1\}$ independently.

Disclosure

An earlier version of this paper was presented at the 10th International Conference on Geometry, Integrability and Quantization 2010, Varna, Bulgaria; see [39].

Conflicts of Interest

The author declares that there are no conflicts of interest.

Acknowledgments

This work was inspired by several discussions with M. Doria, visiting Professor at Condensed Matter Theory Group at University of Antwerp. His ideas and comments are acknowledged as the motivation of this paper. The author also acknowledges financial support from the Condensed Matter Theory Group at University of Antwerp and he is grateful to F. M. Peeters for support through sabbatical year.

References

[1] G. M. Whitesides, "Nanoscience, nanotechnology, and chemistry," *Small*, vol. 1, no. 2, pp. 172–179, 2005.

[2] S. Kruchinin, H. Nagao, and S. Aono, *Modern Aspects of Superconductivity*, World Scientific, New Jersey, USA, 2010.

[3] M. Safronova, D. Budker, D. DeMille, D. F. Kimball, A. Derevianko, and C. W. Clark, "Search for new physics with atoms and molecules," *Reviews of Modern Physics*, vol. 90, no. 2, 2018.

[4] D. Bimberg and U. W. Pohl, "Quantum dots: Promises and accomplishments," *Materials Today*, vol. 14, no. 9, pp. 388–397, 2011.

[5] H. Küpfer, G. Linker, G. Ravikumar et al., "Correlation of the vortex order-disorder transition with the symmetry of the crystal lattice in V_3Si," *Physical Review B: Condensed Matter and Materials Physics*, vol. 67, no. 6, 2003.

[6] E. T. Filby, A. A. Zhukov, P. A. de Groot, M. A. Ghanem, P. N. Bartlett, and V. V. Metlushko, "Shape induced anomalies in vortex pinning and dynamics of superconducting antidot arrays with spherical cavities," *Applied Physics Letters*, vol. 89, no. 9, p. 092503, 2006.

[7] A. K. Geim, I. V. Grigorieva, S. V. Dubonos et al., "Phase transitions in individual sub-micrometre superconductors," *Letter to Nature*, vol. 390, no. 6657, pp. 259–262, 1997.

[8] V. R. Misko, "Recent Advances in Superconductivity and Vortex Matter: Selected Topics," *Reports in Advances of Physical Sciences*, vol. 1, no. 4, p. 1750009, 2017.

[9] B. Xu, M. V. Milošević, and F. M. Peeters, "Magnetic properties of vortex states in spherical superconductors," *Physical Review B: Condensed Matter and Materials Physics*, vol. 77, no. 14, 2008.

[10] M. M. Doria, A. R. Romaguera, and F. M. Peeters, "Effect of the boundary condition on the vortex patterns in mesoscopic three-dimensional superconductors: Disk and sphere," *Physical Review B*, vol. 75, Article ID 064505, 2007.

[11] M. M. Doria, A. R. Romaguera, M. V. Milosevic, and F. M. Peeters, "Threefold onset of vortex loops in superconductors with a magnetic core," *EPL (Europhysics Letters)*, vol. 79, p. 47006, 2007.

[12] B. J. Baelus, D. Sun, and F. M. Peeters, "Vortex structures in mesoscopic superconducting spheres," *Physical Review B: Condensed Matter and Materials Physics*, vol. 75, no. 17, 2007.

[13] G. Zha, S. Zhou, B. Zhu, Y. Shi, and H. Zhao, "Superconducting phase transitions in thin mesoscopic rings with enhanced surface superconductivity," *Physical Review B: Condensed Matter and Materials Physics*, vol. 74, no. 2, 2006.

[14] G. Zha, S. Zhou, B. Zhu, Y. Shi, and H. Zhao, "Charge distribution in thin mesoscopic superconducting rings with enhanced surface superconductivity," *Physical Review B: Condensed Matter and Materials Physics*, vol. 73, 2006.

[15] V. A. Schweigert and F. M. Peeters, "Phase transitions in thin mesoscopic superconducting disks," *Physical Review B: Condensed Matter and Materials Physics*, vol. 57, no. 21, pp. 13817–13832, 1998.

[16] V. A. Schweigert, F. M. Peeters, and P. S. Deo, "Vortex Phase Diagram for Mesoscopic Superconducting Disks," *Physical Review Letters*, vol. 81, no. 13, pp. 2783–2786, 1998.

[17] J. J. Palacios, "Metastability and Paramagnetism in Superconducting Mesoscopic Disks," *Physical Review Letters*, vol. 84, no. 8, pp. 1796–1799, 2000.

[18] J. J. Palacios, "Vortex matter in superconducting mesoscopic disks: Structure, magnetization, and phase transitions," *Physical Review B: Condensed Matter and Materials Physics*, vol. 58, no. 10, pp. R5948–R5951, 1998.

[19] O. Sato, M. Kato, and J. Physics 871, "A variety of vortex state solutions of Ginzburg-Landau equation on superconducting mesoscopic plates," *Journal of Physics: Conference Series (JPCS)*, Article ID 012029, pp. 1–4, 2017.

[20] H. L. Cycon, R. G. Froese, W. Kirsch, and B. Simon, *Schrödinger Operators*, Schrödinger Operators, Berlin, Heidelberg, 1987.

[21] K. Miller and B. Simon, "Quantum magnetic Hamiltonians with remarkable spectral properties," *Physical Review Letters*, vol. 44, no. 25, pp. 1706-1707, 1980.

[22] D. Picca, "The dipolar potential," *Journal of Physics A: Mathematical and General*, vol. 15, no. 9, pp. 2801–2808, 1982.

[23] G. Lévai and B. W. Williams, "The generalized Coulomb problem: an exactly solvable model," *Journal of Physics A: Mathematical and General*, vol. 26, no. 13, pp. 3301–3306, 1993.

[24] A. D. Alhaidari and H. Bahlouli, "Publisher's Note: Electron in the Field of a Molecule with an Electric Dipole Moment," *Physical Review Letters*, vol. 100, no. 12, 2008.

[25] D. Zwillinger, *Handbook of Differential Equations*, Academic Press, San Diego, Calif, USA, 1989.

[26] P. Moon and D. E. Spencer, *Field Theory Handbook*, Springer, Berlin, Germany, 1971.

[27] A. Ronveaux, *Heuns Differential Equations*, University Press, Oxford, England, 1955.

[28] A. Ralko and T. T. Truong, "Heun functions and the energy spectrum of a charged particle on a sphere under a magnetic field and Coulomb force," *Journal of Physics A: Mathematical and General*, vol. 35, no. 45, pp. 9573–9583, 2002.

[29] A. Ralko and T. T. Truong, "Behaviour of three charged particles on a plane under perpendicular magnetic field," *Journal of Physics A: Mathematical and General*, vol. 35, no. 45, pp. 9671–9683, 2002.

[30] A. Debosscher, "Unification of one-dimensional Fokker-Planck equations beyond hypergeometrics: factorizer solution method and eigenvalue schemes," *Physical Review E: Statistical, Nonlinear, and Soft Matter Physics*, vol. 57, no. 1, pp. 252–275, 1998.

[31] A. Y. Kazakov, "The central two-point connection problem for the reduced confluent Heun equation," *Journal of Physics A: Mathematical and General*, vol. 39, no. 10, pp. 2339–2348, 2006.

[32] C. W. Bauer, Z. Ligeti, M. Luke, A. V. Manohar, and M. Trott, "Global analysis of inclusive B decays," *Physical Review D: Particles, Fields, Gravitation and Cosmology*, vol. 70, Article ID 094017, 2004.

[33] D. Bouaziz and M. Bawin, "Regularization of the singular inverse square potential in quantum mechanics with a minimal length," *Physical Review A: Atomic, Molecular and Optical Physics*, vol. 76, Article ID 032112, 2007.

[34] B. Gao, "Solutions of the Schrödinger equation for an attractive $1/r^6$ potential," *Physical Review A*, vol. 58, p. 1728, 1998.

[35] A. A. Stahlhofen, "Susy, Gauss, Heun and physics: a magic square?" *Journal of Physics A: Mathematical and General*, vol. 37, no. 43, pp. 10129–10138, 2004.

[36] M. Swisdak, "Notes on the Dipole Coordinate System," https://arxiv.org/pdf/physics/0606044.pdf.

[37] M. N. Fatkullin and Y. S. Sitnov, "Dipole coordinate system and some of its characteristics," *Geomegnatism and Aeronomy*, vol. 12, pp. 293–295, 1972.

[38] A. Kageyama, T. Sugiyama, and K. Watanabe, "A Note on the Dipole Coordinates," *Computers & Geosciences*, vol. 32, no. 3, pp. 265–269, 2006.

[39] A. Ludu, "Vortex patterns beyond hypergeometric," *Journal of Geometry and Symmetry in Physics*, vol. 26, pp. 85–103, 2012.

[40] A. Ludu, M. V. Milosevic, and F. M. Peeters, "Vortex states in axially symmetric superconductors in applied magnetic field," *Mathematics and Computers in Simulation*, vol. 82, no. 7, pp. 1258–1270, 2012.

[41] A. Ludu, J. Van Deun, M. V. Milosevic, A. Cuyt, and F. M. Peeters, "Analytic treatment of vortex states in cylindrical superconductors in applied axial magnetic field," *Journal of Mathematical Physics*, vol. 51, no. 8, 082903, 29 pages, 2010.

[42] E. Hille, *Ordinary Differential Equations in The Complex Domain*, Willey, New York, NY, USA, 1976.

[43] M. N. Hounkonnou, A. Ronveaux, and K. Sodoga, "Factorization of some confiuent Heun's differential equations," *Applied Mathematics and Computation*, vol. 189, pp. 816–820, 2007.

[44] W. Bühring, "The double confluent Heun equation: characteristic exponent and connection formulae," *Methods and Applications of Analysis*, vol. 1, no. 3, pp. 348–370, 1994.

[45] W. Lay, K. Bay, and S. Y. Slavyanov, "Asymptotic and numeric study of eigenvalues of the double confluent Heun equation," *Journal of Physics A: Mathematical and General*, vol. 31, no. 42, pp. 8521–8531, 1998.

[46] W. Magnus and S. Winkler, *Hill's Equation*, Inderscience Publishers, New York, NY, USA, 1966.

[47] K. Kuiken, "Heun's equation and the hypergeometric equation," *SIAM Journal on Mathematical Analysis*, vol. 10, no. 3, pp. 655–657, 1979.

[48] P. A. Clarkson and P. J. Olver, "Symmetry and the Chazy Equation," *Journal of Differential Equations*, vol. 124, no. 1, pp. 225–246, 1996.

[49] P. A. Becker, "Normalization integrals of orthogonal Heun functions," *Journal of Mathematical Physics*, vol. 38, no. 7, pp. 3692–3699, 1997.

[50] G. Valent, "Heun functions versus elliptic functions," *Mathematical Physics*, 2005.

[51] D. D. B. Figueiredo, "Ince's limits for confluent and double-confluent Heun equations," *Mathematical Physics*, 2005.

[52] P. Malits and J. Math, "On certain Heun functions, associated functions of discrete variables, and applications," *International*

Journal of Mathematics and Mathematical Sciences, vol. 59, p. 3717, 2003.

[53] M. Abramowitz and I. A. Stegun, *Handbook of Mathematical Functions*, National Bureau of Standards, Washington, DC, USA, 1964, http://www.math.sfu.ca/.

[54] G. Arfken in, *Mathematical Methods in Physics*, Academic Press, Orlando, FL, USA, pp. 112-117, 1970.

[55] H. Bateman, "Spheroidal and bipolar coordinates," *Duke Mathematical Journal*, vol. 4, no. 1, pp. 39–50, 1938.

[56] E. H. Lockwood, *A book of curves*, Cambridge University Press, Cambridge, England, pp. 186-190, 1967.

10

Diffusion Process and Reaction on a Surface

M. E. K. Fuziki,[1] M. K. Lenzi,[2] M. A. Ribeiro,[3] A. Novatski,[3] and E. K. Lenzi[3,4]

[1]*Departamento de Engenharia Química, Universidade Tecnologica Federal do Paraná, Ponta Grossa, PR 84016-210, Brazil*
[2]*Departamento de Engenharia Química, Universidade Federal do Paraná, Curitiba, PR 81531-990, Brazil*
[3]*Departamento de Física, Universidade Estadual de Ponta Grossa, Ponta Grossa, PR 87030-900, Brazil*
[4]*National Institute of Science and Technology for Complex Systems, Centro Brasileiro de Pesquisas Física, Rio de Janeiro, RJ 22290-180, Brazil*

Correspondence should be addressed to E. K. Lenzi; eklenzi@uepg.br

Academic Editor: Antonio Scarfone

We investigate the influence of the surface effects on a diffusive process by considering that the particles may be sorbed or desorbed or undergo a reaction process on the surface with the production of a different substance. Our analysis considers a semi-infinite medium, where the particles may diffuse in contact with a surface with active sites. For the surface effects, we consider integrodifferential boundary conditions coupled with a kinetic equation which takes non-Debye relation process into account, allowing the analysis of a broad class of processes. We also consider the presence of the fractional derivatives in the bulk equations. In this scenario, we obtain solutions for the particles in the bulk and on the surface.

1. Introduction

The description of the dynamic processes present on the surface such as adsorption-desorption or reactions and their influence on the diffusion is very important due to the broadness of applications in several fields such as optimization of industrial processes, electronic devices, the action of pharmaceuticals in the organism, and materials science [1]. Particularly in chemical engineering applications of reactions that occur on solid surfaces are of the great interest and advances have recently been achieved by using the heterogeneous catalysis in the pollutants removal [2], with emphasis on oxidative pathways involving the formation of radicals [3–5], and in the biodiesel production [6–8]. In this sense, it is worth not only studying chemical reactions on substrates, but also analyzing different phenomena occurring in a given surface, for example, the adsorption of proteins [9–12], which can be controlled by nanoroughness on the surface of the material [13, 14] or hybridization of DNA [15–17], among others. In these systems, different species may diffuse and react [18–21], which implies considering suitable changes in the diffusion equation or in the boundary conditions to account for the processes of interest [18, 19, 22, 23]. These contexts usually are analyzed by using the standard approaches related to Markovian processes. Therefore, extensions of these approaches become very important in order to consider non-Markovian processes where anomalous diffusion or non-Debye relaxations on the kinetic processes are present.

Here, we analyze the surface effects on a diffusion process by considering one-dimensional semi-infinite media in contact with a surface, which may adsorb (and desorb) particles or absorb the particles to perform a reaction process with the formation of another kind of particle. Figure 1 illustrates the scenario analyzed in this manuscript, where two species (particles or substances) 1 and 2 diffuse in the bulk which is in contact with a surface. Species 1 may be adsorbed or sorbed by the surface. For the first process, species 1 is removed from the bulk by the surface and after some time is desorbed with a characteristic time. In the other case, when species 1 is sorbed, a reaction process may occur with the formation of species 2. In this sense, the reaction only occurs on the surface, for instance, in the presence of an specific catalyst. Thus, species 1 reacts promoting the formation of species 2, in other words, following a first-order irreversible reaction $1 \longrightarrow 2$. This reaction can be described by a kinetic equation [21, 24, 25]. In order to describe these processes, the model considered here is a set of coupled equations that may represent reaction, adsorption, and desorption processes

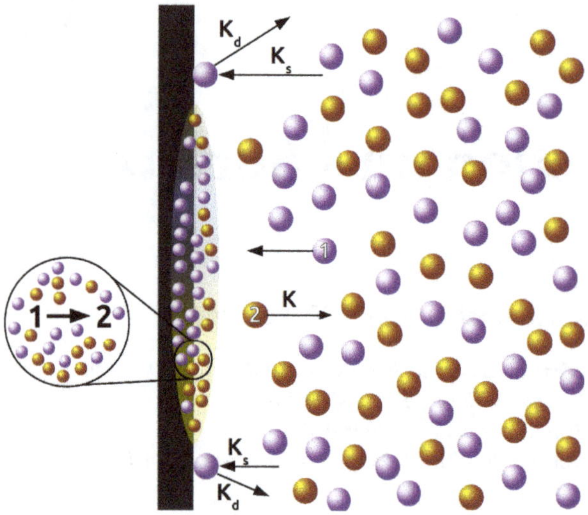

FIGURE 1: Illustration of the interactions occurring between particles and, for example, a catalyst surface. Species 1 in contact with the surface can either be adsorbed or sorbed and by a reaction process produce species 2, i.e., $1 \longrightarrow 2$.

of a substrate on a surface and also the diffusion process in the bulk. These developments are performed in Section 2 by considering generalized diffusion equations [26–40] for the species (1 and 2) in the bulk coupled with kinetic equations as boundary conditions. The discussion and conclusions are presented in Section 3.

2. Diffusion and Surface Effects

Let us start our analysis about the surface effects on the diffusion processes by considering the system governed in the bulk by the following generalized diffusion equations

$$\frac{\partial}{\partial t}\rho_1(x,t) = \mathcal{K}_1 \mathcal{F}_{\alpha_1}\left[\frac{\partial^2}{\partial x^2}\rho_1(x,t)\right], \tag{1}$$

and

$$\frac{\partial}{\partial t}\rho_2(x,t) = \mathcal{K}_2 \mathcal{F}_{\alpha_2}\left[\frac{\partial^2}{\partial x^2}\rho_2(x,t)\right], \tag{2}$$

with $0 \le x < \infty$, where $\mathcal{K}_1$ and $\mathcal{K}_2$ are the generalized diffusion coefficients related to particles 1 and 2, respectively. The quantities $\rho_1(x,t)$ and $\rho_2(x,t)$ represent the density of each particle present in the bulk and $\mathcal{F}_{\alpha_{1(2)}}\{\cdots\}$, in (1) and (2), is the integrodifferential operator:

$$\mathcal{F}_{\alpha_{1(2)}}\{\rho_{1(2)}(x,t)\} = \frac{\partial}{\partial t}\int_0^t dt' k_{\alpha_{1(2)}}(t-t')\rho_{1(2)}(x,t'), \tag{3}$$

where $k_{\alpha_{1(2)}}(t)$ defines how the past history of each system influences the time evolution of $\rho_{1(2)}(x,t)$ and depending on the choice of $k_{\alpha_{1(2)}}(t)$ different fractional differential operators may be obtained. For example, the case

$$k_{\alpha_{1(2)}}(t) = \frac{t^{\alpha_{1(2)}-1}}{\Gamma(\alpha_{1(2)})} \tag{4}$$

corresponds to the well-known Riemann-Liouville fractional operator [41] for $0 < \alpha_{1(2)} < 1$, which has been applied in several situations related to anomalous diffusion [42–45]. The exponential kernel

$$k_{\alpha_{1(2)}}(t) = \frac{\mathcal{R}(\alpha_{1(2)})}{1-\alpha_{1(2)}}e^{-\overline{\alpha}_{1(2)}t}, \tag{5}$$

with $\overline{\alpha}_{1(2)} = \alpha_{1(2)}/(1-\alpha_{1(2)})$ and $\mathcal{R}(\alpha_{1(2)})$ being a normalization factor, corresponds to the fractional operator of Caputo-Fabrizio [46]. Further possibilities for describing the kernel $k_{\alpha_{1(2)}}(t)$ are discussed by Gómez-Aguilar et al. [47], in particular the kernel $k_{\alpha_{1(2)}}(t) = \mathcal{R}(\alpha_{1(2)})E_{\alpha_{1(2)}}(-\overline{\alpha}_{1(2)}t^{\alpha_{1(2)}})/(1-\alpha_{1(2)})$, where $E_\alpha(\ldots)$ is the Mittag-Leffler function [41], which leads us to Atangana-Baleanu fractional operator [48–50]. It has been considered in the Bloch system [51], by resulting in a different behavior from the one obtained in the usual context where differential operators of integer order are considered. These operators, by taking a variable order into account, have been used to extend the Gray-Scott reaction-diffusion model which describe irreversible reaction between two species. In this scenario, the Atangana-Baleanu-Caputo fractional differential operator has shown a faster stabilization behavior and Liouville-Caputo fractional differential operator presented a stronger memory effect [52]. From these choices for $k_{\alpha_{1(2)}}(t)$, we observe that the Riemann-Liouville operator has a singularity at the origin ($t = 0$), while the recently proposed Caputo-Fabrizio and Atangana-Baleanu operators are nonsingular [46–48, 53–55], and the last one may manifest different regimes of diffusion.

On the surface, we consider that the processes are governed by the following equations:

$$\mathcal{K}_1\frac{\partial}{\partial x}\mathcal{F}_{\alpha_1}\{\rho_1(x,t)\}\Big|_{x=0} = \frac{d}{dt}\mathcal{C}(t) + \int_0^t k(t-t')\rho_1(0,t')\,dt' \tag{6}$$

and

$$\mathcal{K}_2\frac{\partial}{\partial x}\mathcal{F}_{\alpha_2}\{\rho_2(x,t)\}\Big|_{x=0} = -\int_0^t k(t-t')\rho_1(0,t')\,dt'. \tag{7}$$

Here $k(t)$ is related to the rate of particle sorption by the surface which by a reaction process produces substance 2. In addition to (6) and (7), we also have the boundary conditions $\partial_x\rho_1(\infty,t) = 0$ and $\partial_x\rho_2(\infty,t) = 0$ to solve (1) and (2). In (6), $\mathcal{C}(t)$ represents the density of particles which are adsorbed by the surface. For the adsorption and desorption processes on the surface, we assume that they may be modeled by the following kinetic equation [56]:

$$\frac{d}{dt}\mathscr{C}(t) = k_s\rho_1(0,t) - \int_0^t k_d\left(t-t'\right)\mathscr{C}\left(t'\right)dt'. \quad (8)$$

In (8), $\rho_1(0,t)$ is the bulk density just in front of the surface, which may be adsorbed by the surface. The parameter k_s is connected to the adsorption phenomenon, being related to a characteristic adsorption time $\tau \propto 1/k_s$, and $k_d(t)$ is a kernel that governs the desorption phenomenon. Thus, the surface density of adsorbed particles depends on the bulk density of particles just in front of the membrane, and on the surface density of particles already sorbed [56]. Equation (8) also extends the usual kinetic equations of first order to situations characterized by unusual relaxations, i.e., non-Debye relaxations for which a nonexponential behavior of the densities can be obtained, depending on the choice of the kernels [56, 57].

From (1), (2), (6), and (7), it is possible to show that

$$\frac{d}{dt}\left(\int_0^\infty \rho_1(x,t)\,dx + \mathscr{C}(t)\right) = -\int_0^t k\left(t-t'\right)\rho_1\left(0,t'\right)dt', \quad (9)$$

and

$$\frac{d}{dt}\left(\int_0^\infty \rho_2(x,t)\,dx\right) = \int_0^t k\left(t-t'\right)\rho_1\left(0,t'\right)dt'. \quad (10)$$

In (9) the term $\int_0^t k(t-t')\rho_1(0,t')dt'$ implies the removal of the particles from the bulk by the surface to promote the production of species 2 by a reaction process. Thus, (11) shows that the mass (number of particles) variation on species 1 is connected to the variations on species 2. In particular, the negative sign shows that the variation of particles of one species produces an opposite variation on the other. The particles produced by a reaction process on the surface are being released from the surface to the bulk. Equations (9) and (10) imply

$$\frac{d}{dt}\left(\int_0^\infty \rho_1(x,t)\,dx + \mathscr{C}(t)\right) = -\frac{d}{dt}\left(\int_0^\infty \rho_2(x,t)\,dx\right), \quad (11)$$

which consequently yields

$$\mathscr{C}(t) + \int_0^\infty \rho_1(x,t)\,dx + \int_0^\infty \rho_2(x,t)\,dx = \text{constant}, \quad (12)$$

i.e., a direct consequence of the conservation of the total number of particles present in the system.

These systems of coupled equations can be solved by using the Laplace transform and the Green function approach. In fact, by applying the Laplace transform ($\mathscr{L}\{\rho_{1(2)}(x,t)\} = \int_0^\infty dt e^{-st}\rho_{1(2)}(x,t) = \overline{\rho}_{1(2)}(x,s)$ and $\mathscr{L}^{-1}\{\overline{\rho}_{1(2)}(x,s)\} = \int_0^\infty dt e^{-st}\overline{\rho}_{1(2)}(x,s) = \rho_{1(2)}(x,t)$) it is possible to simplify the previous equations and obtain that

$$\mathscr{D}_1(s)\frac{\partial^2}{\partial x^2}\overline{\rho}_1(x,s) - s\overline{\rho}_1(x,s) = -\rho_1(x,0), \quad (13)$$

and

$$\mathscr{D}_2(s)\frac{\partial^2}{\partial x^2}\overline{\rho}_2(x,s) - s\overline{\rho}_2(x,s) = -\rho_2(x,0), \quad (14)$$

which are subjected, in the Laplace domain, to the boundary conditions:

$$\mathscr{D}_1(s)\frac{\partial}{\partial x}\overline{\rho}_1(x,s)\Big|_{x=0} = \left[\frac{sk_s}{s+\overline{k}_d(s)} + \overline{k}(s)\right]\overline{\rho}_1(0,s) - \frac{\overline{k}_d(s)}{s+\overline{k}_d(s)}\mathscr{C}(0), \quad (15)$$

$$\mathscr{D}_2(s)\frac{\partial}{\partial x}\overline{\rho}_2(x,s)\Big|_{x=0} = -\overline{k}(s)\overline{\rho}_1(0,s) \quad (16)$$

with $\rho_1(x,0) = \varphi_1(x)$ and $\rho_2(x,0) = \varphi_2(x)$, where $\overline{\mathscr{D}}_{1(2)}(s) = \mathscr{K}_{1(2)}s\overline{k}_{\alpha_{1(2)}}(s)$. By using the Green function approach, the solutions for $\overline{\rho}_1(x,s)$ and $\overline{\rho}_2(x,s)$ can found by considering the previous equations and they are given by

$$\overline{\rho}_1(x,s) = -\int_0^\infty \overline{\mathscr{G}}_1\left(x,x';s\right)\varphi_1\left(x'\right)dx' + \frac{sk_s}{s+\overline{k}_d(s)}\overline{\rho}_1(0,s)\overline{\mathscr{G}}_1(x,0;s) - \frac{\overline{k}_d(s)}{s+\overline{k}_d(s)}\mathscr{C}(0)\overline{\mathscr{G}}_1(x,0;s) \quad (17)$$

and

$$\overline{\rho}_2(x,s) = -\int_0^\infty \overline{\mathscr{G}}_2\left(x,x';s\right)\varphi_2\left(x'\right)dx' - \overline{k}(s)\overline{\mathscr{G}}_2(x,0;s)\overline{\rho}_1(0,s). \quad (18)$$

where $\overline{\mathscr{G}}_1(x,x';s)$ and $\overline{\mathscr{G}}_2(x,x';s)$ correspond to the Green function for each species and $\mathscr{C}(0)$ represents the quantity of particles which may initially be present on the surface. These Green functions are obtained by solving the following equation:

$$\overline{\mathscr{D}}_i(s)\frac{\partial^2}{\partial x^2}\overline{\mathscr{G}}_i\left(x,x';s\right) - s\overline{\mathscr{G}}_i\left(x,x';s\right) = \delta\left(x-x'\right) \quad (19)$$

by taking into account the suitable boundary conditions for each species related to (17) and (18). Thus, the solutions given by (17) and (18) are obtained from the combination of (15), (16), and (19) by taking into account their boundary conditions, following the procedure employed in [58]. It is also interesting to note that the boundary conditions chosen for the Green function were performed in order to evidence the adsorption-desorption process manifested by the surface on the distribution $\rho(x,t)$. For species 1, by performing some calculations, it possible to show that the Green function is given by

$$\begin{aligned}\overline{\mathscr{G}}_1\left(x,x';s\right) &= -\frac{1}{2\sqrt{s\overline{\mathscr{D}}_1(s)}}\left(e^{-\sqrt{s/\overline{\mathscr{D}}_1(s)}|x+x'|}+e^{-\sqrt{s/\overline{\mathscr{D}}_1(s)}|x+x'|}\right) \\ &+\frac{\overline{k}(s)}{\sqrt{s\overline{\mathscr{D}}_1(s)}+\overline{k}(s)}\frac{1}{\sqrt{s\overline{\mathscr{D}}_1(s)}}e^{-\sqrt{s/\overline{\mathscr{D}}_1(s)}|x+x'|},\end{aligned} \tag{20}$$

when the boundary conditions, which are consistent with (17),

$$\overline{\mathscr{D}}_1(s)\frac{\partial}{\partial x}\overline{\mathscr{G}}_1\left(x,x';s\right)\Big|_{x=0}-\overline{k}(s)\overline{\mathscr{G}}_1\left(0,x';s\right)=0 \tag{21}$$

and $\partial_x\overline{\mathscr{G}}_1(x,x';s)|_{x=\infty}=0$ are considered to solve (21). Note that the boundary condition used to solve the equation for the Green function incorporates the reaction on the surface. The term related to the adsorption-desorption process was not incorporated in (21) to be evident on the solution as an additional term. In (20), the first term corresponds to the spreading of the initial condition and the second term shows the influence of the surface on the spreading of the system, which in this case corresponds to a reaction sorption process where species 1 is removed from the system. It is worth mentioning that depending on the choice of $\overline{k}_{\alpha_1}(s)$ (or $k_{\alpha_1}(t)$) the previous Green function may present a stationary solution. In particular, for case $\overline{k}_{\alpha_1}(s)\longrightarrow k_{\alpha_1}=constant$ with $\overline{k}(s)\longrightarrow k=constant$, for $s\longrightarrow 0$ (i.e., for long times), we have, for the Green function, the following stationary behavior:

$$\mathscr{G}_1\left(x,x'\right)=-\frac{1}{2\sqrt{\widetilde{D}_1}}\left(e^{-|x-x'|/\sqrt{\widetilde{D}_1}}-e^{-|x+x'|/\sqrt{\widetilde{D}_1}}\right), \tag{22}$$

which in this case is independent of $\overline{k}(s)$, where $\widetilde{D}_1 = k_{\alpha_1}\mathscr{K}_1 = constant$. This feature is interesting and shows that the surface effects with this characteristic play an important role at intermediate times. For the case $\overline{k}(s)=0$ (or $k(t)=0$), we also have an stationary solution and it is given by

$$\mathscr{G}_1\left(x,x'\right)=-\frac{1}{2\sqrt{\widetilde{D}_1}}\left(e^{-|x+x'|/\sqrt{\widetilde{D}_1}}+e^{-|x+x'|/\sqrt{\widetilde{D}_1}}\right), \tag{23}$$

which is different from (22) due to the boundary conditions used to obtain the Green function. The scenario is typical, for example, of the Caputo-Fabrizio operator, which is characterized by an exponential kernel as mentioned before.

For species 2, we have that

$$\overline{\mathscr{G}}_2\left(x,x';s\right) = -\frac{1}{2\sqrt{s\overline{\mathscr{D}}_2(s)}}\left(e^{-\sqrt{s/\overline{\mathscr{D}}_2(s)}|x+x'|}+e^{-\sqrt{s/\overline{\mathscr{D}}_2(s)}|x+x'|}\right), \tag{24}$$

when the boundary conditions, which are consistent with Eq. (18),

$$\overline{\mathscr{D}}_2(s)\frac{\partial}{\partial x}\overline{\mathscr{G}}_2\left(x,x';s\right)\Big|_{x=0} = \overline{\mathscr{D}}_2(s)\frac{\partial}{\partial x}\overline{\mathscr{G}}_2\left(x,x';s\right)\Big|_{x=\infty}=0 \tag{25}$$

are employed to solve (21). Equation (24) may also present a stationary solution depending on the choice of $\overline{k}_{\alpha_2}(s)$, similar to (20).

In order to perform the inverse of Laplace transform and obtain the time dependent behavior of (20) and (24), we first consider the Riemann-Liouville fractional time derivative and after that the Caputo-Fabrizio fractional time derivative. We also consider that $\overline{k}(s)=k=constant$; i.e., the sorption of species 1 by the surface occurs in a constant rate. For the Riemann-Liouville fractional time derivative, we have that

$$\overline{\mathscr{D}}_{1(2)}(s)=\mathscr{K}_{1(2)}s^{1-\alpha_{1(2)}/2}, \tag{26}$$

and, consequently,

$$\begin{aligned}\mathscr{G}_{1,RL}\left(x,x';t\right) &= -\frac{1}{\sqrt{4\mathscr{K}_1t^{\alpha_1}}}\left(H^{1,0}_{1,1}\left[\frac{\left|x-x'\right|}{\sqrt{\mathscr{K}_1t^{\alpha_1}}}\Bigg|^{(1-\alpha_1/2,\alpha_1/2)}_{(0,1)}\right]\right. \\ &\left.+H^{1,0}_{1,1}\left[\frac{\left|x+x'\right|}{\sqrt{\mathscr{K}_1t^{\alpha_1}}}\Bigg|^{(1-\alpha_1/2,\alpha_1/2)}_{(0,1)}\right]\right)+\frac{k}{\sqrt{4\mathscr{K}_1}} \\ &\cdot\int_0^t\frac{dt'}{\sqrt{t'^{\alpha_1}(t-t')^{\alpha_1}}}E_{\alpha_1',\alpha_1'}\left(-\frac{kt'^{\alpha_1'}}{\sqrt{\mathscr{K}_1}}\right) \\ &\cdot H^{1,0}_{1,1}\left[\frac{\left|x+x'\right|}{\sqrt{\mathscr{K}_1(t-t')^{\alpha_1}}}\Bigg|^{(1-\alpha_1/2,\alpha_1/2)}_{(0,1)}\right]\end{aligned} \tag{27}$$

with $\alpha_1'=1-\alpha_1/2$, where $H^{m,n}_{p,q}[x|^{(a,A)}_{(b,B)}]$ is the Fox H function [59] and $E_{\alpha,\beta}(x)$ is the generalized Mittag-Leffler function [59] (see Figure 2). For the Caputo-Fabrizio fractional time operator, we have that

$$\overline{\mathscr{D}}_{1(2)}(s)=\frac{sD_{1(2)}}{s+\overline{\alpha}_{1(2)}} \tag{28}$$

with $D_{1(2)}=[\mathscr{R}(\alpha_{1(2)})/(1-\alpha_{1(2)})]\mathscr{K}_{1(2)}$. By substituting the previous equation in (20) and performing some calculations with $\overline{k}(s) = k = constant$, we obtain that

$$\begin{aligned}\mathscr{G}_{1,CF}\left(x,x';t\right) &= -\int_0^t \frac{dt'}{\sqrt{4\pi D_1 t'}}\Xi\left(t,t'\right) \\ &\cdot e^{-\overline{\alpha}_1 t'}\left(e^{-(x-x')^2/4D_1 t'} + e^{-(x+x')^2/4D_1 t'}\right) \\ &- \frac{k}{\sqrt{k^2+4D_1\overline{\alpha}_1}}\int_0^t dt'\Xi\left(t,t'\right)e^{-\overline{\alpha}_1 t'}\int_0^{t'} dt'' \\ &\cdot \frac{\Lambda\left(t''\right)}{\sqrt{4\pi D_1\left(t'-t''\right)^3}}e^{-x^2/4D_1(t-t')}\end{aligned} \tag{29}$$

for the Caputo-Fabrizio fractional time operator with $\Xi(t,t') = \overline{\alpha}_1 + \delta(t-t')$:

$$\Lambda(t) = \xi_- erfc\left(\xi_-\sqrt{t}\right)e^{\xi_-^2 t} - \xi_+ erfc\left(\xi_+\sqrt{t}\right)e^{\xi_+^2 t} \tag{30}$$

where $erfc(x)$ is the complementary error function and $\xi_\pm = k/\sqrt{4D_1} \mp \sqrt{k^2/(4D_1)+\overline{\alpha}_1}$. Figure 3 shows the behavior of (29) for different values of α_1. In Figure 4, we present the behavior of (29) for different times in order to show that for long times a stationary state is reached. For (24), we have

$$\begin{aligned}&\mathscr{G}_{2,RL}\left(x,x';t\right) \\ &= -\frac{1}{\sqrt{4\mathscr{K}_2 t^{\overline{\alpha}_2}}}\left(H_{1,1}^{1,0}\left[\frac{\left|x-x'\right|}{\sqrt{\mathscr{K}_2 t^{\overline{\alpha}_2}}}\middle|_{(0,1)}^{(1-\overline{\alpha}_2/2,\alpha_2/2)}\right]\right. \\ &\left. + H_{1,1}^{1,0}\left[\frac{\left|x+x'\right|}{\sqrt{\mathscr{K}_2 t^{\overline{\alpha}_2}}}\middle|_{(0,1)}^{(1-\overline{\alpha}_2/2,\overline{\alpha}_2/2)}\right]\right)\end{aligned} \tag{31}$$

for the Riemann-Liouville operator (see Figure 5) and

$$\begin{aligned}\mathscr{G}_{2,CF}\left(x,x';t\right) &= -\int_0^t \frac{dt'}{\sqrt{4\pi D_2 t'}}\Xi\left(t,t'\right) \\ &\cdot e^{-\alpha t'}\left(e^{-(x-x')^2/4D_2 t'} + e^{-(x+x')^2/4D_2 t'}\right)\end{aligned} \tag{32}$$

for the Caputo-Fabrizio fractional time operator (see Figure 6).

In (17) and (18), the first term promotes the spreading of the initial condition and the other terms represent the influence of the surface on the diffusive process in the bulk. Thus, we need to quantify these terms, in order to obtain the solution for each species. From (17), after some calculations, it is also possible to show that

$$\begin{aligned}&\overline{\rho}_1(0,s) \\ &= -\frac{1}{1-sk_s\overline{\omega}(s)\overline{\mathscr{G}}_1(0,0;s)}\left(\int_0^\infty \overline{\mathscr{G}}_1\left(0,x';s\right)\right. \\ &\left.\cdot\varphi\left(x'\right)dx' + \overline{k}_d(s)\overline{\omega}(s)\overline{\mathscr{G}}_1(0,0;s)\mathscr{C}(0)\right)\end{aligned} \tag{33}$$

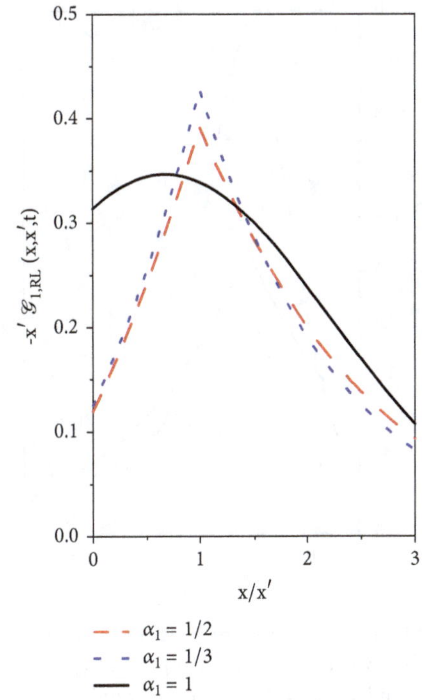

FIGURE 2: This figure illustrates the behavior of (27) by considering different values of α_1. We consider, for simplicity, $\tau_{1,RL} = (x'^2/\mathscr{K}_1)^{1/\alpha_1}$, $k\tau_{1,RL}/x' = 1$, and $t = \tau_{1,RL}$, in arbitrary unities.

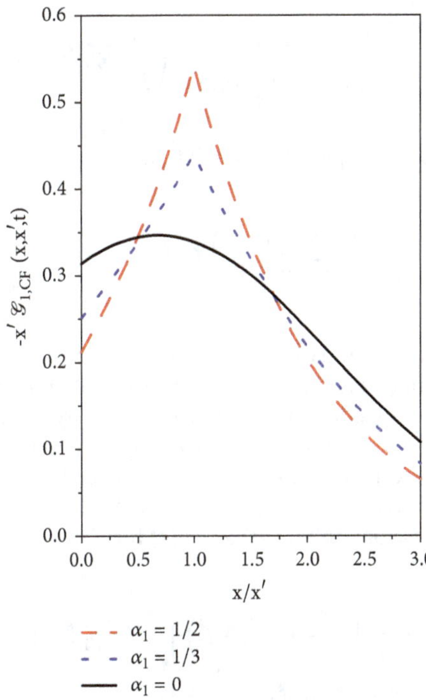

FIGURE 3: This figure illustrates the behavior of (29) by considering different values of α_1. We consider, for simplicity, $\tau_{1,CF} = \sqrt{x'^2/D_1}$, $k\tau_{1,CF}/x' = 1$, $\tau_{1,CF} = 1$, and $\alpha_1 = 1/2$, in arbitrary unities.

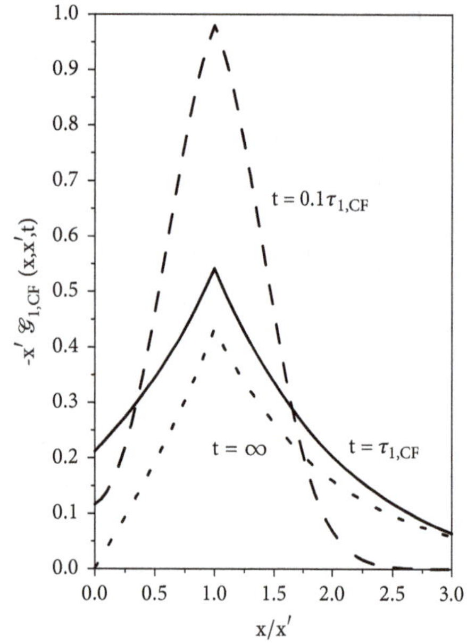

FIGURE 4: This figure illustrates the behavior of (29) by considering different times. We consider, for simplicity, $\tau_{1,CF} = \sqrt{x'^2/D_1}$, $k\tau_{1,CF}/x' = 1$, and $\alpha_1\tau_{1,CF} = 1/2$, in arbitrary unities.

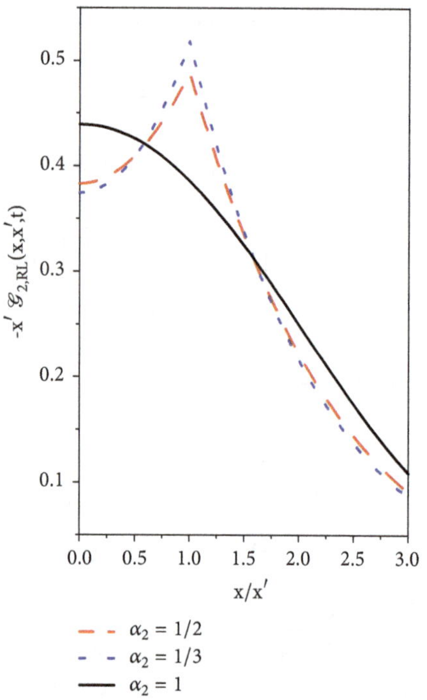

FIGURE 5: This figure illustrates the behavior of (31) by considering different values of α_2. We consider, for simplicity, $\tau_{2,RL} = (x'^2/\mathscr{K}_2)^{1/\alpha_2}$, $k\tau_{2,RL}/x' = 1$, and $t = \tau_{RL,2}$, in arbitrary unities.

where $\overline{\omega}(s) = 1/(s + \overline{k}_d(s))$. By using the previous result, it is possible to show that the quantity of adsorbed particles by the surface is given by

$$\overline{\mathscr{C}}(s) = -\frac{k_s\overline{\omega}(s)}{1 - sk_s\overline{\omega}(s)\overline{\mathscr{G}}_1(0,0;s)}\left(\int_0^\infty \overline{\mathscr{G}}_1(0,x';s) \cdot \varphi(x')\,dx'\right) + \overline{\omega}(s)\left(1 - k_s\overline{k}_d(s) \cdot \frac{\overline{\omega}(s)\overline{\mathscr{G}}_1(0,0;s)}{1 - sk_s\overline{\omega}(s)\overline{\mathscr{G}}_1(0,0;s)}\right)\mathscr{C}(0). \tag{34}$$

By using (33) and (34) and by applying the inverse Laplace transform in (13) and (14), it is possible to obtain $\rho_1(x,t)$ and $\rho_2(x,t)$. We consider, as performed for the Green functions, the fractional operators characterized by (4) and (5) in order to obtain the inverse Laplace transform of (33) and (34). For the first case, i.e., the Riemann-Liouville fractional time derivative, we obtain for (33) and (34) that

$$\rho_{1,RL}(0,t) = \int_0^t \frac{dt'}{t'}\Upsilon_{RL}(t-t')\int_0^\infty dx'\varphi_1(x') \cdot H_{1,1}^{1,0}\left[\frac{|x-x'|}{\sqrt{\mathscr{D}_2 t'^{\alpha}}}\middle|_{(0,1)}^{(0,\alpha/2)}\right] + \left(\Upsilon_{RL}(t) - \int_0^t dt'\Upsilon_{RL}(t-t')\mathscr{I}_{RL}(t')\right)\mathscr{C}_{RL}(0) \tag{35}$$

$$\mathscr{C}_{RL}(t) = \int_0^t dt''\omega(t-t'')\left\{\int_0^{t''}\frac{dt'}{t'}\Upsilon_{RL}(t''-t') \cdot \int_0^\infty dx'\varphi_1(x')H_{1,1}^{1,0}\left[\frac{|x-x'|}{\sqrt{\mathscr{D}_2 t'^{\alpha}}}\middle|_{(0,1)}^{(0,\alpha/2)}\right] + \left((1-k_s)\Upsilon_{RL}(t'') - k_s\int_0^{t''}dt'\Upsilon_{RL}(t''-t')\mathscr{I}_{RL}(t')\right)\mathscr{C}_{RL}(0)\right\}, \tag{36}$$

For the case characterized by (4), i.e., the Caputo-Fabrizio fractional time operator, we obtain that

$$\rho_{1,CF}(0,t) = \int_0^t \frac{dt'}{(4\pi D_1 t')^{3/2}}\Upsilon_{CF}(t-t') \cdot \int_0^\infty dx'\varphi_1(x')e^{-x'^2/4D_1t'} + \left(\Upsilon_{CF}(t) - \int_0^t dt'\Upsilon_{CF}(t-t')\mathscr{I}_1(t')\right)\mathscr{C}_{CF}(0), \tag{37}$$

$$\mathscr{C}_{CF}(t) = \int_0^t dt''\omega\left(t-t''\right)\left\{\int_0^{t''}\frac{dt'}{\left(4\pi D_1 t'\right)^{3/2}}\Upsilon_{CF}\left(t''-t'\right)\int_0^{\infty}dx'\varphi_1\left(x'\right)e^{-x'^2/4D_1t'} + \left(\left(1-k_s\right)\Upsilon_{CF}\left(t''\right) - k_s\int_0^{t''}dt'\Upsilon_{CF}\left(t''-t'\right)\mathscr{I}_{CF}\left(t'\right)\right)\mathscr{C}_{CF}(0)\right\}. \tag{38}$$

The quantities $\Upsilon_{RL(CF)}(t)$ and $\mathscr{I}_{RL(CF)}(t)$ presented in previous equations are defined as follows:

$$\begin{aligned}\Upsilon_{RL(CF)}(t) = {}& -\mathscr{G}_{RL(CF)}(0,0;t) - \sum_{n=1}^{\infty}k_s^n \cdot \int_0^t dt_n\mathscr{I}_{RL(CF)}\left(t-t_n\right)\mathscr{G}_{RL(CF)}\left(0,0;t-t_n\right)\\ &\times\int_0^{t_n}dt_{n-1}\mathscr{I}_{RL(CF)}\left(t_n-t_{n-1}\right)\cdot\mathscr{G}_{RL(CF)}\left(0,0;t_n-t_{n-1}\right)\cdots\\ &\cdot\int_0^{t_2}dt_1\mathscr{I}_{RL(CF)}\left(t_2-t_1\right)\cdot\int_0^{t_1}dt'\mathscr{G}_{RL(CF)}\left(0,0;t_1-t'\right)\mathscr{G}_{RL(CF)}\left(0,0;t'\right)\end{aligned} \tag{39}$$

with

$$\begin{aligned}\mathscr{I}_{RL(CF)}(t) = {}& \delta(t) - \sum_{n=1}^{\infty}(-1)^n\int_0^t dt_n k_d\left(t-t_n\right)\cdot\int_0^{t_n}dt_{n-1}k_d\left(t_n-t_{n-1}\right)\cdots\\ &\cdot\int_0^{t_3}dt_2k_d\left(t_2-t_1\right)\int_0^{t_2}dt_1k_d\left(t_1\right).\end{aligned} \tag{40}$$

By using the previous equations, the inverse of Laplace transform of (17) and (18)

$$\begin{aligned}\rho_{1,RL(CF)}(x,t) = {}& -\int_0^{\infty}\mathscr{G}_{1,RL(CF)}\left(x,x';t\right)\varphi_1\left(x'\right)dx' + k_s\int_0^t dt''\mathscr{I}_{RL(CF)}\left(t-t''\right)\int_0^{t''}\rho_1\left(0,t''-t'\right)\\ &\cdot\mathscr{G}_{1,RL(CF)}\left(x,0;t'\right) - \left(\mathscr{G}_{1,RL(CF)}(x,0;t) - \int_0^t dt'\mathscr{I}_{RL(CF)}\left(t-t'\right)\mathscr{G}_{1,RL(CF)}\left(x,0;t'\right)\right)\mathscr{C}(0)\end{aligned} \tag{41}$$

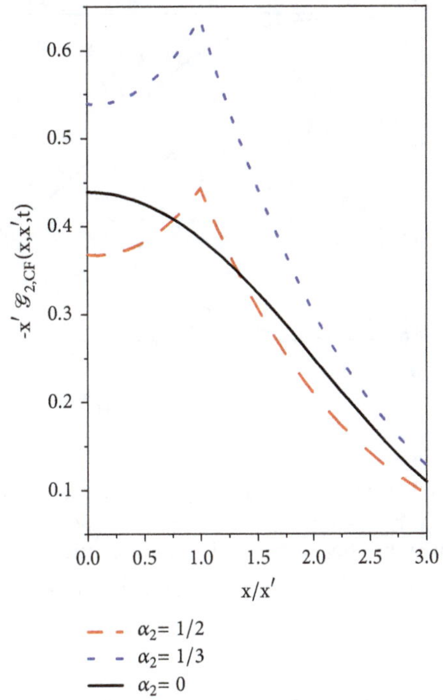

FIGURE 6: This figure illustrates the behavior of (32) by considering different values of α_2. We consider, for simplicity, $\tau_{2,CF} = \sqrt{x'^2/D_2}$, $k\tau_{2,CF}/x' = 1$, $\tau_{2,CF} = 1$, and $\alpha_2 = 1/2$, in arbitrary unities.

and

$$\begin{aligned}\rho_{2,RL(CF)}(x,t) = {}& -\int_0^{\infty}\mathscr{G}_{2,RL(CF)}\left(x,x';t\right)\varphi_2\left(x'\right)dx' - \int_0^t dt''k\left(t-t''\right)\\ &\cdot\int_0^{t''}dt'\mathscr{G}_{2,RL(CF)}\left(x,0;t''-t'\right)\rho_{1,RL(CF)}\left(0,t'\right).\end{aligned} \tag{42}$$

Equations (41) and (42) are the solutions for the scenario analyzed here, which is characterized by a diffusion with processes occurring on a surface. By using these results, it is possible to obtain the survival probability, which is defined as $S_{1(2)}(t) = \int_0^{\infty}dx\rho_{1(2)}(x,t)$ and it is related to the quantity of particles each species present in the bulk. In Figure 7, we illustrate the time behavior of the survival probability for species 1 and 2 by considering different fractional time operators. We observe that for the fractional operator related to the exponential kernel a stationary state is reached. Another interesting quantity to be analyzed is the mean square displacement, i.e., $(\Delta_{1(2)}x)^2 = \langle(x - \langle x\rangle_{1(2)})^2\rangle_{1(2)}$, related to the spreading of the distributions of each species. For example, in absence of adsorption-desorption process ($k_s = 0$ and $\mathscr{C}(0) = 0$) with $k(t) = constant$, species 1 manifests two different regimes, when the Riemann-Liouville fractional time derivative is considered. In particular, for small times $(\Delta_1 x)^2 \sim t^{\gamma}$ and for long times $(\Delta_1 x)^2 \sim t^{\gamma/2}$. In this context, for the Caputo-Fabrizio fractional time operator, we have $(\Delta_1 x)^2 \sim t$ for

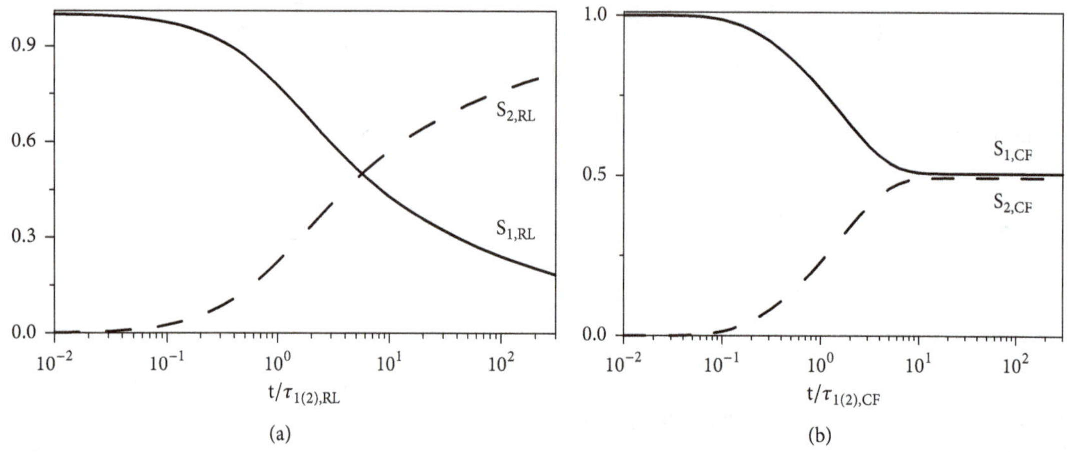

FIGURE 7: (a) illustrates the survival probability for species 1 and 2 by considering the Riemann-Liouville fractional time operator. We consider, for simplicity, $\tau_{1(2),RL} = (x'^2/\mathcal{K}_{1(2)})^{1/\alpha_{1(2)}}$, $k\tau_{1(2),RL}/x' = 1$, $\varphi_1(x) = \delta(x - x')$, $\varphi_2(x) = 0$, and $\alpha_{(1)2} = 1/2$, in arbitrary unities. (b) illustrates the survival probability for species 1 and 2 by considering the Caputo-Fabrizio fractional time operator. We consider, for simplicity, $\tau_{1(2),CF} = \sqrt{x'^2/D_{1(2)}}$, $k\tau_{1(2),CF}/x' = 1$, $\tau_{1(2),CF} = 1$, $\varphi_1(x) = \delta(x - x')$, $\varphi_2(x) = 0$, and $\alpha_{1(2)} = 1/2$, in arbitrary unities.

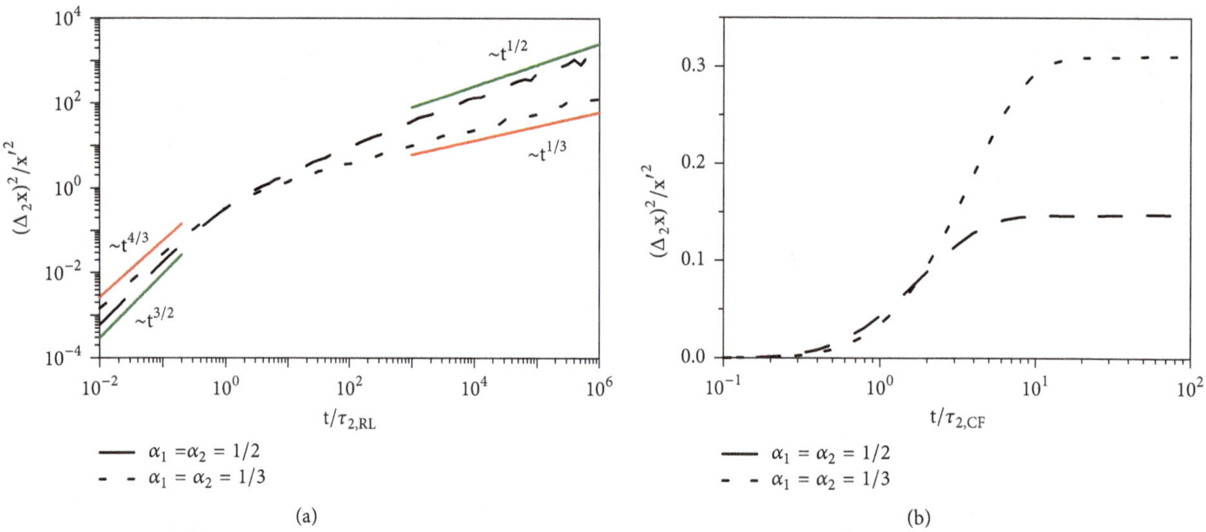

FIGURE 8: (a) illustrates the mean square displacement for species 2 for different values of $\alpha_{1(2)}$ by taking into account the Riemann-Liouville fractional time operator. We consider, for simplicity, $\tau_{1(2),RL} = (x'^2/\mathcal{K}_{1(2)})^{1/\alpha_{1(2)}}$, and $k\tau_{1,CF}/x' = 1$, $\varphi_1(x) = \delta(x-x')$, $\varphi_2(x) = 0$, and $\tau_{1(2),RL} = 1$, in arbitrary unities. (b) illustrates the mean square displacement for species 2 for different values of $\alpha_{1(2)}$ for species 2 by taking into account the Caputo-Fabrizio fractional time operator. We consider, for simplicity, $\tau_{1(2),CF} = \sqrt{x'^2/D_{1(2)}}$, and $k\tau_{1,CF}/x' = 1$, $\varphi_1(x) = \delta(x-x')$, $\varphi_2(x) = 0$, and $\tau_{1(2),CF} = 1$, in arbitrary unities.

small times and $(\Delta_1 x)^2 \sim$ *constant* for long times. For species 2, the behavior of the mean square displacement is illustrated in Figure 8 and, similar to species 1, the Riemann-Liouville fractional time derivative presents two different regimes. For the Caputo-Fabrizio fractional time operator, we observe a superdiffusion (see Figure 8(a)) or a transient (see Figure 8(b)) before reaching a stationary state. The initial behavior observed for species 2 is a consequence of the initial condition considered in this application.

3. Discussions and Conclusions

We have investigated a diffusion process in semi-infinite media in contact with a surface, where the particles may be adsorbed (and desorbed) or sorbed by the surface in order to promote though a reaction process the formation of another species of particle. For particles in the bulk, we have considered generalized diffusion equations, which may be related to a different types of fractional differential operator. In particular, we have considered the Riemann-Liouville and

Caputo-Fabrizio fractional time derivative. For each one of these operators, we have found the solutions and obtained the quantity of particles adsorbed by the surface as a function of time. The reaction process considered here occurs only on the surface and it is irreversible, i.e., $1 \longrightarrow 2$. The particles obtained from the reaction process are released from the surface to the bulk. In the bulk, we have considered that reaction between the particles is absent; i.e., only the diffusion governs the dynamics of the particles in the bulk. The solutions for these cases have been obtained in terms of the Fox H-functions and generalized Mittag-Leffler functions which are usually related to anomalous diffusion. They have exhibited different diffusive behaviors depending on the conditions employed to analyze the system. In particular, for the situation discussed in Figure 8(a), we have verified that species 2 manifests a superdiffusive behavior for small times when the Riemann-Liouville fractional time operator is considered. For long times, the behavior is essentially governed by the diffusion equations and, consequently, subdiffusive behaviors were obtained. In Figure 8(b), we have analyzed the behavior of the mean square displacement for the Caputo-Fabrizio fractional time operator. For small times, we have a transient until the system reaches a stationary solution, i.e., a time independent solution, for long times. The stationary solution manifested by the system, when the Caputo-Fabrizio is employed, may be related to the resetting process connected to this operator as pointed out in [60]. Finally, we hope that the results presented here may be useful to discussion of situations related to anomalous diffusion.

Conflicts of Interest

The authors declare no conflicts of interest.

Acknowledgments

E. K. Lenzi acknowledges CNPq for partial financial support.

References

[1] G. Hariharan and K. Kannan, "Review of wavelet methods for the solution of reaction-diffusion problems in science and engineering," *Applied Mathematical Modelling*, vol. 38, no. 3, pp. 799–813, 2014.

[2] K. Pirkanniemi and M. Sillanpää, "Heterogeneous water phase catalysis as an environmental application: a review," *Chemosphere*, vol. 48, no. 10, pp. 1047–1060, 2002.

[3] D. Wan, W. Li, G. Wang, K. Chen, L. Lu, and Q. Hu, "Adsorption and heterogeneous degradation of rhodamine B on the surface of magnetic bentonite material," *Applied Surface Science*, vol. 349, pp. 988–996, 2015.

[4] W.-D. Oh, Z. Dong, and T.-T. Lim, "Generation of sulfate radical through heterogeneous catalysis for organic contaminants removal: Current development, challenges and prospects," *Applied Catalysis B: Environmental*, vol. 194, pp. 169–201, 2016.

[5] S. Rahim Pouran, A. A. Abdul Raman, and W. M. A. Wan Daud, "Review on the application of modified iron oxides as heterogeneous catalysts in Fenton reactions," *Journal of Cleaner Production*, vol. 64, pp. 24–35, 2014.

[6] S. Semwal, A. K. Arora, R. P. Badoni, and D. K. Tuli, "Biodiesel production using heterogeneous catalysts," *Bioresource Technology*, vol. 102, no. 3, pp. 2151–2161, 2011.

[7] H.-J. Kim, B.-S. Kang, M.-J. Kim et al., "Transesterification of vegetable oil to biodiesel using heterogeneous base catalyst," *Catalysis Today*, vol. 93–95, pp. 315–320, 2004.

[8] A. F. Lee, J. A. Bennett, J. C. Manayil, and K. Wilson, "Heterogeneous catalysis for sustainable biodiesel production via esterification and transesterification," *Chemical Society Reviews*, vol. 43, no. 22, pp. 7887–7916, 2014.

[9] M. Malmsten, *Biopolymers at Interfaces*, vol. 110, CRC Press, 2003.

[10] D. S. Vieira, P. R. G. Fernandes, H. Mukai, R. S. Zola, G. G. Lenzi, and E. K. Lenzi, "Surface roughness influence on CPE parameters in electrolytic cells," *International Journal of Electrochemical Science*, vol. 11, no. 9, pp. 7775–7784, 2016.

[11] V. Hlady and J. Buijs, "Protein adsorption on solid surfaces," *Current Opinion in Biotechnology*, vol. 7, no. 1, pp. 72–77, 1996.

[12] K. Ishihara, H. Nomura, T. Mihara, K. Kurita, Y. Iwasaki, and N. Nakabayashi, "Why do phospholipid polymers reduce protein adsorption?" *Journal of Biomedical Materials Research Part B: Applied Biomaterials*, vol. 39, no. 2, pp. 323–330, 1998.

[13] K. Rechendorff, M. B. Hovgaard, M. Foss, V. P. Zhdanov, and F. Besenbacher, "Enhancement of protein adsorption induced by surface roughness," *Langmuir*, vol. 22, no. 26, pp. 10885–10888, 2006.

[14] M. B. Hovgaard, K. Rechendorff, J. Chevallier, M. Foss, and F. Besenbacher, "Fibronectin adsorption on tantalum: the influence of nanoroughness," *The Journal of Physical Chemistry B*, vol. 112, no. 28, pp. 8241–8249, 2008.

[15] A. Dolatshahi-Pirouz, K. Rechendorff, M. B. Hovgaard, M. Foss, J. Chevallier, and F. Besenbacher, "Bovine serum albumin adsorption on nano-rough platinum surfaces studied by QCM-D," *Colloids and Surfaces B: Biointerfaces*, vol. 66, no. 1, pp. 53–59, 2008.

[16] Y.-P. Ho, M. C. Kung, S. Yang, and T.-H. Wang, "Multiplexed hybridization detection with multicolor colocalization of quantum dot nanoprobes," *Nano Letters*, vol. 5, no. 9, pp. 1693–1697, 2005.

[17] L. C. Brousseau, "Label-Free "Digital Detection" of Single-Molecule DNA Hybridization with a Single Electron Transistor," *Journal of the American Chemical Society*, vol. 128, no. 35, pp. 11346-11347, 2006.

[18] M. Sinder, V. Sokolovsky, and J. Pelleg, "Reversible A⟷B reaction–diffusion process with initially mixed reactants: Boundary layer function approach," *Physica B: Condensed Matter*, vol. 406, no. 15-16, pp. 3042–3049, 2011.

[19] D. ben-Avraham and S. Havlin, *Diffusion and Reactions in Fractals and Disordered Systems*, Cambridge University Press, Cambridge, UK, 2000.

[20] D. Marin, M. A. Ribeiro, H. V. Ribeiro, and E. . Lenzi, "A nonlinear Fokker–Planck equation approach for interacting systems: Anomalous diffusion and Tsallis statistics," *Physics Letters A*, vol. 382, no. 29, pp. 1903–1907, 2018.

[21] E. K. Lenzi, M. A. Ribeiro, M. E. Fuziki, M. K. Lenzi, and H. V. Ribeiro, "Nonlinear diffusion equation with reaction terms: analytical and numerical results," *Applied Mathematics and Computation*, vol. 330, pp. 254–265, 2018.

[22] B. Chopard and M. Droz, *Cellular automata modeling of physical systems. collection aléa-saclay: Monographs and texts in statistical physics*, 1998.

[23] I. Zeldovich, G. I. Barenblatt, V. Librovich, and G. Makhviladze, *Mathematical Theory of Combustion and Explosions*, Consultants Bureau, New York, NY, USA, 1985.

[24] L. G. Arnaut, S. J. Formosinho, and H. Burrows, *Chemical Kinetics: from Molecular Structure to Chemical Reactivity*, Elsevier, 2006.

[25] A. Sapora, P. Cornetti, B. Chiaia, E. K. Lenzi, and L. R. Evangelista, "Nonlocal diffusion in porous media: A spatial fractional approach," *Journal of Engineering Mechanics*, vol. 143, no. 5, 2017.

[26] L. R. Evangelista and E. K. Lenzi, *Fractional Diffusion Equations and Anomalous Diffusion*, Cambridge University Press, 2018.

[27] P. C. Assis, R. P. de Souza, P. C. da Silva, L. R. da Silva, L. S. Lucena, and E. K. Lenzi, "Non-Markovian Fokker-Planck equation: Solutions and first passage time distribution," *Physical Review E: Statistical, Nonlinear, and Soft Matter Physics*, vol. 73, no. 3, 2006.

[28] E. K. Lenzi, R. S. Mendes, J. S. Andrade Jr, L. R. da Silva, and L. S. Lucena, "N-dimensional fractional diffusion equation and green function approach: Spatially dependent diffusion coefficient and external force," *Physical Review E*, vol. 71, no. 5, 2005.

[29] E. K. Lenzi, R. Mendes, K. S. Fa, L. C. Malacarne, and L. Da Silva, "Anomalous diffusion: Fractional Fokker–Planck equation and its solutions," *Journal of Mathematical Physics*, vol. 44, pp. 2179–2185, 2003.

[30] S. A. El-Wakil and M. A. Zahran, "Fractional representation of Fokker-Planck equation," *Chaos, Solitons & Fractals*, vol. 12, no. 10, pp. 1929–1935, 2001.

[31] A. Hanyga, "Multi-dimensional solutions of space-time-fractional diffusion equations," *Proceedings A*, vol. 458, no. 2018, pp. 429–450, 2002.

[32] O. P. Agrawal, "Solution for a fractional diffusion-wave equation defined in a bounded domain," *Nonlinear Dynamics*, vol. 29, no. 1–4, pp. 145–155, 2002.

[33] F. Ren, J. Liang, W. Qiu, X. Wang, Y. Xu, and R. Nigmatullin, "An anomalous diffusion model in an external force fields on fractals," *Physics Letters A*, vol. 312, no. 3-4, pp. 187–197, 2003.

[34] R. Gorenflo, A. Iskenderov, and Y. Luchko, "Mapping between solutions of fractional diffusion-wave equations," *Fractional Calculus & Applied Analysis. An International Journal for Theory and Applications*, vol. 3, no. 1, pp. 75–86, 2000.

[35] B. N. N. Achar and J. W. Hanneken, "Fractional radial diffusion in a cylinder," *Journal of Molecular Liquids*, vol. 114, no. 1-3, pp. 147–151, 2004.

[36] F. Mainardi and G. Pagnini, "The Wright functions as solutions of the time-fractional diffusion equation," *Applied Mathematics and Computation*, vol. 141, no. 1, pp. 51–62, 2003.

[37] G. M. Zaslavsky, "Chaos, fractional kinetics, and anomalous transport," *Physics Reports A Review Section of Physics Letters*, vol. 371, no. 6, pp. 461–580, 2002.

[38] R. Hilfer, *Applications of Fractional Calculus in Physics*, World Scientific, Singapore, 2000.

[39] R. Metzler and J. Klafter, "The restaurant at the end of the random walk: recent developments in the description of anomalous transport by fractional dynamics," *Journal of Physics A: Mathematical and General*, vol. 37, no. 31, pp. R161–R208, 2004.

[40] M. Ashrafuzzaman, "Aptamers as Both Drugs and Drug-Carriers," *BioMed Research International*, vol. 2014, Article ID 697923, 21 pages, 2014.

[41] I. Podlubny, *Fractional Differential Equations: An Introduction to Fractional Derivatives, Fractional Differential Equations, to Methods of Their Solution And Some of Their Applications*, vol. 198, Elsevier, 1998.

[42] R. Metzler and J. Klafter, "The random walk's guide to anomalous diffusion: a fractional dynamics approach," *Physics Reports*, vol. 339, no. 1, pp. 1–77, 2000.

[43] V. R. Voller, "Fractional Stefan problems," *International Journal of Heat and Mass Transfer*, vol. 74, pp. 269–277, 2014.

[44] S. Blasiak, "Time-fractional heat transfer equations in modeling of the non-contacting face seals," *International Journal of Heat and Mass Transfer*, vol. 100, pp. 79–88, 2016.

[45] F. R. G. B. Silva, G. Gonçalves, M. K. Lenzi, and E. K. Lenzi, "An extension of the linear Luikov system equations of heat and mass transfer," *International Journal of Heat and Mass Transfer*, vol. 63, pp. 233–238, 2013.

[46] M. Caputo and M. Fabrizio, "A new definition of fractional derivative without singular kernel," *Progress in Fractional Differentiation and Applications*, vol. 1, no. 2, pp. 73–85, 2015.

[47] J. F. Gómez-Aguilar and A. Atangana, "Fractional Hunter-Saxton equation involving partial operators with bi-order in Riemann-Liouville and Liouville-Caputo sense," *The European Physical Journal Plus*, vol. 132, no. 2, 2017.

[48] A. Atangana and I. Koca, "Model of thin viscous fluid sheet flow within the scope of fractional calculus: fractional derivative with and no singular kernel," *Fundamenta Informaticae*, vol. 151, no. 1-4, pp. 145–159, 2016.

[49] A. Atangana and J. F. Gómez-Aguilar, "Decolonisation of fractional calculus rules: Breaking commutativity and associativity to capture more natural phenomena," *The European Physical Journal Plus*, vol. 133, no. 4, 2018.

[50] A. Atangana, "Non validity of index law in fractional calculus: A fractional differential operator with MARkovian and non-MARkovian properties," *Physica A: Statistical Mechanics and its Applications*, vol. 505, pp. 688–706, 2018.

[51] J. Gómez-Aguilar, "Chaos in a nonlinear Bloch system with Atangana-Baleanu fractional derivatives," in *Numerical Methods for Partial Differential Equations*, pp. 1–23, 2017.

[52] A. Coronel-Escamilla, J. F. Gómez-Aguilar, L. Torres, and R. F. Escobar-Jiménez, "A numerical solution for a variable-order reaction-diffusion model by using fractional derivatives with non-local and non-singular kernel," *Physica A: Statistical Mechanics and its Applications*, vol. 491, pp. 406–424, 2018.

[53] J. Hristov, "Transient heat diffusion with a non-singular fading memory: from the cattaneo constitutive equation with Jeffrey's kernel to the caputo-fabrizio time-fractional derivative," *Thermal Science*, vol. 20, no. 2, pp. 757–762, 2016.

[54] J. F. Gómez-Aguilar, "Space-time fractional diffusion equation using a derivative with nonsingular and regular kernel," *Physica A: Statistical Mechanics and its Applications*, vol. 465, pp. 562–572, 2017.

[55] T. Abdeljawad and D. Baleanu, "Integration by parts and its applications of a new nonlocal fractional derivative with Mittag-Leffler nonsingular kernel," *The Journal of Nonlinear Science and Applications*, vol. 10, no. 03, pp. 1098–1107, 2017.

[56] V. G. Guimarães, H. V. Ribeiro, Q. Li, L. R. Evangelista, E. K. Lenzi, and R. S. Zola, "Unusual diffusing regimes caused by different adsorbing surfaces," *Soft Matter*, vol. 11, no. 9, pp. 1658–1666, 2015.

[57] R. S. Zola, E. K. Lenzi, L. R. Evangelista, and G. Barbero, "Memory effect in the adsorption phenomena of neutral particles," *Physical Review E: Statistical, Nonlinear, and Soft Matter Physics*, vol. 75, no. 4, 2007.

[58] H. W. Wyld, *Mathematical Methods for Physics*, Westview Press, 1999.

[59] A. M. Mathai, R. K. Saxena, and H. J. Haubold, *The H-function : Theory and Applications*, Springer Science & Business Media, New York, NY, USA, 2010.

[60] A. A. Tateishi, H. V. Ribeiro, and E. K. Lenzi, "The role of fractional time-derivative operators on anomalous diffusion," *Frontiers of Physics*, vol. 5, p. 52, 2017.

11

CIP Method of Characteristics for the Solution of Tide Wave Equations

Yafei Nie,[1] Kai Fu,[2] and Xianqing Lv[1]

[1]*Physical Oceanography Laboratory, Ocean University of China, Qingdao 266100, China*
[2]*School of Mathematical Sciences, Ocean University of China, Qingdao 266100, China*

Correspondence should be addressed to Kai Fu; kfu@ouc.edu.cn

Academic Editor: Zhi-Yuan Sun

The CIP-MOC (Constrained Interpolation Profile/Method of Characteristics) is proposed to solve the tide wave equations with large time step size. The bottom topography and bottom friction, which are very important factor for the tidal wave model, are included to the equation of Riemann invariants as the source term. Numerical experiments demonstrate the good performance of the scheme. Compared to traditional semi-implicit (SI) finite difference scheme which is widely used in tidal wave simulation, CIP-MOC has better stability in simulating large gradient water surface change and has the ability to use much longer time step size under the premise of maintaining accuracy. Besides, numerical tests with reflective boundary conditions are carried out by CIP-MOC with large time step size and good results are obtained.

1. Introduction

Tide wave is the water movement caused by the gravitational force of the celestial body. The control equations of tidal wave model are the original shallow water equations including source terms (e.g., bottom friction, bottom topography and horizontal eddy viscosity term) [1]. Therefore, getting the solution of shallow water equations with source terms is of great significance.

In the simulation of ocean tidal waves, Eulerian schemes are widely used, for example, Backhaus [2] and Casulli [3] used semi-implicit scheme (hereafter SI) for the solution of shallow water equations; Lv and Zhang [4] used the semi-implicit scheme to solve tide wave equations and their computational format was used to study bottom friction coefficients [5] and tidal open boundary conditions [6]; the finite volume method is also frequently used [7, 8]. However, since the phase speed of the gravity waves is so fast, time step is strictly limited by CFL condition when using Eulerian schemes [9].

There are growing interests in finding methods that enable using large time steps to solve shallow water equations. Erbes [9] proposed a semi-Lagrangian method of characteristics (MOC) with quadratic interpolation to solve the shallow water equations with a large CFL number, but there was dispersion error produced. Ogata and Yabe [10] applied the CIP method to shallow waters combined with the MOC and showed low dispersion error and low numerical damping even with large CFL number. Toda et al. [11] proposed a new scheme by adopting the Conservative Semi-Lagrangian (CSL) based on the CIP method which shows good conservation. However, applying this method in realistic simulation of ocean tidal wave will meet a difficulty treating source terms including bottom topography, bottom friction, and boundary conditions.

This paper is designated to apply the CIP-MOC [9] to the solution of tide wave equations which are based on shallow water equations and include source terms of bottom topography and friction. The ideal numerical experiment shows that the CIP-MOC method outperforms SI scheme in stability and dispersion errors when using large time step. Problem with reflection boundary condition is also discussed.

This paper is organized as follows. Section 2 presents the CIP-MOC method. In Section 3, numerical experiments are carried out and comparison is made between the CIP and SI. Conclusion is given in Section 4.

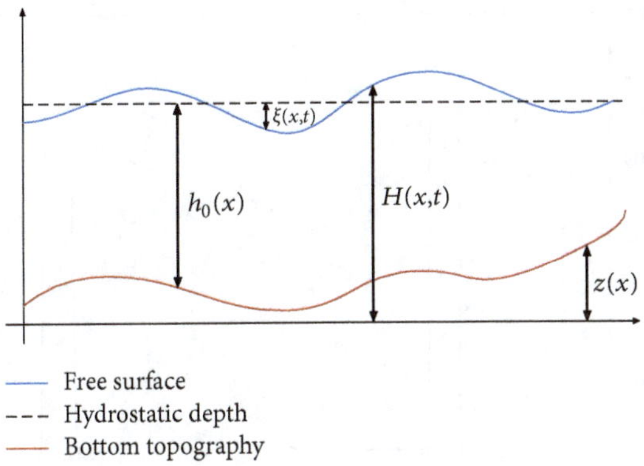

FIGURE 1: Shallow water variables.

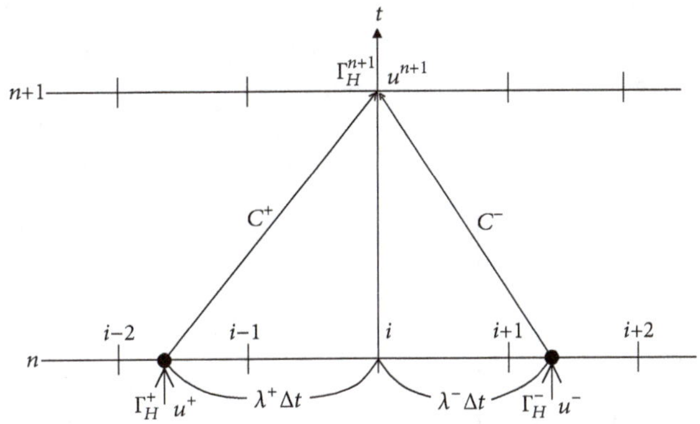

FIGURE 2: Space-time diagram that shows the characteristics $C^{\pm}$. The Riemann invariants consist of $\Gamma_H^{\pm}$ and $u^{\pm}$, $\lambda^{\pm}$ are characteristic speeds.

2. Model and Method

2.1. CIP Method. CIP method is a compact, robust, less diffusive, and high-order scheme in computational fluid dynamics [12]. Compared with other traditional interpolation methods, the internal information of grid is better described by CIP which uses spatial derivative in the description of function distribution in grid cells. We consider the following advection problem:

$$\frac{\partial f}{\partial t} + u\frac{\partial f}{\partial x} = 0, \tag{1}$$

the solution is [13]

$$\begin{aligned} f^{n+1} &= F(x - u\Delta t), \\ \partial_x f^{n+1} &= \frac{dF(x - u\Delta t)}{dx}. \end{aligned} \tag{2}$$

If f and $\partial_x f$ are known at each grid point, the profile between two adjacent points can be approximated by cubic polynomial

$$F_i(X) = a_i X^3 + b_i X^2 + \partial_x f_i X + f_i, \tag{3}$$

where $X = x - x_i$. Consider four constraints

$$\begin{aligned} F(x_i) &= f_i^n, \\ F(x_{iup}) &= f_{iup}^n, \\ \partial_x F(x_i) &= \partial_x f_i^n, \\ \partial_x F(x_{iup}) &= \partial_x f_{iup}^n. \end{aligned} \tag{4}$$

then coefficients can be given as

$$a_i = \frac{(\partial_x f_i + \partial_x f_{iup})}{D^2} + \frac{2(f_i - f_{iup})}{D^3}, \tag{5}$$

$$b_i = \frac{3(f_{iup} - f_i)}{D^2} - \frac{(2\partial_x f_i + \partial_x f_{iup})}{D}, \tag{6}$$

where $\text{sgn}(u_i) = 1\ (u_i \geq 0), -1\ (u_i < 0)$, $iup = i - \text{sgn}(u_i)$, and $D = -\Delta x \cdot \text{sgn}(u_i)$.

Thus, the profile of f and $\partial_x f$ at next time step can be obtained as

$$f_i^{n+1} = F_i(x_i - u_i\Delta t) = a_i\xi^3 + b_i\xi^2 + \partial_x f_i^n \cdot \xi + f_i^n, \tag{7}$$

$$\partial_x f_i^{n+1} = \frac{dF_i(x_i - u_i\Delta t)}{dx} = 3a_i\xi^2 + 2b_i\xi + \partial_x f_i^n, \tag{8}$$

where $\xi = -u_i\Delta t$.

2.2. One-Dimensional Tide Wave Equations. Let bottom friction term be the linear form, then the one-dimensional tide wave equations in a primitive form are written as [14]

$$\begin{aligned} \frac{\partial \zeta}{\partial t} + \frac{\partial[(h_0 + \zeta)u]}{\partial x} &= 0, \\ \frac{\partial u}{\partial t} + u\frac{\partial u}{\partial x} + g\frac{\partial \zeta}{\partial x} + ku &= 0, \end{aligned} \tag{9}$$

where t is time, x is Cartesian coordinate, $h_0(x)$ is undisturbed water depth, $\zeta(x,t)$ is sea surface elevation above the undisturbed sea level, $u(x,t)$ is the flow velocity, k is the bottom friction coefficient, and g is the gravitational acceleration.

To solve (9) with the method of characteristics and preserve the balance between the depth gradient and the bottom effect, the surface gradient method [15] is used here. Replace the variable ζ with the water level $H = \zeta + h_0 + z$, where $z(x)$ represents the bottom topographic height (see Figure 1), we get the equivalent of (9) which can be written as

$$\begin{aligned} \frac{\partial H}{\partial t} + \frac{\partial(Hu)}{\partial x} &= \frac{\partial(uz)}{\partial x}, \\ \frac{\partial u}{\partial t} + u\frac{\partial u}{\partial x} + g\frac{\partial H}{\partial x} + ku &= 0. \end{aligned} \tag{10}$$

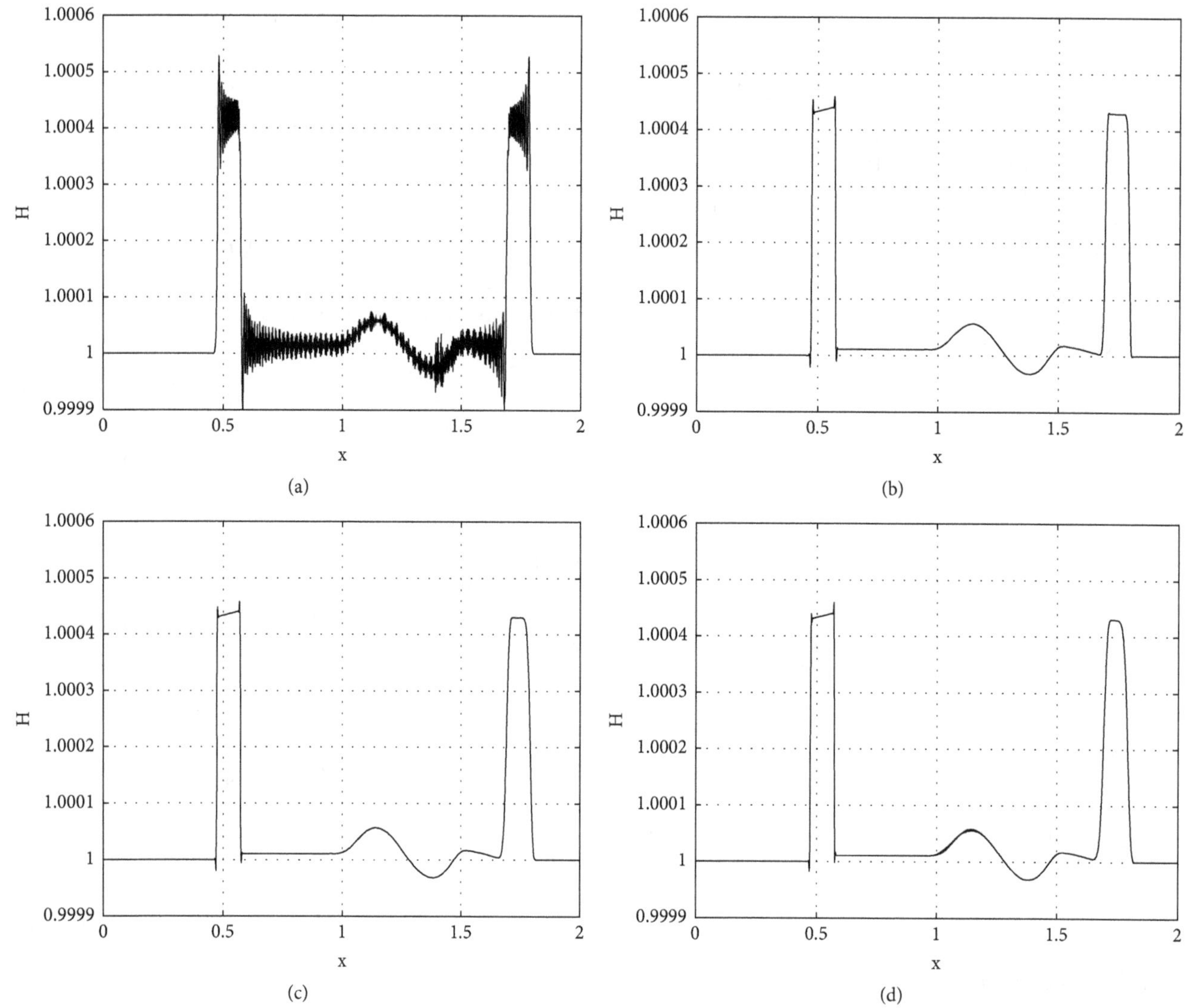

FIGURE 3: A small perturbation of one-dimensional steady state water with a pulse $\zeta = 0.001$. (a) shows SI scheme with CFL=0.63; (b), (c), and (d) are the results of the CIP method with CFL=0.63, 1.57, and 2.61.

In a vector-matrix form, (10) can be rewritten as

$$\frac{\partial}{\partial t}\begin{pmatrix} H \\ u \end{pmatrix} + \begin{pmatrix} u & H \\ g & u \end{pmatrix} \frac{\partial}{\partial x}\begin{pmatrix} H \\ u \end{pmatrix} \equiv \frac{\partial W}{\partial t} + A(W)\frac{\partial W}{\partial x} = \begin{pmatrix} \dfrac{\partial (uz)}{\partial x} \\ -ku \end{pmatrix}. \tag{11}$$

The eigenvalue matrix Λ and eigenvector matrix L of coefficient matrix A are

$$\Lambda = \begin{pmatrix} \lambda^+ & 0 \\ 0 & \lambda^- \end{pmatrix}, \quad L = \begin{pmatrix} \dfrac{\Gamma_H}{g} & -\dfrac{\Gamma_H}{g} \\ 1 & 1 \end{pmatrix}, \tag{12}$$

where $\Gamma_H = \sqrt{gH}$ and $\lambda^{\pm} = u \pm \Gamma_H$. Thus, matrix A can be diagonalized as $L^{-1}AL$, where

$$L^{-1} = \frac{1}{2}\begin{pmatrix} \dfrac{g}{\Gamma_H} & 1 \\ -\dfrac{g}{\Gamma_H} & 1 \end{pmatrix} \tag{13}$$

is the inverse of L. Then (11) can be written as

$$L^{-1}\frac{\partial W}{\partial t} + \left(L^{-1}AL\right)L^{-1}\frac{\partial W}{\partial x} = L^{-1}S, \tag{14}$$

where S represents the right hand side of (11).

Two characteristics about Riemann invariants $R^{\pm} = \Gamma_H \pm u/2$ can be obtained from (14) as

$$\frac{\partial R^{\pm}}{\partial t} + \lambda^{\pm}\frac{\partial R^{\pm}}{\partial x} = \frac{g}{2\Gamma_H}\frac{\partial (uz)}{\partial x} \mp \frac{1}{2}ku. \tag{15}$$

By using CIP method, we have

$$\Gamma_H^{n+1} \pm \frac{u^{n+1}}{2} = \Gamma_H^{\pm} \pm \frac{u^{\pm}}{2} + \frac{g}{2}\int_{t^n}^{t^{n+1}} \frac{1}{\Gamma_H}\frac{\partial (uz)}{\partial x} d\tau^{\pm} \mp \frac{1}{2}k \cdot \Delta t \frac{\left(u^{\pm} + u^{n+1}\right)}{2}, \quad (16)$$

where $\Gamma_H^{\pm}$ and $u^{\pm}$ are calculated by

$$\Gamma_H^{\pm} = \Gamma_{HCIP}\left(x_i - \lambda^{\pm}\Delta t\right), \quad u^{\pm} = u_{CIP}\left(x_i - \lambda^{\pm}\Delta t\right), \quad (17)$$

the subscript CIP in variables means that they are interpolated with the CIP scheme, and the bottom friction term is discretized with the Crank-Nicholson method which is implicit in time. Γ_H and u at time step t^{n+1} can be obtained by solving linear equations (16) (see Figure 2):

$$\Gamma_H^{n+1} = \frac{1}{2}\left[\Gamma_H^{+} + \Gamma_H^{-} + \frac{1}{2}\left(u^{+} - u^{-}\right) + \frac{g}{2}\left(\int_{t^n}^{t^{n+1}} \frac{1}{\Gamma_H}\frac{\partial (uz)}{\partial x} d\tau^{+} + \int_{t^n}^{t^{n+1}} \frac{1}{\Gamma_H}\frac{\partial (uz)}{\partial x} d\tau^{-}\right) - \frac{1}{2}k\Delta t \frac{\left(u^{+} - u^{-}\right)}{2}\right], \quad (18)$$

$$u^{n+1} = \frac{(1/2)\left[u^{+} + u^{-} + 2\left(\Gamma_H^{+} - \Gamma_H^{-}\right) + g\left(\int_{t^n}^{t^{n+1}} (1/\Gamma_H)(\partial (uz)/\partial x)\, d\tau^{+} - \int_{t^n}^{t^{n+1}} (1/\Gamma_H)(\partial (uz)/\partial x)\, d\tau^{-}\right) - (1/2)k\Delta t\left(\left(u^{+} + u^{-}\right)/2\right)\right]}{1 + (1/2)k\Delta t}. \quad (19)$$

Time integration of the bottom topographic term in (18) and (19) can be approximated using $dx^{\pm}/d\tau^{\pm} = u \pm \Gamma_H$ as follows [16]:

$$\int_{t^n}^{t^{n+1}} \frac{1}{\Gamma_H}\frac{\partial (uz)}{\partial x} d\tau^{\pm} \approx -\frac{(uz)^{\pm} - (uz)^n}{\Gamma_H^{\pm}\left(u \pm \Gamma_H\right)^{\pm}}. \quad (20)$$

When applying the CIP method, spatial gradient at the next time step is also needed. Take spatial derivative of (15)

$$\frac{\partial\left(\partial_x R^{\pm}\right)}{\partial t} + \lambda^{\pm}\frac{\partial\left(\partial_x R^{\pm}\right)}{\partial x} = G^{\pm}, \quad (21)$$

$G^{\pm}$ includes the terms related to spatial derivatives of $\lambda^{\pm}$, Γ_H, and u. Considering that the influence of $G^{\pm}$ is not significant in this problem, we neglect $G^{\pm}$ here. Then (21) can be solved as

$$\partial_x\Gamma_H^{n+1} = \frac{1}{2}\left[\partial_x\Gamma_H^{+} + \partial_x\Gamma_H^{-} + \frac{1}{2}\left(\partial_x u^{+} - \partial_x u^{-}\right)\right], \quad (22)$$

$$\partial_x u^{n+1} = \frac{1}{2}\left[\partial_x u^{+} + \partial_x u^{-} + 2\left(\partial_x\Gamma_H^{+} - \partial_x\Gamma_H^{-}\right)\right], \quad (23)$$

then values and derivatives of each value at time step $(n+1)$ are obtained.

2.3. Two-Dimensional Tide Wave Equations. Assuming that pressure is hydrostatic and density is constant, the depth averaged two-dimensional tidal model without horizontal eddy viscosity is given as follows [17]:

$$\begin{aligned}
&\frac{\partial\zeta}{\partial t} + \frac{\partial\left[\left(h_0+\zeta\right)u\right]}{\partial x} + \frac{\partial\left[\left(h_0+\zeta\right)v\right]}{\partial y} = 0,\\
&\frac{\partial u}{\partial t} + u\frac{\partial u}{\partial x} + v\frac{\partial u}{\partial y} - fv + \frac{ku\sqrt{u^2+v^2}}{h_0+\zeta} + g\frac{\partial\zeta}{\partial x} = 0,\\
&\frac{\partial v}{\partial t} + u\frac{\partial v}{\partial x} + v\frac{\partial v}{\partial y} + fu + \frac{kv\sqrt{u^2+v^2}}{h_0+\zeta} + g\frac{\partial\zeta}{\partial y} = 0,
\end{aligned} \quad (24)$$

where $u(x, y, t)$ and $v(x, y, t)$ are x and y components of flow velocity and f represents the Coriolis parameter.

In the same way as one dimension, replace variable ζ with water level $H = \zeta + h_0 + z$; we get the equivalent of (24) which can be written in vector-matrix form as

$$\frac{\partial W}{\partial t} + A(W)\frac{\partial W}{\partial x} + B(W)\frac{\partial W}{\partial y} + S = 0, \quad (25)$$

where

$$W = \begin{pmatrix} H \\ u \\ v \end{pmatrix},$$

$$A(W) = \begin{pmatrix} u & H & 0 \\ g & u & 0 \\ 0 & 0 & u \end{pmatrix},$$

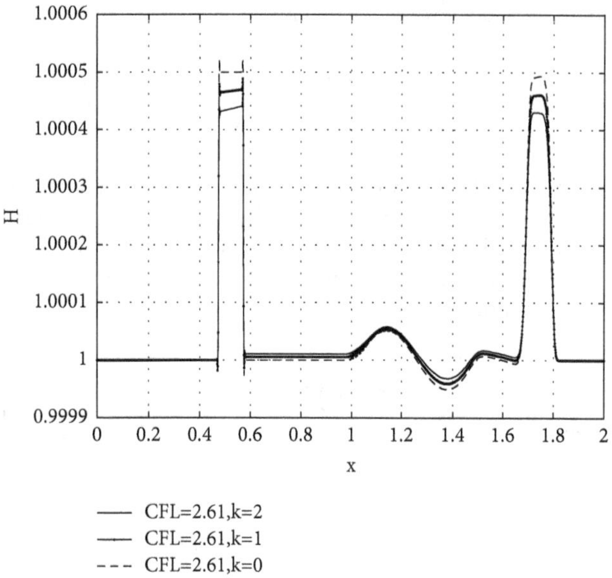

Figure 4: Effect of bottom friction. Results computed by CIP with CFL=2.61 for k=0, 1, and 2.

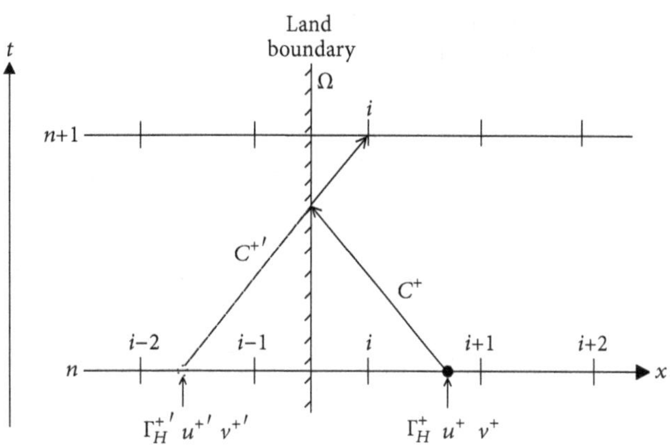

Figure 5: Space-time diagram that shows the reflecting boundary conditions. The filled circle represents the approximated upstream departure point, while the hollow circle represents the virtual upstream point on the shore.

$$B(W)=\begin{pmatrix} v & 0 & H \\ 0 & v & 0 \\ g & 0 & v \end{pmatrix},$$

$$S=\begin{pmatrix} -\left(\dfrac{\partial(uz)}{\partial x}+\dfrac{\partial(vz)}{\partial y}\right) \\ -fv+k\dfrac{u\sqrt{u^2+v^2}}{H-z} \\ fu+k\dfrac{v\sqrt{u^2+v^2}}{H-z} \end{pmatrix}. \tag{26}$$

Since matrices A and B are not commutative; thus, there is no matrix to diagonalize them at the same time. Split (25) into two sequential phases; the source term is divided into two terms added to each directional phase:

$$\frac{\partial W}{\partial t}+A\frac{\partial W}{\partial x}+S_x=0 \quad W^n \longrightarrow W^*, \tag{27}$$

$$\frac{\partial W^*}{\partial t}+B\frac{\partial W^*}{\partial y}+S_y=0 \quad W^* \longrightarrow W^{n+1}, \tag{28}$$

where

$$S_x=\begin{pmatrix} -\dfrac{\partial(uz)}{\partial x} \\ -\dfrac{1}{2}fv+\dfrac{1}{2}k\dfrac{u\sqrt{u^2+v^2}}{H-z} \\ \dfrac{1}{2}fu+\dfrac{1}{2}k\dfrac{v\sqrt{u^2+v^2}}{H-z} \end{pmatrix},$$

$$S_y=\begin{pmatrix} -\dfrac{\partial(vz)}{\partial y} \\ -\dfrac{1}{2}fv+\dfrac{1}{2}k\dfrac{u\sqrt{u^2+v^2}}{H-z} \\ \dfrac{1}{2}fu+\dfrac{1}{2}k\dfrac{v\sqrt{u^2+v^2}}{H-z} \end{pmatrix}. \tag{29}$$

In order to get high accuracy in time, we calculate alternately in the x- and y-directions as follows:

$$W^{n+2}=L_xL_yL_yL_xW^n, \tag{30}$$

where L_x and L_y represent operation of (27) and (28), respectively.

Here we give the solution process in x-direction. Similar to the one-dimensional case, left multiply (27) by the eigenvector matrix L:

$$\begin{aligned} &L^{-1}\frac{\partial W}{\partial t}+\left(L^{-1}AL\right)L^{-1}\frac{\partial W}{\partial x}+L^{-1}S_x \\ &\equiv L^{-1}\frac{\partial W}{\partial t}+\begin{pmatrix} \lambda_x^- & 0 & 0 \\ 0 & \lambda_x^+ & 0 \\ 0 & 0 & u \end{pmatrix}L^{-1}\frac{\partial W}{\partial x}+L^{-1}S_x=0, \end{aligned} \tag{31}$$

where

$$\lambda_x^{\pm}=u\pm\Gamma_H,$$

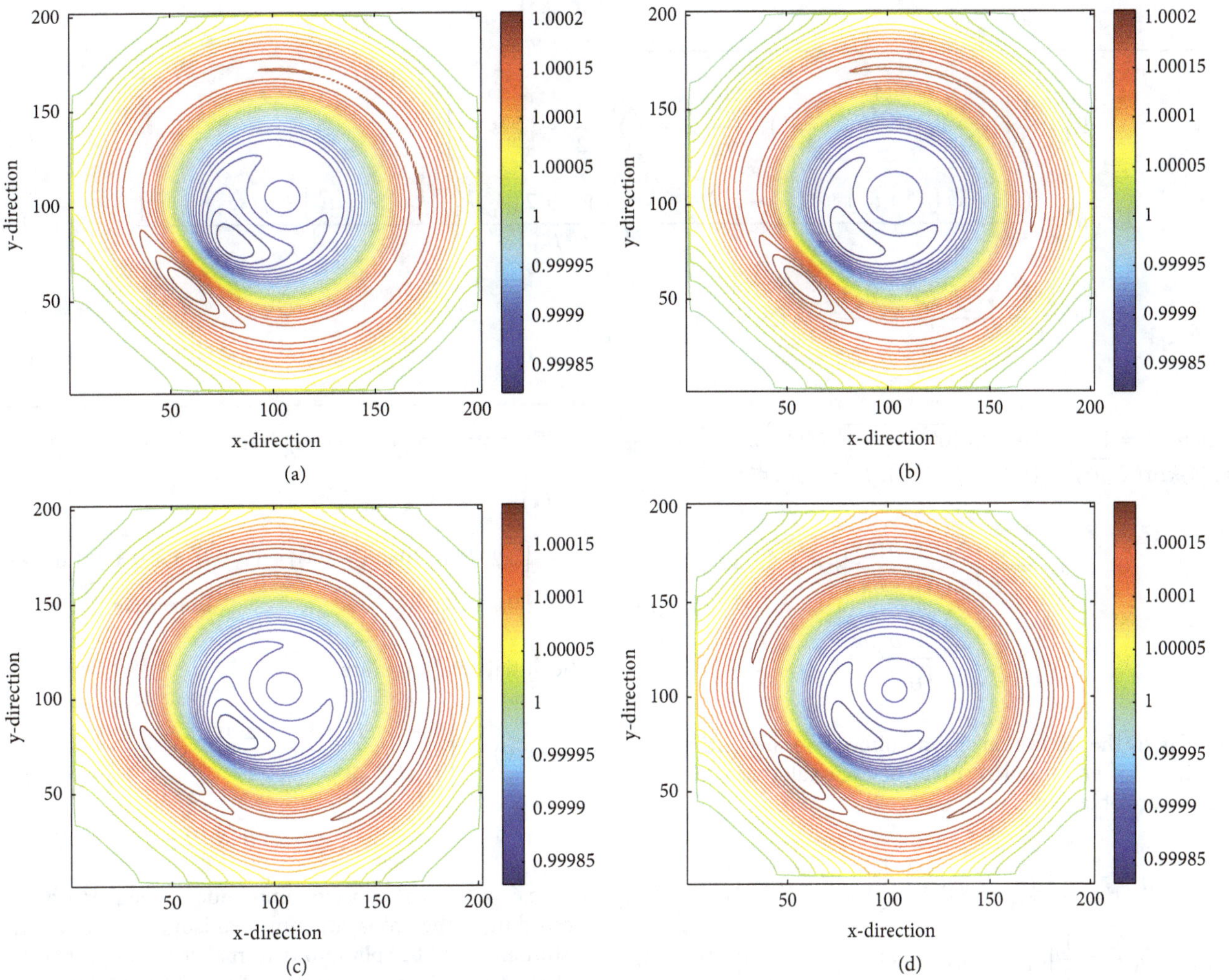

FIGURE 6: Smooth surface wave propagation problem. The contour of the water level H at t=20. (a) is the result of SI with CFL=0.63; (b), (c), and (d) are the results of CIP-MOC with CFL=0.63, 1.42, and 1.90.

$$L = \begin{pmatrix} -\frac{\Gamma_H}{g} & \frac{\Gamma_H}{g} & 0 \\ 1 & 1 & 0 \\ 0 & 0 & 1 \end{pmatrix},$$

$$L^{-1} = \frac{1}{2} \begin{pmatrix} -\frac{g}{\Gamma_H} & 1 & 0 \\ \frac{g}{\Gamma_H} & 1 & 0 \\ 0 & 0 & 2 \end{pmatrix}. \tag{32}$$

Equation (31) leads to the following three equations:

$$\frac{\partial R_x^{\pm}}{\partial t} + \lambda_x^{\pm} \frac{\partial R_x^{\pm}}{\partial x} \mp \frac{1}{4} f v \pm \frac{1}{4} k u \frac{\sqrt{u^2+v^2}}{H-z} - \frac{g}{2\Gamma_H} \frac{\partial (uz)}{\partial x} = 0, \tag{33}$$

$$\frac{\partial v}{\partial t} + u \frac{\partial v}{\partial x} + \frac{1}{2} f u + \frac{1}{2} k v \frac{\sqrt{u^2+v^2}}{H-z} = 0, \tag{34}$$

where $R_x^{\pm} = \Gamma_H \pm u/2$ are the Riemann invariants. These three equations all have the form of convective equations that can be calculated by CIP method. The Coriolis force, bottom topography, and bottom friction term are added to each Riemann invariant along characteristic line. Thus, the Riemann invariants and v can be discretized as follows:

$$\begin{aligned} \Gamma_H^* \pm \frac{1}{2} u^* = {} & \Gamma_H^{\pm} \pm \frac{1}{2} u^{\pm} \pm \frac{1}{4} f \Delta t \frac{\left(v^{\pm} + v^*\right)}{2} \\ & \mp \frac{1}{4} k \Delta t \frac{\left(u^{\pm} + u^*\right)}{2} \frac{\sqrt{(u^n)^2 + (v^n)^2}}{H^n - z} \\ & + \frac{g}{2} \int_{t^n}^{t^{n+1}} \frac{1}{\Gamma_H} \frac{\partial (uz)}{\partial x} d\tau^{\pm}, \end{aligned} \tag{35}$$

$$\begin{aligned} v^* = {} & v^0 - \frac{1}{2} f \Delta t \frac{\left(u^0 + u^*\right)}{2} \\ & - \frac{1}{2} k \Delta t \frac{\left(v^0 + v^*\right)}{2} \frac{\sqrt{(u^n)^2 + (v^n)^2}}{H^n - z}. \end{aligned} \tag{36}$$

So we have

$$\Gamma_H^* = \frac{1}{2}\left[\Gamma_H^+ + \Gamma_H^- + \frac{k_2'}{2}\left(u^+ - u^-\right) + \frac{f'}{2}\left(v^+ - v^-\right) + \frac{g}{2}\left({s_x}^+ + {s_x}^-\right)\right], \quad (37)$$

$$u^* = \frac{(1/2)\left[k_1'\left(u^+ + u^-\right) + 2\left(\Gamma_H^+ - \Gamma_H^-\right) + f'\left(v^+ + v^-\right) + 2\left(f'/k_1'\right)\left[k_2' v^0 - f' u^0\right] + g\left({s_x}^+ - {s_x}^-\right)\right]}{k_1' + f'^2/k_1}, \quad (38)$$

$$v^* = \frac{k_2' v^0 - f'\left(u^0 + u^*\right)}{k_1'}, \quad (39)$$

where $k_1' = 1 + (1/4)k\Delta t(\sqrt{(u^n)^2 + (v^n)^2}/(H^n - z))$, $k_2' = 1 - (1/4)k\Delta t(\sqrt{(u^n)^2 + (v^n)^2}/(H^n - z))$, $f' = (1/4)f\Delta t$,

$$\begin{aligned} {s_x}^+ &= \frac{g}{2}\int_{t^n}^{t^{n+1}} \frac{1}{\Gamma_H}\frac{\partial (uz)}{\partial x} d\tau^+, \\ {s_x}^- &= \frac{g}{2}\int_{t^n}^{t^{n+1}} \frac{1}{\Gamma_H}\frac{\partial (uz)}{\partial x} d\tau^-. \end{aligned} \quad (40)$$

Similar to the derivation in one dimension, neglecting the influence of $G^{\pm}$, the derivatives of each value at time step $(n+1)$ are obtained

$$\partial_x \Gamma_H^* = \frac{1}{2}\left[\partial_x \Gamma_H^+ + \partial_x \Gamma_H^- + \frac{1}{2}\left(\partial_x u^+ - \partial_x u^-\right)\right], \quad (41)$$

$$\partial_x u^* = \frac{1}{2}\left[\partial_x u^+ + \partial_x u^- + 2\left(\partial_x \Gamma_H^+ - \partial_x \Gamma_H^-\right)\right], \quad (42)$$

$$\partial_x v^* = \frac{\partial v^0}{\partial x}. \quad (43)$$

The operation in the x-direction is completed.

The operation in y-direction is essentially the same as the x-direction and the y derivative is computed by linear interpolation

$$\partial_y f_{i,j}^* = \partial_y f_{i,j} - \operatorname{sgn}\left(\lambda_x\right) \cdot \frac{\partial_y f_{i,j} - \partial_y f_{i-\operatorname{sgn}(\lambda_x),j}}{\Delta x}\lambda_x \Delta t, \quad (44)$$

where $\operatorname{sgn}(\lambda_x) = 1\ (\lambda_x \geq 0), -1\ (\lambda_x < 0)$.

3. Numerical Results

3.1. A Small Perturbation of One-Dimensional Steady State Water. To demonstrate that the CIP method can be applied to the computation of ocean tidal waves with bottom topography and bottom friction, we first consider the quasi-stationary test case given in [18]. The difference is that we add the effect of the bottom friction here.

The bottom topography is given by

$$b(x) = \begin{cases} 0.25\left(\cos\left(10\pi\left(x - 1.5\right)\right) + 1\right), & if\ 1.4 \leq x \leq 1.6, \\ 0, & otherwise. \end{cases} \quad (45)$$

The initial conditions are

$$H(x,0) = \begin{cases} 1 + \zeta, & if\ 1.1 \leq x \leq 1.2, \\ 1, & otherwise. \end{cases} \quad (46)$$

$$u(x,0) = 0, \quad k = 2.0,$$

where $\zeta = 0.001$ is constant amplitude of the perturbation. According to the propagation mechanism of wave, the small disturbance will be split into two smaller waves propagating to both sides. However, because of the Gibbs phenomenon, the small pulse of perturbation is a challenge for many numerical schemes [19]. Numerical experiments are carried out with both the CIP and SI schemes; the surface level with $Nx = 2000$ cells at the final time $T = 0.2$ is shown in Figure 3. We keep $\Delta t = T/1000$ as the SI scheme is strictly restricted by the CFL condition, and $\Delta t = T/1000, T/400, T/240$ when using the CIP.

Since flow velocity u and v are far smaller than wave velocity Γ_H, the CFL number is defined as

$$CFL \approx \Gamma_{H\max} \cdot \frac{\Delta t}{\Delta x}. \quad (47)$$

The experimental results are similar to those in [19, 20]. It can be seen that nonphysical oscillations occur when simulating marching wave using the SI scheme, waves are broken in the process of travel, and the amplitude of nonphysical oscillations increases gradually in time, which also means that the scheme does not have the ability to simulate large gradient surface change. However, the shape of the simulated wave is smooth and has no oscillations when using CIP until $CFL = 2.61$. Figure 4 gives the results computed by CIP with CFL=2.61 for k=0, 1, and 2. It can be seen that the separated wave height is less than half of the initial perturbation because of the bottom friction. Numerical experimentsshow that the

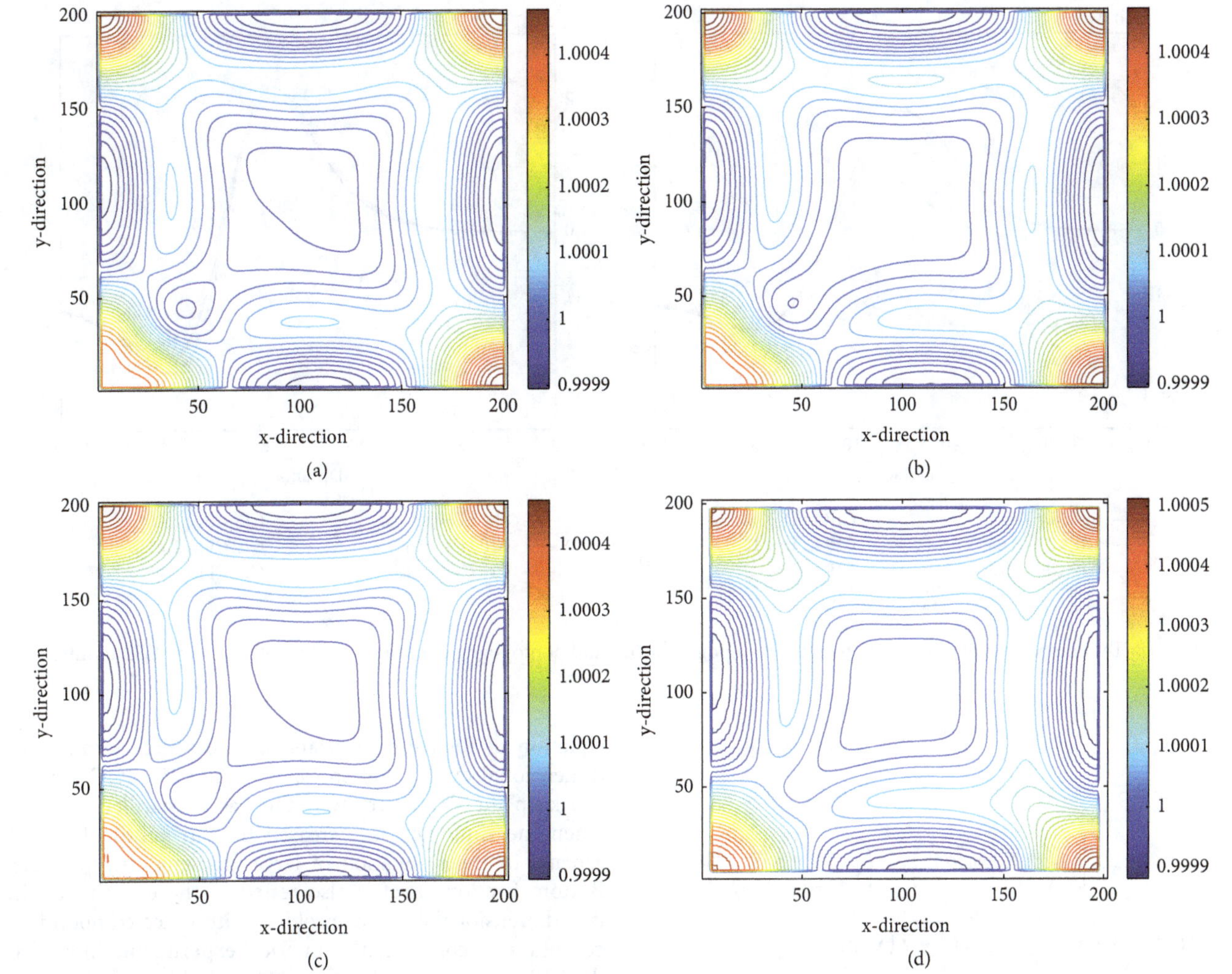

FIGURE 7: Smooth surface wave propagation problem. The contour of the water level H at t=40. (a) is the result of SI with CFL=0.63; (b), (c), and (d) are the results of CIP-MOC with CFL=0.63, 1.42, and 1.90.

CIP method can successfully simulate large gradient water surface with both bottom topography and bottom friction using large time step sizes.

3.2. Smooth Surface Wave Propagation Problem. Since boundary condition plays a significant role in the shallow water flow of an actual marine area, we consider a smooth surface wave propagation problem here with reflective boundary conditions. The conditions for the x-direction are

$$\begin{aligned} \frac{\partial H}{\partial \Omega} &= 0, \\ \frac{\partial v}{\partial \Omega} &= 0, \\ u(\Omega) &= 0. \end{aligned} \tag{48}$$

Since the spatial gradient is also required when using the CIP method, the spatial gradient condition at the boundary is obtained according to (48):

$$\begin{aligned} \partial_x H(\Omega) &= 0, \\ \partial_x v(\Omega) &= 0, \\ \frac{\partial_x u}{\partial \Omega} &= 0. \end{aligned} \tag{49}$$

When the traced upstream departure point on the shore, the corresponding actual upstream departure point in the waters is approximated using reflective boundary conditions (see Figure 5).

The initial conditions are given as follows:

$$\begin{aligned} &\zeta(x, y, 0) \\ &\quad = 0.001 \\ &\quad \cdot \exp\left[-\left(\frac{x - X_{\max}/2}{\sigma_x}\right)^2 - \left(\frac{y - Y_{\max}/2}{\sigma_y}\right)^2\right], \end{aligned}$$

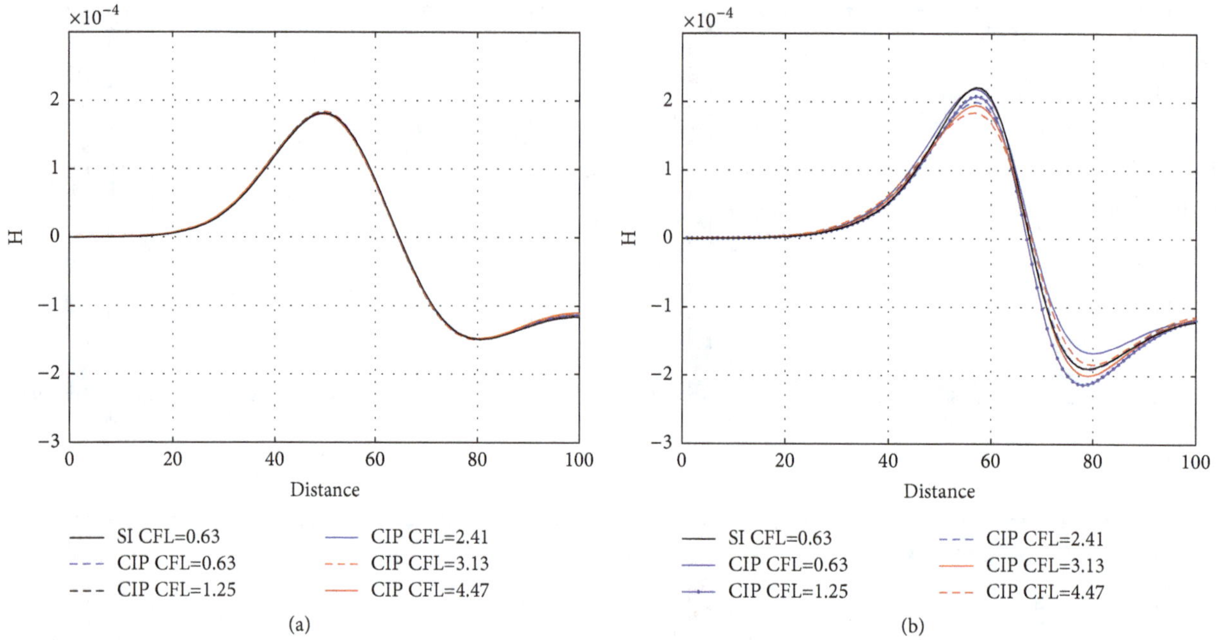

FIGURE 8: The comparison of cross-sections between topographic (a) and nontopographic conditions (b) with different CFL numbers.

$$z(x,y) = 0.5 \cdot \exp\left[-\left(\frac{x - X_{\max}/3}{\sigma_x}\right)^2 - \left(\frac{y - Y_{\max}/3}{\sigma_y}\right)^2\right],$$
$$H(x,y,0) = 1 + \zeta(x,y,0) - z(x,y),$$
$$u(x,y,0) = 0,$$
$$v(x,y,0) = 0, \quad (50)$$

where $\sigma_x = \sigma_y = 25$. Let the bottom friction coefficient $k = 2 \times 10^{-3}$ and the constant Coriolis parameter f be 1×10^{-3}.

The computation domain is $\Omega = [0, 200] \times [0, 200]$, and Δx and Δy are set to 1.0. The contour map of the water surface at t=20 and t=40 is shown in Figures 6 and 7, respectively. It can be seen that at t=20 the contour line is not a complete circle because of the impact of the bottom topography (similar to the one-dimensional case). At $t = 40$, the reflected waves are superimposed on each other and eventually form what is shown in Figure 7. The results show that when time step increases to three times to the SI method, CIP-MOC scheme still simulate the change process of water surface very well, even considering the bottom topography (see Figure 8), boundary, and Coriolis effect.

4. Conclusion

In this paper, we apply the CIP-MOC to the solution of tide wave equations using large time step size. Bottom topography, bottom friction, and Coriolis term are included to the equation of Riemann invariants as the source term. One-dimensional experiment shows that the CIP-MOC scheme can keep the shape very well and has no oscillations even when the time step is increased to four times of the SI scheme in the simulation of large gradient water surface. Bottom friction effect is also tested in the experiment. In two-dimensional case, the problem with reflected boundary conditions is considered. Numerical experiments show that the CIP-MOC scheme is succeeded in simulating the change process of water surface using time step that is three times of the SI method. All the results validated the good stability and low dispersion error of the CIP-MOC method.

Conflicts of Interest

The authors declare that there are no conflicts of interest regarding the publication of this paper.

Acknowledgments

Partial support for this research was provided by the National Key Research and Development Plan through Grant 2016YFC1402304, the Key Research and Development Plan of Shandong Province through Grants 2016ZDJS09A02 and ZR2016AB16, and the National Natural Science Foundation of China through Grant 41606006.

References

[1] J. Pedlosky, *Geophysical fluid dynamics*, Springer Science & Business Media, 2013.

[2] J. O. Backhaus, "A semi-implicit scheme for the shallow water equations for application to shelf sea modelling," *Continental Shelf Research*, vol. 2, no. 4, pp. 243–254, 1982.

[3] V. Casulli, "Semi-implicit finite difference methods for the two-dimensional shallow water equations," *Journal of Computational Physics*, vol. 86, no. 1, pp. 56–74, 1990.

[4] X. Lu and J. Zhang, "Numerical study on spatially varying bottom friction coefficient of a 2D tidal model with adjoint method," *Continental Shelf Research*, vol. 26, no. 16, pp. 1905–1923, 2006.

[5] H. Pan, Z. Guo, and X. Lv, "Inversion of tidal open boundary conditions of the M2 constituent in the bohai and yellow seas," *Journal of Atmospheric and Oceanic Technology*, vol. 34, no. 8, pp. 1661–1672, 2017.

[6] Z. Guo, H. Pan, W. Fan, and X. Lv, "Application of surface spline interpolation in inversion of bottom friction coefficients," *Journal of Atmospheric and Oceanic Technology*, vol. 34, no. 9, pp. 2021–2028, 2017.

[7] R. H. Weisberg and L. Zheng, "A finite volume coastal ocean model simulation of the tide, buoyancy, and wind-driven circulation of Tampa Bay," *Journal of Geophysical Research*, vol. 111, article C01005, 2006.

[8] C. Chen, H. Huang, R. C. Beardsley, H. Liu, Q. Xu, and G. Cowles, "A finite volume numerical approach for coastal ocean circulation studies: comparisons with finite difference models," *Journal of Geophysical Research: Oceans*, vol. 112, no. 3, Article ID C03018, 2007.

[9] G. Erbes, "A Semi-Lagrangian Method of Characteristics for the Shallow-Water Equations," *Monthly Weather Review*, vol. 121, no. 12, pp. 3443–3452, 1993.

[10] Y. Ogata and T. Yabe, "Multi-dimensional semi-lagrangian characteristic approach to the shallow water equations by the CIP method," *International Journal of Computational Engineering Science*, vol. 5, no. 3, pp. 699–730, 2004.

[11] K. Toda, Y. Ogata, and T. Yabe, "Multi-dimensional conservative semi-Lagrangian method of characteristics CIP for the shallow water equations," *Journal of Computational Physics*, vol. 228, no. 13, pp. 4917–4944, 2009.

[12] T. Yabe and T. Aoki, "A universal solver for hyperbolic equations by cubic-polynomial interpolation. I. One-dimensional solver," *Computer Physics Communications*, vol. 66, no. 2-3, pp. 219–232, 1991.

[13] T. Yabe and E. Takei, "A new higher-order Godunov method for general hyperbolic equations," *Journal of the Physical Society of Japan*, vol. 57, no. 8, pp. 2598–2601, 1988.

[14] Y. Anle and L. Fengqi, *Physical Oceanography*, China Ocean University Press, 1992.

[15] J. G. Zhou, D. M. Causon, C. G. Mingham, and D. M. Ingram, "The surface gradient method for the treatment of source terms in the shallow-water equations," *Journal of Computational Physics*, vol. 168, no. 1, pp. 1–25, 2001.

[16] R. Akoh, S. Ii, and F. Xiao, "A CIP/multi-moment finite volume method for shallow water equations with source terms," *International Journal for Numerical Methods in Fluids*, vol. 56, no. 12, pp. 2245–2270, 2008.

[17] R. W. Lardner, "Optimal assimilation of current and surface elevation data in a two-dimensional numerical tidal model," *Applied Mathematical Modelling*, vol. 19, no. 1, pp. 30–38, 1995.

[18] R. J. LeVeque, "Balancing source terms and flux gradients in high-resolution Godunov methods: the quasi-steady wave-propagation algorithm," *Journal of Computational Physics*, vol. 146, no. 1, pp. 346–365, 1998.

[19] Q. Zhu, Z. Gao, W. S. Don, and X. Lv, "Well-balanced hybrid compact-WENO scheme for shallow water equations," *Applied Numerical Mathematics*, vol. 112, pp. 65–78, 2017.

[20] Y. Xing and C.-W. Shu, "High order finite difference WENO schemes with the exact conservation property for the shallow water equations," *Journal of Computational Physics*, vol. 208, no. 1, pp. 206–227, 2005.

12

Existence of Positive Ground State Solution for the Nonlinear Schrödinger-Poisson System with Potentials

Lu Xiao,[1] Haiqin Qian,[1] Mengmeng Qu,[2] and Jun Wang[2]

[1]*School of Management, Jiangsu University, Zhenjiang, Jiangsu, 212013, China*
[2]*Faculty of Science, Jiangsu University, Zhenjiang, Jiangsu, 212013, China*

Correspondence should be addressed to Jun Wang; wangmath2011@126.com

Academic Editor: Pietro d'Avenia

In the present paper we study the existence of positive ground state solutions for the nonautonomous Schrödinger-Poisson system with competing potentials. Under some assumptions for the potentials we prove the existence of positive ground state solution.

1. Introduction

In this paper we are concerned with the existence of solution of the nonautonomous Schrödinger-Poisson system

$$\begin{aligned} -\Delta u + A(x)u + B(x)\phi(x)u &= Q(x)|u|^{p-1}u, \\ & x \in \mathbb{R}^3, \\ -\Delta\phi &= B(x)u^2, \\ & u \in H^1(\mathbb{R}^3), \end{aligned} \tag{1}$$

where $p \in (3,5)$, $A(x)$, $B(x)$, and $Q(x)$ are positive functions such that $\lim_{|x|\longrightarrow\infty} A(x) = a_\infty > 0$, $\lim_{|x|\longrightarrow\infty} B(x) = b_\infty > 0$, and $\lim_{|x|\longrightarrow\infty} Q(x) = q_\infty > 0$. Here $B : \mathbb{R}^3 \longrightarrow \mathbb{R}$ denotes the nonnegative measurable function which represents a nonconstant charge corrector to the density u^2. A and Q are called the potentials of system (1). In the context of the so-called Density Functional Theory, variants of system (1) appear as mean field approximations of quantum many-body systems; see [1–3].

This kind of system also arises in many fields of physics. Indeed, one considers the following system:

$$\begin{aligned} i\hbar\frac{\partial\psi}{\partial t} &= -\frac{\hbar^2}{2m}\Delta\psi + k(x)\psi + B(x)\phi\psi \\ &\quad - Q(x)|\psi|^{p-2}\psi, \quad x \in \mathbb{R}^3,\ t \in \mathbb{R}, \\ -\frac{\hbar^2}{2m}\Delta\phi(x) &= B(x)|\psi|^2, \quad u \in H^1(\mathbb{R}^3), \end{aligned} \tag{2}$$

where i is the imaginary unit, Δ is the Laplacian operator, and $\hbar$ is the Planck constant. A standing wave is a solution of (2) of the form $\psi(x,t) = u(x)e^{i\omega t}$, $\omega > 0$ and $t \in \mathbb{R}$. It is clear that $\psi(x,t)$ solves (2) if and only if $u(x)$ solves the so-called stationary system

$$\begin{aligned} -\frac{\hbar^2}{2m}\Delta u + A(x)u + B(x)\phi(x)u &= Q(x)|u|^{p-2}u, \\ & x \in \mathbb{R}^3, \\ -\frac{\hbar^2}{2m}\Delta\phi &= B(x)u^2, \\ & x \in \mathbb{R}^3, \end{aligned} \tag{3}$$

where $A(x) = k(x) + \omega\hbar$. Here we consider that the case $\hbar$ is constant. Without loss of generality we assume that $\hbar^2/2m = 1$, then system (3) becomes (1). System (1) is also modeled in Abelian Gauge Theories; for instance, see [4–6] and reference therein. In fact, in order to describe the interaction of a nonlinear Schrödinger field with an electromagnetic field $\mathbb{E}-\mathbb{H}$, the gauge potentials $\mathbb{A}-\mathbb{F}$

$$\begin{aligned} \mathbb{A} &: \mathbb{R}^3 \times \mathbb{R} \longrightarrow \mathbb{R}^3, \\ \mathbb{F} &: \mathbb{R}^3 \times \mathbb{R} \longrightarrow \mathbb{R} \end{aligned} \tag{4}$$

are related to $\mathbb{E} - \mathbb{H}$ by the Maxwell equations

$$\mathbb{E} = -\left(\nabla\mathbb{F} + \frac{\partial\mathbb{A}}{\partial t}\right), \quad \mathbb{H} = \nabla\times\mathbb{A}. \tag{5}$$

If we are interested in finding standing waves (solutions of a field equation whose energy travels as a localized packet preserving this localization in time) and we consider the electrostatic case (when $\mathbb{A} = 0$), the Schrödinger field is described by a real function $u : \mathbb{R}^3 \longrightarrow \mathbb{R}$, which represents the matter (see [4, 7]). In this situation we need to consider the stationary states of Schrödinger-Maxwell system (1). The system is also used in quantum electrodynamics, semiconductor theory, nonlinear optics, and plasma physics; for more information on this direction, one can refer to [7–9] and the references therein.

In recent years many papers focus on the existence, multiplicity, and concentration of positive solutions of (3) for the semiclassical case ($\hbar > 0$ sufficiently small). In this framework one is interested not only in existence of solutions but also in their asymptotic behavior as $\hbar \longrightarrow 0$. Typically, the solution tends to concentrate on critical points of $A(B$ or $Q)$. These solutions are called spikes. For more information on this direction. one can refer to [10–14] and the references therein.

In the present paper we are interested in studying the case when $\hbar$ is constant. In [15], the authors studied the existence and nonexistence of solutions of (1) when $A = B = Q = 1$. The existence of the multiple solutions of (1) has been found in the paper [16] in a radial setting. In the paper [17], the author considers that the case $B = 1, V, K$ is radial and satisfies

$$\frac{A_0}{1+|x|^{\alpha}} \le A(x) \le A_1, \quad 0 < Q(x) \le \frac{Q_0}{1+|x|^{\beta}}, \tag{6}$$

where $\alpha \in (0,2], \beta, A_0, A_1$, and Q_0 are positive constants. The author proved the existence of nontrivial positive classical mountain-pass solution of (1). Moreover, some generalizations of the last cases, with $Q(x)|u|^{p-1}u$ replaced by a more generic function $f(x,u)$, were considered in [18, 19]. It is well known that, dealing with system (1), one has to face different kinds of difficulties, which are related to the potential and to the unboundedness of the space $\mathbb{R}^3$. Many early studies were devoted to the autonomous case and to the case in which the coefficients are supposed to be radial (see [16, 20, 21]), just to overcome the lack of compactness-taking advantage of the compact embedding $H_r^1(\mathbb{R}^3) \hookrightarrow L^q(\mathbb{R}^3)$ $(\forall q \in (2,6))$. More recently, many contributions to (1) have also been given looking at cases in which no symmetry assumptions are given on the coefficients appearing in (1); one can refer to the papers [22–24]. Furthermore, for more results on the existence of positive solutions, ground and bound states, one can see [18, 19, 21, 25–33] and references therein. Nearly, the paper [34] proves the existence of bound state solution of (1) under some decay condition on A, B, and Q. Precisely, assume that $A(x) = V_\infty + W(x)$ and $\int_{\mathbb{R}^3} W(x)(|x|e^{2\sqrt{V_\infty}|x|}) < \infty$, $Q(x) = Q_\infty - \beta(x)$ and $\int_{\mathbb{R}^3} \beta(x)(|x|e^{2\sqrt{V_\infty}|x|}) < \infty$, and

$$0 \le B(x) \le ce^{-\sigma|x|}, \quad \text{for } |x| \text{ large enough}, \tag{7}$$

where V_∞ and Q_∞ are positive constants. From this assumption one can easily deduce that $\lim_{|x|\to\infty} B(x) = 0$. In order to get the compactness, the paper [34] studies the case when the limit equation is Scalar Schrödinger equation, i.e.,

$$-\Delta u + a_\infty u = q_\infty |u|^{p-1} u, \quad u \in H^1(\mathbb{R}^3). \tag{8}$$

According to [35, 36], (8) has unique positive solution, and the energy level of any sign-changing solution is strictly greater than $2c_\infty$, where c_∞ is the least energy level for the positive solution of (8). This information is very important for proving the existence of positive bound solution (high energy) of (1.1) in [34].

Motivated by [34, 37], in the present paper we shall study the case when the limit equation is the Schrödinger-Poisson system

$$\begin{aligned} -\Delta u + a_\infty u + b_\infty \phi(x) u &= q_\infty |u|^{p-1} u, \quad \text{in } \mathbb{R}^3, \\ -\Delta\phi &= b_\infty u^2, \quad u \in H^1(\mathbb{R}^3), \end{aligned} \tag{9}$$

and, under some conditions for the A, B, and Q, we prove the existence of positive ground state solution of (1).

In order to state our main results, we shall give some assumptions. For $\lambda \in \mathbb{R}^+$, we define $A(x) = a_\infty + \lambda a(x)$, $B(x) = b_\infty - b(x)$, and $Q(x) = q_\infty + q(x)$. Throughout the paper we need the following conditions.

(A_1) $a_\infty > 1$, $a \in L^{3/2}(\mathbb{R}^3)$, $0 < a < 1$, $a \ne 0$, $\lim_{|x|\to\infty} a(x) = 0$.

(A_2) $b_\infty > 1$, $b \in L^{3/2}(\mathbb{R}^3) \cap L^\infty(\mathbb{R}^3)$, $0 \le b \le 1$, $b \ne 0$, $\lim_{|x|\to\infty} b(x) = 0$.

(A_3) $q_\infty > 1$, $q \in L^\infty(\mathbb{R}^3)$, $0 < q < 1$, $q \ne 0$, $\lim_{|x|\to\infty} q(x) = 0$.

Clearly, system (1) becomes

$$\begin{aligned} &-\Delta u + (a_\infty + \lambda a(x)) u + (b_\infty - b(x)) \phi(x) u \\ &= (q_\infty + q(x)) |u|^{p-1} u, \quad \text{in } \mathbb{R}^3, \\ &-\Delta\phi = (b_\infty - b(x)) u^2, \quad u \in H^1(\mathbb{R}^3), \end{aligned} \tag{10}$$

where $\lambda \in \mathbb{R}^+$ and $p \in (3,5)$. In the following we shall focus on system (10). According to [11, 12], we know that (9) has a positive radial symmetric solution w. Moreover, w is decreasing when the radial coordinate increases. Precisely, there exists a constant $c^* > 0$ such that

$$\lim_{|x|\to+\infty} |w(x)| e^{c^*|x|} = \text{Constant}. \tag{11}$$

Then we have the following main results.

Theorem 1. *Assume that $p \in (3,5)$ and (A_1)-(A_3) hold. Moreover, if one of the following conditions holds*

(1) $\lim_{|x|\to\infty} q(x)e^{(2\tau/(1+\tau))c^*|x|} = \infty$ and $\lim_{|x|\to\infty} q(x)e^{(2\tau/(1+2\tau))c^*|x|} \le c_0$, $\lim_{|x|\to\infty} a(x)|x|^{1+\tau}e^{(2\tau/(1-\tau))c^*|x|} \le c_1$, $\lim_{|x|\to\infty} b(x)e^{(2\tau(1-\tau)/(1+\tau)(1+2\tau))c^*|x|} \ge c_2$, *where* $\tau \in (0,1)$, c_0, c_1, *and* c_2 *are positive constants.*

(2) $\lim_{|x|\to\infty} q(x)e^{(2\tau/(1+\tau))c^*|x|} = \infty$ *and* $\lim_{|x|\to\infty} a(x)e^{((p-1)\tau/(1-\tau))c^*|x|} \le c_3$, *where* $\tau \in (0,1)$ *and* c_3 *is positive constant.*

Then system (10) has a positive ground state solution for each $\lambda \in \mathbb{R}^+$.

Remark 2. (i) From the assumptions on B, we know that $\lim_{|x|\to\infty} B(x) = b_\infty$. Hence we need to consider the limit equation (9) to recover the compactness. This is quite different from the recent work of [34]. The main novelty here is that we shall compare the decay rate of A, B, and Q to recover the compactness and prove the existence of positive ground state solution.

(ii) Note that in condition (1) of Theorem 1, we know that $a(x)$ is decaying faster than $b(x)$. In condition (2) of Theorem 1, we only need the decay condition for $a(x)$ (whatever the decay speed of $b(x)$ to 0 as $|x| \to \infty$) to prove the existence of positive solution. This is different phenomenon compared to [34].

2. Preliminary Results

We define the following notation:

(i) $\|\cdot\|$ is the norm of $H^1(\mathbb{R}^3)$ defined by $\|u\|^2 = \int_{\mathbb{R}^3}(|\nabla u|^2 + a_\infty u^2)$;

(ii) $\|\cdot\|_{D^{1,2}}$ is the norm of $D^{1,2}(\mathbb{R}^3)$ defined by $\|u\|^2_{D^{1,2}} = \int_{\mathbb{R}^3}|\nabla u|^2$;

(iii) $(\cdot,\cdot)$ is the scalar product of $H^1(\mathbb{R}^3)$ defined by $(u,v) = \int_{\mathbb{R}^3}[\nabla u \nabla v + uv]$;

(iv) let C, c^*, c, c_i $(i = 1, \cdots)$ denote different positive constants.

In this part we mainly give some basic knowledge which will be used later. We first show that the second equation of (10) can be solved. We consider, for all $u \in H^1(\mathbb{R}^3)$, the linear functional J_u defined in $D^{1,2}(\mathbb{R}^3)$ by

$$J_u(v) = \int_{\mathbb{R}^3} (b_\infty - b(x)) u^2 v. \tag{12}$$

We infer from condition (A_2) and Hölder inequality that

$$|J_u(v)| \le C |u|^2_{12/5} \|v\|_{D^{1,2}}. \tag{13}$$

By the Lax-Milgram theorem, we know that there exists unique $\phi_u \in D^{1,2}(\mathbb{R}^3)$ such that

$$\int_{\mathbb{R}^3} \nabla\phi_u \nabla v = \int_{\mathbb{R}^3} (b_\infty - b(x)) u^2 v \quad \forall v \in D^{1,2}(\mathbb{R}^3). \tag{14}$$

So, ϕ_u is a weak solution of $-\Delta\phi = u^2$ and the following formula holds

$$\begin{aligned}\phi_u(x) &= \int_{\mathbb{R}^3} \frac{(b_\infty - b(y)) u^2(y)}{4\pi|x-y|} dy \\ &= \frac{1}{4\pi|x|} * ((b_\infty - b) u^2).\end{aligned} \tag{15}$$

Moreover, $\phi_u > 0$ when $u \neq 0$.

We recall the following classical Hardy-Littlewood-Sobolev inequality (see [38, Theorem 4.3]). Assume that $f \in L^p(\mathbb{R}^3)$ and $g \in L^q(\mathbb{R}^3)$. Then one has

$$\int_{\mathbb{R}^3}\int_{\mathbb{R}^3} \frac{f(x) g(y)}{|x-y|^t} dx\, dy \le c(p,q,t) |f|_p |g|_q, \tag{16}$$

where $1 < p, q < \infty$, $0 < t < 3$, and $1/p + 1/q + t/3 = 2$. By (16) we know that

$$\int_{\mathbb{R}^3}\int_{\mathbb{R}^3} \frac{u^2(x) u^2(y)}{|x-y|} dx\, dy \le c |u|^4_{12/5} \le c \|u\|^4. \tag{17}$$

It is well known that solutions of (10) correspond to critical points of the energy functional

$$\begin{aligned}\Phi_\lambda(u) = &\frac{1}{2}\int_{\mathbb{R}^3} (|\nabla u|^2 + (a_\infty + \lambda a(x)) u^2) \\ &+ \frac{1}{4}\int_{\mathbb{R}^3} (b_\infty - b(x)) \phi_u u^2 \\ &- \frac{1}{p+1}\int_{\mathbb{R}^3} (q_\infty + q(x)) |u|^{p+1}.\end{aligned} \tag{18}$$

From (17), we know that Φ_λ is well defined and that

$$\begin{aligned}\Phi'_\lambda(u)[v] = &\int_{\mathbb{R}^3} (\nabla u \nabla v + (a_\infty + \lambda a(x)) uv) \\ &+ \int_{\mathbb{R}^3} (b_\infty - b(x)) \phi_u uv \\ &- \int_{\mathbb{R}^3} (q_\infty + q(x)) |u|^{p-2} uv.\end{aligned} \tag{19}$$

We define the operator $I : H^1(\mathbb{R}^3) \to D^{1,2}(\mathbb{R}^3)$ as

$$I[u] = \phi_u. \tag{20}$$

We infer from [34, Proposition 2.1-2.2] that I has the following properties.

Lemma 3. *(1) I is continuous.*

(2) I maps bounded sets into bounded sets.

(3) $I[tu] = t^2 I[u]$ for all $t \in \mathbb{R}$.

(4) If $u_n \rightharpoonup u \in H^1(\mathbb{R}^3)$, then we have $\int_{\mathbb{R}^3} b(x)\psi_{u_n}(x)u_n^2 \to \int_{\mathbb{R}^3} b(x)\psi_u(x)u^2$ and $\int_{\mathbb{R}^3} b(x)\psi_{u_n}(x)u_n\phi \to \int_{\mathbb{R}^3} b(x)\psi_u(x)u\phi$ for each $\phi \in H^1(\mathbb{R}^3)$, as $n \to \infty$, where $\psi_u(x) = \int_{\mathbb{R}^3}(b(y)u^2(y)/4\pi|x-y|)dy$.

(5) If $u_n \rightharpoonup u \in H^1(\mathbb{R}^3)$, then we have $I(u_n) \to I(u)$ in $D^{1,2}(\mathbb{R}^3)$ and $\int_{\mathbb{R}^3} b_\infty\overline{\phi}_{u_n}(x)u_n\phi \to \int_{\mathbb{R}^3} b_\infty\overline{\phi}_u(x)u\phi$ for each $\phi \in H^1(\mathbb{R}^3)$, as $n \to \infty$, where $\overline{\phi}_u(x) = \int_{\mathbb{R}^3}(b_\infty u^2(y)/4\pi|x-y|)dy$.

Proof. The conclusions (1)-(4) can be proved as in [34, Proposition 2.1-2.2]. Hence we only focus on the proof of (5). Since, by definition of I, for all $u \in D^{1,2}(\mathbb{R}^3)$ we have

$$\|I(u)\|_{D^{1,2}(\mathbb{R}^3)} = \|J_u\|_{\mathscr{L}(D^{1,2}(\mathbb{R}^3),\mathbb{R})} \tag{21}$$

then, in order to prove $I(u_n) \longrightarrow I(u)$ in $D^{1,2}(\mathbb{R}^3)$, it suffices to prove that

$$\|J_{u_n} - J_u\|_{\mathscr{L}(D^{1,2}(\mathbb{R}^3),\mathbb{R})} \longrightarrow 0, \quad \text{as } n \longrightarrow \infty. \tag{22}$$

Let $\varepsilon > 0$ be fixed arbitrarily. Then there exists a positive number $R_\varepsilon > 0$ so large that $|v|_{L^6(B^c_{R_\varepsilon})} < \varepsilon$ for $v \in D^{1,2}(\mathbb{R}^3)$, where $B_{R_\varepsilon} = \{x \in \mathbb{R}^3 : |x| \le R_\varepsilon\}$ and $B^c_{R_\varepsilon} = \mathbb{R}^3 \setminus B_{R_\varepsilon}$. Hence we deduce that

$$\begin{aligned}
\left|J_{u_n}(v) - J_u(v)\right| &= \int_{\mathbb{R}^3} (b_\infty - b(x))\left(u_n^2 - u^2\right) v\,dx \\
&= \int_{B_{R_\varepsilon}} (b_\infty - b(x))\left(u_n^2 - u^2\right) v\,dx \\
&\quad + \int_{B^c_{R_\varepsilon}} (b_\infty - b(x))\left(u_n^2 - u^2\right) v\,dx \\
&\le c|v|_6 |u_n - u|_{L^{12/5}(B_{R_\varepsilon})} |u_n + u|_{L^{12/5}(R_\varepsilon)} \\
&\quad + c|v|_{L^6(B_{R_\varepsilon})^c} \left|u_n^2 - u^2\right|_{6/5} \\
&\le \left(c\varepsilon + c|u_n - u|_{L^{12/5}(B_{R_\varepsilon})} |u_n + u|_{L^{12/5}(R_\varepsilon)}\right) \|v\|_{D^{1,2}}.
\end{aligned} \tag{23}$$

Since $u_n \longrightarrow u$ in $L^p_{loc}(\mathbb{R}^3)$ $(p \in (2,6))$, we know that (22) holds.

Next we prove the later conclusion. For any fixed $R > 0$ large, we infer that for each $x \in \mathbb{R}$

$$\begin{aligned}
\left|\overline{\phi}_{u_n}(x) - \overline{\phi}_u(x)\right| &= \int_{\mathbb{R}^3} b_\infty \frac{u_n^2(y) - u^2(y)}{4\pi|x-y|} dy \\
&= \int_{|x-y|\le 1} b_\infty \frac{u_n^2(y) - u^2(y)}{4\pi|x-y|} dy \\
&\quad + \int_{1\le|x-y|\le R} b_\infty \frac{u_n^2(y) - u^2(y)}{4\pi|x-y|} dy \\
&\quad + \int_{|x-y|\ge R} b_\infty \frac{u_n^2(y) - u^2(y)}{4\pi|x-y|} dy \\
&\le c|u_n - u|_{L^{4/3}(B_1(x))} |u_n + u|_{L^4(B_1(x))} \\
&\quad \cdot \left(\int_{|x-y|\le 1} \frac{1}{|x-y|^2} dy\right) + c|u_n - u|_{L^{8/3}(B_R(x)\setminus B_1(x))} \\
&\quad \cdot |u_n + u|_{L^{8/3}(B_R(x)\setminus B_1(x))} \left(\int_{1\le|x-y|\le R} \frac{1}{|x-y|^4} dy\right) \\
&\quad + c|u_n - u|_{L^{8/3}(B^c_R(x))} |u_n + u|_{L^{8/3}(B^c_R(x))} \\
&\quad \cdot \left(\int_{|x-y|\ge R} \frac{1}{|x-y|^4} dy\right)
\end{aligned} \tag{24}$$

and, hence, we deduce that $\overline{\phi}_{u_n}(x) \longrightarrow \overline{\phi}_u(x)$ a.e., in $x \in \mathbb{R}^3$ and $|\overline{\phi}_{u_n} - \overline{\phi}_u|_{L^\infty(B_{R_0})} \longrightarrow 0$ for each $R_0 > 0$, as $n \longrightarrow \infty$. A direct computation shows that for $R > 0$ large

$$\begin{aligned}
&\left|\int_{\mathbb{R}^3} b_\infty \overline{\phi}_{u_n}(x) u_n \phi - \int_{\mathbb{R}^3} b_\infty \overline{\phi}_u(x) u\phi\right| \\
&= \left|\int_{\mathbb{R}^3} b_\infty \left(\overline{\phi}_{u_n} - \overline{\phi}_u\right) u_n\phi + b_\infty \overline{\phi}_u(x)(u_n - u)\phi\right| \\
&\le c\left|\int_{B_R} \left(\overline{\phi}_{u_n} - \overline{\phi}_u\right) u_n\phi\right| + c\left|\int_{B^c_R} \left(\overline{\phi}_{u_n} - \overline{\phi}_u\right) u_n\phi\right| \\
&\quad + c\left|\int_{B_R} \overline{\phi}_u(x)(u_n - u)\phi\right| \\
&\quad + c\left|\int_{B^c_R} \overline{\phi}_u(x)(u_n - u)\phi\right| \\
&\le c\left|\overline{\phi}_{u_n} - \overline{\phi}\right|_{L^\infty(B_R)} |\phi|_2 |u_n|_2 \\
&\quad + c\left|\overline{\phi}_{u_n} - \overline{\phi}\right|_6 |\phi|_{L^{(12/5)}(B^c_R)} |u_n|_{12/5} \\
&\quad + c\left|\overline{\phi}_u\right|_6 |\phi|_{12/5} |u_n - u|_{L^{12/5}(B_R)} \\
&\quad + c\left|\overline{\phi}_u\right|_{L^6(B^c_R)} |\phi|_{L^{12/5}(B^c_R)} |u_n - u|_{12/5}
\end{aligned} \tag{25}$$

which proves $\int_{\mathbb{R}^3} \overline{\phi}_{u_n}(x) u_n\phi \longrightarrow \int_{\mathbb{R}^3} \overline{\phi}_u(x) u\phi$ for each $\phi \in H^1(\mathbb{R}^3)$, as $n \longrightarrow \infty$. □

It is very easy to verify that, whatever $\lambda \in \mathbb{R}$ is, the function Φ_λ is bounded either from above or from below. Hence, it is convenient to consider Φ_λ restricted to a natural constraint, the Nehari manifold. We set

$$\mathscr{N}_\lambda := \left\{u \in H^1\left(\mathbb{R}^3\right) \setminus \{0\} : \Phi'_\lambda(u)[u] = 0\right\}. \tag{26}$$

Next lemma contains the statement of the main properties of $\mathscr{N}_\lambda$.

Lemma 4. *Assume that (A_1)-(A_3) hold. Then the following conclusions hold.*

(a) $\mathscr{N}_\lambda$ is a C^1 regular manifold diffeomorphic to the sphere of $H^1(\mathbb{R}^3)$.

(b) Φ_λ is bounded from below on $\mathscr{N}_\lambda$ by a positive constant.

(c) u is a free critical point of Φ_λ if and only if u is a critical point of Φ_λ constrained on $\mathscr{N}_\lambda$.

Proof. (a) Let $u \in H^1(\mathbb{R}^3)\setminus\{0\}$ be such that $\|u\| = 1$. We claim that there exists a unique $t \in (0,\infty)$ such that $tu \in \mathscr{N}_\lambda$. Let $f(t) = \Phi_\lambda(tu)$ for $t > 0$. It is easy to verify, by (A_1)-(A_3), that $f(0) = 0$, $f(t) > 0$ for $t > 0$ small and $f(t) < 0$ for

t large. Therefore, $\max_{t>0} f(t)$ is achieved at a $t = t(u) > 0$ so that $f(tu) = 0$ and $t(u)u \in \mathcal{N}_\lambda$. Assume that there exist $t_1 > t_2 > 0$ such that $t_1 u, t_2 u \in \mathcal{N}_\lambda$. Then we see

$$\left(\frac{\|w\|^2 + \int_{\mathbb{R}^3} \lambda a(x))u^2}{t_1^2} - \frac{\|w\|^2 + \int_{\mathbb{R}^3} \lambda a(x))u^2}{t_2^2}\right) = \left(t_1^{p-3} - t_2^{p-3}\right) \int_{\mathbb{R}^3} (q_\infty + q(x)) |u|^{p+1}. \tag{27}$$

This is a contradiction. So, we prove that $f(t)$ admits a unique positive solution $t(u) > 0$ and $t(u)u \in \mathcal{N}_\lambda$. We infer from (16) and (A_1)-(A_3) that

$$\begin{aligned} &c_1 |u|_{p+1}^2 \\ &\le \left(\|u\|^2 + \int_{\mathbb{R}^3} \lambda a(x) u^2 + \int_{\mathbb{R}^3} (b_\infty - b(x)) \phi_u u^2\right) \\ &= \int_{\mathbb{R}^3} (q_\infty + q(x)) |u|^{p+1} \le c_1 |u|_{p+1}^2. \end{aligned} \tag{28}$$

This implies that

$$\|u\| \ge c > 0. \tag{29}$$

Let $G_\lambda(u) := \Phi_\lambda'(u)[u]$. Then $G_\lambda \in C^1(H^1(\mathbb{R}^3), \mathbb{R})$ by the regularity of Φ_λ. Moreover, we infer from (29) that

$$\begin{aligned} &G_\lambda'(u)[u] \\ &= -(p-1) \int_{\mathbb{R}^3} \left(|\nabla u|^2 + (a_\infty + \lambda a(x)) u^2\right) \\ &\quad - (p-3) \int_{\mathbb{R}^3} (b_\infty - b(x)) \phi_u u^2 \le -(p-3) C \\ &< 0. \end{aligned} \tag{30}$$

(b) For all $u \in \mathcal{N}_\lambda$,

$$\begin{aligned} &\Phi_\lambda(u) \\ &= \left(\frac{1}{2} - \frac{1}{p+1}\right) \int_{\mathbb{R}^3} \left(|\nabla u|^2 + (a_\infty + \lambda a(x)) u^2\right) \\ &\quad + \left(\frac{1}{4} - \frac{1}{p+1}\right) \int_{\mathbb{R}^3} (b_\infty - b(x)) \phi_u u^2 \\ &\ge \left(\frac{1}{2} - \frac{1}{p+1}\right) \|u\|^2 \ge c > 0. \end{aligned} \tag{31}$$

Here we use the fact that $3 < p < 5$.

(c) If $u \ne 0$ is a critical point of Φ_λ, then we have $\Phi_\lambda'(u) = 0$, $u \in \mathcal{N}_\lambda$. On the other hand, if u is a critical point of $\Phi_\lambda(u)$ constrained on $\mathcal{N}_\lambda$, then there exists $k \in \mathbb{R}$ such that

$$0 = \Phi_\lambda'(u)[u] = G_\lambda(u) = k G_\lambda'(u)[u]. \tag{32}$$

One infers from (30) that $k = 0$. □

Next we consider the limit functional $\Phi_\infty : H^1(\mathbb{R}^3) \longrightarrow \mathbb{R}$, defined as

$$\begin{aligned} \Phi_\infty(u) = &\frac{1}{2} \int_{\mathbb{R}^3} \left(|\nabla u|^2 + a_\infty u^2\right) + \frac{1}{4} \int_{\mathbb{R}^3} b_\infty \overline{\phi}_u(x) u^2 \\ &- \frac{1}{p+1} \int_{\mathbb{R}^3} q_\infty |u|^{p+1}. \end{aligned} \tag{33}$$

And we consider the corresponding natural constraint

$$\mathcal{N}_\infty := \left\{u \in H^1\left(\mathbb{R}^3\right) \setminus \{0\} : \Phi_\infty'(u)[u] = 0\right\}. \tag{34}$$

Critical points of Φ_∞ are solutions of the limit problem at infinity

$$\begin{aligned} -\Delta u + a_\infty u + b_\infty \overline{\phi}_u(x) u = q_\infty |u|^{p-1} u, \quad x \in \mathbb{R}^3, \\ -\Delta \overline{\phi}_u = b_\infty u^2, \quad u \in H^1\left(\mathbb{R}^3\right), \end{aligned} \tag{35}$$

and, clearly, the conclusions of Lemma 4 hold true for Φ_∞ and $\mathcal{N}_\infty$. Moreover, for any $u \in H^1(\mathbb{R}^3) \setminus \{0\}$, it is easy to see that there exists unique $t(u) > 0$ such that $t(u)u \in \mathcal{N}_\infty$. Set

$$m_\infty := \inf\{\Phi_\infty(u),\ u \in \mathcal{N}_\infty\}. \tag{36}$$

From [11, 12], we deduce that m_∞ is achieved by a positive radially symmetric function w satisfying (11). In what follows, for any $y \in \mathbb{R}^3$, we use the translation symbol $w_y := w(\cdot - y)$. Set

$$m_\lambda := \inf\{\Phi_\lambda(u),\ u \in \mathcal{N}_\lambda\}. \tag{37}$$

Then the following relations of m_λ and m_∞ hold true.

Lemma 5. *Suppose that (A_1)-(A_3) hold. Then for each $\lambda \ge 0$ one has*

$$0 < m_\lambda \le m_\infty. \tag{38}$$

Proof. Let $\lambda \ge 0$ be fixed. The first inequality of (38) is a straight consequence of (31). In order to show the second inequality we should construct a sequence $\{u_n\} \subset \mathcal{N}_\lambda$ and $\lim_n \Phi_\lambda(u_n) = m_\infty$. To accomplish this, we take $\{y_n\}$, with $y_n \in \mathbb{R}^3$, $|y_n| \longrightarrow +\infty$, as $n \longrightarrow +\infty$ and set $u_n = t_n w_{y_n}$, where $w_{y_n} = w(x - y_n)$ and $t_n = t_\lambda(w_{y_n}) > 0$ such that $u_n = t_n w_{y_n} \in \mathcal{N}_\lambda$. Here we recall that w is a radial solution of (35). A direct computation shows that

$$\begin{aligned}\Phi_\lambda(u_n) &= \frac{t_n^2}{2}\left[\|w\|^2 + \int_{\mathbb{R}^3}\lambda a(x) w_{y_n}^2\right] + \frac{t_n^4}{4}\int_{\mathbb{R}^3}\int_{\mathbb{R}^3}\frac{(b_\infty - b(x)) w_{y_n}^2(x)(b_\infty - b(y)) w_{y_n}^2(y)}{4\pi|x-y|}dx\,dy - \frac{q_\infty t_n^{p+1}}{p+1}|w_{y_n}|_{p+1}^{p+1}\\ &\quad - \frac{t_n^{p+1}}{p+1}\int_{\mathbb{R}^3} q(x)|w_{y_n}|^{p+1}\\ &= \frac{t_n^2}{2}\left[\|w\|^2 + \int_{\mathbb{R}^3}\lambda a(x+y_n) w^2\right] - \frac{t_n^{p+1} q_\infty}{p+1}|w|_{p+1}^{p+1} - \frac{t_n^{p+1}}{p+1}\int_{\mathbb{R}^3} q(x+y_n)|w|^{p+1}\\ &\quad + \frac{t_n^4}{4}\int_{\mathbb{R}^3}\int_{\mathbb{R}^3}\frac{(b_\infty - b(x+y_n)) w^2(x)(b_\infty - b(y+y_n)) w^2(y)}{4\pi|x-y|}dx\,dy.\end{aligned} \tag{39}$$

It is clear that

$$\begin{aligned}\lim_{n\to\infty}\int_{\mathbb{R}^3} a(x+y_n) w^2 &= 0,\\ \lim_{n\to\infty}\int_{\mathbb{R}^3} q(x+y_n)|w|^{p+1} &= 0.\end{aligned} \tag{40}$$

We claim that

$$\lim_{n\to\infty}\int_{\mathbb{R}^3}\int_{\mathbb{R}^3}\frac{b(x+y_n) w^2(x) b(y+y_n) w^2(y)}{|x-y|}dx\,dy = 0. \tag{41}$$

In fact, we infer from Hardy-Littlewood-Sobolev inequality (17) that for $R > 0$ large

$$\begin{aligned}&\int_{\mathbb{R}^3}\int_{\mathbb{R}^3}\frac{b(x+y_n) w^2(x) b(y+y_n) w^2(y)}{|x-y|}dx\,dy\\ &\le c\left(\int_{\mathbb{R}^3} b^{6/5}(x+y_n) w^{12/5}(x)\,dx\right)\\ &\le c|w|_6\int_{B_R} b^2(x+y_n)\,dx + c|b|_2|w|_{L^6(B_R^c)}\end{aligned} \tag{42}$$

and, from condition (A_2), we know that the claim holds. On the other hand, we infer from (40)-(41) and $t_n w_{y_n} \in \mathcal{N}_\lambda$ that

$$\begin{aligned}&\|w\|^2 + t_n^2\int_{\mathbb{R}^3} b_\infty\overline{\phi}_w w^2 + o(t_n^2)\\ &= t_n^{p-1} q_\infty|w|_{p+1}^{p+1} + o(t_n^{p-1}).\end{aligned} \tag{43}$$

Thus, we get $\sigma_0 \le t_n \le \sigma_1$ for $\sigma_0, \sigma_1 > 0$. Moreover, we deduce from $w \in \mathcal{N}_\infty$ that

$$\left(\frac{1}{t_n^2} - 1\right)\|w\|^2 = \left(t_n^{p-3} - 1\right)|w|_{p+1}^{p+1} + o(1). \tag{44}$$

Since $p \in (3,5)$, we infer that $t_n \to 1$ as $n \to \infty$. Finally, we let $n \to \infty$ in (39) and obtain

$$\begin{aligned}t_n &\to 1,\\ \Phi_\lambda(u_n) &\to m_\infty.\end{aligned} \tag{45}$$

This finishes the proof. □

By applying the well-known concentration-compactness principle [39] and maximum principle [40], we have the following splitting lemma results. For the details of the proof, one can refer to [23, Lemma 4.1 and Corollary 4.2]

Lemma 6. *If the strict inequality*

$$m_\lambda < m_\infty \tag{46}$$

holds, then m_λ is achieved by a positive function. Furthermore, all the minimizing sequences are relatively compact.

Next we consider the special case $\lambda = 0$.

Lemma 7. *Assume that (A_1)-(A_3) and $\lambda = 0$ hold. Then (10) has a ground state positive solution.*

Proof. Note that

$$\begin{aligned}\Phi_0(u) &= \frac{1}{2}\|u\|^2 + \frac{1}{4}\int_{\mathbb{R}^3}(b_\infty - b(x))\phi_u(x) u^2\\ &\quad - \frac{1}{p+1}\int_{\mathbb{R}^3}(q_\infty + q(x))|u|^{p+1},\end{aligned} \tag{47}$$

and

$$\begin{aligned}\Phi_\infty(u) &= \frac{1}{2}\|u\|^2 + \frac{1}{4}\int_{\mathbb{R}^3} b_\infty\overline{\phi}_u(x) u^2\\ &\quad - \frac{1}{p+1}\int_{\mathbb{R}^3} q_\infty|u|^{p+1},\end{aligned} \tag{48}$$

where $\phi_u(x) = \int_{\mathbb{R}^3}((b_\infty - b(y))u^2(y)/4\pi|x-y|)dy$ and $\overline{\phi}_u(x) = \int_{\mathbb{R}^3}(b_\infty u^2(y)/4\pi|x-y|)dy$. Therefore, we have that

$$m_0 \le m_\infty. \tag{49}$$

From (36) we know that $\Phi_\infty(w) = m_\infty$. Similar to the proof of Lemma 4 (a), we infer that $g(t) = \Phi_0(tw)$ has unique maximum $t_0 > 0$ such that $t_0 w \in \mathcal{N}_0$. Then we infer the following from condition (A_2).

$$\begin{aligned}\Phi_0(t_0 w) &= \frac{t_0^2}{2}\|w\|^2 - \frac{t_0^{p+1}}{p+1}\int_{\mathbb{R}^3}(q_\infty + q(x))|u|^{p+1} + \frac{t_0^4}{4}\\ &\quad\cdot\int_{\mathbb{R}^3}\int_{\mathbb{R}^3}\frac{(b_\infty - b(x)) w^2(x)(b_\infty - b(y)) w^2(y)}{4\pi|x-y|}dx\,dy\\ &< \Phi_\infty(t_0 w)\end{aligned} \tag{50}$$

Moreover, since $w \in \mathcal{N}_\infty$ is the unique maximum point of $g_\infty(t) = \Phi_\infty(tw)$ for $t > 0$, it follows that $\Phi_\infty(t_0 w) \leq \Phi_\infty(w)$. Combining the above arguments, we infer that

$$m_0 \leq \Phi_0 (t_0 w) < \Phi_\infty (t_0 w) \leq \Phi_\infty (w) = m_\infty. \quad (51)$$

From Lemma 6, we know that the conclusion holds. □

3. Proof of Theorem 1

In this section we shall give the proof of Theorem 1. This can be accomplished by the following Lemma.

Lemma 8. *Assume that* (A_1)-(A_3) *hold. Then for each* $\lambda \in \mathbb{R}^+$, *we have* $m_\lambda < m_\infty$.

Proof. We first observe that, by Lemma 7, we know that $m_0 < m_\infty$. So, we consider the case $\lambda > 0$ in the sequel. For fixed $\lambda > 0$, we choose t_n such that $u_n = t_n w_{y_n} \in \mathcal{N}_\lambda$, where y_n and t_n are chosen as in the proof of Lemma 5. Moreover, as in (45), we infer that $0 < \sigma_0 \leq t_n \leq \sigma_1$. Similar to (51), we know that $\Phi_\infty(t_n w) \leq m_\infty$. Thus, we infer that

$$\begin{aligned} m_\lambda &\leq \Phi_\lambda (u_n) = \Phi_\lambda \left(t_n w_{y_n}\right) = \frac{t_n^2}{2} \left[\|w\| + \lambda \int_{\mathbb{R}^3} a(x+y_n) w^2 \right] - \frac{t_n^{p+1}}{p+1} \int_{\mathbb{R}^3} (q_\infty + q(x+y_n)) |w|^{p+1} + \frac{t_n^4}{4} \\ &\cdot \int_{\mathbb{R}^3} \int_{\mathbb{R}^3} \frac{(b_\infty - b(x+y_n)) w^2(x) (b_\infty - b(y+y_n)) w^2(y)}{4\pi |x-y|} dx\, dy = \Phi_\infty (t_n w) + \frac{t_n^2}{2} \left[\lambda \int_{\mathbb{R}^3} a(x+y_n) w^2 + \frac{t_n^2}{2} \right. \\ &\cdot \left. \int_{\mathbb{R}^3} \int_{\mathbb{R}^3} \frac{b(x+y_n) w^2(x) b(y+y_n) w^2(y)}{4\pi |x-y|} dx\, dy - t_n^2 \int_{\mathbb{R}^3} b(x+y_n) \overline{\phi}_w w^2 - \frac{2t_n^{p-1}}{p+1} \int_{\mathbb{R}^3} q(x+y_n) |w|^{p+1} \right] \leq m_\infty \\ &+ \frac{t_n^2}{2} \left[\lambda \int_{\mathbb{R}^3} a(x+y_n) w^2 + \frac{t_n^2}{2} \int_{\mathbb{R}^3} \int_{\mathbb{R}^3} \frac{b(x+y_n) w^2(x) b(y+y_n) w^2(y)}{4\pi |x-y|} dx\, dy - t_n^2 \int_{\mathbb{R}^3} b(x+y_n) \overline{\phi}_w w^2 - \frac{2t_n^{p-1}}{p+1} \right. \\ &\cdot \left. \int_{\mathbb{R}^3} q(x+y_n) |w|^{p+1} \right]. \end{aligned} \quad (52)$$

Since $b \in L^\infty(\mathbb{R}^3)$, we prove the conclusion if we show that, for large n,

$$\begin{aligned} \lambda &\int_{\mathbb{R}^3} a(x+y_n) w^2 \\ &+ c \int_{\mathbb{R}^3} b(x+y_n) w^2(x) \varphi_w(x)\, dx \\ &- c \int_{\mathbb{R}^3} b(x+y_n) \overline{\phi}_w w^2 \\ &- c \int_{\mathbb{R}^3} q(x+y_n) |w|^{p+1} < 0, \end{aligned} \quad (53)$$

where $\varphi_w(x) = \int_{\mathbb{R}^3} (b(y+y_n) w^2(y) / 4\pi |x-y|) dy$. This is equal to proving that, for large n,

$$\begin{aligned} A_1 = \lambda &\int_{\mathbb{R}^3 \setminus B_{\tau|y_n|}} a(x+y_n) w^2 \\ &+ c \int_{\mathbb{R}^3 \setminus B_{\tau|y_n|}} b(x+y_n) w^2(x) \varphi_w(x)\, dx \\ &- c \int_{\mathbb{R}^3 \setminus B_{\tau|y_n|}} b(x+y_n) \overline{\phi}_w w^2 \\ &- c \int_{\mathbb{R}^3 \setminus B_{\tau|y_n|}} q(x+y_n) |w|^{p+1} \\ &< -\lambda \int_{B_{\tau|y_n|}} a(x+y_n) w^2 \\ &- c \int_{B_{\tau|y_n|}} b(x+y_n) w^2(x) \varphi_w(x)\, dx \\ &+ c \int_{B_{\tau|y_n|}} b(x+y_n) \overline{\phi}_w w^2 \\ &+ c \int_{B_{\tau|y_n|}} q(x+y_n) |w|^{p+1} = A_2. \end{aligned} \quad (54)$$

A direct computation shows that for $R > 0$ large

$$\begin{aligned} \varphi_w(x) &= \int_{|x-y|\leq 1} \frac{b(y+y_n) w^2(y)}{4\pi |x-y|} dy \\ &+ \int_{1\leq |x-y| \leq R} \frac{b(y+y_n) w^2(y)}{4\pi |x-y|} dy \\ &+ \int_{|x-y|\geq R} \frac{b(x+y_n) w^2(y)}{4\pi |x-y|} dy \\ &\leq c |w|_4^2 \left(\int_{|x-y|\leq 1} \frac{1}{|x-y|} dy \right) \end{aligned}$$

$$+c|w|_{8/3}^2\left(\left(\int_{1\le|x-y|\le R}\frac{1}{|x-y|}dy\right)^{1/4}\right.$$
$$\left.+c|w|_{8/3}^2\left(\int_{|x-y|\ge R}\frac{1}{|x-y|}dy\right)^{1/4}\right)\le c. \quad (55)$$

Now we are ready to give the estimate of the term A_1. We infer from condition (A_2), (11), and (55) that

$$\begin{aligned}A_1 &\le \int_{\mathbb{R}^3\setminus B_{\tau|y_n|}}\lambda a(x+y_n)w^2 + c\int_{\mathbb{R}^3\setminus B_{\tau|y_n|}}b(x+y_n)\varphi_w w^2 \\ &\le c\left(\left(\int_{\mathbb{R}^3\setminus B_{\tau|y_n|}}|a(x+y_n)|^{3/2}\right)^{2/3} + \int_{\mathbb{R}^3\setminus B_{\tau|y_n|}}|b(x+y_n)|^{3/2}\right)\left[\int_{\mathbb{R}^3\setminus B_{\tau|y_n|}}|w|^6\right]^{1/3} \\ &\le ce^{-2c^*\tau|y_n|}\end{aligned} \quad (56)$$

holds for n large.

Next we give the estimate for the second part A_2. We first consider that the case $a(x)$ decays faster than $b(x)$. By condition (1) of Theorem 1, we know that, for all $M>0$, there exists $n_0\ge 1$ such that, for all $n\ge n_0$ and for all $x\in B_{\tau|y_n|}$,

$$\begin{aligned}&Me^{-2\tau c^*|y_n|}\le q(x+y_n)\le ce^{-(2\tau(1-\tau)/(1+2\tau))c^*|y_n|},\\ &a(x+y_n)\le c(1-\tau)^{-(1+\tau)}|y_n|^{-(1+\tau)}e^{-2\tau c^*|y_n|}\\ &b(x+y_n)\ge ce^{-(2\tau(1-\tau)/(1+2\tau))c^*|y_n|}.\end{aligned} \quad (57)$$

Moreover, by [41, Lemmas 2.3 and 2.6], we know that

$$\overline{\phi}_w(x)=\int_{\mathbb{R}^3}\frac{b_\infty w^2(y)}{4\pi|x-y|}dy\sim\frac{1}{|x|},\quad \text{as } |x|\longrightarrow\infty. \quad (58)$$

Since $|x+y_n|\ge(1-\tau)|y_n|\longrightarrow\infty$ as $n\longrightarrow\infty$, we infer from (58) and (A_2) that

$$\varphi_w(x)=\int_{\mathbb{R}^3}\frac{b(x+y_n)w^2(y)}{4\pi|x-y|}dy\sim o(\overline{\phi}_w) \quad \text{as } n\longrightarrow\infty. \quad (59)$$

Thus, we infer that, for n sufficiently large and for all $x\in B_{\tau|y_n|}$,

$$\frac{cb(x+y_n)\overline{\phi}_w}{q(x+y_n)}-\frac{cb(x+y_n)\varphi_w}{q(x+y_n)}-\frac{\lambda a(x+y_n)}{q(x+y_n)}>0. \quad (60)$$

Hence we deduce that the inequality

$$\begin{aligned}A_2 &\ge \int_{B_{\tau|y_n|}}q(x+y_n)w^2\left[\frac{cb(x+y_n)\overline{\phi}_w}{q(x+y_n)}+cw^{p-1}-\frac{cb(x+y_n)\varphi_w}{q(x+y_n)}-\frac{\lambda a(x+y_n)}{q(x+y_n)}\right]>c\int_{B_{\tau|y_n|}}q(x\\ &\quad +y_n)w^{p+1}>cMe^{-2\tau c^*|y_n|}\int_{B_1}w^{p+1}>cMe^{-2\tau c^*|y_n|}\end{aligned} \quad (61)$$

holds for n sufficiently large. Thus, by the arbitrariness of M, we can conclude that $A_1<A_2$, as desired.

Finally, we consider the remaining case in Theorem 1. By (2) of Theorem 1, for all $\varepsilon>0$ and $M>0$, there exists $n_0\ge 1$ such that, for all $n\ge n_0$ and for all $x\in B_{\tau|y_n|}$,

$$\begin{aligned}&Me^{-2\tau c^*|y_n|}\le q(x+y_n),\\ &a(x+y_n)\le ce^{-(p-1)\tau c^*|y_n|},\end{aligned} \quad (62)$$

where $p\in(3,5)$. Thus we get that, for n sufficiently large and for all $x\in B_{\tau|y_n|}$,

$$cw^{p-1}-\frac{\lambda a(x+y_n)}{q(x+y_n)}>\frac{c}{2}w^{p-1}. \quad (63)$$

Hence, we infer from (59) that

$$\begin{aligned}A_2 &\ge c\int_{B_{\tau|y_n|}}\left(b(x+y_n)w^2(\overline{\phi}_w-\varphi_w)\right)+\int_{B_{\tau|y_n|}}\left(q(x+y_n)w^2\left(cw^{p-1}-\frac{\lambda a(x+y_n)}{q(x+y_n)}\right)\right)\\ &>\frac{c}{2}\int_{B_{\tau|y_n|}}q(x+y_n)w^{p+1}>\frac{c}{2}Me^{-2\tau c^*|y_n|}\int_{B_1}w^{p+1}\\ &>\frac{c}{2}Me^{-2\tau c^*|y_n|}.\end{aligned} \quad (64)$$

Hence, by the arbitrariness of M, we can conclude that $A_1<A_2$. This finishes the proof. □

Proof of Theorem 1. By Lemmas 6 and 8, we know that the conclusions of Theorem 1 hold. □

4. Conclusion

In this paper, the authors prove the existence of positive ground state solutions for the nonautonomous Schrödinger-Poisson system. In condition (1) of Theorem 1, we know that if $a(x)$ decays faster than $b(x)$, we find the existence of positive ground state solution of (10). In condition (2) of Theorem 1, we only need the decay condition for $a(x)$(whatever the decay speed of $b(x)$ to 0 as $|x|\longrightarrow\infty$) to prove the existence of positive solution. This is different phenomenon compared to the previous paper [34].

Conflicts of Interest

The authors declare that they have no conflicts of interest.

Authors' Contributions

The authors contributed equally to this article. They have all read and approved the final manuscript.

Acknowledgments

This work was supported by the National Natural Science Foundation of China (Grants 11571140, 11671077, 71704066), the Fellowship of Outstanding Young Scholars of Jiangsu Province (BK20160063), the Major Projects of the National Social Science Foundation (18VSJ038), the Six Big Talent Peaks Project in Jiangsu Province (XYDXX-015), and the Natural Science Foundation of Jiangsu Province (Nos. BK20150478, BK20170542).

References

[1] W. Bao, N. J. Mauser, and H. P. Stimming, "Effective one particle quantum dynamics of electrons: a numerical study of the Schrödinger-Poisson-X α model," *Communications in Mathematical Sciences*, vol. 1, no. 4, pp. 809–828, 2003.

[2] I. Catto, J. Dolbeault, O. Sanchez, and J. Soler, "Existence of steady states for the Maxwell-Schrödinger-Poisson system: exploring the applicability of the concentration-compactness principle," *Mathematical Models and Methods in Applied Sciences*, vol. 23, no. 10, pp. 1915–1938, 2013.

[3] P.-L. Lions, "Solutions of Hartree-Fock equations for Coulomb systems," *Communications in Mathematical Physics*, vol. 109, no. 1, pp. 33–97, 1987.

[4] T. D'Aprile and D. Mugnai, "Solitary waves for nonlinear Klein-Gordon-MAXwell and Schrödinger-MAXwell equations," *Proceedings of the Royal Society of Edinburgh, Section: A Mathematics*, vol. 134, no. 5, pp. 893–906, 2004.

[5] T. D'Aprile and D. Mugnai, "Non-existence results for the coupled Klein-Gordon-Maxwell equations," *Advanced Nonlinear Studies*, vol. 4, no. 3, pp. 307–322, 2004.

[6] D. Mugnai, "The Schrödinger-Poisson system with positive potential," *Communications in Partial Differential Equations*, vol. 36, no. 7, pp. 1099–1117, 2011.

[7] V. Benci and D. Fortunato, "An eigenvalue problem for the Schrödinger-Maxwell equations," *Topological Methods in Nonlinear Analysis*, vol. 11, no. 2, pp. 283–293, 1998.

[8] V. Benci and D. Fortunato, "Solitary waves of the nonlinear Klein-Gordon equation coupled with the Maxwell equations," *Reviews in Mathematical Physics*, vol. 14, no. 4, pp. 409–420, 2002.

[9] G. Vaira, "Ground states for Schrödinger-Poisson type systems," *Ricerche di Matematica*, vol. 60, no. 2, pp. 263–297, 2011.

[10] X. He, "Multiplicity and concentration of positive solutions for the Schrödinger-Poisson equations," *Zeitschrift für Angewandte Mathematik und Physik*, vol. 5, pp. 869–889, 2011.

[11] J. Wang, L. Tian, J. Xu, and F. Zhang, "Existence of multiple positive solutions for Schrödinger-Poisson systems with critical growth," *Zeitschrift für Angewandte Mathematik und Physik*, vol. 66, no. 5, pp. 2441–2471, 2015.

[12] J. Wang, L. Tian, J. Xu, and F. Zhang, "Existence and concentration of positive ground state solutions for semilinear Schrödinger-Poisson systems," *Advanced Nonlinear Studies*, vol. 13, no. 3, pp. 553–582, 2013.

[13] J. Wang, L. Tian, J. Xu, and F. Zhang, "Existence and concentration of positive solutions for semilinear Schrödinger-Poisson systems in $\mathbb{R}^3$," *Calculus of Variations and Partial Differential Equations*, vol. 48, no. 1-2, pp. 243–273, 2013.

[14] L. Xiao and J. Wang, "Existence of positive solutions for a Schrödinger-Poisson system with critical growth," *Applicable Analysis: An International Journal*, pp. 1–38, 2018.

[15] A. Ambrosetti and D. Ruiz, "Multiple bound states for the Schrödinger-Poisson problem," *Communications in Contemporary Mathematics*, vol. 10, pp. 1–14, 2008.

[16] D. Ruiz, "The Schrödinger-Poisson equation under the effect of a nonlinear local term," *Journal of Functional Analysis*, vol. 237, no. 2, pp. 655–674, 2006.

[17] C. Mercuri, "Positive solutions of nonlinear Schrödinger-Poisson systems with radial potentials vanishing at infinity," *Atti della Accademia nazionale dei Lincei. Rendiconti Lincei. Matematica e applicazioni*, vol. 19, no. 3, pp. 211–227, 2008.

[18] Z. Wang and H.-S. Zhou, "Positive solution for a nonlinear stationary Schrdinger-Poisson system in $\mathbb{R}^3$," *Discrete and Continuous Dynamical Systems - Series A*, vol. 18, no. 4, pp. 809–816, 2007.

[19] L. Zhao and F. Zhao, "On the existence of solutions for the Schrödinger-Poisson equations," *Journal of Mathematical Analysis and Applications*, vol. 346, no. 1, pp. 155–169, 2008.

[20] A. Bahri and Y. Y. Li, "On a minimax procedure for the existence of a positive solution for certain scalar field equations in RN," *Revista Matematica Iberoamericana*, vol. 6, no. 1-2, pp. 1–15, 1990.

[21] A. Azzollini and A. Pomponio, "Ground state solutions for the nonlinear Schrödinger-Maxwell equations," *Journal of Mathematical Analysis and Applications*, vol. 345, no. 1, pp. 90–108, 2008.

[22] A. Ambrosetti, "On Schrödinger-Poisson systems," *Milan Journal of Mathematics*, vol. 76, pp. 257–274, 2008.

[23] G. Cerami and G. Vaira, "Positive solutions for some nonautonomous Schrodinger-Poisson systems," *Journal of Differential Equations*, vol. 248, no. 3, pp. 521–543, 2010.

[24] M. F. Furtado, L. A. Maia, and E. S. Medeiros, in *A Note on The Existence of a Positive Solution for a Non-Autonomous Schrödinger-Poisson System Analysis and Topology in Nonlinear Differential Equations*, vol. 85 of *Progress in Nonlinear Differential Equations and their Applications*, pp. 277–286, Springer, New York, NY, USA, 2014.

[25] A. Azzollini, P. d'Avenia, and A. Pomponio, "On the Schrödinger-Maxwell equations under the effect of a general nonlinear term," *Annales de l'Institut Henri Poincare. Analyse Non Linéaire*, vol. 27, no. 2, pp. 779–791, 2010.

[26] V. Benci and D. Fortunato, "Solitons in Schrödinger-Maxwell equations," *Journal of Fixed Point Theory and Applications*, vol. 15, no. 1, pp. 101–132, 2014.

[27] D. Bonheure and C. Mercuri, "Embedding theorems and existence results for nonlinear Schrödinger-Poisson systems with unbounded and vanishing potentials," *Journal of Differential Equations*, vol. 251, no. 4-5, pp. 1056–1085, 2011.

[28] T. D'Aprile and J. Wei, "Standing waves in the Maxwell-Schrödinger equation and an optimal configuration problem," *Calculus of Variations and Partial Differential Equations*, vol. 25, no. 1, pp. 105–137, 2006.

[29] I. Ianni and G. Vaira, "On concentration of positive bound states for the Schrödinger-Poisson system with potentials," *Advanced Nonlinear Studies*, vol. 8, no. 3, pp. 573–595, 2008.

[30] Y. Jiang and H. Zhou, "Bound states for a stationary nonlinear Schrödinger-Poisson system with signchanging potential in R3," *Acta Mathematica Scientia B*, vol. 29, no. 4, pp. 1095–1104, 2009.

[31] C. Mercuri and T.-M. Tyler, "On a class of nonlinear Schrödinger-Poisson systems involving a nonradial charge density," https://arxiv.org/abs/1805.00964.

[32] E. S. Noussair and J. Wei, "On the effect of the domain geometry on the existence and profile of nodal solution of some singularly perturbed semilinear Dirichlet problem," *Indiana University Mathematics Journal*, vol. 46, no. 4, pp. 1321–1332, 1997.

[33] D. Ruiz, "On the Schrödinger-Poisson-Slater System: behavior of minimizers, radial and nonradial cases," *Archive for Rational Mechanics and Analysis*, vol. 198, no. 1, pp. 349–368, 2010.

[34] G. Cerami and R. Molle, "Positive bound state solution for some Schrödinger-Poisson systems," *Nonlinearity*, vol. 29, no. 10, pp. 3103–3119, 2016.

[35] H. Berestycki and P. Lions, "Nonlinear scalar field equations. I. Existence of a ground state," *Archive for Rational Mechanics and Analysis*, vol. 82, no. 4, pp. 313–345, 1983.

[36] M. K. Kwong, "Uniqueness of positive solutions of $\Delta u - u + u^p = 0$ in $\mathbb{R}^3$," *Archive for Rational Mechanics and Analysis*, vol. 105, no. 3, pp. 243–266, 1989.

[37] G. Cerami and A. Pomponio, "On some scalar field equations with competing coefficients," *International Mathematics Research Notices*, pp. 1–27, 2017.

[38] E.-H. Lieb and M. Loss, *Analysis*, vol. 14 of *Graduate Studies in Mathematics*, American Mathematical Society, Providence, RI, USA, 2nd edition, 2001.

[39] P.-L. Lions, "The concentration-comoactness principle in the calculus of variations. The locally compact case, part 1 and part 2," *Annales de l'Institut Henri Poincaré (C) Analyse Non Linéaire*, vol. 1, no. 4, pp. 223–283, 1984.

[40] M. Willem, *Minimax Theorems*, vol. 24 of *Progress in Nonlinear Differential Equations and their Applications*, Birkhäuser Boston Inc., Boston, Mass, USA, 1996.

[41] Y. Li and W.-M. Ni, "On Conformal Scalar Curvature Equations in RN," *Duke Mathematical Journal*, vol. 57, no. 3, pp. 895–924, 1988.

13

Evaluation of Heat Irreversibility in a Thin Film Flow of Couple Stress Fluid on a Moving Belt

Samuel O. Adesanya,[1,2] **Ramosheuw S. Lebelo**,[2] **and K. C. Moloi**[2]

[1] *Department of Mathematical Sciences, Redeemer's University, Ede, Nigeria*
[2] *Department of Education, Vaal University of Technology, Vanderbijlpark, South Africa*

Correspondence should be addressed to Samuel O. Adesanya; adesanyas@run.edu.ng

Academic Editor: Antonio Scarfone

This article addresses the inherent heat irreversibility in the flow of a couple stress thin film along a moving vertical belt subjected to free and adiabatic surface. Mathematical analysis for the fluid-governing-equations is performed in detail. For maximum thermal performance and efficiency, the present analysis follows the second law of thermodynamics approach for the evaluation of entropy generation rate in the moving film. With this thermodynamic process, the interconnectivity between variables responsible for energy wastage is accounted for in the thermo-fluid equipment. Results of the analysis revealed the fluid properties that contribute more to energy loss and how the exergy of the system can be restored.

1. Introduction

The past few years have witnessed a growing interest in studies related to thin film flow along a moving belt. This has been so due to its wide range of applications in the field of environmental sciences, chemical engineering, and physical sciences. For example, in the design of film evaporator, paper, coating production, etc., by a way of a short literature review on a thin film on moving belt, Siddiqui et al. [1] focused on the film flow of Sisko fluid and Oldroyd 6-constant fluids on the vertically moving belt. Alam et al. [2, 3] described the non-Newtonian Johnson–Segalman film in the case of hydrodynamics and magnetohydrodynamics flow conditions, respectively. In a study by Farooq et al. [4], the exact solution for the Jeffery film flow on a moving belt was obtained. Hameed and Ellahi [5] obtained a numerical approximation for the hydromagnetic flow of Oldroyd 6-constant film by using Chebyshev collocation method. By using the Optimal Homotopy analysis and Lie symmetry group classification both Ene et al. [6] and Hayat et al. [7] derived explicit solution for the Oldroyd 6-constant fluid film. The literature is inexhaustive when it comes to results on a thin film on a moving belt; these interesting results are not limited to [6, 8–13] and the references therein.

All the studies in [1–13] neglect the thermal effect; therefore, the energy transfer associated with heat flow in many mechanical and chemical engineering applications cannot be described. Motivated by Gul et al. [14–16], in which thermal effect on the moving belt is confined to the first law of thermodynamics. Also, Mahmud and Fraser [17] have done second law analysis to conserve exergy of a thermo-fluid setup. They argued that the useful energy destroyed accounts for underperformance and decreased the thermodynamic efficiency of a system. Therefore, the present study addresses both the energy conservation and management from a thermodynamics viewpoint; in other words, both first and second law of thermodynamics that governs energy conservation and transmission are incorporated in the vertically moving thin film for the very first time. Our focus here is to implement the energy evaluation technique as a valuable tool for identifying variables connected with the destruction of available energy for work and a measure of the efficiency of thermal equipment so as to reduce waste especially during fluid coupling in a conveyor belt. It is interesting to note that, a thorough literature search revealed that little or nothing has been done on couple stress fluid thin film on a moving belt despite the fact that the class of fluid is mostly used as lubricants since it accommodates

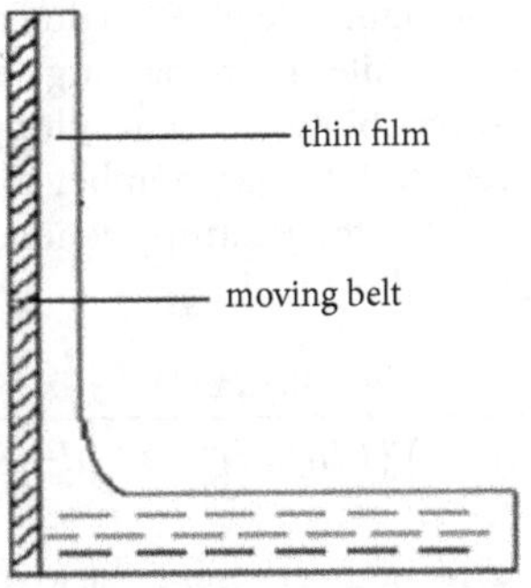

FIGURE 1: Flow geometry.

tiny microstructures of mechanical importance even in the lubrication of pulley bearing-system used for conveyor belts. The presence of these polymer additives in synthetic lubricants cannot be explained by the classical Newtonian model and this challenged was addresses by Stokes [18]. In the next section, the mathematical analysis is presented based on the steady, viscous, and incompressible unidirectional flow conditions. For the sake of brevity, the derivation of the couple stress model used here is described explicitly in ([19–21]) and not included. In Section 3 of the work, the solutions of the dimensionless boundary-value problems are obtained and used to generate the Bejan and entropy generation profiles.

2. Mathematical Analysis

The analysis is focused on a wide belt moving steadily through a container that is filled with couple stress fluid in an upward direction. A thin fluid layer is picked up as the belt passes through the container and the effect of gravitational force g on the thin is considered. The lamina film is assumed to be of uniform thickness δ. As shown in Figure 1, the (x', y') Cartesian coordinate is chosen along the substrate such that the x-axis lies in the transverse direction of the film flow while the y-axis is chosen in an upward direction.

Taking the thermal effect into consideration, the balanced equations for the thin film with constant physical properties are as follows.

Momentum equation:

$$0 = \mu \frac{d^2u'}{dx'^2} - \eta \frac{d^4u'}{dx'^4} - \rho g \tag{1}$$

The appropriate boundary conditions for the moving belt problem are as follows.

$$u'(0) = U,$$
$$0 = \left(\frac{d^2u'}{dx'^2}\right)\bigg|_{x'=0} = \left(\frac{du'}{dx'} - \frac{\eta}{\mu}\frac{d^3u'}{dx'^3}\right)\bigg|_{x'=\delta} = \left(\frac{d^2u'}{dx'^2}\right)\bigg|_{x'=\delta} \tag{2}$$

If the flow is accompanied with thermal effect, then the balanced energy equation for the fluid temperature T is given by

$$0 = k\frac{d^2T}{dx'^2} + \mu\left(\frac{du'}{dx'}\right)^2 + \eta\left(\frac{d^2u'}{dx'^2}\right)^2 \tag{3}$$

with thermal conditions

$$T(0) = T_a,$$
$$\left(\frac{dT}{dx'}\right)\bigg|_{x'=\delta} = 0. \tag{4}$$

As a result of the inherent irreversibility in many heat transfer processes, the momentum and energy exchange on the couple stress thin film on the moving belt will ultimately lead to a nonequilibrium situation; therefore there will be a continuous generation of entropy, E_G, due to heat transfer and fluid friction within the moving film. The rate at which exergy is depleted which results in underperformance is given by

$$E_G = \frac{k}{T_a^2}\left(\frac{dT}{dx'}\right)^2 + \frac{\mu}{T_a}\left(\frac{du'}{dx'}\right)^2 + \frac{\eta}{T_a}\left(\frac{d^2u'}{dx'^2}\right)^2 \tag{5}$$

with

$$u = \frac{u'}{U},$$
$$x = \frac{x'}{\delta},$$
$$G = \frac{\rho g \delta^2}{\mu U},$$
$$\kappa^2 = \frac{\eta}{\delta^2 \mu},$$
$$\lambda = \frac{\mu U^2}{k T_a}, \tag{6}$$
$$Ns = \frac{\delta^2 E_G}{k},$$
$$\theta = \frac{T - T_a}{T_a}.$$

Using (6), (1)-(2) give the Boundary-Value Problem (BVP)

$$0 = \frac{d^2u}{dx^2} - \kappa^2\frac{d^4u}{dx^4} - G;$$
$$u(0) = 1, \tag{7}$$
$$0 = \left(\frac{d^2u}{dx^2}\right)\bigg|_{x=0} = \left(\frac{du}{dx} - \kappa^2\frac{d^3u}{dx^3}\right)\bigg|_{x=1} = \left(\frac{d^2u}{dx^2}\right)\bigg|_{x=1}$$

and the dimensionless form (3)-(4) gives the following.

$$0 = \frac{d^2\theta}{dx^2} + \lambda\left(\left(\frac{du}{dx}\right)^2 + \kappa^2\left(\frac{d^2u}{dx^2}\right)^2\right);$$
$$\theta(0) = 0, \tag{8}$$
$$\frac{d\theta(1)}{dx} = 0$$

Using the same analysis in (6), we have the dimensionless entropy generation rate (Ns) as follows.

$$Ns = \left(\frac{d\theta}{dx}\right)^2 + \lambda\left(\left(\frac{du}{dx}\right)^2 + \kappa^2\left(\frac{d^2u}{dx^2}\right)^2\right) \tag{9}$$

The first term on the right side of (9) is due to Heat Transfer Irreversibility (HTI) while the remaining terms are due to Fluid Friction Irreversibility (FFI) resulting from viscous dissipation. As a result, the Bejan number can be defined by the ratio of HTI with the total entropy generation, Ns, on the thin film. In other words

$$Be = \frac{(d\theta/dx)^2}{(d\theta/dx)^2 + \lambda\left((du/dx)^2 + \kappa^2 (d^2u/dx^2)^2\right)}. \tag{10}$$

The coefficients μ, u', η, ρ represent the dynamic viscosity, dimensional velocity, couple stress, and density, respectively, and λ, θ, k, T_a viscous heating parameter, dimensionless entropy generation rate, thermal conductivity of the fluid, and referenced temperature. The exact solution of the BVP (7) gives

$$u(x) = -\frac{e^{-x/\kappa}\left(-2e^{1/\kappa+x/\kappa} - 2e^{x/\kappa} + 2e^{1/\kappa+x/\kappa}Gx + 2e^{x/\kappa}Gx - e^{1/\kappa+x/\kappa}Gx^2 - e^{x/\kappa}Gx^2 - 2e^{1/\kappa+x/\kappa}G\kappa^2 + 2e^{1/\kappa}G\kappa^2 - 2e^{x/\kappa}G\kappa^2 + 2e^{2x/\kappa}G\kappa^2\right)}{2\left(1+e^{1/\kappa}\right)} \tag{11}$$

and also (8) admits

$$\begin{aligned}
\theta(x) = {} & \frac{1}{12\left(1+e^{1/\kappa}\right)^2} e^{-2x/\kappa} G^2 \left(8e^{1/\kappa+2x/\kappa}x\right. \\
& + 4e^{2/\kappa+2x/\kappa}x + 4e^{2x/\kappa}x - 12e^{1/\kappa+2x/\kappa}x^2 \\
& - 6e^{2/\kappa+2x/\kappa}x^2 - 6e^{2x/\kappa}x^2 + 8e^{1/\kappa+2x/\kappa}x^3 \\
& + 4e^{2/\kappa+2x/\kappa}x^3 + 4e^{2x/\kappa}x^3 - 2e^{1/\kappa+2x/\kappa}x^4 \\
& - e^{2/\kappa+2x/\kappa}x^4 - e^{2x/\kappa}x^4 + 24e^{1/\kappa+2x/\kappa}x\kappa^2 \\
& + 12e^{2/\kappa+2x/\kappa}x\kappa^2 + 12e^{2x/\kappa}x\kappa^2 - 12e^{1/\kappa+2x/\kappa}x^2\kappa^2 \\
& - 6e^{2/\kappa+2x/\kappa}x^2\kappa^2 - 6e^{2x/\kappa}x^2\kappa^2 + 24e^{1/\kappa+x/\kappa}\kappa^3 \\
& + 24e^{2/\kappa+x/\kappa}\kappa^3 - 24e^{2/\kappa+2x/\kappa}\kappa^3 - 24e^{1/\kappa+3x/\kappa}\kappa^3 \\
& + 24e^{2x/\kappa}\kappa^3 - 24e^{3x/\kappa}\kappa^3 - 24e^{1/\kappa+x/\kappa}x\kappa^3 \\
& - 24e^{2/\kappa+x/\kappa}x\kappa^3 + 12e^{2/\kappa+2x/\kappa}x\kappa^3 + 24e^{1/\kappa+3x/\kappa}x\kappa^3 \\
& - 12e^{2x/\kappa}x\kappa^3 + 24e^{3x/\kappa}x\kappa^3 - 24e^{1/\kappa+x/\kappa}\kappa^4 \\
& - 24e^{2/\kappa+x/\kappa}\kappa^4 + 48e^{1/\kappa+2x/\kappa}\kappa^4 + 30e^{2/\kappa+2x/\kappa}\kappa^4 \\
& - 24e^{1/\kappa+3x/\kappa}\kappa^4 - 6e^{2/\kappa}\kappa^4 + 30e^{2x/\kappa}\kappa^4 - 24e^{3x/\kappa}\kappa^4 \\
& \left. - 6e^{4x/\kappa}\kappa^4\right)\lambda.
\end{aligned} \tag{12}$$

The exact solutions are used to determine the entropy generation rate (9) and heat irreversibility ratio in (10). The skin friction and the rate of heat transfer along the belt are defined as

$$\tau_w = \left(\frac{du'}{dx'} - \frac{\eta}{\mu}\frac{d^3u'}{dx'^3}\right)\bigg|_{x=0},$$
$$q = -k\frac{dT}{dx}\bigg|_{x=0} \tag{13}$$

and the dimensionless form becomes

$$Sf = \frac{\delta\tau_w}{U} = \left(\frac{du}{dx} - \kappa^2\frac{d^3u}{dx^3}\right)\bigg|_{x=0},$$
$$Nu = \frac{q\delta}{kT_a} = -\frac{d\theta}{dx}\bigg|_{x=0} \tag{14}$$

3. Results and Discussion

In the section, the responses of the exact solution to a variation of parameters for the velocity, temperature, and entropy generation are presented graphically as a function of the spatial variable. The two limiting cases of the couple stress parameter are shown in Figure 2; evidently, as the couple stress parameter approaches zero, the Newtonian class is fully recovered, while as it becomes infinite, the fluid behaves like plastic. Therefore in this work, the couple stress parameter is carefully chosen in between the limiting cases. In Figures 3(a)–3(d), the drainage problem is presented. In Figure 3(a), the flow is observed to increase with increasing values of the couple stress parameter, since the increase in the couple stress parameter implies a decrease in the dynamic viscosity of the fluid and thus an increase in the molecular distance between the fluid particles. Figure 3(b) displayed an asymmetrical heat flow from the isothermal wall to the adiabatic surface; this

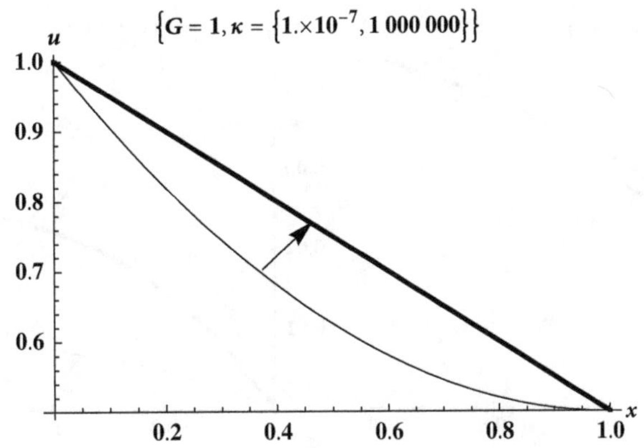

FIGURE 2: Drainage velocity against the limiting cases of the couple stress parameter.

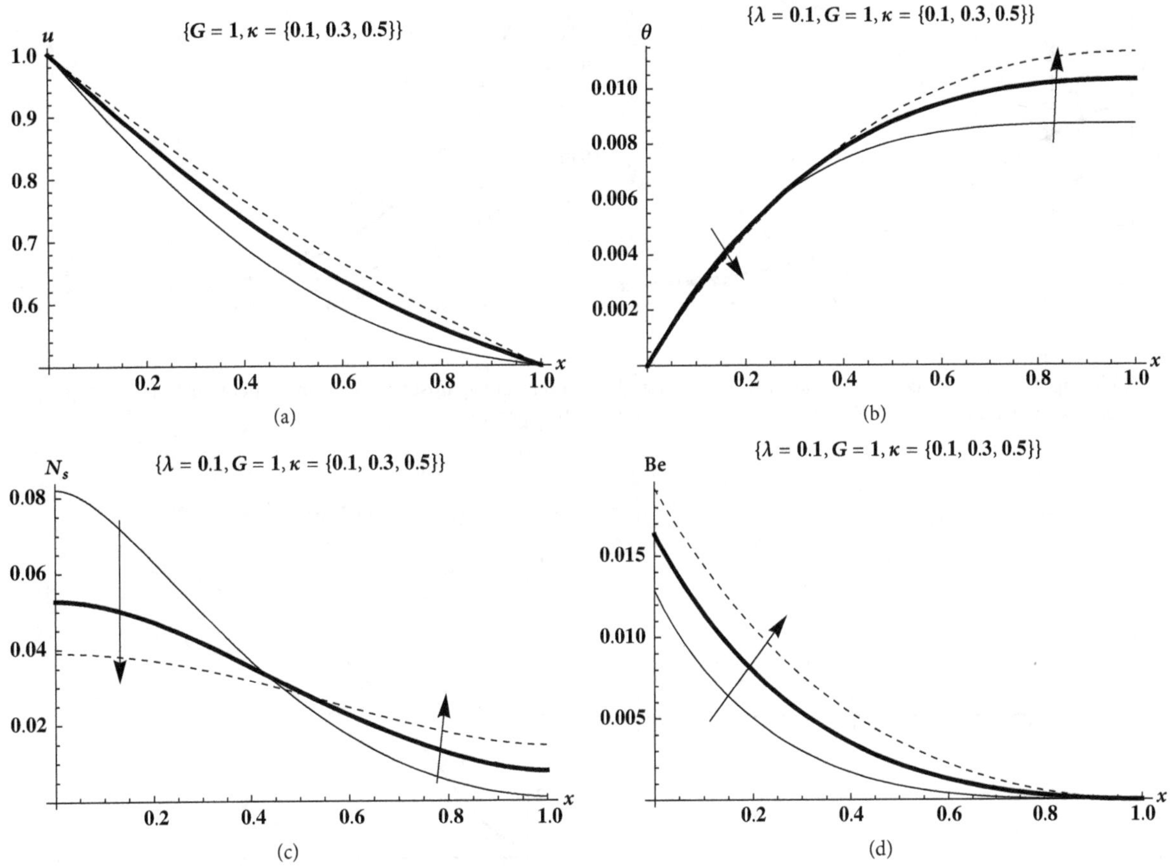

FIGURE 3: (a) drainage velocity against couple stress parameter, (b) temperature against couple stress parameter, (c) entropy generation against couple stress parameter, (d) Bejan number against couple stress parameter.

shows that the temperature profile is well behaved. Moreover, the rate of heat transfer at the free end increases significantly due to the migration of heat from cold region to the cold ambient air. Figure 3(c) revealed the contribution of the couple stress parameter to the entropy generation profile in the moving thin film flow. The result here shows that entropy at the cold isothermal vertical wall is decreasing with increasing values of the couple stress parameter while the reverse trend is observed at the free surface. Figure 3(d) represents the nonequilibrium in the irreversible process; at small values of the couple stress parameter, FFI is seen to contribute more to the heat irreversibility; this dominance declines across the vertical length as seen in the plot. Interestingly as the couple stress parameter increases, the dynamic viscosity of the fluid declines and the impact of the HTI is seen. Result in Figure 4(a) depicts the lifting effect when the fluid moves against gravitational force. Naturally, the gravitational pull decreases the flow velocity as the dynamic

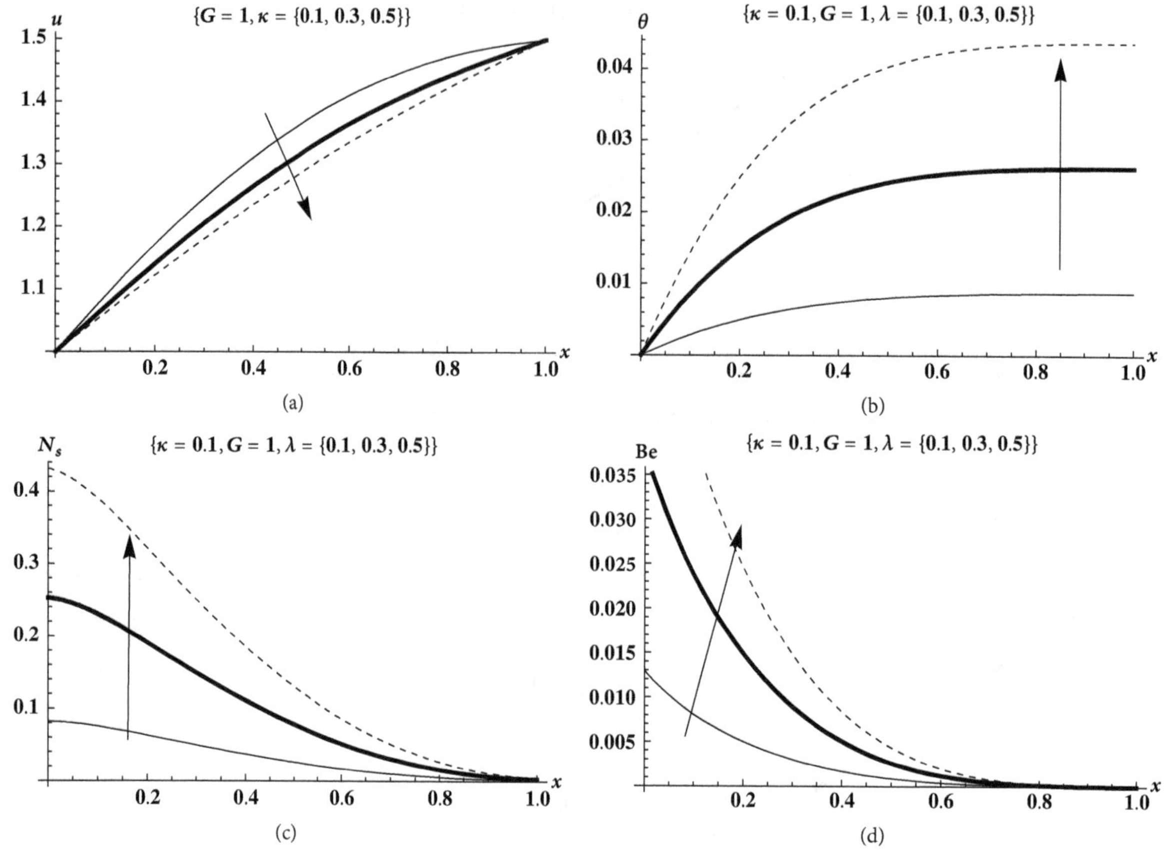

FIGURE 4: (a) lifting flow velocity against couple stress parameter, (b) temperature against viscous heating parameter, (c) entropy generation against viscous heating parameter, (d) Bejan number against viscous heating parameter.

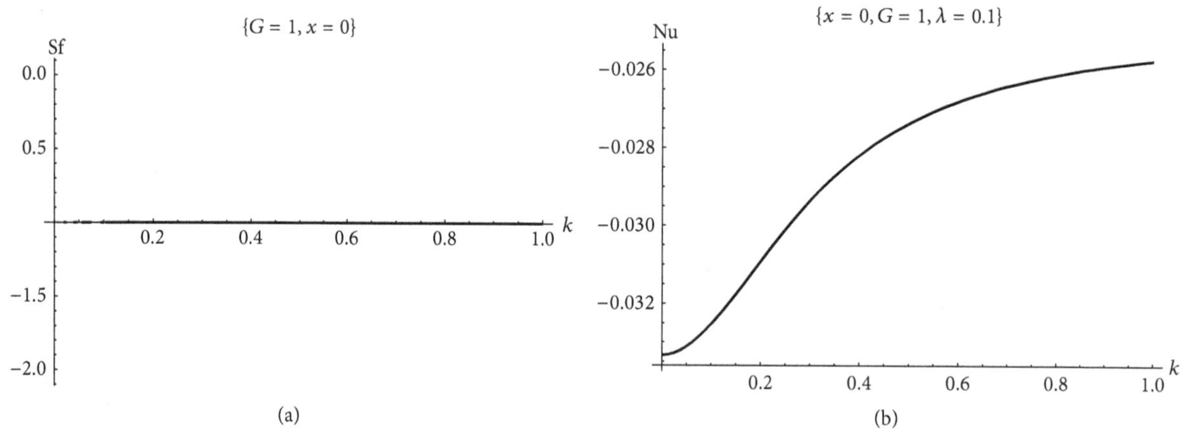

FIGURE 5: (a) Skin friction for lifting flow against couple stress parameter, (b) heat transfer rate for lifting flow against couple stress parameter.

viscosity of the fluid decreases. Figure 4(b) confirmed that, as the viscous heating increases, the fluid temperature is expected to increase as shown in the plot. This contributes to the total heat energy, and entropy is expected to increase the kinetic energy of the fluid particles as presented in Figure 4(c) and the HTI dominates over FFI in the moving film as shown in Figure 4(d). Result in Figure 5(a) represents a constant variation of couple stress parameter with skin friction while the rate of heat transfer along the moving belt increases with increasing value of the couple stress parameter as shown in Figure 5(b).

4. Concluding Remarks

The inclusion of couple stresses into the vertically moving thin film is the major contribution of this study. The exact solutions of the velocity and temperature fields are obtained and used to derive the entropy generation rate and heat irreversibility ratio. The result of the computation shows the following:

(i) Drainage and lifting problems are opposing phenomena.

(ii) Couple stress parameter contributes to entropy generation only at the adiabatic surface

(iii) Entropy generation decreases with couple stress parameter at the isothermal cold surface

(iv) Viscous heating of the fluid enhances the entropy generation rate

Conflicts of Interest

The authors declare that they have no conflicts of interest.

References

[1] A. M. Siddiqui, M. Ahmed, and Q. K. Ghori, "Thin film flow of non-Newtonian fluids on a moving belt," *Chaos, Solitons & Fractals*, vol. 33, no. 3, pp. 1006–1016, 2007.

[2] M. K. Alam, M. T. Rahim, T. Haroon, S. Islam, and A. M. Siddiqui, "Solution of steady thin film flow of Johnson-Segalman fluid on a vertical moving belt for lifting and drainage problems using Adomian decomposition method," *Applied Mathematics and Computation*, vol. 218, no. 21, pp. 10413–10428, 2012.

[3] M. K. Alam, A. M. Siddiqui, M. T. Rahim, and S. Islam, "Thin-film flow of magnetohydrodynamic (MHD) Johnson-Segalman fluid on vertical surfaces using the Adomian decomposition method," *Applied Mathematics and Computation*, vol. 219, no. 8, pp. 3956–3974, 2012.

[4] A. Farooq, B. Batiha, and A. M. Siddiqui, "Lifting of a Jeffrey fluid on a vertical belt under the simultaneous effects of magnetic field and wall slip conditions," *International Journal of Advanced Mathematical Sciences*, vol. 1, no. 2, 2013.

[5] M. Hameed and R. Ellahi, "Thin film flow of non-Newtonian MHD fluid on a vertically moving belt," *International Journal for Numerical Methods in Fluids*, vol. 66, no. 11, pp. 1409–1419, 2011.

[6] R.-D. Ene, V. Marinca, and V. B. Marinca, "Thin film flow of an Oldroyd 6-constant fluid over a moving belt: An analytic approximate solution," *Open Physics*, vol. 14, no. 1, pp. 44–64, 2016.

[7] T. Hayat, R. Ellahi, and F. M. Mahomed, "Exact solution of a thin film flow of an Oldroyd 6-constant fluid over a moving belt," *Communications in Nonlinear Science and Numerical Simulation*, vol. 14, no. 1, pp. 133–139, 2009.

[8] A. M. Siddiqui, A. A. Farooq, M. A. Rana, and B. S. Babcock, "Variational iteration and Adomian decomposition methods for the nonlinear thin film flow of a Jeffrey-six constant fluid on a vertically moving belt," *International Journal of Applied Nonlinear Science*, vol. 1, no. 3, pp. 275–288, 2014.

[9] M. Moosavi, M. Momeni, T. Tavangar, R. Mohammadyari, and M. Rahimi-Esbo, "Variational iteration method for flow of non-Newtonian fluid on a moving belt and in a collector," *Alexandria Engineering Journal*, vol. 55, no. 2, pp. 1775–1783, 2016.

[10] M. Sajid, T. Hayat, and S. Asghar, "Comparison between the HAM and HPM solutions of thin film flows of non-Newtonian fluids on a moving belt," *Nonlinear Dynamics*, vol. 50, no. 1-2, pp. 27–35, 2007.

[11] Majeed Ahmed AL-Jawary, Ghassan Hasan Radhi, and Jure Ravnik, "Daftardar-Jafari method for solving nonlinear thin film flow problem," *Arab Journal of Basic and Applied Sciences*, vol. 25, no. 1, pp. 20–27, 2018.

[12] A. M. Siddiqui, A. A. Farooq, T. Haroon, and M. A. Rana, "A study of thin film flow of an Oldroyd 8-constant fluid on a moving belt using variational iteration approach," *Canadian Journal of Physics*, vol. 92, no. 10, pp. 1196–1202, 2014.

[13] S. Abelman and S. N. N. Nguetchue, "Application of the homotopy analysis method (HAM) to thin film flow of a second-grade fluid on a vertically moving belt," *Chemical Engineering Communications*, vol. 199, no. 10, pp. 1298–1319, 2012.

[14] Taza Gul, Rehan Ali Shah, Saeed Islam, and Muhammad Arif, "MHD Thin Film Flows of a Third Grade Fluid on a Vertical Belt with Slip Boundary Conditions," *Journal of Applied Mathematics*, vol. 2013, Article ID 707286, 14 pages, 2013.

[15] T. Gul, F. Ghani, S. Islam et al., "Unsteady thin film flow of a fourth grade fluid over a vertical moving and oscillating belt," *Propulsion and Power Research*, vol. 5, no. 3, pp. 223–235, 2016.

[16] T. Gul, S. Islam, R. A. Shah et al., "Heat Transfer Analysis of MHD Thin Film Flow of an Unsteady Second Grade Fluid Past a Vertical Oscillating Belt," *PLoS ONE*, vol. 9, no. 11, p. e103843, 2014.

[17] S. Mahmud and R. Andrew Fraser, "Flow, thermal, and entropy generation characteristics inside a porous channel with viscous dissipation," *International Journal of Thermal Sciences*, vol. 44, no. 1, pp. 21–32, 2005.

[18] V. K. Stokes, "Couple stresses in fluids," *Physics of Fluids*, vol. 9, no. 9, pp. 1709–1715, 1966.

[19] N. A. Khan, H. Khan, and S. Anwar Ali, "Exact solutions for MHD flow of couple stress fluid with heat transfer," *Journal of the Egyptian Mathematical Society*, vol. 24, no. 1, pp. 125–129, 2016.

[20] G. Janardhana Reddy, M. Kumar, B. Kethireddy, and A. J. Chamkha, "Colloidal study of unsteady magnetohydrodynamic couple stress fluid flow over an isothermal vertical flat plate with entropy heat generation," *Journal of Molecular Liquids*, vol. 252, pp. 169–179, 2018.

[21] N. T. M. Eldabe, A. A. Hassan, and M. A. A. Mohamed, "Effect of couple stresses on the MHD of a Non-Newtonian unsteady flow between two parallel porous plates," *Zeitschrift fur Naturforschung - Section A Journal of Physical Sciences*, vol. 58, no. 4, pp. 204–210, 2003.

14

The Global Existence and Uniqueness of the Classical Solution with the Periodic Initial Value Problem for One-Dimension Klein-Gordon-Zakharov Equations

Cong Sun[1] and **Lixia Li**[2]

[1] *School of Applied Mathematics, Jilin University of Finance and Economics, Changchun, Jilin 130117, China*
[2] *College of Mathematics and Physics, Bohai University, Liaoning 121013, China*

Correspondence should be addressed to Cong Sun; 3211519@163.com

Academic Editor: Zhi-Yuan Sun

Through applying Galerkin method, we establish the approximating solution for one-dimension Klein-Gordon-Zakharov equations and obtain the local classical solution. By applying integral estimates, we also obtain the existence and uniqueness of the global classical solution of Klein-Gordon-Zakharov equations.

1. Introduction

The Klein-Gordon-Zakharov equations are used to show the interaction between Langmuir wave and ion wave in the plasma, where

$$\begin{aligned} u_{tt} - u_{xx} + u + nu &= 0, \\ n_{tt} - n_{xx} &= \left(|u|^2\right)_{xx}, \end{aligned} \tag{1}$$

with periodic initial condition

$$\begin{aligned} u(x,0) &= u_0, \\ n(x,0) &= n_0. \end{aligned} \tag{2}$$

Here, $u(x,t)$ (complex function) denotes the biggest moment scale component produced by electron in electric field. $n(x,t)$ (real function) denotes the speed of deviations between the ion at any position and that at equilibrium position. $x \in R$; $0 \le t \le T$.

Inspired by B. L. Guo's idea and method [1], our paper focuses on the existence and uniqueness of the local and global classical solution of the periodic initial value problem for the above one-dimension Klein-Gordon-Zakharov equations. In past decades, many scholars devoted their researches to Zakharov system and have obtained some important results [2–4].

In [5], H. Pecher has researched the following Klein-Gordon equation:

$$\phi_{tt} - \Delta\phi + \phi = \phi\,|\phi|^{p-1}, \tag{3}$$

where $\phi = \phi(x,t)$, $x \in R^n$, $t \ge 0$. Given the existence of the Cauchy problem of (3), H. A. Levine in [6] adopted potential well method and concave function method to study the blowing up of the solution of the Cauchy problem of this equation. Moreover, in [7], W. A. Strauss introduced a compactness lemma, which proved existence of the solitary wave of these equations. In [4], Ozawa, Tsutaya, and Tsutsumi studied the following Klein-Gordon-Zakharov system:

$$\begin{aligned} \phi_{tt} - \Delta\phi + \phi &= -\psi\phi, \\ \psi_{tt} - c^2\Delta\psi &= \Delta\,|\phi|^2, \end{aligned} \tag{4}$$

with initial condition

$$\begin{aligned} \phi(x,0) &= \phi_0, \\ \phi_t(x,0) &= \phi_1, \\ \psi(x,0) &= \psi_0, \\ \psi_t(x,0) &= \psi_1. \end{aligned} \tag{5}$$

Through applying harmonic analysis method and contracting mapping principle, they obtained that the Cauchy problem (4)-(5) had local well-posedness about t in energy space $H^1(R^3) \times L^2(R^3)$. Furthermore, based on conservation of energy, they also obtained global existence and uniqueness with small initial value of above Cauchy problem.

Furthermore, in [8], T. C. Wang et al considered the following Klein-Gordon-Zakharov equations:

$$\begin{aligned} u_{tt} - u_{xx} + u + nu + |u|^2 u &= 0, \\ n_{tt} - n_{xx} &= \left(|u|^2\right)_{xx}. \end{aligned} \tag{6}$$

By applying Leray-Schauder fixed-point theorem, they proved the existence of different solutions of system (6).

Recently, Z. Y. Zhang et al. studied [9] the following Klein-Gordon-Zakharov equations:

$$\begin{aligned} u_{tt} - u_{xx} + u + \alpha nu &= 0, \\ n_{tt} - n_{xx} &= \beta\left(|u|^2\right)_{xx}. \end{aligned} \tag{7}$$

With different parameter conditions, they adopted the bifurcation method and dynamical systems approach to research bifurcation analysis and dynamic behaviour of travelling wave solutions of (7).

Besides, in [10], by combining the trigonometric function series method and the exp-function method, Y. Zhang et al. considered the travelling wave solutions of (7).

Now, let us describe our results. Inspired by [1], we first introduced the following equations:

$$n_t = \phi_{xx}, \tag{8}$$

$$\phi_t = n + |u|^2, \tag{9}$$

$$u_t = \psi_{xx}, \tag{10}$$

$$\psi_{xxt} = u_{xx} - u - nu, \tag{11}$$

with periodic initial data

$$\begin{aligned} n(x,0) &= n_0, \\ \phi(x,0) &= \phi_0, \\ u(x,0) &= u_0, \\ \psi(x,0) &= \psi_0. \end{aligned} \tag{12}$$

Moreover, $n(x,t)$, $\phi(x,t)$, $u(x,t)$, $\psi(x,t)$, $n(x,0)$, $\phi(x,0)$, $u(x,0)$, and $\psi(x,0)$ are periodic functions with L about $x(-\infty < x < +\infty,\ L > 0,\ 0 \le t \le T)$. Firstly, applying Galerkin method, we structure the approximate solution of (8)–(11); the aim is to obtain the local classical solution. Next, we employ this method of integral estimates to prove the existence and uniqueness of the global solution of (1)-(2).

We divide this paper into four parts. In Section 2, we establish a priori estimate of (8)–(11). In Section 3, we give the existence of local and global smooth solutions of the above equations. Lastly, in Section 4, we prove existence and uniqueness of the global classical solution of (1)-(2).

Notation. In complex Hilbert function space, inner product and standard norm are defined as follows:

$$\begin{aligned} (f,g) &= \int_0^L f(x)\overline{g(x)}\,dx, \\ \|f\|_{L_2} = \|f\| &= \int_0^L |f|^2\,dx, \end{aligned} \tag{13}$$

where $\overline{g(x)}$ is conjugate function of $g(x)$.

H^s is complex function space with generalized differentiable function $D^k u (k \le s) \in L^2$, and the norm is defined as follows:

$$\|f\|_s^2 = \sum_{k\le s} \left\|D^k f\right\|_{L^2}^2. \tag{14}$$

$L^s(0,T;H^s)$ denotes function space about complex function of $f(x,t)$. Moreover, for any $0 \le t \le T$, $f(x,t) \in H^s$, and $\sup_{0\le t\le T}\|f\|_{H^s}$ is bound. Particularly,

$$\|f(x,t)\|_{L^\infty} = \sup|f(x,t)|. \tag{15}$$

Furthermore, $C_0, C_1, C_2, \ldots, D_0, D_1, D_2, \ldots, E_0, E_1, E_2, \ldots$ are estimate constants, which relate to initial conditions.

2. A Priori Estimate of Integrating

Lemma 1. *If $u_0 \in L^2$, $|\psi_{xx}| \le C_0|u|$, and $E_0 = \|u_0\|^2$, then $\|u\|^2 \le C_1 E_0$.*

Proof. Equation (10) makes inner product with u and takes real part; we have

$$\begin{aligned} \int_0^L \frac{1}{2}(uu_t)\,dx &= \int_0^L \psi_{xx}\overline{u}\,dx \le \int_0^L |\psi_{xx}|\,|\overline{u}|\,dx, \\ \frac{1}{2}\frac{d}{dt}\|u\|^2 &\le C_0\|u\|^2. \end{aligned} \tag{16}$$

By Gronwall inequality, we can obtain that

$$\|u\|^2 \le C_1 E_0. \tag{17}$$

□

Lemma 2. *If $u_{0t}, u_{0x}, u_0, n_{0x}, \phi_{0x}, \psi_{0x} \in L^2$, then*

$$\begin{aligned} &\|u_t\|^2 + \|u_x\|^2 + \|u\|^2 + \int_0^L n|u|^2\,dx + \frac{1}{2}\int_0^L \phi_x^2\,dx \\ &+ \frac{1}{2}\int_0^L n^2\,dx = M_1, \end{aligned} \tag{18}$$

where

$$\|u_{0t}\|^2 + \|u_{0x}\|^2 + \|u_0\|^2 + \int_0^L n_0 |u_0|^2 dx + \frac{1}{2}\int_0^L \phi_{0x}^2 dx + \frac{1}{2}\int_0^L n_0^2 dx = M_1. \quad (19)$$

Proof. Taking inner product (11) with u_t and making real part, we have that

$$\int_0^L \psi_{xxt}\overline{u_t}\, dx = \int_0^L u_{tt}\overline{u_t}\, dx = \frac{1}{2}\frac{d}{dt}\int_0^L |u_t|^2 dx = \frac{1}{2}\frac{d}{dt}\|u_t\|^2, \quad (20)$$

$$\int_0^L u_{xx}\overline{u_t}\, dx = -\frac{1}{2}\frac{d}{dt}\int_0^L |u_x|^2 dx = -\frac{1}{2}\frac{d}{dt}\|u_x\|^2, \quad (21)$$

$$\int_0^L u\overline{u_t}\, dx = \frac{1}{2}\frac{d}{dt}\int_0^L |u|^2 dx = \frac{1}{2}\frac{d}{dt}\|u\|^2, \quad (22)$$

$$\int_0^L nu\overline{u_t}\, dx = \frac{1}{2}\frac{d}{dt}n\|u\|^2 dx + \frac{1}{4}\frac{d}{dt}\int_0^L \phi_x^2 dx + \frac{1}{4}\frac{d}{dt}\int_0^L n^2 dx. \quad (23)$$

Combining (20)–(23) yields

$$\frac{d}{dt}\left(\|u_t\|^2 + \|u_x\|^2 + \|u\|^2 + \int_0^L n|u|^2 dx + \frac{1}{2}\int_0^L \phi_x^2 dx + \frac{1}{2}\int_0^L n^2 dx\right) = 0. \quad (24)$$

Therefore,

$$\|u_{0t}\|^2 + \|u_{0x}\|^2 + \|u_0\|^2 + \int_0^L n_0 |u_0|^2 dx + \frac{1}{2}\int_0^L \phi_{0x}^2 dx + \frac{1}{2}\int_0^L n_0^2 dx = M_1. \quad (25)$$

□

Lemma 3. *If $u_0 \in H^1$, $\phi_0 \in H^1$, and $n_0 \in L^2$, then*

$$\|u\|_{H^1}^2 \le E_1, \quad \|n\|^2 \le E_2, \quad \|\phi\|_{H^1} \le E_3, \quad \|\phi_t\|^2 \le E_4, \quad (26)$$

where E_1, E_2, E_3, and E_4 are constants, which only relate to $\|n_0\|$, $\|u_0\|_{H^1}$, $\|\phi_0\|_{H^1}$, and T.

Proof. From (18), we can easily obtain that

$$\|u_x\|^2 + \|u\|^2 \le M_1. \quad (27)$$

So that,

$$\|u\|_{H_1} \le E_1. \quad (28)$$

By (9) and (18), we have

$$\int_0^L n^2 dx = \int_0^L (\phi_t - |u|)^2 dx \le 2\int_0^L \left(\phi_t^2 + |u|^4\right) dx \le 2|M_1|. \quad (29)$$

Hence,

$$\|n\|^2 \le 2M_1 = E_2, \quad \|\phi_t\|^2 \le E_4. \quad (30)$$

Furthermore, (9) takes inner product with ϕ; we get

$$\frac{d}{dt}\|\phi\|^2 \le \left[\|\phi\|^2 + 2\left(\|n\|^2 + \|u^2\|^2\right)\right] \le \|\phi\|^2 + E_2 + M_1 \le E_5\|\phi\|^2. \quad (31)$$

By Gronwall inequality, we have

$$\|\phi\|^2 \le \|\phi_0\|^2 e^T. \quad (32)$$

Hence, in combination with (18), we can deduce that

$$\|\phi\|_{H^1} \le E_3. \quad (33)$$

□

Lemma 4. *If $u_0 \in H^1$, $u_{0t} \in H^2$, $u_{0x} \in H^2$, and $n_0 \in H^1$, then*

$$\|u_t\|^2 + \|u_x\|^2 + \|u\|^2 \le E_5. \quad (34)$$

Proof. The first equation of (1) makes inner product with u_t and takes real part; we have that

$$\begin{aligned}\frac{d}{dt}\left(\|u_t\|^2 + \|u_x\|^1 2 + \|u\|^2\right) &\le \frac{1}{2}\left(\int_0^T |n|^4 dx + \int_0^T |u|^4 dx + 2\|u_t\|^2\right) \le \frac{1}{2}\|n\|^2\|n\|_{L^\infty}^2 + \frac{1}{2}\|u\|^2 \\ &\cdot \|u\|_{L^\infty}^2 + \|u_t\|^2 \le \left[\left(\alpha_1^2\|n_x\|^2 + c_1^2\|n\|^2\right)\|n\|^2 \right. \\ &\left. + \left(\alpha_2^2\|u\|^2 + c_2^2\|u_x\|^2\right)\|u\|^2 + \|u_t\|^2\right] \le B_1\left(\|u_t\|^2 \right. \\ &\left. + \|u_x\|^2 + \|u\|^2\right) + B_2,\end{aligned} \quad (35)$$

where $B_1 = \max\{1, \alpha_2^2 C_1 E_0, c_2^2 C_1 E_0\}$ and $B_2 = \alpha_1^2 E_1 \|n_x\|^2$.

So that, by Gronwall inequality, we obtain that

$$\|u_t\|^2 + \|u_x\|^2 + \|u\|^2 \le E_5. \quad (36)$$

□

Corollary 5. $\|u\|_{L^\infty}^2 \le E_{51}$, $\|u_t\|_{L^\infty}^2 \le E_{52}$, $\|u_x\|_{L^\infty}^2 \le E_{53}$, *and* $\|n\|_{L^\infty} \le E_{54}$.

Lemma 6. *If $n_{0tt}, n_{0xt}, u_{0tt}, u_{0xx} \in L^2$, then one has*

$$\|n_{tt}\|^2 + \|n_{xt}\|^2 + \|u_{tt}\|^2 + \|u_{xx}\|^2 \le E_6. \quad (37)$$

Proof. We differentiate t with respect to the second equation of (1); we obtain that

$$n_{ttt} - n_{xxt} = \left(|u|^2\right)_{xxt}. \quad (38)$$

Equation (38) takes inner product with n_{tt}; we have that

$$\frac{d}{dt}\|n_{tt}\|^2 + \frac{d}{dt}\|n_{xt}\|^2 \le \|n_{tt}\|^2 + \left\|\left(|u|^2\right)_{xxt}\right\|^2. \quad (39)$$

Furthermore,

$$\begin{aligned} \left\||u|^2_{xxt}\right\| &= \|4u_{tx}\overline{u_x} + 2u_{txx}\overline{u} + 2u_{xx}\overline{u_t}\| \\ &\le 4\|\overline{u_x}\|\|u_{tx}\|_{L^\infty} + 2\|u_{txx}\|\|\overline{u}\|_{L^\infty} \\ &\quad + 2\|u_{xx}\|\|u_t\|_{L^\infty} \\ &\le 4\frac{c_{11}\|u\|}{1-\alpha_{12}}\|u_{tx}\|_{L^\infty} + 2\sqrt{E_{51}}\|u_{txx}\| \\ &\quad + 2\sqrt{E_{52}}\|u_{xx}\| \\ &\le 4\frac{c_{11}\sqrt{C_1E_0}}{1-\alpha_{12}}\left(c_{12}\sqrt{E_5} + \alpha_{13}\|u_{txx}\|\right) \\ &\quad + 2\sqrt{E_{51}}\|u_{txx}\| + 2\sqrt{E_{52}}\|u_{xx}\| \\ &\le 4\frac{c_{11}\sqrt{C_1E_0}}{1-\alpha_{12}}c_{12}\sqrt{E_5} \\ &\quad + \frac{4c_{11}\sqrt{C_1E_0}\alpha_{13}}{1-\alpha_{12}}\frac{c_{13}}{1-\alpha_{14}}\|u_{xx}\| \\ &\quad + 2\sqrt{E_{51}}\frac{c_{13}\|u_{xx}\|}{1-\alpha_{14}} + 2\|u_{xx}\|\sqrt{E_{52}} \\ &\le B_3\left(1 + \|u_{xx}\|\right), \end{aligned} \quad (40)$$

where $B_3 = \max\{4c_{11}\sqrt{C_1E_0}c_{12}\sqrt{E_5}/(1-\alpha_{12}), 4c_{11}c_{13}\sqrt{C_1E_0}\alpha_{13}/(1-\alpha_{12})(1-\alpha_{14}) + 2\sqrt{E_{51}}c_{13}/(1-\alpha_{14}) + 2\sqrt{E_{52}}\}$.

So,

$$\left\||u|^2_{xxt}\right\|^2 \le B_3^2\left(1 + \|u_{xx}\|^2\right). \quad (41)$$

Differentiating (1) with respect to t, we have that

$$u_{ttt} - u_{xxt} + u_t + nu_t + n_t u = 0. \quad (42)$$

Next, (42), respectively, makes inner product with u_{xx} and u_{tt} and takes real part; we can obtain that

$$\begin{aligned} \frac{d}{dt}\|u_{xx}\|^2 &\le \|u_{xx}\|^2 + \|u_{ttt}\|^2 + 2\|u_t\|^2_{L^\infty}\|u_{xx}\|^2 \\ &\quad + \|u_t\|_{L^\infty}\left(\|n\|^2 + \|u_x\|^2\right) \\ &\quad + \left(\|n_{tx}\|^2 + \|u\|^2\right)\|u_x\|^2 \\ &\le \|u_{xx}\|^2 + \frac{c_{14}^2}{1-\alpha_{14}^2}\|u_{tt}\|^2 + 2E_5\|u_{xx}\|^2 \\ &\quad + E_5\|n_{tx}\|^2 + E_5^{3/2} + \sqrt{E_5}M_1 + E_5C_1E_0 \\ &\le B_4\left(\|u_{xx}\|^2 + \|u_{tt}\|^2 + \|n_{tx}\|^2 + 1\right), \end{aligned} \quad (43)$$

$$\begin{aligned} \frac{d}{dt}\|u_{tt}\|^2 &\le \|\overline{u_{xxt}}\|^2 + 2\|u_{tt}\|^2 \\ &\quad + \|n\|_{L^\infty}\left(\|u_t\|^2 + \|u_{tt}\|^2\right) \\ &\quad + \|n_t\|_{L^\infty}\left(\|u\|^2 + \|u_{tt}\|^2\right) \\ &\le \frac{c_{15}^2}{(1-\alpha_{15})^2}\|u_{xx}\|^2 + \left(3 + \sqrt{E_2}\right)\|u_{tt}\|^2 \\ &\quad + \alpha_{16}\|n_{tt}\|^2 + C_1E_0 + \sqrt{E_2E_5} + c_{16}^2E_2 \\ &\le B_5\left(\|u_{xx}\|^2 + \|u_{tt}\|^2 + \|n_{tt}\|^2 + 1\right). \end{aligned} \quad (44)$$

Here,

$$\begin{aligned} B_4 &= \max\left\{1 + 2E_5, \frac{c_{14}^2}{(1-\alpha_{14})^2}, E_5, E_5^{3/2} + \sqrt{E_5}M_1 \right. \\ &\quad \left. + E_5C_1E_0\right\}, \\ B_5 &= \max\left\{\frac{c_{15}^2}{(1-\alpha_{15})^2}, 3 + \sqrt{E_2}, C_1E_0 + \sqrt{E_2E_5} \right. \\ &\quad \left. + c_{16}^2E_2\right\}. \end{aligned} \quad (45)$$

Combining (39), (41), (43), and (44) yields

$$\begin{aligned} &\frac{d}{dt}\left(\|n_{tt}\|^2 + \|n_{xt}\|^2 + \|u_{tt}\|^2 + \|u_{xx}\|^2\right) \\ &\quad \le B_6\left(\|n_{tt}\|^2 + \|n_{xt}\|^2 + \|u_{tt}\|^2 + \|u_{xx}\|^2 + 1\right). \end{aligned} \quad (46)$$

Therefore, applying Gronwall inequality, we can obtain

$$\|n_{tt}\|^2 + \|n_{xt}\|^2 + \|u_{tt}\|^2 + \|u_{xx}\|^2 \le E_6. \quad (47)$$

□

Lemma 7. *If* $\|n_{0ttt}\|^2 + \|n_{0xtt}\|^2 + \|u_{0ttt}\|^2 \le E_7$, *then*

$$\|n_{ttt}\|^2 + \|n_{xtt}\|^2 + \|u_{ttt}\|^2 \le E_8. \quad (48)$$

Proof. Differentiating (8) with respect to t thrice and differentiating (9) with respect to x and t twice and eliminating ϕ_{xxtt}, we have

$$n_{tttt} - n_{xxtt} = |u|^2_{xxtt}. \quad (49)$$

Furthermore, also differentiating (10) with respect to t thrice and differentiating (11) with respect to t twice and eliminating ψ_{xxtt}, we can obtain

$$u_{tttt} - u_{xxtt} + u_{tt} + 2n_t u_t + n u_{tt} + n_{tt} u = 0. \quad (50)$$

Equation (49) makes inner product with n_{ttt}; then

$$\frac{d}{dt}\left(\|n_{ttt}\|^2 + \|n_{xtt}\|^2\right) \le 4\|n_{ttt}\|^2 + \left\||u|^2_{xxtt}\right\|^2. \quad (51)$$

Furthermore,

$$\begin{aligned} \left\||u|^2_{xxtt}\right\| \le{}& 2\|\overline{u}\|_{L^\infty}\|u_{ttxx}\| + 4\|\overline{u_x}\|_{L^\infty}\|u_{ttx}\| + \|u_{tt}\|^2 \\ &+ \|u_{xx}\|^2 + 4\|u_{txx}\|\|\overline{u_t}\|_{L^\infty} + 4\|u_{tx}\|^2, \end{aligned} \quad (52)$$

$$\begin{aligned} \|u_{ttx}\| &\le c_{17}\|u_{tt}\| + \alpha_{17}\|u_{ttxx}\|, \\ \|u_{xxt}\| &\le c_{18}\|u_{xx}\| + \alpha_{18}\|u_{ttxx}\|, \end{aligned} \quad (53)$$

$$\|u_{tttt}\| \le \frac{c_{19}}{1-\alpha_{19}}\|u_{ttt}\|, \quad (54)$$

$$\begin{aligned} \|u_{xxtt}\| \le{}& \|u_{tttt}\| + \|u_{tt}\| + \|n_t\|^2 + \|u_t\|^2 + \frac{1}{2}\|n\|^2 \\ &+ \frac{1}{2}\|u_{tt}\|^2 + \frac{1}{2}\|u\|^2 + \frac{1}{2}\|n_{tt}\|^2 \\ \le{}& \frac{c_{19}}{1-\alpha_{19}}\|u_{ttt}\| + \sqrt{E_6} + c_{20}^2 E_2 + \alpha_{20}^2\|n_{ttt}\|^2 \\ &+ \frac{1}{2}E_2 + \sqrt{E_5} + \frac{1}{2}C_1 E_0 + E_6. \end{aligned} \quad (55)$$

Due to Lemmas 4 and 6, Corollary 5, and (53) and (54), we have that

$$\begin{aligned} \left\||u|^2_{xxtt}\right\| \le{}& \left(2\sqrt{E_{51}} + 4\sqrt{E_{53}}\alpha_{18} + 4\sqrt{E_{52}}\alpha_{17}\right)\|u_{ttxx}\| \\ &+ 4\sqrt{E_{52}}c_{17}\sqrt{E_6} + 2\sqrt{E_6} + 4\sqrt{E_{53}}c_{18}\sqrt{E_6} \\ \le{}& B_7\|u_{ttt}\| + B_8\|n_{ttt}\|^2 + B_9, \end{aligned} \quad (56)$$

where

$$\begin{aligned} B_7 &= \left(2\sqrt{E_{51}} + 4\sqrt{E_{53}}\alpha_{18} + 4\sqrt{E_{52}}\alpha_{17}\right)\frac{c_{19}}{1-\alpha_{19}}, \\ B_8 &= \left(2\sqrt{E_{51}} + 4\sqrt{E_{53}}\alpha_{18} + 4\sqrt{E_{52}}\alpha_{17}\right)\alpha_{20}^2, \\ B_9 &= \left(2\sqrt{E_{51}} + 4\sqrt{E_{53}}\alpha_{18} + 4\sqrt{E_{52}}\alpha_{17}\right) \\ &\quad \cdot \left(\sqrt{E_6} + c_{20}^2 E_2 + \frac{1}{2}E_2 + \sqrt{E_5} + \frac{1}{2}C_1 E_0 + E_6\right) \\ &\quad + 4\sqrt{E_{52}}c_{17}\sqrt{E_6} + 2\sqrt{E_6} + 4\sqrt{E_{53}}c_{18}\sqrt{E_6}. \end{aligned} \quad (57)$$

Equation (50) makes inner product with u_{ttt} and takes real part; we have that

$$\begin{aligned} \frac{1}{2}\frac{d}{dt}\|u_{ttt}\|^2 \le{}& \frac{1}{2}\left(\|u_{xxtt}\|^2 + \|u_{ttt}\|^2\right) \\ &+ \frac{1}{2}\left(\|u_{tt}\|^2 + \|u_{ttt}\|^2\right) \\ &+ \|u_t\|_{L^\infty}\left(\|n_t\|^2 + \|u_{ttt}\|^2\right) \\ &+ \frac{1}{2}\|n\|_{L^\infty}\left(\|u_{tt}\|^2 + \|u_{ttt}\|^2\right) \\ &+ \frac{1}{2}\|u\|_{L^\infty}\left(\|n_{tt}\| + \|u_{ttt}\|^2\right). \end{aligned} \quad (58)$$

Combining (51), (56), and (58) and Corollary 5 and applying Gronwall inequality, Lemma 7 can easily be proven. □

3. Existence of Local Classical Solution

Firstly, we apply Galerkin method to establish approximating solution for (8)–(11) as follows:

$$\begin{aligned} n^m &= n^m(x,t) = \sum_{j=1}^m n_{jm}(t) W_j(x), \\ \phi^m &= \phi^m(x,t) = \sum_{j=1}^m \varphi_{jm}(t) W_j(x), \\ u^m &= u^m(x,t) = \sum_{j=1}^m u_{jm}(t) W_j(x), \\ \psi^m &= \psi^m(x,t) = \sum_{j=1}^m u_{jm}(t) W_j(x). \end{aligned} \quad (59)$$

Moreover, n^m, ϕ^m, u^m, and ψ^m satisfy the following equations and initial condition:

$$\left(n_t^m, W_j\right) - \left(\phi_{xx}^m, W_j\right) = 0, \tag{60}$$

$$\left(\phi_t^m, W_j\right) - \left(n^m + |u^m|^2, W_j\right) = 0, \tag{61}$$

$$\left(u_t^m, W_j\right) - \left(\psi_{xx}^m, W_j\right) = 0, \tag{62}$$

$$\left(\psi_{xxt}^m, W_j\right) - \left(u_{xx}^m, W_j\right) + \left(u^m, W_j\right) + \left(n^m u^m, W_j\right) = 0. \tag{63}$$

$$\left(n^m(x,0), W_j\right) = \left(n_0, W_j\right), \tag{64}$$

$$\left(\phi^m(x,0), W_j\right) = \left(\phi_0, W_j\right), \tag{65}$$

$$\left(u^m(x,0), W_j\right) = \left(u_0, W_j\right), \tag{66}$$

$$\left(\psi^m(x,0), W_j\right) = \left(\psi_0, W_j\right). \tag{67}$$

Next, by implementing integration estimates for the approximation solution of (60)–(63), we establish local classical solution of (60)–(63).

Lemma 8. *If $u^m(x,0) \in L^2$, $|\psi_{xx}^m| \le C$, and $\|u^m(x,0)\| = E_9$, then we have*

$$\|u^m\| \le CE_9. \tag{68}$$

Proof. Equation (62) multiplies $\overline{u_j^m}$ and makes summation for $j = 1, 2, \ldots,$ n; we obtain

$$(u_t^m, u^m) - (\psi_{xx}^m, u^m) = 0. \tag{69}$$

Taking real part, we have that

$$\frac{d}{dt}\|u^m\|^2 \le C\|u^m\|^2. \tag{70}$$

Therefore, by Gronwall inequality, we have $\|u^m(t)\| \le CE_9$. □

Lemma 9. *If $n^m(x,0) \in L^2$, $n_x^m(x,0) \in L^2$, $n_t^m(x,0) \in L^2$, $\phi_x^m(x,0) \in L^2$, and $\|u^m\|_{L^\infty} \le E_{91}$, then*

$$\|n_x^m\|^2 + \|\phi^m\|^2 + \|n_t^m\|^2 \le E_{10}. \tag{71}$$

Proof. Equation (60) multiplies $\overline{n_j^m}$ and makes summation for $j = 1, 2, \ldots,$ n; we can obtain

$$(n_t^m, n^m) = \frac{1}{2}\frac{d}{dt}\|n^m\|^2, \tag{72}$$

$$(\phi_{xx}^m, n^m) \le \|\phi_x^m\|^2 + \|n_x^m\|^2. \tag{73}$$

Furthermore, (60) differentiates with respect to t and (61) differentiates with respect to x twice; we have that

$$\left(n_{tt}^m, w_j\right) - \left(n_{xx}^m + |u^m|_{xx}^2, w_j\right) = 0. \tag{74}$$

Next, (74) multiplies $\overline{n_{jt}^m}$ and makes summation for $j = 1, 2, \ldots,$ n; it is easy to obtain the following equation:

$$(n_{tt}^m, n_t^m) - (n_{xx}^m + |u^m|_{xx}, n_t^m) = 0. \tag{75}$$

So, we obtain that

$$(n_{tt}^m, n_t^m) = \frac{1}{2}\frac{d}{dt}\|n_t^m\|^2, \tag{76}$$

$$-(n_{xx}^m, n_t^m) = \frac{1}{2}\frac{d}{dt}\|n_x^m\|^2. \tag{77}$$

$$\left|\left(|u^m|_{xx}^2, n_t^m\right)\right| \le \frac{1}{2}\left(\|n_t^m\|^2 + 4\left\||u_x^m|^2 + u_{xx}^m\overline{u^m}\right\|^2\right). \tag{78}$$

Furthermore,

$$\begin{aligned}\left\||u_x^m|^2 + u_{xx}^m\overline{u^m}\right\| &\le \left\||u_x^m|^2\right\| + \left\|\overline{u^m}\right\|_{L^\infty}\|u_{xx}^m\| \\ &\le \left(c_{18}\|u^m\| + \alpha_{19}\|u_{xx}^m\|\right)\|u_x^m\| \\ &\quad + \left(c_{19}\left\|\overline{u^m}\right\| + \alpha_{20}\|u_{xx}^m\|\right)\|u_{xx}^m\|,\end{aligned} \tag{79}$$

$$\|u_{xx}^m\| \le \frac{c_{20}}{1-\alpha_{21}}\|u_x^m\|. \tag{80}$$

Therefore,

$$\left\||u_x^m|^2 + u_{xx}^m\overline{u^m}\right\|^2 \le D_1 + (1 + D_2)\|u_x^m\|^2, \tag{81}$$

where $D_1 = (c_{18}CE_9 + (c_{20}/(1-\alpha_{21}))c_{19}CE_9)^2$ and $D_2 = (\alpha_{19}(c_{20}/(1-\alpha_{21})) + \alpha_{20}(c_{20}^2/(1-\alpha_{21})^2))^2$.

Equation (61) multiplies $(-\lambda_j)$; we have that

$$\left(\phi_t^m, w_{jxx}\right) - \left(n^m + |u^m|^2, w_{jxx}\right) = 0. \tag{82}$$

Equation (82) multiplies $\phi_{jm}(t)$; then

$$(\phi_t^m, \phi_{xx}^m) - \left(n^m + |u^m|^2, \phi_{xx}^m\right) = 0. \tag{83}$$

Furthermore,

$$|(\phi_t^m, \phi_{xx}^m)| = \frac{1}{2}\frac{d}{dt}\|\phi_x\|^2, \tag{84}$$

$$\begin{aligned}\left|\left(n^m + |u^m|^2, \phi_{xx}^m\right)\right| &= \int_0^{2\pi}\left(n^m + |u^m|^2\right)_x \phi_x\, dx \\ &\le \frac{1}{2}\left[\|n_x^m\|^2 + \|\phi_x^m\|^2 + 4\|u^m\|_{L^\infty}\left(\|n_x^m\|^2 + \left\|\overline{u_x^m}\right\|^2\right)\right] \\ &\le \frac{1}{2}\left[\|n_x^m\|^2 + \|\phi_x^m\|^2 + 4E_{91}\|n_x^m\|^2 + 4\|u_x^m\|^2\right].\end{aligned} \tag{85}$$

So,

$$\frac{d}{dt}\|\phi_x^m\|^2 \le D_3\left(\|n_x^m\|^2 + \|\phi_x^m\|^2 + \|u_x^m\|^2\right), \tag{86}$$

where $D_3 = \max\{4, 4E_{91}\}$.

Combining (76), (77), (78), (81), (84), and (86), we can obtain that

$$\frac{d}{dt}\left(\|n_x^m\|^2 + \|n_t^m\|^2 + \|\phi_x^m\|^2\right) \leq D_4\left(1 + \|n_t^m\|^2 + \|n_t^m\|^2 + \|\phi_x^m\|^2 + \|u_x^m\|^2\right), \tag{87}$$

where $D_4 = \max\{1, 2D_3 + 2, 8D_1, 8(1 + D_2) + 2D_3\}$.

So, by Gronwall inequality, we have that

$$\|n_x^m\|^2 + \|\phi^m\|^2 + \|n_t^m\|^2 \leq E_{10}. \tag{88}$$

□

Lemma 10. *$\|\phi_t^m\|^2 \leq E_{11}$, $\|\phi_{xx}^m\| \leq E_{12}$, $\|u_{tt}^m\| \leq E_{13}$, and $\|u_{xx}^m\| \leq E_{14}$.*

Proof. The proof is similar to that of Lemma 9. For details, see Lemma 9. □

Definition 11. If $n(x,t) \in L^\infty(0, T_0, H^0)$, $\phi(x,t) \in L^\infty(0, T_0, H^2)$, $u(x,t \in L^\infty(0, T_0, H^2))$, $\psi(x,t) \in L^\infty(0, T_0, H^3)$, and $n_t, \phi_t, u_t, \psi_t \in L^\infty(0, T_0, H^0)$. Moreover, they satisfy the following integral identity:

$$\begin{aligned} (n_t, v) &= (\phi_{xx}, v) \\ (\phi_t, v) &= \left(n + |u|^2, v\right) \\ (u_t, v) &= (\psi_{xx}, v) \\ (\psi_{xxt}, v) &= (u_{xx} - u - nu, v), \end{aligned} \tag{89}$$

and

$$\begin{aligned} (n(x,0), v) &= (n_0(x), v) \\ (\phi(x,0), v) &= (\phi_0(x), v) \\ (u(x,0), v) &= (u_0(x), v) \\ (\psi(x,0), v) &= (\psi_0(x), v). \end{aligned} \tag{90}$$

Then, n, ϕ, u, and ψ are local generalized solutions of (60)–(63). Besides, n, ϕ, u, and ψ are periodic functions with L about x.

Theorem 12. *If $n^m(x,0) \in L^2$, $n_x^m(x,0) \in L^2$, $n_t^m(x,0) \in L^2$, $\phi_x^m(x,0) \in L^2$, $\|u^m\|_{L^\infty} \leq E_{91}$, n_0, ϕ_0, u_0, and ψ_0 are period functions with L about x. Then, (60)–(63) have local generalized solution.*

Proof. By Lemmas 8, 9, and 10, $\|n_x^m\|$, $\|n_t^m\|$, $\|\phi_x^m\|$, $\|u_{xx}^m\|$, and $\|u_{tt}^m\|$ have uniform boundedness. So, we can select a subsequence and satisfy $n_t^m \xrightarrow{weak} n_t$, $\phi_x^m \xrightarrow{weak} \phi_x$, $u_{tt}^m \xrightarrow{weak} u_{tt}$, and $u_{xx}^m \xrightarrow{weak} u_{xx}$. Moreover, there exists subsequences, $n^m \xrightarrow{strong} n$, $\phi^m \xrightarrow{strong} \phi$, and $u^m \xrightarrow{strong} u$.

In (60)–(63), let $m \to \infty$; we have

$$\left(n_t, W_j\right) - \left(\phi_{xx}, W_j\right) = 0, \tag{91}$$

$$\left(\phi_t, W_j\right) - \left(n + |u|^2, W_j\right) = 0, \tag{92}$$

$$\left(u_t, W_j\right) - \left(\psi_{xx}, W_j\right) = 0, \tag{93}$$

$$\left(\psi_{xxt}, W_j\right) - \left(u_{xx}, W_j\right) + \left(u, W_j\right) + \left(nu, W_j\right) = 0. \tag{94}$$

Moreover $W_j \in H^2$ and it is dense in H^0, W_j approximates to v. We complete the proof of this theorem. Similarly, we also can prove (64)–(67). □

Lemma 13. *If $\|n_{tt}^m(x,0)\|$, $\|n_{xt}^m(x,0)\|$, $\|u_{xx}^m(x,0)\|$, and $\|u_{tt}^m(x,0)\|$ are bound, then*

$$\|n_{tt}^m\|^2 + \|n_{xt}^m\|^2 + \|u_{xx}^m\|^2 + \|u_{tt}^m\|^2 \leq E_{10}. \tag{95}$$

Proof. Equation (74) differentiates with respect to t. Next, it multiplies $\overline{n_{jtt}^m}$ and makes summation for $j = 1, 2, \ldots, n$. We can easily obtain the following equation:

$$\left(n_{ttt}^m, n_{tt}^m\right) - \left(n_{xxt}^m + |u^m|_{xxt}^2, n_{tt}^m\right) = 0. \tag{96}$$

Furthermore,

$$\left(n_{ttt}^m, n_{tt}^m\right) = \frac{1}{2}\frac{d}{dt}\|n_{tt}^m\|^2, \tag{97}$$

$$\left(n_{xxt}^m, n_{tt}^m\right) = -\frac{1}{2}\frac{d}{dt}\|n_{xt}^m\|^2, \tag{98}$$

$$\left(|u^m|_{xxt}, n_{tt}^m\right) \leq \frac{1}{2}\left(\||u^m|_{xxt}\|^2 + \|n_{tt}\|^2\right). \tag{99}$$

Meanwhile, by CoóojieB inequality,

$$\begin{aligned} \||u^m|_{xxt}\| &= \left\|4u_{tx}^m\overline{u_x^m} + 2u_{txx}^m\overline{u^m} + 2u_{xx}^m\overline{u_t}\right\| \\ &\leq 4\left\|\overline{u_x^m}\right\|\|u_{tx}^m\|_{L^\infty} + 2\|u_{ttx}^m\|\|u^m\|_{L^\infty} + 2\|u_{xx}^m\|\|u_t^m\|_{L^\infty} \\ &\leq 4\left(c_{22}\left\|\overline{u^m}\right\| + \alpha_{22}\left\|\overline{u_{xx}^m}\right\|\right)\left(c_{23}\|u_t^m\| + \alpha_{23}\|u_{xxt}^m\|\right) \\ &\quad + 2\|u_{ttx}^m\|\left(c_{24}\left\|\overline{u^m}\right\| + \alpha_{24}\left\|\overline{u_{xx}^m}\right\|\right) \\ &\quad + 2\|u_{xx}^m\|\left(c_{25}\left\|\overline{u^m}\right\| + \alpha_{25}\|u_{tt}^m\|\right) \\ &\leq D_5 + D_6\|u_{tt}^m\|^2 + D_7\|u_{xx}^m\|^2. \end{aligned} \tag{100}$$

Here,

$$\begin{aligned} D_5 &= 4c_{22}^2C^2E_9^2 + \frac{c_{27}}{1 - \alpha_{27}}\left(c_{24}CE_9\right)^2 + \left(c_{25}CE_9\right)^2, \\ D_6 &= 4c_{23}^2\left(\frac{c_{28}}{1 - \alpha_{28}}\right)^2 + \frac{c_{27}}{1 - \alpha_{27}} + \alpha_{25}^2, \\ D_7 &= 4\alpha_{22}^2 + 4\alpha_{23}\left(\frac{c_{26}}{1 - \alpha_{26}}\right)^2. \end{aligned} \tag{101}$$

Moreover, (62) differentiates with respect to t twice, multiplies $\overline{u^m_{jtt}}$, and makes summation for $j = 1, 2, \ldots, n$; we can obtain that

$$(u^m_{ttt}, u^m_{tt}) - (\psi^m_{xxtt}, u^m_{tt}) = 0, \quad (102)$$

$$(u^m_{ttt}, u^m_{tt}) = \frac{1}{2}\frac{d}{dt}\|u\|^2, \quad (103)$$

$$|(\psi_{xxtt}, u^m_{tt})| \le \frac{1}{2}\left(\|\psi^m_{xxtt}\|^2 + \|u^m_{tt}\|^2\right). \quad (104)$$

Next, (62) differentiates with respect to x twice, multiplies $(-\lambda)$, and makes summation for $j = 1, 2, \ldots, n$; we have that

$$(u^m_{txx}, u^m_{xx}) - (\psi_{xxxx}, u^m_{xx}) = 0, \quad (105)$$

$$(u^m_{txx}, u^m_{xx}) = \frac{1}{2}\frac{d}{dt}\|u^m_{xx}\|^2, \quad (106)$$

$$|(\psi_{xxxx}, u^m_{xx})| \le \frac{1}{2}\left(\|u^m_{xx}\|^2 + \|\psi^m_{xxxx}\|^2\right), \quad (107)$$

and (60) differentiates with respect to t and x; we can obtain that

$$(n^m_{ttx}, W_j) - (\phi^m_{xxxt}, W_j) = 0, \quad (108)$$

where $(n^m_{ttx}, W_{jx}) - (\phi^m_{xxxt}, W_{jx}) = 0$ multiplies $\overline{n^m_{jt}}$ and makes summation for $j = 1, 2, \ldots, n$; we can obtain that

$$(n^m_{ttx}, n_{xt}) - (\phi^m_{xxxt}, n_{xt}) = 0. \quad (109)$$

Taking real part, we have that

$$(n^m_{ttx}, n^m_{xt}) = \frac{1}{2}\frac{d}{dt}\|n^m_{tx}\|^2, \quad (110)$$

$$|(\psi^m_{xxxt}, n^m_{xt})| \le \frac{1}{2}\left(\|n^m_{tx}\|^2 + \|\psi^m_{xxxt}\|^2\right). \quad (111)$$

By (97), (98), (99), (100), (103), (104), (107), (110), and (111) and applying Gronwall inequality, we can obtain the result of Lemma 13. □

Corollary 14. *If $\|u^m_{tx}(x,0)\|$, $\|\psi^m_{tx}(x,0)\|$, and $\|n^m_{xx}(x,0)\|$ are bound, then there exist T_0, $0 \le t \le T_0$, $\|u^m_{tx}(t)\| \le E_{11}$, $\|\psi^m_{tx}(t)\| \le E_{12}$, and $\|n^m_{xx}(t)\| \le E_{13}$.*

Theorem 15. *Assume that the conditions of Lemmas 8, 9, 10, and 13 are satisfied, then initial value problem (1) has local classical solution.*

Proof. The idea and method are similar to Theorem 3.2 of [1]. For details, please see this paper. □

4. Initial Value Function and Its Derivative Estimate in L^2

In this section, we apply the idea and method of [1] to discuss boundedness of $\|n^m_t(0)\|$, $\|\phi^m_t(0)\|$, $\|u^m_t(0)\|$, and $\|\psi^m_{xxt}\|$.

Assume that $n_0 \in H^6$, $\phi_0 \in H^4$, $u_0 \in H^4$, and $\psi_0 \in H^4$, and take n^m_0, ϕ^m_0, u^m_0, and ψ^m_0; they satisfy $n^m_0 \xrightarrow{H^6} n_0$, $\phi^m_0 \xrightarrow{H^4} \phi_0$, $u^m_0 \xrightarrow{H^4}$, and $\psi^m_0 \xrightarrow{H^4} \psi_0$.

Due to (64), (65), (66), and (67) and above assumption, it is easy to deduce that $\|n^m_0\|_{H^6}$, $\|\phi^m_0\|_{H^4}$, $\|u^m_0\|_{H^4}$, and $\|\psi^m_0\|_{H^4}$ have uniform boundedness about m.

Next, we denote an orthogonal projection operator $P \in H^0$, which is spanned by $\{W_1, W_2, \ldots, W_m\}$. By (60), (61), (62), and (63), we have that

$$\begin{aligned} P\left(n^m_t(0) - \phi^m_{xx}(0)\right) &= 0 \\ P\left(\phi^m_t(0)\right) - \left(n^m(0) + |u^m(0)|^2\right) &= 0 \\ P\left(u^m_t(0) - \psi^m_{xx}(0)\right) &= 0 \\ P\left(\psi^m_{xxt}(0) - u^m_{xx}(0) + u^m(0) + u^m(0)\, n^m(0)\right) &= 0, \end{aligned} \quad (112)$$

and by $\|PY\| \le \|Y\|$, we have

$$\begin{aligned} \|n^m_t(0)\| &\le \|\phi^m_{xx}(0)\|, \\ \|\phi^m_t(0)\| &\le \|\phi^m_t(0)\| + \|n^m(0)\| + \left\||u^m(0)|^2\right\|, \\ \|u^m_t(0)\| &\le \|\psi^m_{xx}\|, \\ \|\psi^m_{xxt}(0)\| &\le \|u^m_{xx}(0)\| + \|u^m_{xx}(0)\| + \|u^m(0)\| \\ &\quad + \|u^m(0)\|\,\|n^m(0)\|. \end{aligned} \quad (113)$$

So, we obtain that $\|n^m_{0t}\|$, $\|\phi^m_{0t}\|$, $\|u^m_{0t}\|$, and $\|\psi^m_{0xxt}\|$ have uniform boundedness about m.

Similarly, it is not difficult to obtain some higher-order derivative estimation, including n^m_{0t}, ϕ^m_{0t}, u^m_{0t}, and ψ^m_{0xxt}. For details, we do not discuss. We have the following Lemma.

Lemma 16. *Assume that $n_0 \in H^6$, $\phi_0 \in H^4$, $u_0 \in H^4$, $\psi_0 \in H^4$, then $\|n^m_t(0)\|$, $\|\phi^m_t(0)\|$, $\|u^m_t(0)\|$, $\|\psi^m_{xxt}(0)\|$, $\|n^m_{tt}(0)\|$, $\|\phi^m_{tt}(0)\|$, $\|u^m_{tt}(0)\|$, $\|\psi^m_{xxtt}(0)\|$, $\|n^m_{ttt}(0)\|$, $\|n^m_{ttx}(0)\|$, $\|n^m_{txx}(0)\|$, $\|\phi^m_{ttt}(0)\|$, $\|\phi^m_{ttx}(0)\|$, $\|\phi^m_{txx}(0)\|$, $\|u^m_{tt}(0)\|$, $\|u^m_{ttt}(0)\|$, $\|u^m_{txx}(0)\|$, $\|u^m_{ttx}(0)\|$, $\|\psi_{xxttt}(0)\|$, and $\|\psi^m_{xxtt}(0)\|$ have uniform boundedness about m.*

5. Existence and Uniqueness of the Global Classical Solution

Theorem 17. *Assume that $n_0 \in H^6$, $\phi_0 \in H^4$, $u_0 \in H^4$, $\psi_0 \in H^4$, and they are periodic functions with L about x; then initial value problem (1) has global classical solution.*

Proof. The idea and method are similar to Theorem 4.1 of [1]. For details, please see this paper. □

Theorem 18. *Global classical solution of (1)-(2) is unique.*

Proof. Assume that u_1, n_1 and u_2, n_2 are both classical solutions with the same periodic initial value of (1)-(2). Let

$V = u_1 - u_2$ and $W = n_1 - n_2$; by (1)-(2), we obtain the following equations about V and W:

$$V_{tt} - V_{xx} + V + n_1 V + W u_2 = 0, \quad (114)$$

$$W_{tt} - W_{xx} = V_{xx}\overline{u_1} + u_{2xx}V + 2\left(V_x\overline{u_{1x}} + u_{2x}\overline{V_x}\right) + \overline{u_{1xx}}V + u_2\overline{V_{xx}}, \quad (115)$$

where V is a complex function and W is a real function.

Equation (114) makes inner product with V_t and takes real part; we have that

$$(V_{tt}, V_t) = \frac{1}{2}\frac{d}{dt}\|V_t\|^2, \quad (116)$$

$$(V_{tx}, V_t) = -\frac{1}{2}\frac{d}{dt}\|V_x\|^2, \quad (117)$$

$$\begin{aligned}|(n_1 V, V_t)| &\le \frac{1}{2}\|n_1\|_{L^\infty}\left(\|V\|^2 + \|V_t\|^2\right)\\ &\le \frac{1}{2}E_{54}\left(\|V\|^2 + \|V_t\|^2\right),\end{aligned} \quad (118)$$

$$(V, V_t) = \frac{1}{2}\frac{d}{dt}\|V\|^2, \quad (119)$$

$$\begin{aligned}(W u_2, V_t) &\le \frac{1}{2}\|u_2\|_{L^\infty}\left(\|W\|^2 + \|V_t\|^2\right),\\ &\le \frac{1}{2}E_{51}\left(\|W\|^2 + \|V_t\|^2\right).\end{aligned} \quad (120)$$

Meanwhile, (115) makes inner product with W_t and takes real part; we have that

$$(W_{tt}, W_t) = \frac{1}{2}\frac{d}{dt}\|W_t\|^2, \quad (121)$$

$$(W_{xx}, W_t) = -\frac{1}{2}\frac{d}{dt}\|W_x\|^2, \quad (122)$$

$$\begin{aligned}|(V_{xx}\overline{u_1}, W_t)| &\le \frac{1}{2}\|\overline{u_1}\|_{L^\infty}\left(\|V_{xx}\|^2 + \|W_{tt}\|^2\right)\\ &\le \frac{1}{2}E_{51}\left[\left(\frac{c_{28}}{1-\alpha_{23}}\right)^2\|V_x\|^2 + \|W_t\|^2\right],\end{aligned} \quad (123)$$

$$\begin{aligned}|(u_{2xx}V, W_t)| &\le \frac{1}{2}\|u_{2xx}\|_{L^\infty}\left(\|V\|^2 + \|W_t\|^2\right)\\ &\le \frac{1}{2}\left(c_{28}\|u_2\| + \alpha_{23}\|u_{2xxx}\|\right)\left(\|V\|^2 + \|W_t\|^2\right)\\ &\le \frac{1}{2}\left(c_{28}\sqrt{C_1E_0} + \alpha_{23}\sqrt{E_8}\right)\left(\|V\|^2 + \|W_t\|^2\right),\end{aligned} \quad (124)$$

$$\begin{aligned}|(V_x\overline{u_{1x}}, W_t)| &\le \frac{1}{2}\|u_{1x}\|_{L^\infty}\left(\|V_x\|^2 + \|W_t\|^2\right)\\ &\le \frac{1}{2}\left(c_{29}\|u_1\| + \alpha_{24}\|u_{1xx}\|\right)\left(\|V_x\|^2 + \|W_t\|^2\right)\\ &\le \frac{1}{2}\left(c_{29}\sqrt{C_1E_0} + \alpha_{24}\sqrt{E_6}\right)\left(\|V_x\|^2 + \|W_t\|^2\right)\end{aligned} \quad (125)$$

$$\begin{aligned}|(u_{2x}\overline{V_x}, W_t)| &\le \frac{1}{2}\|u_{2x}\|_{L^\infty}\left(\|V_x\|^2 + \|W_t\|^2\right)\\ &\le \frac{1}{2}\left(c_{30}\|u_2\| + \alpha_{25}\|u_{2xx}\|\right)\left(\|V_x\|^2 + \|W_t\|^2\right)\\ &\le \frac{1}{2}\left(c_{30}\sqrt{C_1E_0} + \alpha_{25}\sqrt{E_6}\right)\left(\|V_x\|^2 + \|W_t\|^2\right),\end{aligned} \quad (126)$$

$$\begin{aligned}|(\overline{u_{1xx}}V, W_t)| &\le \frac{1}{2}\|\overline{u_{1xx}}\|_{L^\infty}\left(\|V\|^2 + \|W_t\|^2\right)\\ &\le \frac{1}{2}\left(c_{31}\|u_1\| + \alpha_{26}\|u_{1xxx}\|\right)\left(\|V\|^2 + \|W_t\|^2\right)\\ &\le \frac{1}{2}\left(c_{31}\sqrt{C_1E_0} + \alpha_{26}\sqrt{E_8}\right)\left(\|V\|^2 + \|W_t\|^2\right),\end{aligned} \quad (127)$$

$$\begin{aligned}|(u_2\overline{V_{xx}}, W_t)| &\le \frac{1}{2}\|u_2\|_{L^\infty}\left(\|V_{xx}\|^2 + \|W_t\|^2\right)\\ &\le \frac{1}{2}E_{51}\left[\left(\frac{c_{28}}{1-\alpha_{23}}\right)^2\|V_x\|^2 + \|W_t\|^2\right).\end{aligned} \quad (128)$$

Combining (117)–(128), we have that

$$\begin{aligned}&\frac{d}{dt}\left(\|V_t\|^2 + \|V_x\|^2 + \|V\|^2 + \|W_t\|^2 + \|W_x\|^2\right)\\ &\le D_8\left(\|V_t\|^2 + \|V_x\|^2 + \|V\|^2 + \|W_t\|^2\right),\end{aligned} \quad (129)$$

where

$$\begin{aligned}D_8 = \max\Bigg\{&6E_{54}, 6E_{51}, 12E_{51}\left(\frac{c_{28}}{1-\alpha_{23}}\right)^2,\\ &6\left(c_{28}\sqrt{C_1E_0} + \alpha_{23}\sqrt{E_8}\right),\\ &6\left(c_{29}\sqrt{C_1E_0} + \alpha_{24}\sqrt{E_6}\right),\\ &12\left(c_{30}\sqrt{C_1E_0} + \alpha_{25}\sqrt{E_6}\right),\\ &6\left(C_{31}\sqrt{C_1E_0} + \alpha_{26}\sqrt{E_8}\right)\Bigg\}.\end{aligned} \quad (130)$$

Integrating (130) from $[0, T]$, we obtain that

$$\begin{aligned}&\left(\|V_t(T)\|^2 + \|V_x(T)\|^2 + \|V(T)\|^2 + \|W_t(T)\|^2\right.\\ &\left.+ \|W_x(T)\|^2\right)e^{-D_8T} \le 0.\end{aligned} \quad (131)$$

Hence, $u_1 = u_2$ and $n_1 = n_2$. □

Conflicts of Interest

The authors declare that there are no conflicts of interest regarding the publication of this paper.

Acknowledgments

This paper is funded by the new entry doctoral research project of Jilin University of Finance and Economics (2017B2013).

References

[1] B. L. Guo and L. J. Shen, "The existence and uniqueness of the classical solution to the periodic initial value problem for the Zakharov equation," *Acta Mathematicae Applicatae Sinica*, vol. 5, no. 3, pp. 310–324, 1982 (Chinese).

[2] F. Oliveira, "Stability of the solitons for the one-dimensional Zakharov-Rubenchik equation," *Physica D: Nonlinear Phenomena*, vol. 175, no. 3-4, pp. 220–240, 2003.

[3] T. Ozawa, K. Tsutaya, and Y. Tsutsumi, *Normal Form and Global Solutions for the Klein-Gordon- Zakharov Equation*, University of Tokyo, 1994.

[4] T. Ozawa, K. Tsutaya, and Y. Tsutsumi, "Well-posedness in energy space for the Cauchy problem of the Klein-Gordon-Zakharov equations with different propagation speeds in three space dimensions," *Mathematische Annalen*, vol. 313, no. 1, pp. 127–140, 1999.

[5] H. Pecher, "L^p Abschatzungen and klassiche Losungen fur nichtlieare wellengenichungen. I," *Mathematische Zeitschrift*, vol. 150, no. 2, pp. 159–183, 1976.

[6] H. A. Levine, "Instability and nonexistence of global solution to nonlinear wave equation of the form Putt=-Au-F(u)," *Transactions of the American Mathematical Society*, vol. 192, pp. 1–21, 1974.

[7] W. A. Strauss, "Existence of solitary waves in higher dimensions," *Communications in Mathematical Physics*, vol. 55, no. 2, pp. 149–162, 1977.

[8] T. Wang, J. Chen, and L. Zhang, "Conservative difference methods for the Klein-Gordon-Zakharov equations," *Journal of Computational and Applied Mathematics*, vol. 205, no. 1, pp. 430–452, 2007.

[9] Z. Zhang, F.-L. Xia, and X.-P. Li, "Bifurcation analysis and the travelling wave solutions of the Klein-Gordon-Zakharov equations," *Pramana—Journal of Physics*, vol. 80, no. 1, pp. 41–59, 2013.

[10] Z. Y. Zhang, Y. H. Zhang, X. Y. Gan, and D. M. Yu, "A note on exact travelling wave solutions for the klein -gordon-zakharov equations," *Zeitschriftfü Naturforschung*, vol. A67, no. 167, 2012.

15

Minimization of Stress State of a Hub of Friction Pair

Vagif M. Mirsalimov[1,2] **and Parvana E. Akhundova**[1,3]

[1]*Azerbaijan Technical University, H. Javid 25, AZ1073 Baku, Azerbaijan*
[2]*Institute of Mathematics and Mechanics of the Academy of Sciences of Azerbaijan, B. Vahabzade 9, AZ1141 Baku, Azerbaijan*
[3]*Azerbaijan State University of Economics, Istiglaliyyat 6, AZ1001 Baku, Azerbaijan*

Correspondence should be addressed to Vagif M. Mirsalimov; vagif.mirsalimov@imm.az

Academic Editor: Eugen Radu

The function of displacements of external contour points of a friction pair hub that could provide minimization of stress state of a hub was determined on the basis of minimax criterion. The problem is to decrease stress state at that place where it is important. The rough friction surface model is used. To solve a problem of optimal design of friction unit the closed system of algebraic equations is constructed. Increase of serviceability of friction pair parts may be controlled by design-engineering methods, in particular by geometry of triboconjugation elements. Minimization of maximum circumferential stress on contact surface of friction unit is of great importance in the design stage for increasing the serviceability of friction pair. The obtained function of displacements of the hub's external contour points provides the serviceability of friction pair elements. The calculation of friction pair for oil-well sucker-rod pumps is considered as an example.

1. Introduction

Operation efficiency of a friction pair of machines depends in considerable degree on stress state of the friction pairs. For example, typical operational failure of hubs of oil-well sucker-rod pump is appearance of plastic deformations on the internal contour. When operating, the plunger rubs against the hub's surface. The surface layer of the hub's metal is heating. In the process of operation of the friction pair "a hub-a plunger" under repeatedly reciprocating motion of the plunger there arises force interaction between contacting surfaces of the hub and plunger; there occur friction forces that cause wear of mating materials. In its parts there arises stress-strain state caused by the action of force and thermal loads.

According to classic theories of strength, in the simplest case maximum normal stress is responsible for failure of friction pair parts. Consequently, the value of maximum normal peripheral stress achieved in the material may be considered responsible for strength failure of friction pair materials. Increase of serviceability of friction pair parts may be controlled by design-engineering methods, in particular by geometry of triboconjugation elements. At present there are no solutions of tribomechanics problems on construction geometry of the surface of friction pair parts such that the stress field created by it prevented failure or occurrence of irreversible deformations of the materials of contact pair elements. Minimization of maximum circumferential stress on contact surface of friction unit is of great importance in the design stage for increasing serviceability of friction pair. Obviously, the lower the stress state of the hub, the higher its operation life. To improve the efficiency of details of friction pair the optimal design is important [1–11].

The goal of the study is to develop a mathematical model for a hub-plunger pair that allows determination of optimal function of displacements of the points of the external contour of the hub under given operation modes of the plunger.

2. Problem Statement

It is known that real treated surfaces are not absolutely smooth but always have irregularities of technological character. Such micro- or macroscopic irregularities form the rough surface. Despite the very small sizes of the irregularities, they affect the different service properties of

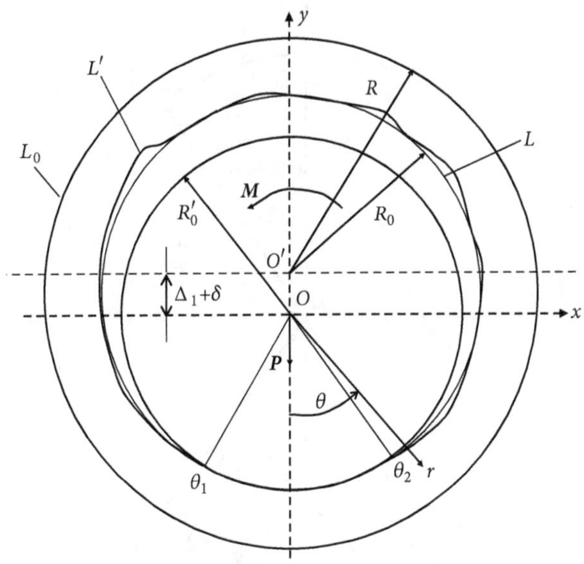

FIGURE 1: Design diagram of contact problem for hub-plunger friction pair.

triboconjugation [12–15]. In the operation process of friction pair, on the internal surface of the hub, on the area of contact with the plunger, there acts surface thermal source caused by external friction of the plunger against the hub's wall in the course of repeatedly reciprocating motion of the plunger. As a result of such interaction, there happen friction forces reducing to wear of mating materials and increase of hub's and plunger's temperature. We adopt the parameters of the function of displacements of the hub's external surface points as controlling variables.

As a mathematical model of the problem on reduction of hub's stress level we assume differential equations of thermoelasticity.

To determine the stress distribution and contact pressure in the elements of friction pair, a wear-contact problem on pressing the plunger into the hub's surface must be considered.

Let in some unknown area a plunger be pressed into the hub internal surface. The shear modulus and Poisson's ratio of the plunger and hub are various.

It is considered that on the external surface the hub has some displacements. The function of these displacements is unknown beforehand and is to be defined. It is assumed that the plane strain conditions are fulfilled. We simulate the hub and plunger by an isotropic elastic homogeneous body. It is assumed that the operation modes of the friction unit at which residual deformations arise are inadmissible. The loading conditions are considered as quasistatic.

Assign the hub of a friction pair to polar system of coordinates $r\theta$ having chosen the origin at the center of concentrical circles L, L_0 of radii R_0 and R, respectively (Figure 1). It is assumed that the internal contour L' of the hub is close to circular one. Let us consider some realization of the hub and plunger's rough internal surface.

The boundary of the internal contour L' we represent is in the form $r = \rho(\theta)$, $\rho = R_0 + \varepsilon H(\theta)$, where $\varepsilon = R^0_{\max}/R_0$ is a small parameter; $R^0_{\max}$ is the greatest irregularity (hollow) of unevenness of the internal contour profile from the circle $r = R_0$.

We assume without loss of generality that the function $H(\theta)$ may be represented in the form of Fourier series

$$H(\theta) = \sum_{k=0}^{n} \left(a_k^0 \cos k\theta + b_k^0 \sin k\theta\right). \quad (1)$$

Similarly, the plunger's external contour is close to circular one and may be represented in the following form:

$$\begin{gathered} r = \rho_1(\theta), \\ \rho_1 = R_0' + \varepsilon H_1(\theta), \\ H_1(\theta) = \sum_{k=0}^{n} \left(a_k^1 \cos k\theta + b_k^1 \sin k\theta\right). \end{gathered} \quad (2)$$

The concentrated force

$$P = R_0 \int_{\theta_1}^{\theta_2} (\cos\theta - f \sin\theta)\, p(\theta)\, d\theta \quad (3)$$

pressing plunger to the boundary of the hub's internal contour is applied at the plunger's center.

The pressure $p(\theta)$ is asymmetrically distributed on the contact area and creates a moment with respect to the plunger

$$M = -R_0^2 \int_{\theta_1}^{\theta_2} \tau_{r_\theta}\, d\theta = f R_0^2 \int_{\theta_1}^{\theta_2} p(\theta)\, d\theta. \quad (4)$$

The contact pressure $p(\theta)$ is unknown beforehand.

It is assumed that the wear of friction pair is of abrasive character. It is required to determine the function of displacements of the points of the hub external contour L_0 under given functions $H(\theta)$ and $H_1(\theta)$ at which minimization of stress state in the hub occurs.

The condition relating the displacements of the plunger and hub is of the form [16–18]

$$v_1 + v_2 = \delta(\theta), \quad \theta_1 \le \theta \le \theta_2, \quad (5)$$

where v_1 and v_2 are the normal components of displacements; $\delta(\theta)$ is the penetration of the points of the hub and plunger surfaces determined by the form of the internal hub surface and the plunger surface and also by the value of the pressing force **P**; $\theta_2 - \theta_1$ is the value of the angle (area) of contact.

In the contact area, in addition to normal pressure tangential stress $\tau_{r\theta}$ acts, connected with contact pressure $p(\theta, t)$ by the Amonton-Coulomb law. The tangential forces (friction forces) work toward heat release in the contact area. The total amount of heat per unit time is proportional to power of friction forces. The amount of heat flux released at the point of the contact area with the coordinate θ will be equal to $Q(\theta, t) = Vfp(\theta, t)$, where V is the average speed of displacement of the plunger with respect to the hub over a period; f is the friction pair coefficient. Total amount of heat

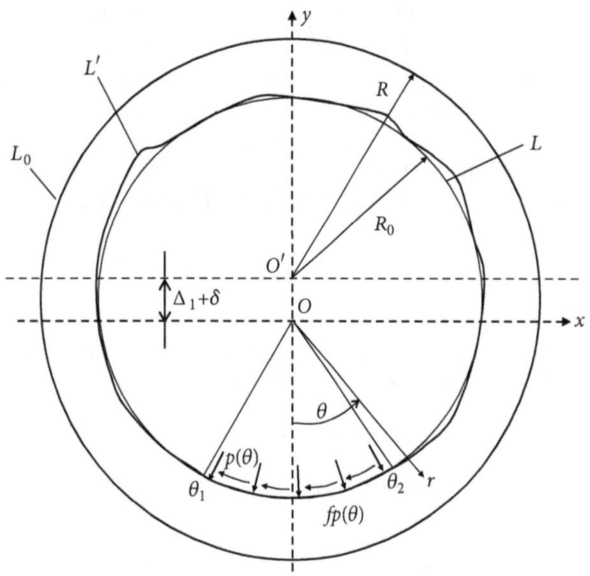

FIGURE 2: Design diagram of contact problem for hub.

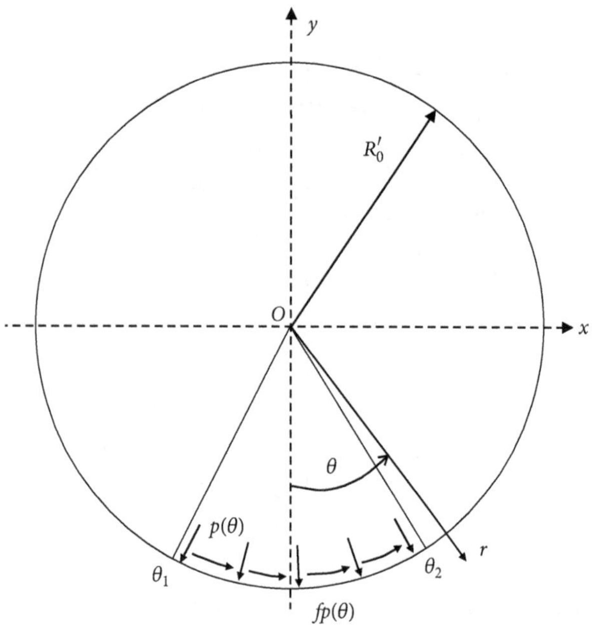

FIGURE 3: Design diagram of contact problem for plunger.

$Q(\theta,t)$ will be consumed in the following way: $Q = Q_b + Q_1$, where Q_b is the heat entering the hub; Q_1 is the heat entering the plunger.

As the plunger's motion frequency is rather great, we consider the problem of determination of temperature stationary.

For displacements of the points of the hub's friction we have $v_1 = v_{1e} + v_{1r} + v_{1w}$, where v_{1e} are thermoelastic displacements of the hub's contact surface; v_{1r}, v_{1w} are displacements caused by crumpling of microregularities and wear of the hub's surface, respectively. Similarly, for the displacements of the plunger's surface we have $v_2 = v_{2e} + v_{2r} + v_{2w}$.

The rate of change of the displacements of the surface under hub's and plunger's wear will be [17]

$$\frac{dv_{kw}}{dt} = K^{(k)} p(\theta,t), \quad k = 1,2, \tag{6}$$

where $K^{(k)}$ is the coefficient of wear of the hub and plunger's material ($k = 1,2$), respectively.

To determine the displacements v_{1e}, v_{1r} and v_{2e}, v_{2r} it is necessary to solve the following thermoelasticity problems for the hub (Figure 2) and the plunger (Figure 3), respectively:

$$\Delta T = 0,$$

$$\text{for } r = \rho(\theta) \quad A_{T_1}\lambda\frac{\partial T}{\partial n} - A_{T_2}\alpha_1 T = -Q_*, \tag{7}$$

$$\text{for } r = R \quad \lambda\frac{\partial T}{\partial r} + \alpha_2 T = 0, \tag{8}$$

$$\text{for } r = \rho(\theta),$$

$$\sigma_n = -p(\theta),$$

$$\tau_{nt} = -fp(\theta)$$

on the contact area,

$$\sigma_n = 0,$$

$$\tau_{nt} = 0$$

out of the contact area,

$$\text{for } r = R, \quad v_r - iv_\theta = g(\theta);$$

$$\Delta T_1 = 0,$$

$$\text{for } r = \rho_1(\theta)$$

$$\lambda_1\frac{\partial T_1}{\partial n} = -Q(\theta)$$

on the contact area,

(9)

$$\lambda_1\frac{\partial T_1}{\partial n} + \alpha_1 T_1 = 0 \quad \text{out of the contact area,} \tag{10}$$

$$\text{for } r = \rho_1(\theta),$$

$$\sigma_n = -p(\theta);$$

$$\tau_{nt} = -fp(\theta)$$

on the contact area, (11)

$$\sigma_n = 0,$$

$$\tau_{nt} = 0$$

out of the contact area.

Here Δ is the Laplace operator; T is overtemperature of the hub; A_{T_1} is heat-absorbing surface; A_{T_2} is cooling surface; $Q_*(\theta) = Q_b(\theta) = \alpha_{m.n.}Vfp(\theta)$ on the contact area; $Q_* = 0$ out of the contact area; λ, λ_1 are the thermal conductivity of the hub's material and plunger's material, respectively; α_1, α_2 are the heat transfer coefficients between the internal and external surface of the hub and medium, respectively; n, t are natural coordinates; v_r, v_θ are radial and tangential components of displacements vector; σ_n, σ_t, τ_{nt} are stress tensor components; $g(\theta)$ is the sought-for function of displacements of the points of the hub's external contour: $i^2 = -1$.

For intensity of the surface heat source, on the friction area we have $Q_1(\theta) = \alpha_{m.n.1} fVp(\theta)$; $\alpha_{m.n.1}$ is the heat partition coefficient for the plunger.

The contact pressure $p(\theta)$ is to be determined in the process of solution of minimization problem. To solve the stated problem, the wear-contact problem on pressing the plunger into the surface of the hub and optimization problem should be solved jointly.

Without loss of generality of the optimization problem, we assume that the sought-for function $g(\theta)$ of displacements of the points of the external contour L_0 may be represented in the form of the Fourier series

$$g(\theta) = \sum_{k=0}^{\infty} (a_k^* \cos k\theta + b_k^* \sin k\theta). \quad (12)$$

The unknown quantities θ_1 and θ_2 are the ends of the contact area of the plunger and the hub. To find these quantities we use the condition [19] that pressure $p(\theta)$ continuously passes into zero, when the point θ leaves the contact area:

$$p(\theta_1) = 0, \quad p(\theta_2) = 0. \quad (13)$$

To find the functions of displacements of the points of the external contour $g(\theta)$, we should complement the problem statement with an additional condition (with a criterion that allows determining the function $g(\theta)$). As a criterion for determining the function of displacements of the points of the hub external contour (function $g(\theta)$) we adopt the condition of minimization of the maximum value of circumferential stresses in the hub's material. Minimization of the maximum value of the stresses σ_θ of the hub will boost the serviceability of the friction pair.

Consequently, the coefficients a_k^*, b_k^* of the function $g(\theta)$ should be managed so that minimization of the maximum value of the stress σ_θ in the hub is provided. This additional condition allows determining the sought-for function $g(\theta)$.

3. Solution Method

We look for temperature functions, stresses, displacements, and contact pressure in the hub and plunger in the form of expansions in small parameter wherein for simplification we ignore the members containing ε higher than first degree. We get the values of temperature, stress tensor components, and displacements for $r = \rho(\theta)$ (similarly for $r = \rho_1(\theta)$ as well) expanding in series the expressions for temperature, stresses, and displacements in the neighborhood of $r = R_0$. Each of approximations satisfies the system of differential equations of plane thermoelasticity. Using the perturbations method, in light of the foregoing we arrive at the sequence of boundary value problems of thermoelasticity for the hub.

In zero approximation,

$$\Delta T^{(0)} = 0, \quad (14)$$

$$\text{for } r = R_0\text{:} \quad A_{T_1} \lambda \frac{\partial T^{(0)}}{\partial r} - A_{T_2} \alpha_1 T^{(0)} = -Q_*^{(0)}, \quad (15)$$

$$\text{for } r = R\text{:} \quad \lambda \frac{\partial T^{(0)}}{\partial r} + \alpha_2 T^{(0)} = 0,$$

$$\text{for } r = R_0\text{:} \quad \sigma_r^{(0)} = -p^{(0)}(\theta), \quad \tau_{r\theta}^{(0)} = -fp^{(0)}(\theta) \quad \text{on the contact area,} \quad (16)$$

$$\sigma_r^{(0)} = 0, \quad \tau_{r\theta}^{(0)} = 0 \quad \text{out of the contact area,} \quad (17)$$

$$\text{for } r = R\text{:} \quad v_r^{(0)} - iv_\theta^{(0)} = g^{(0)}(\theta); \quad (18)$$

in the first approximation,

$$\Delta T^{(1)} = 0 \quad (19)$$

$$\text{for } r = R_0\text{:} \quad A_{T_1} \lambda \frac{\partial T^{(1)}}{\partial r} - A_{T_2} \alpha_1 T^{(1)} = \left[A_{T_2} \alpha_1 \frac{\partial T^{(0)}}{\partial r} - A_{T_1} \lambda \frac{\partial^2 T^{(0)}}{\partial r^2} \right] H(\theta) - Q_*^{(1)}(\theta), \quad (20)$$

$$\text{for } r = R\text{:} \quad \lambda \frac{\partial T^{(1)}}{\partial r} + \alpha_2 T^{(1)} = 0, \quad (21)$$

$$\text{for } r = R_0\text{:}$$

$$\sigma_r^{(1)} = N - p^{(1)}(\theta),$$

$$\tau_{r\theta}^{(1)} = T_t - fp^{(1)}(\theta)$$

on the contact area,

(22)

$$\sigma_r^{(1)} = N$$

$$\tau_{r\theta}^{(1)} = T_t$$

out of the contact area, (23)

for $r = R$:

$$v_r^{(1)} - iv_\theta^{(1)} = g^{(1)}(\theta).$$

Hence for $r = R_0$

$$N = -H(\theta)\frac{\partial\sigma_r^{(0)}}{\partial r} + 2\tau_{r\theta}^{(0)}\cdot\frac{1}{R_0}\frac{dH(\theta)}{d\theta},$$

$$T_t = \left(\sigma_\theta^{(0)} - \sigma_r^{(0)}\right)\frac{1}{R_0}\frac{dH(\theta)}{d(\theta)} - H(\theta)\frac{\partial\tau_{r\theta}^{(0)}}{\partial r}. \quad (24)$$

The boundary conditions for the plunger in each approximation may be written in the same way.

The solution of a boundary value problem for conductivity theory is sought by the method of separation of variables. The temperature is sought in the form of a product of two functions, one of which depends on the variable r, and the other depends only on the polar angle θ. We find the distribution of hub's overtemperature $T = T^{(0)} + \varepsilon T^{(1)} + \ldots$ in the form

$$T^{(0)} = C_{10} + C_{20}\ln r + \sum_{k=1}^{\infty}\left(C_{10}^{(k)}r^k + C_{20}^{(k)}r^{-k}\right)\cos k\theta$$

$$+ \sum_{k=1}^{\infty}\left(A_{10}^{(k)}r^k + A_{20}^{(k)}r^{-k}\right)\sin k\theta,$$

$$T^{(1)} = C_{11} + C_{21}\ln r + \sum_{k=1}^{\infty}\left(C_{11}^{(k)}r^k + C_{21}^{(k)}r^{-k}\right)\cos k\theta$$

$$+ \sum_{k=1}^{\infty}\left(A_{11}^{(k)}r^k + A_{21}^{(k)}r^{-k}\right)\sin k\theta. \quad (25)$$

The constants C_{10}, C_{20}, $C_{10}^{(k)}$, $C_{20}^{(k)}$, $A_{10}^{(k)}$, $A_{20}^{(k)}$ are determined from boundary conditions (15) of the problem in a zero approximation. The coefficients $C_{11}, C_{21}, C_{11}^{(k)}, C_{21}^{(k)}, A_{11}^{(k)}$, $A_{21}^{(k)}$ are found, respectively, from boundary conditions (20) of the problem in the first approximation.

To solve the thermoelasticity problem the thermoelastic potential of displacements [20] is used in each approximation. In the present problem the thermoelastic potentials of displacements in zero and first approximations are determined by solving the following differential equations:

$$\Delta F^{(0)} = \beta T^{(0)},$$

$$\Delta F^{(1)} = \beta T^{(1)},$$

$$\beta = \frac{1+\mu}{1-\mu}\alpha, \quad (26)$$

where μ is Poisson's ratio of the plunger.

We look for the solution of (26) in the form

$$F^{(0)} = \sum_{n=0}^{\infty}\left[f_n^{(0)}(r)\cos n\theta + f_n^{*(0)}(r)\sin n\theta\right],$$

$$F^{(1)} = \sum_{n=0}^{\infty}\left[f_n^{(1)}(r)\cos n\theta + f_n^{*(1)}(r)\sin n\theta\right]. \quad (27)$$

Using the functions $F^{(0)}(r,\theta)$ and $F^{(1)}(r,\theta)$ we satisfy differential equations (26), respectively. Applying the method of separation of variables, we obtain ordinary differential equations for the functions $f_n^{(0)}(r)$, $f_n^{*(0)}(r)$, $f_n^{(1)}(r)$, and $f_n^{*(1)}(r)$.

$$\frac{d^2f_n^0}{dr^2} + \frac{1}{2}\frac{df_n^0}{dr} - \frac{n^2}{r^2}f_n^0 = \beta F_n^0,$$

$$\frac{d^2f_n^{0*}}{dr^2} + \frac{1}{2}\frac{df_n^{0*}}{dr} - \frac{n^2}{r^2}f_n^{0*} = \beta F_n^{0*}. \quad (28)$$

Particular solutions of differential equations are sought by the method of variation of constants

$$f_n^0 = \beta\left[-\ln r\int_{R_0}^{r}\rho F_0^0(\rho)\,d\rho + \int_r^R\rho F_0^0(\rho)\ln\rho d\rho\right],$$

$$f_n^0 = -\frac{\beta}{2n}\left[r^n\int_r^R F_n^0(\rho)\,\rho^{1-n}d\rho\right.$$

$$\left.+ r^{-n}\int_{R_0}^{r}\rho F_n^0(\rho)\,\rho^{1+n}d\rho\right], \quad (29)$$

$$f_n^{0*} = -\frac{\beta}{2n}\left[r^n\int_r^R F_n^{0*}(\rho)\,\rho^{1-n}d\rho\right.$$

$$\left.+ r^{-n}\int_{R_0}^{r}\rho F_n^{0*}(\rho)\,\rho^{1+n}d\rho\right].$$

After defining the thermoelastic potential of displacements in a zero approximation for the hub, by the known formulas [20], we calculate the approximate thermoelastic potential of the stresses $\overline{\sigma}_r^{(0)}, \overline{\sigma}_\theta^{(0)}, \overline{\tau}_{r\theta}^{(0)}$ and displacements $\overline{\upsilon}_r^{(0)}$, $\overline{\upsilon}_\theta^{(0)}$ in the hub

$$\overline{\sigma}_r^{(0)} = -2G\left\{\frac{1}{r}\sum_{n=0}^{\infty}\left[\frac{\partial f_n^0}{\partial r}\cos n\theta + \frac{\partial f_n^{0*}}{\partial r}\sin n\theta\right] + \frac{1}{r^2}\sum_{n=0}^{\infty}\left(-n^2\right)\left[f_n^0\cos n\theta + f_n^{0*}\sin n\theta\right]\right\},$$

$$\overline{\sigma}_\theta^{(0)} = -2G\sum_{n=0}^{\infty}\left[\frac{\partial^2 f_n^0}{\partial r^2}\cos n\theta + \frac{\partial^2 f_n^{0*}}{\partial r^2}\sin n\theta\right],$$

$$\overline{\tau}_{r\theta}^{(0)} = 2G\left\{\left(-\frac{1}{r^2}\right)\sum_{n=0}^{\infty}n\left[f_n^{0*}\cos n\theta - f_n^0\sin n\theta\right] + \frac{1}{r}\sum_{n=0}^{\infty}n\left[\frac{\partial f_n^{0*}}{\partial r}\cos n\theta - \frac{\partial f_n^0}{\partial r}\sin n\theta\right]\right\}, \quad (30)$$

$$\overline{\upsilon}_r^{(0)} = \sum_{n=0}^{\infty}\left[\frac{\partial f_n^0}{\partial r}\cos n\theta + \frac{\partial f_n^{0*}}{\partial r}\sin n\theta\right],$$

$$\overline{\upsilon}_\theta^{(0)} = \frac{1}{r}\sum_{n=0}^{\infty}\left[(-n)\left(f_n^0\sin n\theta - f_n^{0*}\cos n\theta\right)\right].$$

Here G is shear modulus of the plunger.

The found stresses and displacements will not satisfy boundary conditions (16)-(18). Thus, it is necessary to determine for the hub the second stress-strain states $\overline{\overline{\sigma}}_r^{(0)}, \overline{\overline{\sigma}}_\theta^{(0)}, \overline{\overline{\tau}}_{r\theta}^{(0)}$, $\overline{\overline{\upsilon}}_r^{(0)}, \overline{\overline{\upsilon}}_\theta^{(0)}$ such that boundary conditions (16)-(18) are fulfilled. To obtain the second stress-strain state in the hub, we have the following conditions:

$$\text{for } r = R_0: \quad \overline{\overline{\sigma}}_r^{(0)} = -p^{(0)}(\theta) - \overline{\sigma}_r^{(0)}, \quad (31)$$

$$\overline{\overline{\tau}}_{r\theta}^{(0)} = -fp^{(0)}(\theta) - \overline{\tau}_{r\theta}^{(0)} \quad \text{on the contact area}, \quad (32)$$

$$\overline{\overline{\sigma}}_r^{(0)} = -\overline{\sigma}_r^{(0)}, \quad \overline{\overline{\tau}}_{r\theta}^{(0)} = -\overline{\tau}_{r\theta}^{(0)} \quad \text{out of the contact area}, \quad (33)$$

$$\text{for } r = R: \quad \overline{\overline{\upsilon}}_r^{(0)} - i\overline{\overline{\upsilon}}_\theta^{(0)} = g^{(0)}(\theta) - \left(\overline{\upsilon}_r^{(0)} - i\overline{\upsilon}_\theta^{(0)}\right).$$

By means of the Kolosov-Muskhelishvili formulas [19], we can write the boundary conditions of problem (31)-(33) in the form of a boundary value problem for determining two complex functions $\Phi^{(0)}(z)$ and $\Psi^{(0)}(z)$ for the hub

$$\Phi^{(0)}(\tau_0) + \overline{\Phi^{(0)}(\tau_0)} - e^{2i\theta}\left[\overline{\tau}_0\Phi^{(0)\prime}(\tau_0) + \Psi^{(0)}(\tau_0)\right] = X^{(0)}(\theta), \quad (34)$$

$$\Phi^{(0)}(\tau) - k_b\overline{\Phi^{(0)}(\tau)} - e^{2i\theta}\left[\overline{\tau}\Phi^{(0)\prime}(\tau) + \Psi^{(0)}(\tau)\right] = 2G\left\{g^{(0)\prime}(\theta) - \left(\overline{\upsilon}_r^{(0)} - i\overline{\upsilon}_\theta^{(0)}\right)'\right\}. \quad (35)$$

Here

$$\tau_0 = R_0\exp(i\theta); \quad \tau = R\exp(i\theta); \quad k_b = 3 - 4\mu,$$
$$X^{(0)}(\theta) = \begin{cases} -(1-if)\,p^{(0)}(\theta) - \left(\sigma_r^{(0)} - \sigma_{r\theta}^{(0)}\right) & \text{on the contact area} \\ -\left(\overline{\sigma}_r^{(0)} - i\overline{\tau}_{r\theta}^{(0)}\right) & \text{out of the contact area.} \end{cases} \quad (36)$$

We look for complex potentials $\Phi^{(0)}(z)$ and $\Psi^{(0)}(z)$ in the form

$$\Phi^{(0)}(z) = \sum_{k=-\infty}^{\infty} a_k^{(0)} z^k, \quad \Psi^{(0)}(z) = \sum_{k=-\infty}^{\infty} b_k^{(0)} z^k. \quad (37)$$

To solve boundary value problem (34) and (35) with regard to analytic functions $\Phi^{(0)}(z)$ and $\Psi^{(0)}(z)$, we use the method of power series. For that the right hand sides of conditions (34) and (35) expand into Fourier series

$$X^{(0)}(\theta) = \sum_{k=-\infty}^{\infty} A_k^{(0)} e^{ik\theta}, \quad g^{(0)\prime}(\theta) - \left(\overline{\upsilon}_r^{(0)} - i\overline{\upsilon}_\theta^{(0)}\right)' = \sum_{k=-\infty}^{\infty} F_k^{(0)} e^{ik\theta}. \quad (38)$$

Requiring that functions (37) should satisfy boundary conditions (34) and (35), after some transformations we get an infinite linear algebraic system with respect to unknown coefficients $a_k^{(0)}, b_k^{(0)}$ whose solution is written in the form

$$a_0^{(0)} = \frac{A_0^{(0)}R_0^2 - F_0^{(0)}R^2}{2R_0^2 - (1-k_b)R^2}, \quad a_{-1}^{(0)} = \frac{A_1^{(0)}R_0}{1+k_b}, \quad (39)$$

$$b_{-2}^{(0)}R_0^{-2} = 2a_0^{(0)} - A_0^{(0)}, \quad b_{-1}^{(0)} = -\frac{k_b\overline{A}_1^{(0)}R_0}{1+k_b}, \quad (40)$$

$$a_k^{(1)} = \frac{(1+k)\left(R^2 - R_0^2\right)M_k - \overline{M}_k\left(R_0^{-2k+2} + k_bR^{-2k+2}\right)}{\left(1-k^2\right)\left(R^2-R_0^2\right)^2 - \left(R_0^{-2k+2} + k_bR^{-2k+2}\right)\left(R_0^{2k+2} + k_bR^{2k+2}\right)} \quad (k = \pm 2, \pm 3, \ldots),$$

$$M_k = F_k^{(0)}R^{-k+2} - A_k^{(0)}R_0^{-k+2},$$

$$a_1^{(0)} = \frac{2A_1^{(0)}\left(R^2 - R_0^2\right)R_0}{\left(1-k_b\right)\left(R_0^4 + k_b R^4\right)} - \frac{\overline{M}_{-1}}{R_0^4 + k_b R^4},$$

$$b_{k+2}^{(0)} R_0^{k+2} = (1-k)\, a_k^{(0)} R_0^k + \overline{a}_k^{(0)} R_0^{-k} - A_k^{(0)}. \tag{41}$$

The right sides of these formulas (39)-(41) contain the expansion coefficients of the contact pressure $p^{(0)}(\theta)$ and the functions of displacement at the points of the external contour

$$g^{(0)'}(\theta) = a_0^{*(0)} + \sum_{k=1}^{\infty}\left(a_k^{*(0)} \cos k\theta + b_k^{*(0)} \sin k\theta\right). \tag{42}$$

Now by means of complex potentials $\Phi^{(0)}(z)$, $\Psi^{(0)}(z)$, the Kolosov-Muskhelishvili formulas, and integration of kinetic equation (6) of wear of the hub's material in a zero approximation, we find the displacement $v_1^{(0)}$ of the hub's contact surface. In the same way we find the solution of thermoelasticity problem for the plunger in a zero approximation. Using this solution and kinetic equation of wear of the plunger's material in a zero approximation, we find the displacement $v_2^{(0)}$ of the plunger's contact surface. We substitute the found values $v_1^{(0)}$ and $v_2^{(0)}$ in the main contact equation (5) in a zero approximation.

For algebraization of the main equation, we look for the unknown functions of contact pressure in a zero approximation in the form of expansions

$$p^{(0)}(\theta,t) = p_0^0(\theta) + t p_1^0(\theta) + \ldots;$$

$$p_0^0(\theta) = A_0^0 + \sum_{k=1}^{\infty}\left(A_k^0 \cos k\theta + B_k^0 \sin k\theta\right); \tag{43}$$

$$p_1^0(\theta) = A_0^1 + \sum_{k=1}^{\infty}\left(A_k^1 \cos k\theta + B_k^1 \sin k\theta\right)\ldots \tag{44}$$

Substituting relation (43) and (44) in the main contact equation in a zero approximation, we find functional equations for successive determination of $p_0^0(\theta)$, $p_1^0(\theta)$, etc. To construct an algebraic system and to find A_k, B_k we equate the coefficients under identical trigonometric functions in the left and right hand sides of the functional equation of the contact problem. We get an infinite algebraic system with respect to A_k^0 $(k = 0,1,2,\ldots)$, B_k^0 $(k = 1,2,\ldots)$ and A_k^1, B_k^1, etc. The quantities θ_1 and θ_2 are unknown. Because of this the system of equations is nonlinear. To determine the quantities θ_1 and θ_2 $(\theta_1 = \theta_1^0 + \varepsilon\theta_1^1 + \ldots;\ \theta_2 = \theta_2^0 + \varepsilon\theta_2^1 + \ldots)$ we have condition (13). We can represent these equations in the following form:

for zero approximation

$$p^{(0)}\left(\theta_1^0\right) = 0,$$

$$p^{(0)}\left(\theta_2^0\right) = 0,$$

for first approximation

$$p^{(1)}\left(\theta_1^1\right) = 0,$$

$$p^{(1)}\left(\theta_2^1\right) = 0. \tag{45}$$

The right hand sides of the infinite algebraic systems contain the unknown coefficients $a_k^{*(0)}$, $b_k^{*(0)}$ of expansions of functions $g^{(0)}(\theta, a_k^{*(0)}, b_k^{*(0)})$. Under the known function $g^{(0)}(\theta, a_k^{*(0)}, b_k^{*(0)})$ the obtained systems allow determining the contact pressure, temperature, stress-strain state, and wear of friction pair elements by numerical calculations. To construct the missing equations $g^{(0)}(\theta, a_k^{*(0)}, b_k^{*(0)})$ the minimization of the maximum value of the hub's circumferential stress on the friction surface is required:

$$\min_{\eta^0 \in C}\ \max_{\theta\in[\theta_1,\theta_2]} \sigma_\theta^{(0)s}\left(\eta^0,\theta\right) \tag{46}$$

under restrictions associated with bearing capacity and heat-resistance of the friction pair

$$p_{\max} \le [p],$$

$$(pV)_{\max} \le [pV], \tag{47}$$

and also $\sigma_{\theta\max}^0 \le [\sigma]$, where $[p]$ is the admissible specific load on contact surface; $[pV]$ is the admissible heat-resistance of the pair; $[\sigma]$ is the admissible stress for material of the hub and is determined experimentally. C is the constraints set. The design parameters $\eta = (a_k^{*(0)}, b_k^{*(0)})$ are nonnegative.

For the function $\sigma_\theta^{(0)s}$ maximum value in the hub's inner surface is found:

$$\sigma_\theta^{(0)s} = \sigma_\theta^{(0)s}\left(\theta_{\max}\right) = \sigma_{\theta\max}^{(0)s}\left(a_k^{*(0)}, b_k^{*(0)}\right), \tag{48}$$

where value $\theta_{\max}$ is solution of equation

$$\frac{d\sigma_\theta^{(0)s}}{d\theta} = 0. \tag{49}$$

The maximum of the function $\sigma_\theta^{(0)s}$ is found by the usual methods of differential calculus.

As the circumferential stress $\sigma_\theta^{(0)s}$ (control quality index) and $\sigma_{\theta\max}^{(0)s}$ linearly depend on sought-for coefficients of functions $g^{(0)}(\theta, a_k^{*(0)}, b_k^{*(0)})$, the solution of the considered optimization problem in the zero approximation is reduced to a linear programming problem. Thus, finding such non-negative values of variables α_k^*, β_k^*, a_k, b_k and $a_k^{*(0)}$, $b_k^{*(0)}$ that satisfy the obtained system of equations (constraints) and also reduce to minimum the linear function $\sigma_{\theta\max}^{(0)s}$ is required.

After determining the sought-for quantities of zero approximation we can pass to construction of the solution of the problem in the first approximation. Based on the obtained solution for $r = R_0$ we determine the functions N and T_t.

By means of known formulas [20], by the found thermoelastic potential $F^{(1)}(r,\theta)$, we find the stresses $\overline{\sigma}_r^{(1)}$, $\overline{\sigma}_\theta^{(1)}$, $\overline{\tau}_{r\theta}^{(1)}$ and displacements $\overline{v}_r^{(1)}$, $\overline{v}_\theta^{(1)}$ for the hub. The found stress and displacement component do not satisfy the boundary conditions (22) and (23) in the first approximation. Consequently, it is necessary to find the second stress-strain state $\overline{\overline{\sigma}}_r^{(1)}$, $\overline{\overline{\sigma}}_\theta^{(1)}$, $\overline{\overline{\tau}}_{r\theta}^{(1)}$, $\overline{\overline{v}}_r^{(1)}$, $\overline{\overline{v}}_\theta^{(1)}$ for the hub in the first approximation. We can write the boundary conditions for finding $\overline{\overline{\sigma}}_r^{(1)}$, $\overline{\overline{\sigma}}_\theta^{(1)}$, $\overline{\overline{\tau}}_{r\theta}^{(1)}$, $\overline{\overline{v}}_r^{(1)}$,

TABLE 1: The values of Fourier coefficients for an optimal function of displacements of hub's external contour points that allows the minimization of the stress state (mm).

a_0^*	a_1^*	a_2^*	a_3^*	a_4^*	a_5^*	a_6^*	a_7^*	a_8^*	a_9^*
0.1123	0.0879	0.0745	0.0617	0.0528	0.0419	0.0403	0.0315	0.0284	0.0201
	b_1^*	b_2^*	b_3^*	b_4^*	b_5^*	b_6^*	b_7^*	b_8^*	b_9^*
	0.0892	0.0764	0.0583	0.0519	0.0423	0.0374	0.0285	0.0223	0.0154

$\overline{\overline{v}}_\theta^{(1)}$ in the form of a boundary value problem for finding complex potentials $\Phi^{(1)}(z)$ and $\Psi^{(1)}(z)$ that we look for in the form of power series as in a zero approximation with obvious changes.

The further course of solution is as in a zero approximation. The right hand side of the system for determining the quantities $a_k^{(1)}$, $b_k^{(1)}$ contains the sought-for $p^{(1)}(\theta)$ and the functions of displacements at the points of the external contour

$$g^{(1)'} = a_0^{*(1)} + \sum_{k=1}^{\infty} a_k^{*(1)} \cos k\theta + b_k^{*(1)} \sin k\theta. \quad (50)$$

As in the zero approximation, the obtained systems of equations permit expressing the coefficients $a_k^{(1)}$, $b_k^{(1)}$ by the coefficients $a_k^{*(1)}$, $b_k^{*(1)}$ of the function of displacements at the points of the hub's external contour in the first approximation.

A thermoelasticity problem for the plunger in the first approximation is solved in the same way.

In the first approximation the resolving equation of the contact problem is reduced to the algebraic form as in the zero approximation. For that we represent the sought-for functions of contact pressure in the form

$$\begin{aligned} p^{(1)}(\theta,t) &= p_0^1(\theta) + tp_1^1(\theta) + \ldots; \\ p_0^1(\theta) &= A_{0,0}^1 + \sum_{k=1}^{\infty}\left(A_{k,0}^1 \cos k\theta + B_{k,0}^1 \sin k\theta\right); \end{aligned} \quad (51)$$

$$p_1^1(\theta) = A_{0,1}^1 + \sum_{k=1}^{\infty}\left(A_{k,1}^1 \cos k\theta + B_{k,1}^1 \sin k\theta\right)\ldots \quad (52)$$

Substituting relation (51) and (52) in the main contact equation in a first approximation, we find functional equations for successive determination of $p_0^1(\theta)$, $p_1^1(\theta)$, etc. To construct an algebraic system and to find A_k^1, B_k^1 we equate the coefficients under identical trigonometric functions in the left and right hand sides of the functional equation of the contact problem. We get infinite linear algebraic systems with respect to $A_{0,0}^1$, $A_{k,0}^1$, $B_{k,0}^1$ and $A_{0,1}^1$, $A_{k,1}^1$, $B_{k,1}^1$ $(k = 1, 2, \ldots)$, etc. The quantities θ_1^1, θ_2^1 are unknown and the system of equations becomes nonlinear. The obtained system of equations in the first approximation is not closed because of the unknown function $g^{(1)}(\theta)$. The obtained system of equations in zero and first approximations allows finding contact pressure, temperature distribution, stress-strain state, and wear of the hub and plunger of friction pair by numerical calculations under the given functions $g^{(0)}(\theta)$ and $g^{(1)}(\theta)$.

To construct the missing equations we require minimizing the maximum value of the stress σ_θ on the hub's friction surface

$$\sigma_{\theta\max} \longrightarrow \min. \quad (53)$$

with constraints involving load bearing capacity and heat-resistance of the pair and also $\sigma_{\theta\max} \leq [\sigma]$, where σ is admissible stress for the hub's material and is determined experimentally.

For circumferential stress in the hub, for $r = \rho(\theta)$, we have

$$\sigma_\theta = \sigma_{\theta|r=R_0}^{(0)} + \varepsilon\left[H(\theta)\frac{\partial\sigma^{(0)}}{\partial r} + \sigma_\theta^{(1)}\right]_{|r=R_0} \quad (54)$$

For the function $\sigma_\theta(\theta)$ we find its maximum value. The coefficients a_k^*, b_k^* of the sought-for function $g(\theta)$ should be managed so that minimization $\sigma_{\theta\max}$ (minimax criterion) is provided. As the stress $\sigma_{\theta\max}$ (control quality index) linearly depends on sought-for coefficients of the function $g(\theta)$, then the stated minimization problem is reduced to a linear programming problem.

4. Analysis of Simulation Results

To realize the obtained mathematical model, numerical methods for solving linear programming problems can be used. In the minimization problem, the simplex algorithm is found to be the most effective method. In the considered problem there are many free parameters. These are different thermophysical and mechanical characteristics of materials, quality parameters of the surface of the hub's internal contour and plunger's external contour, geometrical sizes of the plunger and hub, and movement rate of the plunger. The numerical calculation is performed by the method of successive approximations [21] and simplex algorithm.

As a numerical example of the application of the mathematical model obtained in Section 3 the minimization was carried out for friction pair of oil-well sucker-rod pump. The calculations were performed for the hub of the pump.

The results of calculations of the expansion coefficients of displacements functions $g(\theta)$ at initial time $t = 0$ are cited in Table 1 for a rough internal contour described by a stationary random function with a zero mean value and known variance. As an example we accepted the following: $2R_0 = 57$ mm; $2R = 73$ mm; $2R_0' = 56.7$ mm; $E = 1.8 \cdot 10^5$ MPa; $E_1 = 2.1 \cdot 10^5$ MPa; $f = 0.2$; $\mu = 0.25$; $\mu = 0.3$; $K^{(1)} = 2 \cdot 10^{-8}$; $K^{(2)} = 2.5 \cdot 10^{-9}$; $\Delta_1 = R_0 - R_0' = 0.15$ mm; $V = 0.4$ m/s.

After solving of the algebraic systems the contact pressure was calculated as a function of polar angle for different values

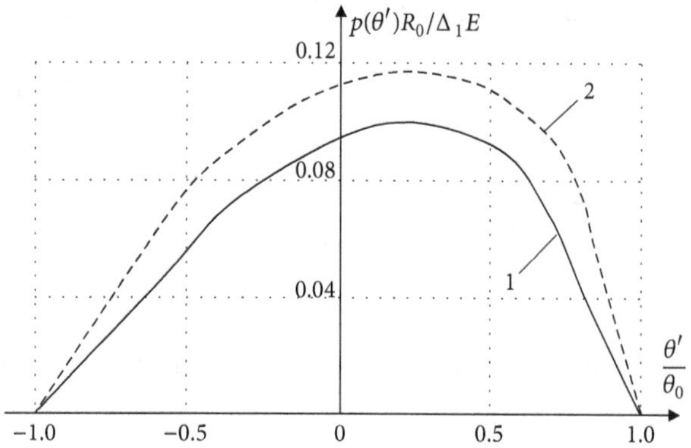

FIGURE 4: Dependence of the contact pressure on the values of polar angle for $V = 0.5$ m/s.

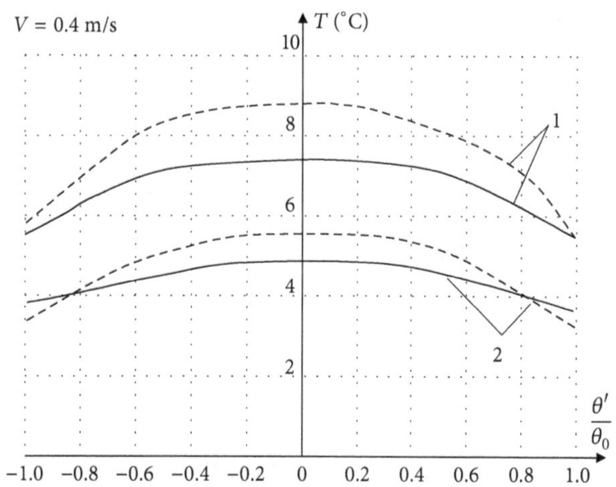

FIGURE 5: Temperature change in the hub for one stroke of the plunger for velocity of plunger motion $V = 0.4$ m/s.

of the free parameters. The contact pressure for the borehole sucker-rod oil pumps depending on the values of polar angle $\theta' = \theta - \theta_+$ $(\theta_+ = (1/2)(\theta_2 + \theta_1)$, $\theta_0 = (1/2)(\theta_2 - \theta_1))$ are represented in Figure 4 for the velocity of plunger motion $V = 0.5$ m/s. Here curve 1 is corresponding to the optimal solution and curve 2 is corresponding to case when the function of the displacements of points of the hub external contour is equal to zero. The large values of contact pressure, as a rule, are in the middle of the contact surface depending on the friction coefficient and the contact angle. The existence of friction forces in the contact area results in displacement of the diagram of contact pressure distribution to direction opposite to the action of the moment.

With known contact pressure it is possible to calculate the temperature distribution and abrasive wear of the friction pair details. The temperature change for different depths on the hub thickness was studied. The temperature change in the hub for one plunger stroke for different velocity of plunger motion is represented in Figures 5-6. Here solid and dashed lines, respectively, correspond to the optimal solution and to the case when the function of displacements of the points of the external contour of the hub is equal to zero. Curve 1

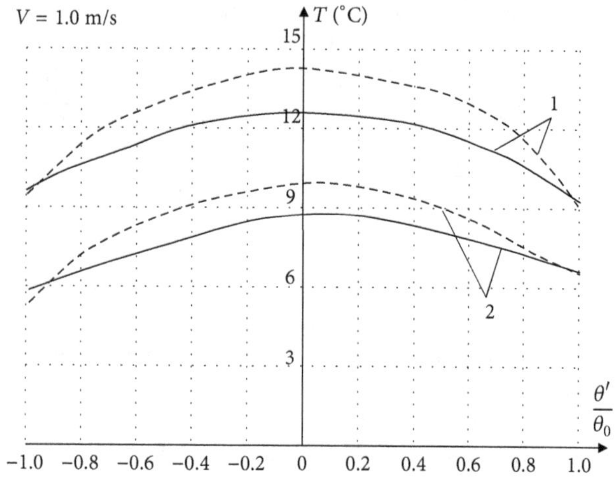

FIGURE 6: Temperature change in the hub for one stroke of the plunger for velocity of plunger motion $V = 1.0$ m/s.

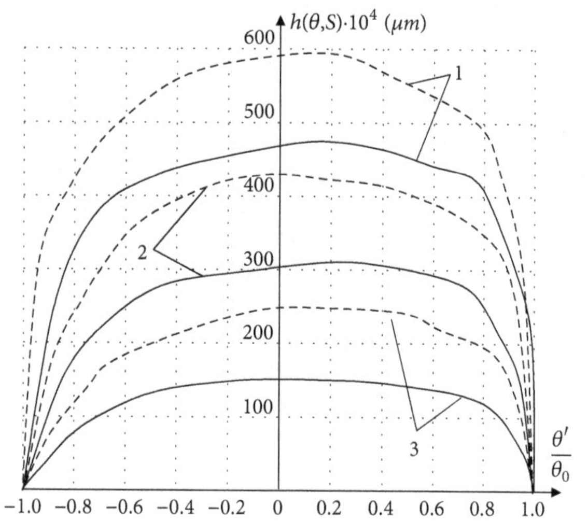

FIGURE 7: Abrasive wear of the hub surface for one stroke of the plunger.

is corresponding to the surface temperature of the hub and curve 2 is corresponding to temperature on depth of 2 mm.

The results of calculations of the abrasive wear of the hub surface for one plunger stroke are shown in Figure 7 for different velocity of the plunger motion. Curves 1-3 correspond to the velocity $V = 0.2$, 0.5, and 1.0 m/s, respectively.

The calculation results show that at low values of the contact pressure the wear of the hub along the contact zone has uneven character. With increasing of the contact pressure the hub wear along the length of the contact zone tends to leveling and mainly depends on wear (friction) path. Because the friction coefficient of the pair has significant effect on the wear of the hub the dependence of maximum wear on the friction coefficient was calculated.

In Figure 8 the dependence of maximum wear is shown for friction path corresponding to ten work hours of the pump. Curve 1 corresponds to the optimal solution and curve

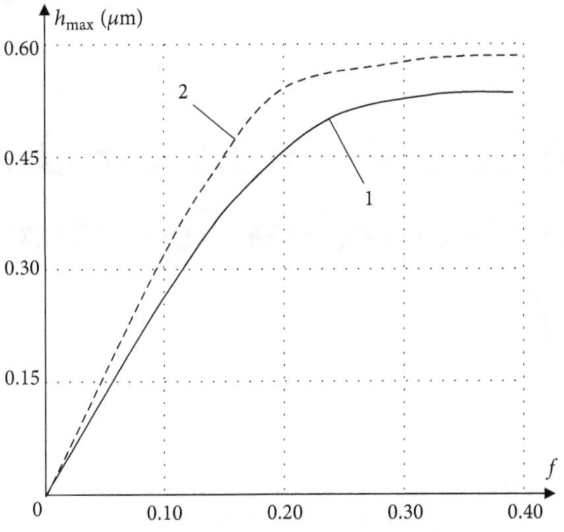

FIGURE 8: Dependence of maximum wear for friction path corresponding to ten work hours of the pump.

2 corresponds to case when the function of displacements of the points of the hub external contour is equal to zero.

Due to the wear the surface microgeometry of the plunger and hub will vary. The relations for the radial wear obtained in the study allow determination of the variation of friction surface for a given moment of time.

5. Conclusions

It is shown that by deriving the function of displacements of the external contour points of a hub of friction pair, one can control (minimize) the stress state distribution in the hub. This will increase serviceability of the pair "a hub-a plunger". The suggested method of minimization of stress state of a friction pair may be extended to other constructions of friction units.

Conflicts of Interest

The authors declare that there are no conflicts of interest regarding the publication of this paper.

References

[1] G. H. Gadzhiev and V. M. Mirsalimov, "Minimizing wear of the internal surface of a split cylinder sleeve in a contact pair," *Journal of Friction and Wear*, vol. 25, no. 3, pp. 231–237, 2004.

[2] G. H. Gadzhiev and V. M. Mirsalimov, "Optimum design of jointed cylinder-ram pairs," *Journal of Friction and Wear*, vol. 25, no. 5, pp. 466–473, 2004.

[3] V. M. Mirsalimov, "An inverse wear contact problem for a friction couple," *Journal of Machinery Manufacture and Reliability*, vol. 37, no. 1, pp. 53–59, 2008.

[4] V. M. Mirsalimov, "Inverse problem of fracture mechanics for a disk fitted onto a rotating shaft," *Journal of Applied Mechanics and Technical Physics*, vol. 50, no. 4, pp. 712–719, 2009.

[5] V. M. Mirsalimov, "Simulation of bridged crack closure in a contact pair bushing," *Mechanics of Solids*, vol. 44, no. 2, pp. 232–243, 2009.

[6] Z. Mróz and I. Páczelt, "Analysis of thermo-elastic wear problems," *Journal of Thermal Stresses*, vol. 34, no. 5-6, pp. 569–606, 2011.

[7] V. M. Mirsalimov and F. E. Veliyev, "Inverse problem of failure mechanics for a drawing die strengthened with a holder," *Acta Polytechnica Hungarica*, vol. 10, no. 1, pp. 121–138, 2013.

[8] F. Meng, C. Labergere, P. Lafon, M. Duguy, and L. Daniel, "Multi-objective optimization based on meta-models of an aeronautical hub including the ductile damage constraint," *International Journal of Damage Mechanics*, vol. 23, no. 8, pp. 1055–1076, 2014.

[9] V. M. Mirsalimov and P. E. Akhundova, "Minimization of contact pressure for hub–shaft friction pair," *Journal of Friction and Wear*, vol. 36, no. 5, pp. 404–408, 2015.

[10] V. M. Mirsalimov and P. E. Aknundova, "Minimization of abrasive wear for the internal surface of the hub of a friction pair," *Journal of Friction and Wear*, vol. 37, no. 5, pp. 424–429, 2016.

[11] F. J. Szabó, "Multidisciplinary optimization of journal bearings, using a RVA evolutionary type optimization algorithm," *Acta Polytechnica Hungarica*, vol. 13, no. 7, pp. 181–195, 2016.

[12] T. R. Thomas, *Rough surface*, Longman, London, 1982.

[13] B. Bhushan, "Contact mechanics of rough surfaces in tribology: Multiple asperity contact," *Tribology Letters*, vol. 4, no. 1, pp. 1–35, 1998.

[14] G. Carbone and F. Bottiglione, "Contact mechanics of rough surfaces: a comparison between theories," *Meccanica*, vol. 46, no. 3, pp. 557–565, 2011.

[15] Á. Czifra and S. Horváth, "Complex microtopography analysis in sliding friction of steel-ferodo material pair," *Meccanica*, vol. 46, no. 3, pp. 609–616, 2011.

[16] K. L. Johnson, *Contact Mechanics*, Cambridge University Press, Cambridge, 1985.

[17] I. G. Goryacheva, *Contact mechanics in tribology*, vol. 61, Kluwer Academic Publishers, Dordrecht, 1998.

[18] I. G. Goryacheva, *Mechanics of Frictional Interaction*, Moscow, Nauka, 2001, in Russian.

[19] N. I. Muskhelishvili, *Some Basic Problems of the Mathematical Theory of Elasticity*, Kluwer, Amsterdam, 1977.

[20] H. Parkus, *Instationäre wärmespannungen*, Springer-Verlag, Vienna, 1959.

[21] V. M. Mirsalimov, *Non-one-dimensional Elastoplastic Problems*, Nauka, Moscow, Russian, 1987.

16

Some Analytical Solitary Wave Solutions for the Generalized q-Deformed Sinh-Gordon Equation: $\partial^2\theta/\partial z\partial\xi = \alpha[\sinh_q(\beta\theta^{\gamma})]^p - \delta$

H. Eleuch [1,2]

[1] Institute for Quantum Science and Engineering, Texas A&M University, College Station, TX 77843, USA
[2] Department of Applied Sciences and Mathematics, College of Arts and Sciences, Abu Dhabi University, Abu Dhabi, UAE

Correspondence should be addressed to H. Eleuch; hichemeleuch@yahoo.fr

Academic Editor: Boris G. Konopelchenko

We introduce the generalized q-deformed Sinh-Gordon equation and derive analytical soliton solutions for some sets of parameters. This new defined equation could be useful for modeling physical systems with violated symmetries.

1. Introduction

Dynamical models are the cornerstone of several fields of science and uneven covered in the literature [1–9]. They represent an important research area in mathematics and theoretical physics. Such models are usually based on nonlinear ordinary or partial differential equations. Nonlinearity has generated crucial effects and phenomena not only in the macroscopical systems but also in the microscopical systems governed by quantum physics. We can cite as examples Rogue waves [10–12], bifurcation [13, 14], bistability [15–18], chaos [19–22], phase transition [23–26], lasing [27–31], superradiance [32–34], correlations [35–37], squeezing [38–41], and solitons [42–48]. A particular interest is focused on solitons due to their potential new application perspectives in several fields of physics and engineering [49–56]. Soliton is a propagating wave solution that conserves its shape and has a particle-like behavior. This propriety could be explained in general by a balance between dispersion and nonlinearity. Several famous nonlinear partial differential equations generate solitons [43, 57–61] such as Korteweg-de Vries equation, nonlinear Schrödinger equation, Kadomtsev-Petviashvili equation, Sin-Gordon, and Sinh-Gordon equations. The Sinh-Gordon equation has several applications such as in surface theory [62], crystal lattice formation [63], and string dynamics in curved space-time [64]. When the q-deformed hyperbolic function, proposed in the ninetieth of the last century by Arai [65, 66], is included in the dynamical system, the symmetry of the system is destroyed and consequently the symmetry of the solution. Recently, several solutions for Schrödinger equation and Dirac equation with q-deformed hyperbolic potential are generated [67–72]. q-deformed functions are very promising for modeling atom-trapping potentials or statistical distributions in Bose-Einstein condensates [73, 74] as well as for exploring vibrational spectra of diatomic molecules [75, 76].

In this work, we propose to analyze the propagating wave solutions for a more general form of the Sinh-Gordon equation, the generalized q-deformed Sinh-Gordon equation. As far as we know, this equation is introduced for the first time. We derive for some sets of parameters analytical soliton solutions. The generalized q-deformed Sinh-Gordon equation will open the door for conceiving models of physical systems where the symmetry is absent or violated.

2. Preliminaries

Let us first define the generalized q-deformed Sinh-Gordon equation as

$$\frac{\partial^2\theta}{\partial z\partial\xi} = \alpha\left[\sinh_q\left(\beta\theta^{\gamma}\right)\right]^p - \delta, \tag{1}$$

where $\alpha, \gamma, p \in \mathbb{R}^*$, $\delta \in \mathbb{R}$, and $\beta \in \mathbb{R}^+$. For $\delta/\alpha \geqslant 0$, (1) admits a trivial constant solution

$$\theta(z,\xi) = \left[\frac{1}{\beta}\ln\left(\left(\frac{\delta}{\alpha}\right)^{1/p} + \sqrt{\left(\frac{\delta}{\alpha}\right)^{2/p} + q}\right)\right]^{1/\gamma}. \tag{2}$$

It is useful to mention that any first-order polynome ($Po_1(s) = a_1 s + a_0$; with $a_1 \neq 0$) in z or in ξ or any combination of two arbitrary first-order polynomes ($\alpha_0 Po_1(z) + \alpha_1 po_1(\xi)$) can not be a solution to the generalized q-deformed Sinh-Gordon equation (1). Equation (1) is also equivalent to

$$\frac{\partial^2 (\theta_1^{\gamma_1})}{\partial z \partial \xi} = \alpha \left[\sinh_q (\beta \theta_1)\right]^p - \delta, \tag{3}$$

where $\theta_1 = \theta^\gamma$ and $\gamma_1 = 1/\gamma$. $\sinh_q$ is the Arai q-deformed function defined by

$$\sinh_q (x) = \frac{e^x - qe^{-x}}{2}, \tag{4}$$

and we have also

$$\cosh_q (x) = \frac{e^x + qe^{-x}}{2}, \tag{5}$$

$$\sec h_q (x) = \frac{1}{\cosh_q (x)}, \tag{6}$$

$$\tanh_q (x) = \frac{e^x - qe^{-x}}{e^x + qe^{-x}} = \frac{\sinh_q (x)}{\cosh_q (x)}, \tag{7}$$

where $0 < q \leq 1$. For $q = 1$ we get the standard sinh, cosh, sech, and tanh functions. We list below some simple and useful relations for q-deformed functions

$$\left(\cosh_q (x)\right)^2 - \left(\sinh_q (x)\right)^2 = q, \tag{8}$$

$$\sinh_q (-x) = -q \sinh_{1/q} (x), \tag{9}$$

$$\cosh_q (-x) = q \cosh_{1/q} (x), \tag{10}$$

$$\frac{d}{dx} \sinh_q (x) = \cosh_q (x), \tag{11}$$

$$\frac{d}{dx} \cosh_q (x) = \sinh_q (x), \tag{12}$$

$$\frac{d}{dx} \tanh_q (x) = \frac{q}{\left(\cosh_q (x)\right)^2} = 1 - \tanh_q^2 (x), \tag{13}$$

$$\frac{d}{dx} \sec h_q (x) = -\tanh_q (x) \sec h_q (x), \tag{14}$$

$$\frac{d}{dx} \arcsin h_q (x) = \frac{1}{\sqrt{q + x^2}}, \tag{15}$$

$$\frac{d}{dx} \arctan h_q (x) = \frac{1}{1 - x^2}, \tag{16}$$

where $\arctan h_q$ and $\operatorname{arcsinh}_q$ are, respectively, the inverse functions of $\tanh_q$ and $\sinh_q$. Since $\tanh_q(x)$ and $\sinh_q(x)$ vanish at $x = (1/2)\ln(q)$ then

$$\begin{aligned} \operatorname{arcsinh}_q (x) &= \int_0^x \frac{dx'}{\sqrt{q + (x')^2}} + \frac{1}{2} \ln (q) \\ &= \ln \left[x + \sqrt{x^2 + q}\right], \end{aligned} \tag{17}$$

$$\begin{aligned} \operatorname{arctanh}_q (x) &= \int_0^x \frac{dx'}{1 - (x')^2} + \frac{1}{2} \ln (q) \\ &= \frac{1}{2} \ln \left[\frac{q(1+x)}{1-x}\right]; \quad |x| < 1. \end{aligned} \tag{18}$$

It is worth mentioning that the generalized q-deformed Sinh-Gordon equation (1) with $\delta = 0$

$$\frac{\partial^2 \theta}{\partial z \partial \xi} = \alpha \left[\sinh_q (\beta \theta^\gamma)\right]^p \tag{19}$$

can be easily transformed to the generalized q-deformed Sin-Gordon equation defined as

$$\frac{\partial^2 \theta}{\partial z \partial \xi} = \alpha \left[\sin_q (\beta \theta^\gamma)\right]^p, \tag{20}$$

by changing $\theta \longrightarrow (i)^{1/\gamma}\theta$ and $z \longrightarrow (i)^{(1/\gamma - p)} z$.

Let us return to (1), by defining the following new normalized coordinates:

$$z_1 = \beta^{1/2\gamma} \sqrt{|\alpha|} z, \tag{21}$$

$$\xi_1 = \beta^{1/2\gamma} \sqrt{|\alpha|} \xi, \tag{22}$$

and using

$$u = \beta^{1/\gamma} \theta, \tag{23}$$

(1) becomes

$$\frac{\partial^2 u}{\partial z_1 \partial \xi_1} = \pm \left(\left[\sinh_q (u^\gamma)\right]^p - \delta_1\right). \tag{24}$$

Here $\pm$ is defined by the sign of α and $\delta_1 = \delta/|\alpha|$. Using the standard transformation

$$x = a z_1 + \frac{\xi_1}{a} \tag{25}$$

and

$$t = a z_1 - \frac{\xi_1}{a}, \tag{26}$$

with a is an arbitrary constant, we obtain then

$$\frac{\partial^2 u}{\partial x^2} - \frac{\partial^2 u}{\partial t^2} = \left[\sinh_q (u^\gamma)\right]^p - \delta_1. \tag{27}$$

This equation is obtained for positive α, and for negative α we obtain the same equation as (27) by interchanging $x \longrightarrow t$ and $t \longrightarrow x$. Equation (27) has two conservative quantities, namely, the total energy E and the momentum P [77, 78] defined by

$$E = \int_{-\infty}^{\infty} \left(\frac{\partial^2 u}{\partial x^2} - \frac{\partial^2 u}{\partial t^2} \right) - F(u)\, dx, \quad (28)$$

and

$$P = \frac{1}{2} \int_{-\infty}^{\infty} \left(\frac{\partial u}{\partial t} \right) \left(\frac{\partial u}{\partial x} \right) dx, \quad (29)$$

where

$$F(u) = \int \left[\sinh_q (u^\gamma) \right]^p du - \delta_1. \quad (30)$$

3. Traveling Wave Solution

In this section we explore the traveling wave solutions of the generalized q-deformed Sinh-Gordon equation (27). We first define a new moving coordinate

$$\eta = \frac{1}{\sqrt{1-c^2}} (x - ct), \quad (31)$$

so

$$\frac{\partial}{\partial t} = -\frac{c}{\sqrt{1-c^2}} \frac{\partial}{\partial \eta}, \qquad \frac{\partial}{\partial x} = \frac{1}{\sqrt{1-c^2}} \frac{\partial}{\partial \eta}. \quad (32)$$

c can be interpreted as the speed of the traveling wave in the space-time $(x-t)$. Note that by defining $\eta_1 = x \pm t$ we get the trivial constant solution for $\delta_1 \geqslant 0$

$$u(\eta_1) = \left[\ln \left(\delta_1^{1/p} + \sqrt{\delta_1^{2/p} + q} \right) \right]^{1/\gamma}. \quad (33)$$

Here we take $c \neq \pm 1$; the traveling wave in the moving frame η verifies

$$\frac{d^2 u(\eta)}{d\eta^2} = \left[\sinh_q (u^\gamma(\eta)) \right]^p - \delta_1. \quad (34)$$

It is worth mentioning that, for $p = 1, \gamma = 1, \delta_1 = 0$, and $q = 1$, (27) describes the standard traveling wave of Sinh-Gordon equation. Let us consider now the following particular cases.

3.1. Special Case (1): The Deformed Sinh-Gordon Equation ($p = 1, \gamma = 1$, and $\delta_1 = 0$). For $= 1, \delta_1 = 0$, and $\gamma = 1$ (in this case (27) can be identified as a deformed Sinh-Gordon equation). We can multiply both side of (34) by $du(\eta)/d\eta$ and after the integration we get

$$\frac{1}{2} \left[\frac{du(\eta)}{d\eta} \right]^2 = \cosh_q (u(\eta)) + c_1', \quad (35)$$

so we have an implicit equation for the traveling wave

$$\int^u \frac{du'}{\sqrt{2\cosh_q (u') + c_1}} = \pm \left(\eta + c_2' \right), \quad (36)$$

which can be written also as

$$H(u) \pm \eta = C, \quad (37)$$

where

$$H(u) = \int^u \frac{du'}{\sqrt{2\cosh_q (u') + c_1}}. \quad (38)$$

C and c_1 are free parameters. We can derive general explicit solutions for the considered q-deformed Sinh-Gordon equation. In fact

$$H(u) = \int^u \frac{du'}{\sqrt{e^{u'} + q e^{-u'} + c_1}} = \int^{e^u} \frac{dy}{\sqrt{y^3 + qy + c_1 y^2}}, \quad (39)$$

and let us first remind the definition of the Weierstrass Elliptic function $\wp(x)$:

$$x = \int_{\wp(x;g_2;g_3)}^{\infty} \frac{dw}{\sqrt{4w^3 - g_2 w - g_3}}. \quad (40)$$

so we need to transform the expression of $H(u)$ and in particular the polynomial $P_3(y) = y^3 + c_1 y^2 + qy$. By defining $y_1 = y + c_1/3$ the polynomial $P_3(y)$ will be transformed to $(1/4)(4y_1^3 + 4((3q - c_1^2)/3)y_1 + 4(2c_1^3/27 - qc_1/3))$, where $g_2 = -4((3q - c_1^2)/3)$ and $g_3 = -4(2c_1^3/27 - qc_1/3)$, and we have then

$$H(u) = \int^{e^u + c_1/3} \frac{2dy_1}{\sqrt{4y_1^3 - g_2 y_1 - g_3}}. \quad (41)$$

Finally, we obtain the general expression of the soliton in the moving frame η

$$u(\eta) = \ln \left[\wp \left(\pm \frac{1}{2} \eta + c_2; g_2; g_3 \right) - \frac{c_1}{3} \right], \quad (42)$$

with

$$g_2 = 4 \left(\frac{c_1^2 - 3q}{3} \right), \quad (43)$$

$$g_3 = 4 \left(\frac{qc_1}{3} - \frac{2c_1^3}{27} \right). \quad (44)$$

c_1 and c_2 are free parameters. Let us now study another case.

3.2. Special Case (2): ($p = -2$, $\gamma = 1$, and $\delta_1 = 0$). In this section we shall derive analytical soliton solutions for the case

$$\frac{d^2u(\eta)}{d\eta^2} = \frac{1}{\left[\sinh_q(u(\eta))\right]^2}. \quad (45)$$

After multiplication of (45) by $du(\eta)/d\eta$ and integration, we have

$$\frac{du(\eta)}{d\eta} = \pm\sqrt{\frac{4}{(q - e^{2u(\eta)})} + c_1}, \quad (46)$$

where c_1 is a free parameter.

Equation (46) can be integrated and we obtain then the following general implicit solutions:

$$\begin{aligned} 0 = {} & \sqrt{\frac{q}{4 + c_1 q}}\ln\left[c_1 q^2 + 4q\right. \\ & + \sqrt{(c_1 q^2 + 4q)(e^{2u(\eta)} - q)(c_1 e^{2u(\eta)} - c_1 q - 4)} \\ & \left. - (2 + c_1 q)e^{2u(\eta)}\right] + \frac{1}{\sqrt{c_1}}\ln\left[2 + c_1 q - c_1 e^{2x}\right. \\ & \left. - \sqrt{c_1 (e^{2u(\eta)} - q)(c_1 e^{2u(\eta)} - c_1 q - 4)}\right] \\ & - \sqrt{\frac{4q}{4 + c_1 q}}u(\eta) \pm 2\eta + c_2. \end{aligned} \quad (47)$$

It is worth mentioning that this equation is valid only for $c_1 \neq -4/q$ and $c_1 \neq 0$.

By choosing $c_1 = 0$ or $c_1 = -4/q$ we can generate from direct integration of (46) two other sets of implicit solutions, namely,

$$\begin{aligned} & \sqrt{q - e^{2u(\eta)}} - \sqrt{q}\,\mathrm{arctanh}\left(\sqrt{\frac{q - e^{2u(\eta)}}{q}}\right) \pm 2\eta + c_2 \\ & = 0, \end{aligned} \quad (48)$$

and

$$\begin{aligned} & \sqrt{qe^{-2u(\eta)} - 1} + \arctan\left[\frac{1}{\sqrt{qe^{-2u(\eta)} - 1}}\right] \pm \frac{2}{\sqrt{q}}\eta + c_2 \\ & = 0. \end{aligned} \quad (49)$$

3.3. Special Case (3): ($p = 2$, $\gamma = 1$, and $\delta_1 = -q/2$). Another interesting case to study is for ($p = 2$, $\gamma = 1$, and $\delta_1 = -q/2$), the traveling wave equation is then

$$\begin{aligned} \frac{d^2u(\eta)}{d\eta^2} & = \left[\sinh_q(u(\eta))\right]^2 + \frac{q}{2} \\ & = \frac{1}{2}\left[\cosh_{q^2}(2u(\eta))\right]. \end{aligned} \quad (50)$$

Let us first mention that $u = cst$ is not a solution for (50). By using the transformation

$$v(\eta) = e^{2u(\eta)}, \quad (51)$$

we get

$$2v(\eta)v''(\eta) - 2v'^2 - v(\eta)^3 - q^2 v(\eta) = 0. \quad (52)$$

In order to solve this nonlinear second-order equation we define

$$w(v) = \left[\frac{dv(\eta)}{d\eta}\right]^2, \quad (53)$$

so

$$\frac{d^2v(\eta)}{d\eta^2} = \frac{1}{2}\frac{dw}{dv}(v), \quad (54)$$

and (52) will be transformed to

$$v\frac{dw}{dv}(v) - 2w(v) = v^3 + q^2 v. \quad (55)$$

This is a first-order inhomogeneous linear differential equation with the general solution

$$w(v) = v^3 + c_1' v^2 - q^2 v. \quad (56)$$

Here c_1' is an arbitrary constant. From definition (53) and (56) the expression of the soliton in the implicit form is then

$$\int^{v(\eta)} \frac{dv}{\sqrt{v^3 + c_1 v^2 - q^2 v}} = \pm\eta + C, \quad (57)$$

where C and c_1 are free parameters. We can write

$$\begin{aligned} & \int^{v(\eta)} \frac{dv}{\sqrt{v^3 + c_1 v^2 - q^2 v}} \\ & = -\int_{v(\eta)+c_1/3} \frac{2dv_1}{\sqrt{4v_1^3 - g_4 v_1 - g_5}}, \end{aligned} \quad (58)$$

where

$$g_4 = 4\left(\frac{c_1^2 + 3q^2}{3}\right), \quad (59)$$

$$g_5 = -4\left(\frac{q^2 c_1}{3} + \frac{2c_1^3}{27}\right), \quad (60)$$

The expression of the solution for (52) is then

$$v(\eta) = \wp\left(\pm\frac{1}{2}\eta + c_2; g_4; g_5\right) - \frac{c_1}{3}, \quad (61)$$

and consequently the general expression of the soliton in the moving frame $\eta = (1/\sqrt{1 - c^2})(x - ct)$ is

$$u(\eta) = \ln\left[\sqrt{\wp\left(\pm\frac{1}{2}\eta + c_2; g_4; g_5\right) - \frac{c_1}{3}}\right], \quad (62)$$

where c_1 and c_2 are free parameters.

4. Conclusion

We have introduced a generalized q-deformed Sinh-Gordon equation defined by (1) ($\partial^2\theta/\partial z\partial\xi = \alpha[\sinh_q(\beta\theta^\gamma)]^p - \delta$). We have derived general explicit analytical soliton solutions for two sets of parameters, namely, ($p = 1$, $\gamma = 1$, and $\delta = 0$) and ($p = 2$, $\gamma = 1$, and $\delta = -(q/2)|\alpha|$). Furthermore, we have given general implicit soliton solutions for the case: $p = -2$, $\gamma = 1$, and $\delta = 0$. Future investigations will be focusing on generating more soliton solutions using algebraic methods such as Tanh-method [79], rational expansion method [80], F–expansion method [81], and auxiliary method [82]. The general study of the traveling wave solutions for the generalized q-deformed Sinh-Gordon equation or its asymptotic behavior as well as the analysis of some other proprieties will be for sure very interesting. The usefulness of the generalized q-deformed Sinh-Gordon equation will not be limited to the fields of applied mathematics and mathematical physics but also will open the door for applications in physics where the symmetry of the studied system is absent or violated.

Conflicts of Interest

The author declares that they have no conflicts of interest.

References

[1] S. H. Strogatz, *Nonlinear Dynamics and Chaos: With Applications to Physics, Biology, Chemistry, and Engineering*, Westview Press, Boulder, Colorado, 2015.

[2] J. Wang, "Landscape and flux theory of non-equilibrium dynamical systems with application to biology," *Advances in Physics*, vol. 64, no. 1, pp. 1–137, 2015.

[3] V. Jaksic, D. P. Mandic, K. Ryan, B. Basu, V. Pakrashi, and R. Soc, "A comprehensive study of the delay vector variance method for quantification of nonlinearity in dynamical systems," *Royal Society Open Science*, vol. 3, Article ID 150493, 2016.

[4] M. Savescu, K. R. Khan, R. W. Kohl, L. Moraru, A. Yildirim, and A. Biswas, "Optical soliton perturbation with improved nonlinear Schrödinger's equation in nano fibers," *Journal of Nanoelectronics and Optoelectronics*, vol. 8, pp. 208–220, 2013.

[5] R. Sassaman and A. Biswas, "Topological and non-topological solitons of the generalized Klein-Gordon equations," *Applied Mathematics and Computation*, vol. 215, no. 1, pp. 212–220, 2009.

[6] A. L. Fabian, R. Kohl, and A. Biswas, "Perturbation of topological solitons due to sine-Gordon equation and its type," *Communications in Nonlinear Science and Numerical Simulation*, vol. 14, no. 4, pp. 1227–1244, 2009.

[7] B. Geranmehr and S. R. Nekoo, "Nonlinear suboptimal control of fully coupled non-affine six-DOF autonomous underwater vehicle using the state-dependent Riccati equation," *Ocean Engineering*, vol. 96, pp. 248–257, 2015.

[8] X.-Y. Gao, "Looking at a nonlinear inhomogeneous optical fiber through the generalized higher-order variable-coefficient Hirota equation," *Applied Mathematics Letters*, vol. 73, pp. 143–149, 2017.

[9] R. Duan, "Global solutions for a one-dimensional compressible micropolar fluid model with zero heat conductivity," *Journal of Mathematical Analysis and Applications*, vol. 463, no. 2, pp. 477–495, 2018.

[10] Z. Du, B. Tian, H.-P. Chai, Y. Sun, and X.-H. Zhao, "Rogue waves for the coupled variable-coefficient fourth-order nonlinear Schrödinger equations in an inhomogeneous optical fiber," *Chaos, Solitons & Fractals*, vol. 10, pp. 90–98, 2018.

[11] L. Liu, B. Tian, Y.-Q. Yuan, and Z. Du, "Dark-bright solitons and semirational rogue waves for the coupled Sasa-Satsuma equations," *Physical Review E, Covering Statistical, Nonlinear, Biological, and Soft Matter Physics*, vol. 97, Article ID 052217, 2018.

[12] X.-Y. Wu, B. Tian, L. Liu, and Y. Sun, "Rogue waves for a variable-coefficient Kadomtsev–Petviashvili equation in fluid mechanics," *Computers & Mathematics with Applications*, vol. 76, no. 2, pp. 215–223, 2018.

[13] Y. A. Kuznetsov, *Elements of Applied Bifurcation Theory*, vol. 112 of *Applied Mathematics Science*, Springer, New York, NY, USA, 2nd edition, 1998.

[14] B. Zhilinski, "Quantum bifurcations," in *Encyclopedia of Complexity and Systems Science*, p. 7135, Springer-Verlag, Berlin, Germany, 2009.

[15] N. N. Rosanov, *Spatial Hysteresis and Optical Patterns*, Springer Series in Synergetics, Springer-Verlag, Berlin, Germany, 2002.

[16] A. Baas, J. P. Karr, H. Eleuch, and E. Giacobino, "Optical bistability in semiconductor microcavities," *Physical Review A: Atomic, Molecular and Optical Physics*, vol. 69, no. 2, Article ID 023809, 2004.

[17] E. A. Sete, H. Eleuch, and S. Das, "Controllable nonlinear effects in an optomechanical resonator containing a quantum well," *Physical Review A: Atomic, Molecular and Optical Physics*, vol. 85, no. 4, Article ID 043824, 2012.

[18] E. A. Sete and H. Eleuch, "Anomalous optical bistability and robust entanglement of mechanical oscillators using two-photon coherence," *Journal of the Optical Society of America B*, vol. 32, no. 5, pp. 971–982, 2015.

[19] J.-P. Eckmann and D. Ruelle, "Ergodic theory of chaos and strange attractors," *Reviews of Modern Physics*, vol. 57, no. 3, pp. 617–656, 1985.

[20] M. Sieber and F. Steiner, "Quantum chaos in the hyperbola billiard," *Physics Letters A*, vol. 148, no. 8-9, pp. 415–420, 1990.

[21] H. Eleuch and A. Prasad, "Chaos and regularity in semiconductor microcavities," *Physics Letters A*, vol. 376, no. 26-27, pp. 1970–1977, 2012.

[22] H. Eleuch, A. Prasad, and I. Rotter, "Enhancement of photon intensity in forced coupled quantum wells inside a semiconductor microcavity," *Physical Review E: Statistical, Nonlinear, and Soft Matter Physics*, vol. 87, no. 2, Article ID 022916, 2013.

[23] V. G. Ivancevic and T. T. Ivancevic, *Complex Nonlinearity: Chaos, Phase Transitions, Topology Change and Path Integrals*, Springer, New York, NY, USA, 2008.

[24] H. K. Janssen, "On the nonequilibrium phase transition in reaction-diffusion systems with an absorbing stationary state," *Zeitschrift für Physik B Condensed Matter*, vol. 42, no. 2, pp. 151–154, 1981.

[25] D. G. Joshi and M. Vojta, "Nonlinear bond-operator theory and 1/d expansion for coupled-dimer magnets. II. Antiferromagnetic phase and quantum phase transition," *Physical Review B: Condensed Matter and Materials Physics*, vol. 91, no. 9, Article ID 094405, 2015.

[26] S. Shrivastava, K. H. Kang, and M. F. Schneider, "Solitary shock waves and adiabatic phase transition in lipid interfaces and nerves," *Physical Review E: Statistical, Nonlinear, and Soft Matter Physics*, vol. 91, no. 1, Article ID 012715, 2015.

[27] H. Haken, *Laser Theory*, Springer-Verlag, New York, NY, USA, 1983.

[28] R. Buschlinger, M. Lorke, and U. Peschel, "Light-matter interaction and lasing in semiconductor nanowires: A combined finite-difference time-domain and semiconductor Bloch equation approach," *Physical Review B: Condensed Matter and Materials Physics*, vol. 91, no. 4, Article ID 045203, 2015.

[29] R. Buschlinger, M. Lorke, and U. Peschel, "Erratum: Light-matter interaction and lasing in semiconductor nanowires: A combined finite-difference time-domain and semiconductor Bloch equation approach [Phys. Rev. B 91, 045203 (2015)]," *Physical Review B: Condensed Matter and Materials Physics*, vol. 91, no. 15, Article ID 159903, 2015.

[30] M. Eghbalpour, R. Karimi, S. Batebi, and H. R. Soleimani, "Effect of electron-spin relaxation on optical bistability and lasing without population inversion in a three-level v type quantum system," *Optik - International Journal for Light and Electron Optics*, vol. 127, no. 5, pp. 2525–2530, 2016.

[31] E. A. Sete, A. A. Svidzinsky, Y. V. Rostovtsev et al., "Using quantum coherence to generate gain in the XUV and X-ray: gain-swept superradiance and lasing without inversion," *IEEE Journal of Selected Topics in Quantum Electronics*, vol. 18, no. 1, pp. 541–553, 2012.

[32] M. Gross and S. Haroche, "Superradiance: An essay on the theory of collective spontaneous emission," *Physics Reports*, vol. 93, no. 5, pp. 301–396, 1982.

[33] E. A. Sete, A. A. Svidzinsky, H. Eleuch, Z. Yang, R. D. Nevels, and M. O. Scully, "Correlated spontaneous emission on the Danube," *Journal of Modern Optics*, vol. 57, no. 14-15, pp. 1311–1330, 2010.

[34] C. Svetina, N. Mahne, L. Raimondi et al., "Photon transport of the superradiant TeraFERMI THz beamline at the FERMI free-electron laser," *Journal of Synchrotron Radiation*, vol. 23, pp. 106–110, 2016.

[35] P. L. Lai and C. Fyfe, "Kernel and nonlinear canonical correlation analysis," *International Journal of Neural Systems*, vol. 10, 377, no. 5, p. 365, 2000.

[36] K. Berrada, M. El Baz, H. Eleuch, and Y. Hassouni, "A comparative study of negativity and concurrence based on spin coherent states," *International Journal of Modern Physics C*, vol. 21, no. 3, pp. 291–305, 2010.

[37] H. Eleuch, S. Guérin, and H. R. Jauslin, "Effects of an environment on a cavity-quantum-electrodynamics system controlled by bichromatic adiabatic passage," *Physical Review A: Atomic, Molecular and Optical Physics*, vol. 85, no. 1, Article ID 013830, 2012.

[38] K. Kasai, G. Jiangrui, and C. Fabre, "Observation of squeezing using cascaded nonlinearity," *EPL (Europhysics Letters)*, vol. 40, no. 1, article 25, 1997.

[39] P. D. Drummond, *Quantum Squeezing*, Springer Series on Atomic, Optical, and Plasma Physics, Springer, New York, NY, USA, 2010.

[40] Y. A. Korkmaz and C. Bulutay, "Nuclear spin squeezing via electric quadrupole interaction," *Physical Review A: Atomic, Molecular and Optical Physics*, vol. 93, no. 1, Article ID 013812, 2016.

[41] E. Giacobino, J.-P. Karr, G. Messin, H. Eleuch, and A. Baas, "Quantum optical effects in semiconductor microcavities," *Comptes Rendus Physique*, vol. 3, no. 1, pp. 41–52, 2002.

[42] X.-H. Zhao, B. Tian, J. Chai, X.-Y. Wu, and Y.-J. Guo, "Multi-soliton interaction of a generalized Schrödinger-Boussinesq system in a magnetized plasma," *The European Physical Journal Plus*, vol. 132, article 192, 2017.

[43] E. Infeld and G. Rowlands, *Nonlinear Waves, Solitons and Chaos*, Cambridge University Press, Cambridge, UK, 2000.

[44] H. Eleuch and R. Bennaceur, "An optical soliton pair among absorbing three-level atoms," *Journal of Optics A: Pure and Applied Optics*, vol. 5, no. 5, pp. 528–533, 2003.

[45] D. S. Kharenko, A. E. Bednyakova, E. V. Podivilov, M. P. Fedoruk, A. Apolonski, and S. A. Babin, "Cascaded generation of coherent Raman dissipative solitons," *Optics Expresss*, vol. 41, no. 1, pp. 175–178, 2016.

[46] A. Biswas, D. Milovic, and A. Ranasinghe, "Solitary waves of Boussinesq equation in a power law media," *Communications in Nonlinear Science and Numerical Simulation*, vol. 14, no. 11, pp. 3738–3742, 2009.

[47] G. Ebadi, A. H. Kara, M. D. Petkovic, A. Yildirim, and A. Biswas, "Solitons and conserved quantities of the Ito equation," *Proceedings of the Romanian Academy - Series A: Mathematics, Physics, Technical Sciences, Information Science*, vol. 13, no. 3, pp. 215–224, 2012.

[48] R. Sassaman, A. Heidari, and A. Biswas, "Topological and non-topological solitons of nonlinear Klein-Gordon equations by He's semi-inverse variational principle," *Journal of The Franklin Institute*, vol. 347, no. 7, pp. 1148–1157, 2010.

[49] Y. S. Kivshara and B. Luther-Davies, "Dark optical solitons: physics and applications," *Physics Reports*, vol. 298, no. 2-3, pp. 81–197, 1998.

[50] M. J. Ablowitz, G. Biondini, and L. A. Ostrovsky, "Optical solitons: perspectives and applications," *Chaos: An Interdisciplinary Journal of Nonlinear Science*, vol. 10, no. 3, pp. 471–474, 2000.

[51] N. Boutabba, H. Eleuch, and H. Bouchriha, "Thermal bath effect on soliton propagation in three-level atomic system," *Synthetic Metals*, vol. 159, no. 13, pp. 1239–1243, 2009.

[52] R. Hao, "Optical soliton control in inhomogeneous nonlinear media with the parity-time symmetric potentials," *Optics Communications*, vol. 338, pp. 265–268, 2015.

[53] A. Biswas, "Dispersion-managed solitons in optical fibres," *Journal of Optics A: Pure and Applied Optics*, vol. 4, no. 1, article 84, 2002.

[54] M. Savescu, A. H. Bhrawy, A. A. Alshaery et al., "Optical solitons in nonlinear directional couplers with spatio-temporal dispersion," *Journal of Modern Optics*, vol. 61, no. 5, pp. 441–458, 2014.

[55] J. V. Guzman, E. M. Hilal, A. A. Alshaery et al., "Thirring optical solitons with spatio-temporal dispersion," *Proceedings of The Romanian Academy, Series A*, vol. 16, no. 1, pp. 41–46, 2015.

[56] H. Triki, A. H. Kara, A. Bhrawy, and A. Biswas, "Soliton Solution and Conservation Law of Gear-Grimshaw Model for Shallow Water Waves," *Acta Physica Polonica A*, vol. 125, no. 5, pp. 1099–1106, 2014.

[57] A. H. Bhrawy, M. A. Abdelkawy, and A. Biswas, "Cnoidal and snoidal wave solutions to coupled nonlinear wave equations by the extended Jacobi's elliptic function method," *Communications in Nonlinear Science and Numerical Simulation*, vol. 18, no. 4, pp. 915–925, 2013.

[58] A. H. Bhrawy, M. A. Abdelkawy, S. Kumar, and A. Biswas, "Solitons and other solutions to Kadomtsev-Petviashvili equation of B-type," *Romanian Journal of Physics*, vol. 58, no. 7-8, pp. 729–748, 2013.

[59] G. Ebadi, N. Y. Fard, A. H. Bhrawy et al., "Solitons and other solutions to the (3+1)-dimensional extended Kadomtsev-Petviashviliequation with power law nonlinearity," *Romanian Reports in Physics*, vol. 65, no. 1, pp. 27–62, 2013.

[60] A. Biswas, A. H. Bhrawy, M. A. Abdelkawy, A. A. Alshaery, and E. M. Hilal, "Symbolic computation of some nonlinear fractional differential equations," *Romanian Reports in Physics*, vol. 59, no. 5-6, pp. 433–442, 2014.

[61] A. Sardar, K. Ali, S. Tahir et al., "Dispersive optical solitons in nanofibers with Schrödinger-Hirota equation," *Journal of Nanoelectronics and Optoelectronics*, vol. 11, no. 3, pp. 382–387, 2016.

[62] L. Hauswirth, R. Sa Earp, and E. Toubiana, "Associate and conjugate minimal immersions in $M \times R$," *The Tohoku Mathematical Journal*, vol. 60, no. 2, pp. 267–286, 2008.

[63] M. J. Ablowitz and P. A. Clarkson, *Solitons, Nonlinear Evolution Equations and Inverse Scattering*, Cambridge University Press, Cambridge, UK, 1991.

[64] A. L. Larsen and N. Sánchez, "sinh-GORdon, cosh-GORdon, and Liouville equations for strings and multistrings in constant curvature spacetimes," *Physical Review D: Particles, Fields, Gravitation and Cosmology*, vol. 54, no. 4, pp. 2801–2807, 1996.

[65] A. Arai, "Exactly solvable supersymmetric quantum mechanics," *Journal of Mathematical Analysis and Applications*, vol. 158, no. 1, pp. 63–79, 1991.

[66] A. Arai, "Exact solutions of multi-component nonlinear Schrödinger and Klein-Gordon equations in two-dimensional space-time," *Journal of Physics A: Mathematical and General*, vol. 34, no. 20, pp. 4281–4288, 2001.

[67] B. J. Falaye, K. J. Oyewumi, and M. Abbas, "Exact solution of Schrödinger equation with q-deformed quantum potentials using Nikiforov—Uvarov method," *Chinese Physics B*, vol. 22, no. 11, Article ID 110301, 2013.

[68] H. Yılmaz, D. Demirhan, and F. Büyükkılıç, "Solution of the Schrödinger equation for two q-deformed potentials by the SWKB method," *Journal of Mathematical Chemistry*, vol. 47, article 539, 2010.

[69] M. Sebawe Abdalla, H. Eleuch, and T. Barakat, "Exact analytical solutions of the wave function for some q-deformed potentials," *Reports on Mathematical Physics*, vol. 71, no. 2, pp. 217–229, 2013.

[70] M. Sebawe Abdalla and H. Eleuch, "Exact analytic solutions of the Schrödinger equations for some modified q-deformed potentials," *Journal of Applied Physics*, vol. 115, no. 23, Article ID 234906, 2014.

[71] M. Eshghi and H. Mehraban, "Solution of the Dirac equation with position-dependent mass for q-parameter modified pöschl-teller and coulomb-like tensor potential," *Few-Body Systems*, vol. 52, no. 1-2, pp. 41–47, 2012.

[72] A. Kurniawan, A. Suparmi, and C. Cari, "Approximate analytical solution of the Dirac equation with q-deformed hyperbolic Pöschl–Teller potential and trigonometric Scarf II non-central potential," *Chinese Physics B*, vol. 24, no. 3, Article ID 030302, 2015.

[73] Q. J. Zeng, Z. Cheng, and J. H. Yuan, "Bose–Einstein condensation of a q-deformed boson system in a harmonic potential trap," *Physica A: Statistical Mechanics and its Applications*, vol. 3, pp. 563–571, 391.

[74] Y. Shu, J. Chen, and L. Chen, "Bose-Einstein condensation of a q-deformed ideal Bose gas," *Physics Letters A*, vol. 292, no. 6, pp. 309–314, 2002.

[75] S. M. Ikhdair, "Rotation and vibration of diatomic molecule in the spatially-dependent mass Schrödinger equation with generalized q-deformed Morse potential," *Chemical Physics*, vol. 361, no. 1-2, pp. 9–17, 2009.

[76] D. Bonatsos, E. N. Argyres, and P. P. Raychev, "$SU_q(1,1)$ description of vibrational molecular spectra," *Journal of Physics A: Mathematical and General*, vol. 24, no. 8, pp. L403–L408, 1991.

[77] M. Dehghan and A. Ghesmati, "Application of the dual reciprocity boundary integral equation technique to solve the nonlinear Klein-Gordon equation," *Computer Physics Communications*, vol. 181, no. 8, pp. 1410–1418, 2010.

[78] S. Jiménez and L. Vázquez, "Analysis of four numerical schemes for a nonlinear Klein-Gordon equation," *Applied Mathematics and Computation*, vol. 35, no. 1, pp. 61–94, 1990.

[79] A.-M. Wazwaz, "The tanh method: exact solutions of the sine-GORdon and the sinh-GORdon equations," *Applied Mathematics and Computation*, vol. 167, no. 2, pp. 1196–1210, 2005.

[80] Q. Wang, Y. Chen, and H. Zhang, "A new Riccati equation rational expansion method and its application to (2+1)-dimensional Burgers equation," *Chaos, Solitons & Fractals*, vol. 25, no. 5, pp. 1019–1028, 2005.

[81] J. L. Zhang, M. L. Wang, Y. M. Wang, and Z. D. Fang, "The improved F-expansion method and its applications," *Physics Letters A*, vol. 350, no. 1-2, pp. 103–109, 2006.

[82] A. Sirendaoreji, "Auxiliary equation method and new solutions of Klein-Gordon equations," *Chaos, Solitons & Fractals*, vol. 31, no. 4, pp. 943–950, 2007.

17

On the Second-Order Quantum (p, q)-Difference Equations with Separated Boundary Conditions

Chanon Promsakon,[1] Nattapong Kamsrisuk,[1] Sotiris K. Ntouyas,[2,3] and Jessada Tariboon[1]

[1] *Intelligent and Nonlinear Dynamic Innovations Research Center, Department of Mathematics, Faculty of Applied Science, King Mongkut's University of Technology North Bangkok, Bangkok 10800, Thailand*
[2] *Department of Mathematics, University of Ioannina, 451 10 Ioannina, Greece*
[3] *Nonlinear Analysis and Applied Mathematics (NAAM)-Research Group, Department of Mathematics, Faculty of Science, King Abdulaziz University, P.O. Box 80203, Jeddah 21589, Saudi Arabia*

Correspondence should be addressed to Jessada Tariboon; jessada.t@sci.kmutnb.ac.th

Academic Editor: Antonio Scarfone

In this paper, we investigate the existence and uniqueness of solutions for a boundary value problem for second-order quantum (p, q)-difference equations with separated boundary conditions, by using classical fixed point theorems. Examples illustrating the main results are also presented.

1. Introduction

Quantum calculus or q-calculus is known as the study of calculus without limits. The study of q-calculus was initiated by Euler on studying infinite series. In 1910, Jackson [1, 2] was the first one, who establised the q-derivative or q-difference operator for a function f on $[0, \infty)$ by

$$D_q f(t) = \frac{f(t) - f(qt)}{(1-q)t}, \quad t \neq 0, \qquad (1)$$

$$\text{and } D_q f(0) = \lim_{t \to 0} D_q f(t), \quad t = 0,$$

and gave its properties. The q-integral of a function f on $[0, \infty)$ is defined by

$$\int_0^t f(s)\, d_q s = (1-q)t \sum_{n=0}^{\infty} q^n f(q^n t), \qquad (2)$$

provided that infinite series converges. The general theory of linear quantum difference equations was published in 1912 by Carmichael [3]. Details of its basic notions, results, and methods can be found in the text [4]. For other papers on the subject, see [5–7]. In recent years, the topic has been attracting the attention of several researchers and a variety of new results can be found in [8–13] and the references therein. In addition, the classical quantum calculus was generalized to q_k-calculus by Tariboon and Ntouyas [14]. For details, we refer to the recent monograph [15].

There are some applications of q-calculus and difference equations to molecular problems in physics. In 1967, Finkelstein [16] studied behaviors of hydrogen atoms by using Schrödinger equation and q-calculus. In [17], the author investigated the q-field theory. The q-Coulomb problem and q-hydrogen atom were studied by [18–21]. In addition, Yang-Mills theories and also q-Yang-Mills equation were developed by [22–24]. The theory of quantum group applied to vibration and rotation molecules with q-algebra and q-Heisemberg algebra technique was established in [25–27]. The subject of elementary particle physics and chemical physics using q-calculus was investigated in [28–31]. The string theory involving q-calculus was studied in [32].

The subject of (p, q)-calculus first appeared in quantum algebras which contained two quantum numbers p and

q introduced by [33]. For some recent results see [34–41] and the references cited therein. Recently in [42], the authors initiated the study on boundary value problems for (p,q)-difference equations by investigating the existence and uniqueness of solution for first-order quantum (p,q)-difference equation subject to a nonlocal condition of the form

$$D_{p,q}x(t) = f(t, x(pt)), \quad t \in \left[0, \frac{T}{p}\right], \tag{3}$$

$$x(0) = \alpha x(T) + \sum_{i=1}^{m} \beta_i \int_0^{\eta_i} x(s)\, d_{p_i,q_i}s, \tag{4}$$

where $0 < q < p \le 1$, $0 < q_i < p_i \le 1$, $i = 1, 2, \ldots, m$, are quantum numbers, $D_{p,q}$ is (p,q)-difference operator, $f \in C([0, T/p] \times \mathbb{R}, \mathbb{R})$, $T > 0$, α, β_i, $i = 1, 2, \ldots, m$, are given constants, and the points $\eta_i \in [0, p_i T]$, $i = 1, 2, \ldots, m$. They proved existence and uniqueness results for problem (3)-(4) by using the classical fixed point theorems, such as Banach's fixed point theorem, Boyd and Wong fixed point theorem for nonlinear contractions, and Leray-Schauder nonlinear alternative.

In this paper, we continue the study on (p,q) boundary value problems, by considering the following second-order quantum (p,q)-difference equation with separated boundary conditions

$$\begin{gathered} D_{p,q}^2 x(t) = f(t, x(p^2 t)), \quad t \in \left[0, \frac{T}{p^2}\right], \\ \alpha_1 x(0) + \alpha_2 D_{p,q}x(0) = \alpha_3, \\ \beta_1 x(T) + \beta_2 D_{p,q}x\left(\frac{T}{p}\right) = \beta_3, \end{gathered} \tag{5}$$

where $0 < q < p \le 1$ are two given quantum numbers, $D_{p,q}^2$ is the second-order (p,q)-difference operator, and $f \in C([0, T/p^2] \times \mathbb{R}, \mathbb{R})$, $T > 0$, α_i, β_i, $i = 1, 2, 3$, are given real constants. Some new existence results are established by using Schaefer fixed point theorem, Krasnoselskii's fixed point theorem, and Lelay-Schauder's nonlinear alternative. In addition the uniqueness of solutions is established via Banach contraction mapping principle.

The paper is organized as follows: In Section 2, we recall some definitions and basic facts from (p,q)-calculus. The main existence and uniqueness results are given in Section 3. Examples illustrating the obtained results are presented in Section 4.

2. Preliminaries

In this section, we recall some basic concepts of (p,q)-calculus (see [41]). The (p,q)-number is defined by

$$[n]_{p,q} = \frac{p^n - q^n}{p - q}, \tag{6}$$

where $0 < q < p \le 1$. For each $k, n \in \mathbb{N}$, $n \ge k \ge 0$, the (p,q)-factorial and (p,q)-binomial are defined by

$$[n]_{p,q}! = \prod_{k=1}^{n} [k]_{p,q}, \quad n \ge 1, \ [0]_{p,q}! = 1, \tag{7}$$

and

$$\begin{bmatrix} n \\ k \end{bmatrix}_{p,q} = \frac{[n]_{p,q}!}{[n-k]_{p,q}!\,[k]_{p,q}!}, \tag{8}$$

respectively.

Let $f : [0, T] \longrightarrow \mathbb{R}$. The (p,q)-derivative of function f is defined as

$$D_{p,q}f(t) = \frac{f(pt) - f(qt)}{(p-q)t}, \quad t \ne 0, \tag{9}$$

and $D_{p,q}f(0) = \lim_{t \to 0} D_{p,q}f(t)$. Observe that the function $D_{p,q}f(t)$ is defined on $[0, T/p]$ which is extended from $[0, T]$ of a function $f(t)$. In addition, we say that f is (p,q)-differentiable on $[0, T/p]$ provided that $D_{p,q}f(t)$ exists for all $t \in [0, T/p]$.

Let a function $f : [0, T] \longrightarrow \mathbb{R}$. Then the (p,q)-integral of f is defined by

$$\int_0^t f(s)\, d_{p,q}s = (p-q)t \sum_{n=0}^{\infty} \frac{q^n}{p^{n+1}} f\left(\frac{q^n}{p^{n+1}} t\right), \tag{10}$$

provided that the right hand side converges. Note that the domain of function $\int_0^t f(s) d_{p,q}s$ is $[0, pT]$ which shrinks from $[0, T]$ of a function $f(t)$ since $0 < p \le 1$.

In the following theorems we collect the basic properties of (p,q)-differentiation and (p,q)-integration, respectively. See [37].

Theorem 1 (see [37]). *Suppose that $f, g : [0, T] \longrightarrow \mathbb{R}$ are (p,q)-differentiable on $[0, T/p]$. Then*

(a) $f + g : [0, T] \longrightarrow \mathbb{R}$ is (p,q)-differentiable on $[0, T/p]$, and

$$D_{p,q}(f(t) + g(t)) = D_{p,q}f(t) + D_{p,q}g(t). \tag{11}$$

(b) $\lambda f : [0, T] \longrightarrow \mathbb{R}$ is (p,q)-differentiable on $[0, T/p]$ for any constant λ, and

$$D_{p,q}(\lambda f)(t) = \lambda D_{p,q}f(t). \tag{12}$$

(c) $fg : [0, T] \longrightarrow \mathbb{R}$ is (p,q)-differentiable on $[0, T/p]$, and

$$\begin{gathered} D_{p,q}(fg)(t) = f(pt) D_{p,q}g(t) + g(qt) D_{p,q}f(t). \\ D_{p,q}(fg)(t) = g(pt) D_{p,q}f(t) + f(qt) D_{p,q}g(t). \end{gathered} \tag{13}$$

(d) If $g(t) \ne 0$, then f/g is (p,q)-differentiable on $[0, T/p]$ with

$$\begin{gathered} D_{p,q}\left(\frac{f}{g}\right)(t) = \frac{g(qt) D_{p,q}f(t) - f(qt) D_{p,q}g(t)}{g(pt)\, g(qt)}. \\ D_{p,q}\left(\frac{f}{g}\right)(t) = \frac{g(pt) D_{p,q}f(t) - f(pt) D_{p,q}g(t)}{g(pt)\, g(qt)}. \end{gathered} \tag{14}$$

Theorem 2 (see [37]). *Let $f, g : [0,T] \longrightarrow \mathbb{R}$ be continuous functions and constants $0 < q < p \leq 1$, $a, b \in [0,T]$. The following formulas hold:*

(a) The (p,q)-integration by parts is given by

$$\int_a^b f(pt)\, D_{p,q} g(t)\, d_{p,q}t = f(t)\, g(t)\Big|_a^b - \int_a^b g(qt)\, D_{p,q} f(t)\, d_{p,q}t. \quad (15)$$

(b) $D_{p,q}\int_0^t f(s)d_{p,q}s = f(t)$.

(c) $\int_0^t D_{p,q}f(s)d_{p,q}s = f(t) - f(0)$.

(d) $\int_a^t D_{p,q}f(s)d_{p,q}s = f(t) - f(a)$ where $a \in (0,t)$.

Theorem 3 (see [42]). *Let a function $f : [0,T] \longrightarrow \mathbb{R}$ and constants $0 < q < p \leq 1$. Then for $t \in [0, p^2T]$, we have*

$$\int_0^t \int_0^s f(r)\, d_{p,q}r\, d_{p,q}s = \frac{1}{p}\int_0^t (t - qs)\, f\left(\frac{1}{p}s\right) d_{p,q}s. \quad (16)$$

The second-order (p,q)-difference of a function $f : [0,T] \longrightarrow \mathbb{R}$ can be expressed as

$$\begin{aligned} D_{p,q}^2 f(t) &= D_{p,q}\left(D_{p,q}f\right)(t) \\ &= D_{p,q}\left(\frac{f(pt) - f(qt)}{(p-q)t}\right) \\ &= \frac{f(p^2t) - f(pqt)}{p(p-q)^2 t^2} - \frac{f(pqt) - f(q^2t)}{q(p-q)^2 t^2} \\ &= \frac{qf(p^2t) - qf(pqt) - pf(pqt) + pf(q^2t)}{pq(p-q)^2 t^2} \\ &= \frac{qf(p^2t) - (p+q)f(pqt) + pf(q^2t)}{pq(p-q)^2 t^2}, \\ &\qquad t \in \left[0, \frac{T}{p^2}\right]. \end{aligned} \quad (17)$$

Now, for $h : [0, T/p^2] \longrightarrow \mathbb{R}$, we transform the linear second-order (p,q)-difference equation with separated boundary condition

$$\begin{aligned} &D_{p,q}^2 x(t) = h(t), \quad t \in \left[0, \frac{T}{p^2}\right], \\ &\alpha_1 x(0) + \alpha_2 D_{p,q}x(0) = \alpha_3, \\ &\beta_1 x(T) + \beta_2 D_{p,q}x\left(\frac{T}{p}\right) = \beta_3, \end{aligned} \quad (18)$$

into an integral equation which will be used to define the solution of the boundary value problem (5). For convenience we put constants $\Lambda = \alpha_1(\beta_1 T + \beta_2) - \alpha_2\beta_1 \neq 0$,

$$\begin{aligned} \Lambda_1 &= \frac{\alpha_3(\beta_1 T + \beta_2) - \alpha_2\beta_3}{\Lambda}, \\ \Lambda_2 &= \frac{\alpha_1\beta_3 - \alpha_3\beta_1}{\Lambda}. \end{aligned} \quad (19)$$

Lemma 4. *The unique solution of linear problem (18) is given by*

$$\begin{aligned} x(t) &= \Lambda_1 + \Lambda_2 t \\ &+ \frac{(\alpha_2 - \alpha_1 t)\beta_1}{p\Lambda}\int_0^T (T - qs)\, h\left(\frac{s}{p}\right) d_{p,q}s \\ &+ \frac{(\alpha_2 - \alpha_1 t)\beta_2}{\Lambda}\int_0^{T/p} h(s)\, d_{p,q}s \\ &+ \frac{1}{p}\int_0^t (t - qs)\, h\left(\frac{s}{p}\right) d_{p,q}s. \end{aligned} \quad (20)$$

Proof. Applying (p,q)-integration to the first equation of (18), we have

$$D_{p,q}x(t) - D_{p,q}x(0) = \int_0^t h(s)\, d_{p,q}s, \quad t \in \left[0, \frac{T}{p}\right]. \quad (21)$$

Taking (p,q)-integration to (21) and using Theorem 3, we have

$$\begin{aligned} x(t) &= x(0) + D_{p,q}x(0)\, t \\ &+ \frac{1}{p}\int_0^t (t - qs)\, h\left(\frac{s}{p}\right) d_{p,q}s, \quad t \in [0,T]. \end{aligned} \quad (22)$$

For convenience, we set constants $x(0) = A$, $D_{p,q}x(0) = B$ and then

$$x(t) = A + Bt + \frac{1}{p}\int_0^t (t - qs)\, h\left(\frac{s}{p}\right) d_{p,q}s. \quad (23)$$

In particular for $t = T$ in (23), we have

$$x(T) = A + BT + \frac{1}{p}\int_0^T (T - qs)\, h\left(\frac{s}{p}\right) d_{p,q}s. \quad (24)$$

Using (21) and by replacing $t = T/p$, we obtain

$$D_{p,q}x\left(\frac{T}{p}\right) = B + \int_0^{T/p} h(s)\, d_{p,q}s. \quad (25)$$

From the first boundary condition of (18) and the above setting constants, we get

$$\alpha_1 A + \alpha_2 B = \alpha_3. \quad (26)$$

The second boundary condition of (18) with (24) and (25) yields

$$\begin{aligned} &\beta_1\left(A + BT + \frac{1}{p}\int_0^T (T - qs)\, h\left(\frac{s}{p}\right) d_{p,q}s\right) \\ &+ \beta_2\left(B + \int_0^{T/p} h(s)\, d_{p,q}s\right) = \beta_3. \end{aligned} \quad (27)$$

Solving (26) and (27) for constants A and B, it follows that

$$A = \Lambda_1 + \frac{1}{\Lambda}\left(\frac{\alpha_2\beta_1}{p}\int_0^T (T-qs)\,h\left(\frac{s}{p}\right)d_{p,q}s + \alpha_2\beta_2\int_0^{T/p} h(s)\,d_{p,q}s\right) \tag{28}$$

and

$$B = \Lambda_2 - \frac{1}{\Lambda}\left(\frac{\alpha_1\beta_1}{p}\int_0^T (T-qs)\,h\left(\frac{s}{p}\right)d_{p,q}s + \alpha_1\beta_2\int_0^{T/p} h(s)\,d_{p,q}s\right). \tag{29}$$

Substituting constants A and B into (23), we obtain the unique solution in (20) of linear problem (18). The proof is completed. □

3. Main Results

Let $\mathscr{C} := C([0,T],\mathbb{R})$ denote the Banach space of all continuous functions from $[0,T]$ to $\mathbb{R}$ endowed with sup-norm $\|x\| = \sup\{|x(t)| : t \in [0,T]\}$. In view of Lemma 4 we define an operator $\mathscr{A} : \mathscr{C} \longrightarrow \mathscr{C}$ by

$$\begin{aligned}(\mathscr{A}x)(t) &= \Lambda_1 + \Lambda_2 t + \frac{(\alpha_2-\alpha_1 t)\beta_1}{p\Lambda}\int_0^T (T-qs)\,f(s,x(ps))\,d_{p,q}s \\ &\quad + \frac{(\alpha_2-\alpha_1 t)\beta_2}{\Lambda}\int_0^{T/p} f\left(s,x\left(p^2s\right)\right)d_{p,q}s + \frac{1}{p}\int_0^t (t-qs)\,f(s,x(ps))\,d_{p,q}s.\end{aligned} \tag{30}$$

It should be noticed that problem (5) has solutions if and only if the operator $\mathscr{A}$ of equation $x = \mathscr{A}x$ has fixed points. Our first result is an existence theorem for separated boundary value problem of quantum (p,q)-difference equation (5) by using Schaefer's fixed point theorem.

Theorem 5. *Assume that*

(H_1) *$f : [0,T/p^2] \times \mathbb{R} \longrightarrow \mathbb{R}$ is a continuous function and there exists a constant $N > 0$ such that*

$$\left|f\left(t,x\left(p^2t\right)\right)\right| \le N, \tag{31}$$

for each $t \in [0,T/p^2]$ and all $x \in \mathbb{R}$.

Then the separated boundary value problem (5) has at least one solution on $[0,T]$.

Proof. Schaefer's fixed point theorem is used to prove that the operator $\mathscr{A}$ defined by (30) has at least one fixed point. So, we divide the proof into four steps.

Step 1 (continuity of $\mathscr{A}$). Let $\{x_n\}$ be a sequence of function such that $x_n \longrightarrow x$ on $[0,T/p^2]$. Given that f is continuous function on $[0,T/p^2]$, we have

$$f\left(t,x_n\left(p^2t\right)\right) \longrightarrow f\left(t,x\left(p^2t\right)\right). \tag{32}$$

Therefore, we obtain

$$\begin{aligned}|(\mathscr{A}x_n)(t) - (\mathscr{A}x(t))| &\le \frac{(|\alpha_2|+|\alpha_1|T)|\beta_1|}{p|\Lambda|}\cdot\int_0^T (T-qs)\,|f(s,x_n(ps)) - f(s,x(ps))|\,d_{p,q}s \\ &\quad + \frac{(|\alpha_2|+|\alpha_1|T)|\beta_2|}{|\Lambda|}\cdot\int_0^{T/p}\left|f\left(s,x_n\left(p^2s\right)\right) - f\left(s,x\left(p^2s\right)\right)\right|d_{p,q}s + \frac{1}{p} \\ &\quad \cdot\int_0^t (t-qs)\,|f(s,x_n(ps)) - f(s,x(ps))|\,d_{p,q}s,\end{aligned} \tag{33}$$

which implies that $\|\mathscr{A}x_n - \mathscr{A}x\| \longrightarrow 0$ as $n \longrightarrow \infty$. Hence the operator $\mathscr{A}$ is continuous.

Step 2 ($\mathscr{A}$ maps bounded sets into bounded sets in $[0,T]$). Choosing $r > 0$, we define a bounded ball B_r as $B_r = \{x \in \mathscr{C} : \|x\| \le r\}$. Then, for any $x \in B_r$, we have

$$\begin{aligned}|(\mathscr{A}x)(t)| &\le |\Lambda_1| + |\Lambda_2 t| + \frac{(|\alpha_2|+|\alpha_1|T)|\beta_1|}{p|\Lambda|}\cdot\int_0^T (T-qs)\,|f(s,x(ps))|\,d_{p,q}s \\ &\quad + \frac{(|\alpha_2|+|\alpha_1|T)|\beta_2|}{|\Lambda|}\int_0^{T/p}\left|f\left(s,x\left(p^2s\right)\right)\right|d_{p,q}s \\ &\quad + \frac{1}{p}\int_0^t (t-qs)\,|f(s,x(ps))|\,d_{p,q}s \le |\Lambda_1| + |\Lambda_2 T| \\ &\quad + \frac{(|\alpha_2|+|\alpha_1|T)|\beta_1|}{p|\Lambda|}\int_0^T (T-qs)\,N d_{p,q}s \\ &\quad + \frac{(|\alpha_2|+|\alpha_1|T)|\beta_2|}{|\Lambda|}\int_0^{T/p} N d_{p,q}s + \frac{1}{p} \\ &\quad \cdot\int_0^T (T-qs)\,N d_{p,q}s = |\Lambda_1| + |\Lambda_2 T| \\ &\quad + \frac{(|\alpha_2|+|\alpha_1|T)T^2|\beta_1|}{(p+q)|\Lambda|}N + \frac{(|\alpha_2|+|\alpha_1|T)T|\beta_2|}{p|\Lambda|} \\ &\quad \cdot N + \frac{\mathrm{T}^2}{p+q}N := \rho\end{aligned} \tag{34}$$

This means that $\|\mathscr{A}x\| \le \rho$. Therefore, the set $\mathscr{A}B_r$ is uniformly bounded.

Step 3 ($\mathscr{A}$ maps bounded sets into equicontinuous sets of $[0,T]$). Let $\tau_1,\tau_2 \in [0,T]$ with $\tau_1 < \tau_2$ be two points and B_r be a bounded ball in $\mathscr{C}$. Then for any $x \in B_r$, we get

$$\begin{aligned}
&|(\mathscr{A}x)(\tau_2)-(\mathscr{A}x)(\tau_1)| \le (\tau_2-\tau_1)|\Lambda_2| \\
&+\frac{(\tau_2-\tau_1)|\alpha_1\beta_1|}{p|\Lambda|}\int_0^T (T-qs)|f(s,x(ps))|\,d_{p,q}s \\
&+\frac{(\tau_2-\tau_1)|\alpha_1\beta_2|}{|\Lambda|}\int_0^{T/p}\left|f\left(s,x_n\left(p^2s\right)\right)\right| d_{p,q}s \\
&+\frac{1}{p}\left|\int_0^{\tau_2}(\tau_2-qs)f(s,x(ps))\,d_{p,q}s\right. \\
&\left.-\int_0^{\tau_1}(\tau_1-qs)f(s,x(ps))\,d_{p,q}s\right| \le (\tau_2-\tau_1)|\Lambda_2| \\
&+\frac{(\tau_2-\tau_1)|\alpha_1\beta_1|}{p|\Lambda|}\int_0^T (T-qs)|f(s,x(ps))|\,d_{p,q}s \\
&+\frac{(\tau_2-\tau_1)|\alpha_1\beta_2|}{|\Lambda|}\int_0^{T/p}\left|f\left(s,x_n\left(p^2s\right)\right)\right| d_{p,q}s \\
&+\frac{1}{p}\left|\int_0^{\tau_1}(\tau_2-\tau_1)f(s,x(ps))\,d_{p,q}s\right. \\
&\left.+\int_{\tau_1}^{\tau_2}(\tau_2-qs)f(s,x(ps))\,d_{p,q}s\right| \le (\tau_2-\tau_1)|\Lambda_2| \\
&+\frac{(\tau_2-\tau_1)|\alpha_1\beta_1|}{p|\Lambda|}\int_0^T (T-qs)|f(s,x(ps))|\,d_{p,q}s \\
&+\frac{(\tau_2-\tau_1)|\alpha_1\beta_2|}{|\Lambda|}\int_0^{T/p}\left|f\left(s,x_n\left(p^2s\right)\right)\right| d_{p,q}s \\
&+\frac{N}{p}\left\{\int_0^{\tau_1}(\tau_2-\tau_1)\,d_{p,q}s+\int_{\tau_1}^{\tau_2}(\tau_2-qs)\,d_{p,q}s\right\}.
\end{aligned} \tag{35}$$

As $\tau_1 \longrightarrow \tau_2$, the right-hand side of the above inequality (which is independent of x) tends to zero. This shows that the set $\mathscr{A}B_r$ is equicontinuous set. From a consequence of Steps 1 to 3, together with the Arzelá-Ascoli theorem, we deduce that the operator $\mathscr{A}:\mathscr{C}\longrightarrow\mathscr{C}$ is completely continuous.

Step 4. Finally, we show that the set

$$E=\{x\in\mathscr{C}: x=\theta\mathscr{A}x \text{ for some } 0<\theta<1\} \tag{36}$$

is bounded.

Let $x \in E$ be a solution of problem (5). Then $x(t) = \theta(\mathscr{A}x)(t)$ for some $0<\theta<1$. Hence, for each $t\in[0,T]$, by the method of computation in Step 2, we obtain

$$\|\mathscr{A}x\| \le \rho. \tag{37}$$

This gives that the set E is bounded. By applying Schaefer's fixed point theorem, we get that $\mathscr{A}$ has at least one fixed point which is a solution of the second-order quantum (p,q)-difference equation with separated boundary value problem (5) on $[0,T]$. The proof is completed. □

The second existence of a unique solution is based on Banach's contraction mapping principle. For convenience we set a positive constant

$$\Phi=\frac{(|\alpha_2|+|\alpha_1|T)T}{|\Lambda|}\left\{\frac{|\beta_1|T}{p+q}+\frac{|\beta_2|}{p}\right\}+\frac{T^2}{p+q}. \tag{38}$$

Theorem 6. *Assume that the function $f:[0,T/p^2]\times\mathbb{R}\longrightarrow\mathbb{R}$ is continuous and satisfying the assumption:*

(H_2) There exists constant $L>0$ such that $|f(t,x)-f(t,y)|\le L|x-y|$ for each $t\in[0,T/p^2]$ and $x,y\in\mathbb{R}$.

If

$$L\Phi<1, \tag{39}$$

where Φ is defined by (38), then the boundary value problem of quantum (p,q)-difference equation (5) has a unique solution on $[0,T]$.

Proof. The existence of a unique fixed point of operator equation $x=\mathscr{A}x$, where the operator $\mathscr{A}$ is defined by (30), is proved by using Banach's contraction mapping principle. Let us define a ball $B_r=\{x\in\mathscr{C}:\|x\|\le r\}$ with the value r satisfying

$$r\ge\frac{|\Lambda_1|+|\Lambda_2T|+M\Phi}{1-L\Phi}, \tag{40}$$

where $M=\sup_{t\in[0,T/p^2]}|f(t,0)|$. Note that the above inequality is well defined as $L\Phi<1$. Next, we will show that $\mathscr{A}B_r\subset B_r$. For any $x\in B_r$, we have

$$\begin{aligned}
&|(\mathscr{A}x)(t)| \le \sup_{t\in[0,T/p^2]}\left\{|\Lambda_1|+|\Lambda_2|t+\frac{|(\alpha_2-\alpha_1t)\beta_1|}{p|\Lambda|}\right. \\
&\cdot\int_0^T(T-qs)|f(s,x(ps))|\,d_{p,q}s \\
&+\frac{|(\alpha_2-\alpha_1t)\beta_2|}{|\Lambda|}\int_0^{T/p}\left|f\left(s,x_n\left(p^2s\right)\right)\right| d_{p,q}s+\frac{1}{p} \\
&\left.\cdot\int_0^T(T-qs)|f(s,x(ps))|\,d_{p,q}s\right\} \le |\Lambda_1| \\
&+|\Lambda_2|T+\frac{(|\alpha_2|+|\alpha_1|T)|\beta_1|}{|p\Lambda|}\int_0^T(T-qs) \\
&\cdot\left(|f(s,x(ps))-f(s,0)|+|f(s,0)|\right)d_{p,q}s \\
&+\frac{(|\alpha_2|+|\alpha_1|T)|\beta_2|}{|\Lambda|} \\
&\cdot\int_0^{T/p}\left(\left|f\left(s,x\left(p^2s\right)\right)-f(s,0)\right|\right. \\
&\left.+|f(s,0)|\right)d_{p,q}s+\frac{1}{p}\int_0^T(T-qs) \\
&\cdot\left(|f(s,x(ps))-f(s,0)|+|f(s,0)|\right)d_{p,q}s
\end{aligned}$$

$$\leq |\Lambda_1| + |\Lambda_2| T + (L\|x\| + M) \cdot \left\{ \frac{(|\alpha_2| + |\alpha_1| T) |\beta_1|}{p |\Lambda|} \int_0^T (T - qs) \, d_{p,q}s + \frac{(|\alpha_2| + |\alpha_1| T) |\beta_2|}{|\Lambda|} \int_0^{T/p} d_{p,q}s + \frac{1}{p} \cdot \int_0^T (T - qs) \, d_{p,q}s \right\} \leq |\Lambda_1| + |\Lambda_2| T + (Lr + M) \left\{ \frac{(|\alpha_2| + |\alpha_1| T) T^2 |\beta_1|}{(p+q) |\Lambda|} + \frac{(|\alpha_2| + |\alpha_1| T) T |\beta_2|}{p |\Lambda|} + \frac{T^2}{p+q} \right\} = |\Lambda_1| + |\Lambda_2| T + (Lr + M) \Phi \leq r, \tag{41}$$

which implies that $\|\mathscr{A}x\| < r$. Therefore, we have $\mathscr{A}B_r \subset B_r$. Finally we shall show that $\mathscr{A}$ is a contraction. For any $x, y \in \mathscr{C}$ and for each $t \in [0, T]$, we obtain

$$|\mathscr{A}x(t) - \mathscr{A}y(t)| \leq \frac{(|\alpha_2| + |\alpha_1| T) |\beta_1|}{|p\Lambda|} \int_0^T (T - qs) \cdot (|f(s, x(ps)) - f(s, y(ps))|) \, d_{p,q}s + \frac{(|\alpha_2| + |\alpha_1| T) |\beta_2|}{|\Lambda|} \int_0^{T/p} (|f(s, x(p^2 s)) - f(s, y(p^2 s))|) \, d_{p,q}s + \frac{1}{p} \int_0^T (T - qs) \cdot (|f(s, x(ps)) - f(s, x(ps))|) \, d_{p,q}s \leq \|x - y\| \left\{ \frac{(|\alpha_2| + |\alpha_1| T) T^2 |\beta_1|}{(p+q) |\Lambda|} + \frac{(|\alpha_2| + |\alpha_1| T) T |\beta_2|}{p |\Lambda|} + \frac{T^2}{p+q} \right\} = L\Phi \|x - y\|. \tag{42}$$

Therefore, we have $\|\mathscr{A}x - \mathscr{A}y\| \leq L\Phi \|x - y\|$. As $L\Phi < 1$, then the operator $\mathscr{A}$ is a contraction. Hence, by the Banach contraction mapping principle, it follows that the operator $\mathscr{A}$ has a unique fixed point which is the unique solution of problem (5) on $[0, T]$. This completes the proof. □

The third existence theorem is based on Lelay-Schauder's nonlinear alternative.

Lemma 7 (nonlinear alternative for single-value maps [43]). *Let E be a Banach space, C be a closed, convex subset of E, U be an open subset of C, and $0 \in U$. Suppose that $A : \overline{U} \longrightarrow C$ is continuous and compact (that is, $A(\overline{U})$ is a relatively compact subset of C map). Then either*

(i) *A has a fixed point in $\overline{U}$, or*

(ii) *there is a $x \in \partial U$ (the boundary of U in C) and $\lambda \in (0, 1)$ with $x = \lambda A(x)$.*

Theorem 8. *Assume that $f : [0, T/p^2] \times \mathbb{R} \longrightarrow \mathbb{R}$ is a continuous function. In addition we suppose that*

(H_3) *there exist a continuous nondecreasing function $\psi : [0, \infty) \longrightarrow (0, \infty)$ and a function $z \in C([0, T/p^2], \mathbb{R}^+)$ such that*

$$|f(t, x)| \leq z(t) \psi(\|x\|) \quad \text{for each } (t, x) \in \left[0, \frac{T}{p^2}\right] \times \mathbb{R}; \tag{43}$$

(H_4) *there exists a constant $K > 0$ such that*

$$\frac{K}{|\Lambda_1| + |\Lambda_2| T + \|z\| \psi(K) \Phi} > 1. \tag{44}$$

Then the separated boundary value problem (5) has at least one solution on $[0, T]$.

Proof. As in the proof of Theorem 5, the operator $\mathscr{A} : \mathscr{C} \longrightarrow \mathscr{C}$ is completely continuous. The result will follow from the Lelay-Schauder nonlinear alternative (Lemma 7) once we have proved the boundedness of the set of all solutions to equations $x = \theta\mathscr{A}x$ for $\theta \in [0, 1]$.

Let x be a solution of problem (5). Then, from (H_3), we have

$$\|x\| \leq |\Lambda_1| + |\Lambda_2 T| + \|z\| \psi(\|x\|) \cdot \left\{ \frac{(|\alpha_2| + |\alpha_1| T) T^2 |\beta_1|}{(p+q) |\Lambda|} + \frac{(|\alpha_2| + |\alpha_1| T) T |\beta_2|}{p |\Lambda|} + \frac{T^2}{p+q} \right\} = |\Lambda_1| + |\Lambda_2 T| + \|z\| \psi(\|x\|) \Phi. \tag{45}$$

Thus, we obtain

$$\frac{\|x\|}{|\Lambda_1| + |\Lambda_2| T + \|z\| \psi(\|x\|) \Phi} \leq 1. \tag{46}$$

In view of (H_4), there exists a positive constant K such that $\|x\| \neq K$. Let us define the set

$$U = \{x \in \mathscr{C} : \|x\| < K\}. \tag{47}$$

Note that the operator $\mathscr{A} : \overline{U} \longrightarrow \mathscr{C}$ is continuous and completely continuous. From the choice of U, there is no $x \in \partial U$ such that $x = \theta\mathscr{A}x$ for some $\theta \in (0, 1)$. Therefore, by applying the nonlinear alternative of Leray-Schauder type, we can conclude that the operator $\mathscr{A}$ has at least one fixed point in $\overline{U}$, which is a solution of the quantum (p, q)-difference boundary value problem (5) on $[0, T]$. The proof is completed. □

The final existence theorem is established by using Krasoselskii's fixed point theorem.

Lemma 9 (Krasnoselskii's fixed point theorem [44]). *Let M be a closed, bounded, convex, and nonemtry subset of a Banach space X. Let A, B be the operators such that (a) $Ax + By \in M$ whenever $x, y \in M$; (b) A is compact and continuous; (c) B is a contraction mapping. Then there exists $z \in M$ such that $z = Az + Bz$.*

Theorem 10. *Assume that $f : [0, T/p^2] \times \mathbb{R} \longrightarrow \mathbb{R}$ is a continuous function satisfying the assumption (H_2). In addition we suppose that*

(H_5) $|f(t,x)| \leq \kappa(t), \forall (t,x) \in [0, T/p^2] \times \mathbb{R}$ and $\kappa \in C([0, T/p^2], \mathbb{R}^+)$.

If

$$\left(\frac{T^2}{p+q}\right) L < 1, \tag{48}$$

then the separated boundary value problem of quantum (p,q)-difference equation (5) has at least one solution on $[0,T]$.

Proof. Let us define $\sup_{t\in[0,T/p^2]}|\kappa(t)| = \|\kappa\|$ and choose a suitable constant $\overline{r}$ as

$$\overline{r} \geq |\Lambda_1| + |\Lambda_2| T + \|\kappa\| \Phi, \tag{49}$$

where Φ is defined by (38). Furthermore, we define the operators $\mathscr{P}$ and $\mathscr{Q}$ on $B_{\overline{r}} = \{x \in \mathscr{C} : \|x\| \leq \overline{r}\}$ by

$$\begin{aligned}
(\mathscr{P}x)(t) &= \frac{(\alpha_2 - \alpha_1 t)\beta_1}{p\Lambda} \int_0^T (T - qs) f(s, x(ps)) \, d_{p,q}s \\
&\quad + \frac{(\alpha_2 - \alpha_1 t)\beta_2}{\Lambda} \int_0^{T/p} f(s, x(p^2 s)) \, d_{p,q}s,
\end{aligned} \tag{50}$$

$$(\mathscr{Q}y)(t) = \Lambda_1 + \Lambda_2 t + \frac{1}{p}\int_0^t (t - qs) f(s, y(ps)) \, d_{p,q}s.$$

Observe that $\mathscr{P}x + \mathscr{Q}x = \mathscr{A}x$. For $x, y \in B_{\overline{r}}$, we have

$$\begin{aligned}
\|\mathscr{P}x + \mathscr{Q}y\| &\leq |\Lambda_1| + |\Lambda_2 T| + \|\kappa\| \\
&\quad \cdot \left\{\frac{(|\alpha_2| + |\alpha_1| T) T^2 |\beta_1|}{(p+q)|\Lambda|} + \frac{(|\alpha_2| + |\alpha_1| T) T |\beta_2|}{p|\Lambda|} \right. \\
&\quad \left. + \frac{T^2}{p+q}\right\} = |\Lambda_1| + |\Lambda_2 T| + \|\kappa\| \Phi \leq \overline{r},
\end{aligned} \tag{51}$$

which yields $\mathscr{P}x + \mathscr{Q}y \in B_{\overline{r}}$. Therefore, the condition (a) of Lemma 9 is satisfied. It follows from the assumption (H_2) together with (48) that $\mathscr{Q}$ is a contraction mapping. Hence the condition (c) of Lemma 9 is fulfilled. Now we will show that the condition (b) of Lemma 9 is satisfied. Given that function f is continuous, we get that the operator $\mathscr{P}$ is continuous. It is easy to verify that

$$\begin{aligned}
\|\mathscr{P}x\| &\leq \|\kappa\| \left\{\frac{(|\alpha_2| + |\alpha_1| T) T^2 |\beta_1|}{(p+q)|\Lambda|}\right. \\
&\quad \left. + \frac{(|\alpha_2| + |\alpha_1| T) T |\beta_2|}{p|\Lambda|}\right\}.
\end{aligned} \tag{52}$$

Thus, the set $\mathscr{P}(B_r)$ is uniformly bounded. Next we prove the compactness of the operator $\mathscr{P}$. Give $\sup_{(t,x)\in[0,T/p^2]\times B_{\overline{r}}}|f(t,x)| = \overline{f} < \infty$ and let $t_1, t_2 \in [0,T]$ with $t_1 < t_2$. Then we get

$$\begin{aligned}
&|(\mathscr{P}x)(t_2) - (\mathscr{P}x(t_1))| \\
&\leq \frac{|\alpha_1\beta_1|(t_2 - t_1)}{p|\Lambda|}\int_0^T (T - qs)|f(s, x(ps))| \, d_{p,q}s \\
&\quad + \frac{|\alpha_1\beta_2|(t_2 - t_1)}{|\Lambda|}\int_0^{T/p} |f(s, x(p^2 s))| \, d_{p,q}s \\
&\leq \overline{f}\left\{\frac{|\alpha_1\beta_1| T^2}{(p+q)|\Lambda|} + \frac{|\alpha_1\beta_2| T}{|\Lambda|}\right\}(t_2 - t_1),
\end{aligned} \tag{53}$$

which is independent of x and tends to zero as $t_1 \longrightarrow t_2$. So, the set $\mathscr{P}(B_r)$ is equicontinuous. By the Arzelá-Ascoli theorem, $\mathscr{P}$ is compact on $B_{\overline{r}}$. Thus all the assumptions of Lemma 9 are satisfied. Therefore the separated boundary value problem of quantum (p,q)-difference equation (5) has at least one solution on $[0,T]$. The proof is completed. □

4. Examples

Example 1. Consider the following separated boundary value problem for second-order quantum (p,q)-difference equation of the form

$$\begin{aligned}
&D^2_{1/2,1/5} x(t) = f\left(t, x\left(\frac{t}{4}\right)\right), \quad t \in [0,4], \\
&\frac{5}{6} x(0) + \frac{4}{5} D_{p,q} x(0) = \frac{1}{3}, \\
&\frac{1}{4} x(1) + \frac{5}{6} D_{p,q} x(2) = \frac{3}{5}.
\end{aligned} \tag{54}$$

Set constants $p = 1/2, q = 1/5, T = 1, \alpha_1 = 5/6, \alpha_2 = 4/5$, $\alpha_3 = 1/3$, $\beta_1 = 1/4$, $\beta_2 = 5/6$, and $\beta_3 = 3/5$. Then we can find constants as $\Lambda = 0.7027777778$, $\Lambda_1 = -0.1691699605$, $\Lambda_2 = 0.5928853755$, and $\Phi = 6.132128741$.

(i) Let $f : [0,4] \times \mathbb{R} \longrightarrow \mathbb{R}$ by

$$f\left(t, x\left(\frac{t}{4}\right)\right) = \frac{5t^2}{(1 + x^2(t/4))} + \frac{t^2 \sin t}{3t + 1}. \tag{55}$$

It is easy to see that $|f(t,x)| \leq 5.25 = N$ which satisfies (H_1). Thus, by Theorem 5, the separated boundary value problem (54) with f given by (55) has at least one solution on $[0,1]$.

(ii) Given $f : [0,4] \times \mathbb{R} \longrightarrow \mathbb{R}$ with

$$f\left(t, x\left(\frac{t}{4}\right)\right) = \frac{|\cos t|}{18}\left(\frac{x^2(t/4) + 2|x(t/4)|}{1 + |x(t/4)|}\right). \tag{56}$$

it follows that

$$|f(t,x) - f(t,y)| \leq \frac{1}{9}|x - y|. \quad (57)$$

Therefore, the condition (H_2) is satisfied with $L = 1/9$. Indeed, we have $L\Phi = 0.6813476379 < 1$. Hence, by Theorem 6, the separated boundary value problem of quantum (p,q)-difference equation (54) has a unique solution on $[0,1]$.

Example 2. Consider the following separated boundary value problem for second-order quantum (p,q)-difference equation of the form

$$D^2_{1/3,1/6}x(t) = g\left(t, x\left(\frac{t}{9}\right)\right),$$

$$t \in \left[0, \frac{9}{2}\right],$$

$$\frac{7}{3}x(0) + \frac{5}{6}D_{p,q}x(0) = \frac{4}{7}, \quad (58)$$

$$\frac{2}{5}x\left(\frac{1}{2}\right) + \frac{7}{2}D_{p,q}x\left(\frac{3}{2}\right) = \frac{6}{5}.$$

Set constants $p = 1/3$, $q = 1/6$, $T = 1/2$, $\alpha_1 = 7/3$, $\alpha_2 = 5/6$, $\alpha_3 = 4/7$, $\beta_1 = 2/5$, $\beta_2 = 7/2$, and $\beta_3 = 6/5$. Then we can compute that $\Lambda = 8.3$, $\Lambda_1 = 0.1342512908$, $\Lambda_2 = 0.3098106712$, and $\Phi = 1.813253012$.

(i) Let $g : [0, 9/2] \times \mathbb{R} \longrightarrow \mathbb{R}$ be a nonlinear function defined by

$$g\left(t, x\left(\frac{t}{9}\right)\right) = \frac{1}{(t+2)^3}\left(\frac{x^4(t/9)}{1 + x^2(t/9)} + 2\right). \quad (59)$$

Choosing $z(t) = 1/(t+2)^3$ and $\psi(x) = x^2 + 2$, we notice that the condition (H_3) of Theorem 8 is satisfied. Furthermore, we can find that $\|z\| = 1/8$. Then there exists a constant K such that $K \in (0.9447899357, 3.467170197)$. Therefore, the separated boundary value problem (58) with g given by (59) has at least one solution on $[0, 1/2]$.

(ii) Let a function $g : [0, 9/2] \times \mathbb{R} \longrightarrow \mathbb{R}$ be defined by

$$g\left(t, x\left(\frac{t}{9}\right)\right) = \left(\frac{3e^{-t}}{2 + t^2} \cdot \frac{|x(t/9)|}{|x(t/9)| + 1}\right) + \frac{t}{2}. \quad (60)$$

It follows that

$$|g(t,x)| \leq \frac{3e^{-t}}{2 + t^2} + \frac{t}{2}. \quad (61)$$

Then the function g defined in (60) satisfies condition (H_5) with $\kappa(t) = ((3e^{-t})/(2+t^2)) + (t/2)$. In addition, the condition (H_2) is satisfied with $L = 3/2$ and

$$\left(\frac{T^2}{p+q}\right)L = 0.75 < 1. \quad (62)$$

Hence, by applying Theorem 10, the separated boundary value problem for second-order quantum (p,q)-difference equation (58) with g defined in (60) has at least one solution on $[0, 1/2]$.

Remark 3. In Example 2, Case (ii), although the function $g(t,x)$ satisfies condition (H_2), Theorem 6 cannot be applied, since $L\Phi = 2.719879518 > 1$.

Conflicts of Interest

The authors declare that there are no conflicts of interest regarding the publication of this paper.

Acknowledgments

This research was funded by King Mongkut's University of Technology North Bangkok, Contract no. KMUTNB-60-ART-074.

References

[1] F. H. Jackson, "On q-difference integrals," *The Quarterly Journal of Pure and Applied Mathematics*, vol. 41, pp. 193–203, 1910.

[2] F. H. Jackson, "q-difference equations," *American Journal of Mathematics*, vol. 32, no. 4, pp. 305–314, 1910.

[3] R. D. Carmichael, "The general theory of Linear q-difference equations," *The American Journal of Mathematics*, vol. 34, no. 2, pp. 147–168, 1912.

[4] V. Kac and P. Cheung, *Quantum Calculus*, Springer, New York, NY, USA, 2002.

[5] T. Mason, "On properties of the solutions of linear q-difference equations with entire function coefficient," *The American Journal of Mathematics*, vol. 37, no. 4, pp. 439–444, 1915.

[6] C. R. Adams, "On the linear ordinary q-difference equation," *Annals of Mathematics: Second Series*, vol. 30, no. 1–4, pp. 195–205, 1928.

[7] T. Ernst, *The history of q-calculus and a new method, UUDMReport2000:16*, Department of Mathematics, UppsalaUniversity, 2000.

[8] B. Ahmad and S. K. Ntouyas, "Boundary Value Problems for q-Difference Inclusions," *Abstract and Applied Analysis*, vol. 2011, Article ID 292860, 15 pages, 2011.

[9] A. Dobrogowska and A. Odzijewicz, "Second order q-difference equations solvable by factorization method," *Journal of Computational and Applied Mathematics*, vol. 193, no. 1, pp. 319–346, 2006.

[10] M. El-Shahed and H. A. Hassan, "Positive solutions of q-difference equation," *Proceedings of the American Mathematical Society*, vol. 138, no. 5, pp. 1733–1738, 2010.

[11] R. A. C. Ferreira, "Nontrivial solutions for fractional q-difference boundary value problems," *Electronic Journal of Qualitative Theory of Differential Equations*, vol. 70, pp. 1–10, 2010.

[12] G. Gasper and M. Rahman, *Basic Hypergeometric Series*, vol. 35 of *Encyclopedia of Mathematics and its Applications*, Cambridge University Press, Cambridge, UK, 1990.

[13] J. Ma and J. Yang, "Existence of solutions for multi-point boundary value problem of fractional q-difference equation," *Electronic Journal of Qualitative Theory of Differential Equations*, no. 92, pp. 1–10, 2011.

[14] J. Tariboon and S. K. Ntouyas, "Quantum calculus on finite intervals and applications to impulsive difference equations," *Advances in Difference Equations*, vol. 2013, article no. 282, 2013.

[15] B. Ahmad, S. K. Ntouyas, and J. Tariboon, *Quantum Calculus: New Concepts, Impulsive IVPs and BVPs, Inequalities*, vol. 4 of *Trends in Abstract and Applied Analysis*, World Scientific Publishing, Hackensack, NJ, USA, 2016.

[16] R. J. Finkelstein, "Symmetry group of the hydrogen atom," *Journal of Mathematical Physics*, vol. 8, no. 3, pp. 443–449, 1967.

[17] R. J. Finkelstein, "*q*-field theory," *Letters in Mathematical Physics*, vol. 34, no. 2, pp. 169–176, 1995.

[18] F.-L. Chan and R. J. Finkelstein, "*q*-deformation of the Coulomb problem," *Journal of Mathematical Physics*, vol. 35, no. 7, pp. 3273–3284, 1994.

[19] A. C. Cadavid and R. J. Finkelstein, "The q-Coulomb problem in configuration space," *Journal of Mathematical Physics*, vol. 37, no. 8, pp. 3675–3683, 1996.

[20] R. J. Finkelstein, "The q-Coulomb problem," *Journal of Mathematical Physics*, vol. 37, no. 6, pp. 2628–2636, 1996.

[21] J. Feigenbaum and P. G. Freund, "A q-deformation of the Coulomb problem," *Journal of Mathematical Physics*, vol. 37, no. 4, pp. 1602–1616, 1996.

[22] H. Cheng, *Canonical Quantization of Yang-Mills Theories*, Perspectives in Mathematical Physics, International Press, 1996.

[23] R. J. Finkelstein, "q-gravity," *Letters in Mathematical Physics*, vol. 38, no. 1, pp. 53–62, 1996.

[24] M. Kamata and A. Nakamula, "One-parameter family of selfdual solutions in classical Yang-Mills theory," *Physics Letters B: Particle Physics, Nuclear Physics and Cosmology*, vol. 463, no. 2-4, pp. 257–262, 1999.

[25] T. Negadi and M. Kibler, "A q-deformed Aufbau Prinzip," *Journal of Physics A: Mathematical and General*, vol. 25, no. 4, pp. L157–L160, 1992.

[26] L. P. Marinova, P. P. Raychev, and J. Maruani, "Molecular backbending in AgH and its description in terms of q-algebras," *Molecular Physics. An International Journal in the Field of Chemical Physics*, vol. 82, no. 6, pp. 1115–1129, 1994.

[27] M. R-Monteiro, L. M. C. S. Rodrigues, and S. Wulck, "Quantum algebraic nature of the phonon spectrum in ^{4}He," *Physical Review Letters*, vol. 76, no. 7, pp. 1098–1100, 1996.

[28] A. M. Gavrilik, "q-Serre relations in $U_q(u_n)$ and q-deformed meson mass sum rules," *Journal of Physics A: Mathematical and General*, vol. 27, no. 3, pp. L91–L94, 1994.

[29] R. J. Finkelstein, "q gauge theory," *International Journal of Modern Physics A*, vol. 11, no. 4, pp. 733–746, 1996.

[30] G. Kaniadakis, A. Lavagno, and P. Quarati, "Kinetic model for q-deformed bosons and fermions," *Physics Letters A*, vol. 227, no. 3-4, pp. 227–231, 1997.

[31] K. N. Ilinski, G. V. Kalinin, and A. S. Stepanenko, "q-functional field theory for particles with exotic statistics," *Physics Letters A*, vol. 232, no. 6, pp. 399–408, 1997.

[32] W. Siegel, *Introduction to String Field Theory*, vol. 8 of *Advanced Series in Mathematical Physics*, World Scientific Publishing, Teaneck, NJ, USA, 1988.

[33] R. Chakrabarti and R. Jagannathan, "A (p,q)-oscillator realization of two-parameter quantum algebras," *Journal of Physics A: Mathematical and General*, vol. 24, no. 24, pp. 5683–5701, 1991.

[34] R. Jagannathan and K. Srinivasa Rao, "Two-parameter quantum algebras, twin-basic number, and associated generalized hypergeometric series," *Differential Equations & Applications*, vol. 27, 2006.

[35] V. Sahai and S. Yadav, "Representations of two parameter quantum algebras and p,q-special functions," *Journal of Mathematical Analysis and Applications*, vol. 335, no. 1, pp. 268–279, 2007.

[36] I. Burban, "Two-parameter deformation of the oscillator algebra and (p,q)-analog of two-dimensional conformal field theory," *Journal of Nonlinear Mathematical Physics*, vol. 2, no. 3-4, pp. 384–391, 1995.

[37] P. N. Sadjang, "On the fundamental theorem of (p, q)-calculus and some (p, q)-Taylor formulas," https://arxiv.org/abs/1309.3934, 2015.

[38] M. N. Hounkonnou and J. D. Kyemba, "R(p,q)-calculus: differentiation and integration," *SUT Journal of Mathematics*, vol. 49, no. 2, pp. 145–167, 2013.

[39] M. Mursaleen, K. J. Ansari, and A. Khan, "On (p,q)-analogue of Bernstein operators," *Applied Mathematics and Computation*, vol. 266, pp. 874–882, 2015.

[40] S. Araci, U. G. Duran, M. Acikgoz, and H. M. Srivastava, "A certain (p,q)-derivative operator and associated divided differences," *Journal of Inequalities and Applications*, vol. 2016, article no. 301, 2016.

[41] U. Duran, *Post Quantum Calculus, [Master Thesis]*, University of Gaziantep, 2016.

[42] N. Kamsrisuk, C. Promsakon, S. K. Ntouyas, and J. Tariboon, "Nonlocal boundary value problems for (p,q)-difference equations," *Differential Equations & Applications*, vol. 10, no. 2, pp. 183–195, 2018.

[43] A. Granas and J. Dugundji, *Fixed Point Theory and Applications*, Springer, New York, NY, USA, 2003.

[44] M. A. Krasnosel'skii, "Two remarks on the method of successive approximations," *Uspekhi Matematicheskikh Nauk*, vol. 10, no. 1, pp. 123–127, 1955.

18

The Application of Minimal Length in Klein-Gordon Equation with Hulthen Potential using Asymptotic Iteration Method

Isnaini Lilis Elviyanti, Beta Nur Pratiwi, A. Suparmi, and C. Cari

Physics Department, Graduate Program, Sebelas Maret University, Indonesia

Correspondence should be addressed to A. Suparmi; soeparmi@staff.uns.ac.id

Academic Editor: Yao-Zhong Zhang

The application of minimal length formalism in Klein-Gordon equation with Hulthen potential was studied in the case of scalar potential that was equal to vector potential. The approximate solution was used to solve the Klein-Gordon equation within the minimal length formalism. The relativistic energy and wave functions of Klein-Gordon equation were obtained by using the Asymptotic Iteration Method. By using the Matlab software, the relativistic energies were calculated numerically. The unnormalized wave functions were expressed in hypergeometric terms. The results showed the relativistic energy increased by the increase of the minimal length parameter. The unnormalized wave function amplitude increased for the larger minimal length parameter.

1. Introduction

The relativistic effect gives the correction in the nonrelativistic quantum mechanics by applying the strong potential field in the particles dynamic. The particles dynamic in relativistic effect can be described by using Klein-Gordon equation, particularly for a zero spin particle. The Klein-Gordon equation is formed by two potentials coupling which are the four-vector potential ($V(r)$) and scalar potential ($S(r)$). From these two potentials, the Klein-Gordon equation has two framework conditions, which are as follows: if the scalar potential was equal to vector potential ($S(r) = V(r)$) and if the scalar potential was equal to minus vector potential ($S(r) = -V(r)$). This condition appears in nuclear and high energy physics problem [1–3]. Some of researchers have investigated both these conditions with the certainty vector potential. The main case of that research was how to reduce the Klein-Gordon equation into the Schrodinger-like equation, so we can solve it by using the certainty suitable methods. The methods which are usually used are such as Supersymmetric Quantum Mechanics (SUSY) [4], Nikiforov-Uvarov [1, 5], and Asymptotic Iteration Method [6, 7]. Various potentials are used to explain the dynamic particle, such as the harmonic potential [8], Makarov potential [2], Hulthen potential [1, 5], Kratzer potential [6], and Trigonometric Poschl-Teller potential [7].

The particles dynamic in quantum mechanics corresponds to the position and momentum of particles. And, as we have known, the study of commutation relations between position and momentum operators is explained using Heisenberg uncertainty principle [9], which is given by

$$\left[\widehat{X}, \widehat{P}\right] \geq i\hbar \tag{1}$$

where $\widehat{X}$ is position operator, $\widehat{P}$ is momentum operator, i is imaginer number, and $\hbar = h/2\pi$ with h being Planck constant. The presence of a quantum gravity on quantum mechanics has the consequence of the existence of a minimal observable distance on the order of the Planck length. Therefore the Heisenberg uncertainty principle gets additional correction due to the presence of a quantum gravity, which is well known as Generalized Uncertainty Principle (GUP) [10, 11], given by

$$\left[\widehat{X}, \widehat{P}\right] \geq i\hbar\left(1 + \alpha_{ML}P^2\right) \tag{2}$$

where α_{ML} is minimal length parameter that has value $0 \leq \alpha_{ML} \leq 1$ and P is magnitude of the momentum [11]. When the energy is much smaller than the Planck mass, α_{ML} goes to zero and we recover Heisenberg uncertainty principle [12].

In 2009, Jana and Roy have solved Klein-Gordon equation in the presence of minimal length for scalar potential was equal to vector potential using Algebraic Approach [13]. The Klein-Gordon equation in the presence of a minimal length is solved by using Feynman for scalar potential was equal to vector potential [14]. In addition, the hypergeometric method is used to solve Klein-Gordon equation using hyperbolic cotangent potential [15] and Asymptotic Iteration Method in trigonometric cotangent potential [16, 17].

In this paper, we solved the minimal length formalism in radial part of Klein-Gordon equation for the condition $S(r) = V(r)$ with Hulthen potential by using approximate solution. The approximate solution is used by Chabab et al. to solve the Bohr Mottelson Hamiltonian in the presence of a minimal length formalism by introducing the new wave function [11]. The minimal length formalism in the Klein-Gordon equation is reduced into second-order differential equation. The relativistic energy equation and wave functions of Klein-Gordon equation are obtained by using Asymptotic Iteration Method.

The study is organized as follows. In Section 2, the approximate form of Klein-Gordon equation within minimal length formalism is presented. The Hulthen potential is introduced in Section 3. We describe Asymptotic Iteration Method in Section 4. The result and discussion are presented in Section 5. At last, conclusion is given in Section 6.

2. Approximate Form of Klein-Gordon Equation within Minimal Length Formalism

Generalized Uncertainty Principle is called the minimal length that is deformed from the commutation relations between position and momentum operators in quantum mechanics. In (2), the commutation relations can be rewritten as follows [9, 12]:

$$\widehat{X}_i = \widehat{x}_i \tag{3}$$

$$\widehat{P}_i = \left(1 + \alpha_{ML}\widehat{p}^2\right)\widehat{p}_i \tag{4}$$

where $\widehat{P}_i$ and $\widehat{p}_i$ are momentum operators at high and low energy, respectively. The magnitude of $\widehat{p}_i$ is expressed by p. The occurrence of minimal length is in string theory, black hole, quantum gravity, and noncommutative geometry, which yield new correction to Heisenberg uncertainty principle and imply a finite minimal uncertainty in position measurements, e.g., at the Planck scale [14].

The general Klein-Gordon equation with scalar potential $S(r)$ and vector potential $V(r)$ is given as

$$\begin{aligned}&(E - V(r))^2 \psi(r,\theta,\varphi)\\&= \left[P^2c^2 + \left(M_oc^2 + S(r)\right)^2\right]\psi(r,\theta,\varphi)\end{aligned} \tag{5}$$

where E is relativistic energy and M_o is rest mass. By setting $S(r) = V(r)$ in (5) and substituting (4) into (5), with $\widehat{p} = -i\hbar\nabla$ and $c = \hbar = 1$ (natural unit), we have

$$\begin{aligned}&-\left(\Delta - 2\alpha_{ML}\Delta^2\right)\psi(r,\theta,\varphi)\\&-\begin{pmatrix} E^2 - M_o{}^2 \\ -(E+M_o)V(r)\end{pmatrix}\psi(r,\theta,\varphi) = 0\end{aligned} \tag{6}$$

and here we have set $V(r) \to (1/2)V(r)$.

Accordingly, it would be natural to scale the potential term in (6), so that the nonrelativistic energy is reproduced [18]. The new wave function that is used to get the approximate form of Klein-Gordon equation is given [11]:

$$\psi(r,\theta,\varphi) = (1 + 2\alpha_{ML}\Delta)\phi(r,\theta,\varphi) \tag{7}$$

Equation (7) is modification of (19) in [11]. The modification is proposed to eliminate quadratic of Laplacian factor such that we get (20) in [11]. By substituting (7) into (6), we obtain

$$\left[-\Delta - \frac{\begin{pmatrix} E^2-M_o{}^2 \\ -(E+M_o)V(r)\end{pmatrix}}{\left(1 + 2\alpha_{ML}\begin{pmatrix} E^2-M_o{}^2 \\ -(E+M_o) \\ V(r)\end{pmatrix}\right)}\right]\phi(r,\theta,\varphi) = 0 \tag{8}$$

Equation (8) is the minimal length formalism in Klein-Gordon equation within the approximate form. The component Δ^3 is eliminated due to the value of α_{ML}^2 which goes to zero, and value of α_{ML} is very small. Here, we have used the properties that Δ is scalar differential operator. If Δ operates to scalar fields ϕ at a point (r,θ,φ), $\Delta\phi$ will result in another scalar field. Here, $\Delta\phi$ is also called scalar Laplacian. Inserting (7) into (6) leads us to do the multiplication operation among Laplacian. By using the property that the multiplication operation is commutative for scalar differential operator with constant coefficient, then the multiplication operator among Laplacian also has commutative properties. Therefore, here we can eliminate Δ^2 when (7) is inserted into (6). To get simple solution of (8), binomial expansion is used for small α_{ML}; then (8) becomes

$$\begin{aligned}&\Delta\phi(r,\theta,\varphi)\\&+\begin{bmatrix} \left(E^2 - M_o{}^2\right) \\ -(E+M_o)V(r) \\ -2\alpha_{ML}\begin{pmatrix} E^2 - M_o{}^2 \\ -(E+M_o)V(r)\end{pmatrix}^2\end{bmatrix}\phi(r,\theta,\varphi) = 0\end{aligned} \tag{9}$$

Equation (9) is obtained by setting $\alpha_{ML}{}^2$ which goes to zero, so $\alpha_{ML}{}^2$ is ignored. Applying spherical Laplacian operator, as

$$\begin{aligned}\Delta &= \frac{1}{r^2}\frac{\partial}{\partial r}r^2\frac{\partial}{\partial r}\\&+\frac{1}{r^2}\left(\frac{1}{\sin\theta}\frac{\partial}{\partial\theta}\sin\theta\frac{\partial}{\partial\theta} + \frac{1}{\sin^2\theta}\frac{\partial^2}{\partial\varphi^2}\right)\end{aligned} \tag{10}$$

into (9), we use variable separable method by setting the new wave function $\phi(r,\theta,\varphi) = R(r)\Theta(\theta)\Phi(\varphi)$, and we have a polar part and radial part of Klein-Gordon equation in the presence of minimal length. The polar part is given:

$$-\begin{bmatrix}\dfrac{1}{\Theta(\theta)}\dfrac{1}{\sin\theta}\dfrac{\partial}{\partial\theta}\sin\theta\dfrac{\partial\Theta(\theta)}{\partial\theta} \\ +\dfrac{1}{\Phi(\varphi)}\dfrac{1}{\sin^2\theta}\dfrac{\partial^2\Phi(\varphi)}{\partial\varphi^2}\end{bmatrix} = \lambda \tag{11}$$

and the radial part is as follows:

$$\frac{1}{r^2}\frac{\partial}{\partial r}r^2\frac{\partial R(r)}{\partial r} + \left[\begin{array}{c}\left(\begin{array}{c}E^2 - M_o^{\,2} \\ -(E+M_o)V(r)\end{array}\right) \\ -2\alpha_{ML}\left(\begin{array}{c}(E^2-M_o^{\,2})^2 \\ -2(E^2-M_o^{\,2}) \\ (E+M_o)V(r) \\ +(E+M_o)^2V^2(r)\end{array}\right)\end{array}\right]R(r) = \frac{\lambda}{r^2}R(r) \quad (12)$$

where λ is a constant of variable separable method which corresponds to angular momentum (L). By applying $R(r) = U_H(r)/r$ and $\lambda = L(L+1)$ into (12), it yields

$$\frac{d^2U_H(r)}{dr^2} - \frac{L(L+1)}{r^2}U_H(r) + \left[\begin{array}{c}\left(\begin{array}{c}E^2 - M_o^{\,2} \\ -(E+M_o)V(r)\end{array}\right) \\ -2\alpha_{ML}\left(\begin{array}{c}(E^2-M_o^{\,2})^2 \\ -2(E^2-M_o^{\,2})(E+M_o)V(r) \\ +(E+M_o)^2V^2(r)\end{array}\right)\end{array}\right]U_H(r) = 0 \quad (13)$$

Equation (13) is the minimal length formalism in Klein-Gordon equation with Hulthen potential in the form of one-dimensional Schrodinger-like equation.

3. Asymptotic Iteration Method

Asymptotic Iteration Method is method to solve the second-order differential equation in form [19–21]

$$y_n''(u) = \lambda_0(u)\,y_n'(u) + s_0(u)\,y_n(u) \quad (14)$$

where $\lambda_0(u) \neq 0$ and $s_0(u)$ are the coefficients of a differential equation and n is a quantum number. To obtain solution, we derive (14).

$$y_n^{z+1}(u) = \lambda_{z-1}(u)\,y_n'(u) + s_{z-1}(u)\,y_n(u) \quad (15)$$

and here

$$\begin{aligned}\lambda_z(u) &= \lambda'_{z-1}(u) + s_{z-1}(u) + \lambda_0(u)\lambda_{z-1}(u) \\ s_z(u) &= s'_{z-1}(u) + s_0(u)\lambda_{z-1}(u), \\ z &= 1, 2, 3, \ldots\end{aligned} \quad (16)$$

The eigenvalue is obtained from the quantization condition which is given by

$$\Delta_z(u) = \lambda_z(u)\,s_{z-1}(u) - \lambda_{z-1}(u)\,s_z(u) = 0 \quad (17)$$

To obtain the wave function, (14) is reduced into the formalism, as follows

$$y_n''(u) = \left\{\begin{array}{c}2\left(\dfrac{au^{N+1}}{1-bu^{N+2}} - \dfrac{t+1}{u}\right)y_n'(u) \\ -\dfrac{wu^N}{1-bu^{N+2}}y_n(u)\end{array}\right\} \quad (18)$$

Equation (18) is one-dimensional Schrodinger-like equation that has solution which is expressed in hypergeometric term, given as

$$\begin{aligned}y_n(u) &= (-1)^n C'(N+2)^n(\sigma)_n \\ &\quad {}_2F_1\left(-n, \rho+n; \sigma; bu^{N+2}\right)\end{aligned} \quad (19)$$

where

$$\begin{aligned}(\sigma)_n &= \frac{\Gamma(\sigma+n)}{\Gamma(\sigma)}; \\ \sigma &= \frac{2t+N+3}{N+2}; \\ \rho &= \frac{(2t+1)b+2a}{(N+2)b}\end{aligned} \quad (20)$$

C' is normalization constant and ${}_2F_1$ is a hypergeometric function. The unnormalized wave functions of Klein-Gordon equation are obtained by using (19)-(20) [19–21].

4. Hulthen Potential

The Hulthen potential is one of short range potentials in physics. The Hulthen potential has been used in particle physics, atomic physics, nuclear physics, solid-state physics, and chemical physics. The Hulthen wave functions have been used in solid-state physics problems [22]. The Hulthen-like wave functions have been found to investigate atomic problems [22]. The general Hulthen potential is given by [23]

$$V(r) = -V_H\omega_H\frac{e^{-2\omega_H r}}{1-e^{-2\omega_H r}} \quad (21)$$

where ω_H is a screening parameter and V_H is potential depth. The value of screening potential is 0.025 for low screening and 0.15 for high screening [22]. To get simple solution, (21) was changed in hyperbolic trigonometric term [23], given as

$$V(r) = \frac{V_H\omega_H}{2}(1 - \coth\omega_H r) \quad (22)$$

By setting $V_H = 7$ and $\omega_H = 0.1$, the visualization of Hulthen potential is expressed in Figure 1.

Figure 1 shows the visualization of Hulthen potential in r function. The Hulthen potential for different value of r is approximately from 0 until 0.05μ/eV (natural unit). The Hulthen potential has negative value in a very small value of r, while the Hulthen potential inclines to be constant for higher value of r.

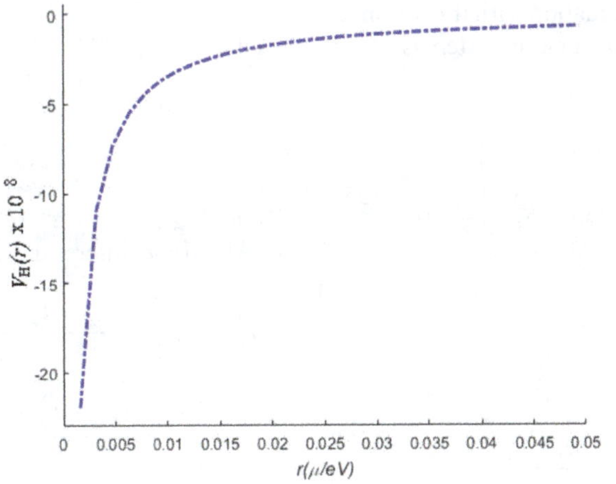

FIGURE 1: The visualization of Hulthen potential.

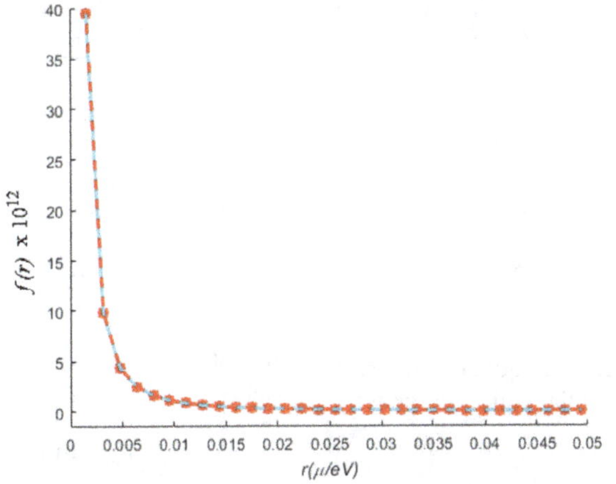

FIGURE 2: The visualization of approximation, with $\omega_H = 0.1$ and value of r from 0 until 0.05 μ/eV (natural unit).

5. Result and Discussion

Equation (13) can not be solved exactly unless we use the approximation to the function of $1/r^2$. The approximation of function $1/r^2$ is given as [23]

$$\frac{1}{r^2} = \frac{\omega_H^2}{\sinh^2 \omega_H r} = f(r) \tag{23}$$

for small value of ω_H or $\omega_H r \ll 1$. The visualization of that approximation is expressed in Figure 2.

Figure 2 shows the red line as a function of $1/r^2$ and light blue line as a function of $\omega_H^2/\sinh^2(\omega_H r)$. It is seen that the two lines overlap with each other; then the centrifugal term $1/r^2$ is approximated by $\omega_H^2/\sinh^2(\omega_H r)$ also as in [23].To obtain the exact solution of (13) the approximate term of $1/r^2$ in (23) is inserted into (13) and together with (22); then we get

$$\begin{aligned}
&\frac{d^2 U_H(r)}{dr^2} - \frac{\left(\omega_H^2 L(L+1) + \alpha_{ML}(E+M_o)^2\left((V_H\omega_H)^2/2\right)\right)}{\sinh^2 \omega_H r} U_H(r) \\
&+ \left[\begin{array}{l} + \left(\begin{array}{c} (E+M_o)\dfrac{V_H\omega_H}{2} + \alpha_{ML}(E+M_o)^2 (V_H\omega_H)^2 \\ -2\alpha_{ML}\left(E^2 - M_o^2\right)(E+M_o) V_H\omega_H \end{array} \right) \coth \omega_H r \\ + \left(\begin{array}{c} \left(E^2 - M_o^2\right) - (E+M_o)\dfrac{V_H\omega_H}{2} \\ +2\alpha_{ML}\left(E^2-M_o^2\right)(E+M_o)V_H\omega_H \\ -\alpha_{ML}(E+M_o)^2(V_H\omega_H)^2 - 2\alpha_{ML}\left(E^2-M_o^2\right)^2 \end{array} \right) \end{array} \right] U_H(r) = 0
\end{aligned} \tag{24}$$

Equation (24) is the Klein-Gordon equation with the minimal length for Hulthen potential which can be rewritten as

$$\frac{d^2U_H(r)}{dr^2} - \left[\frac{\nu_{PH}(\nu_{PH}-1)}{\sinh^2\omega_H r} - 2q_{PH}\coth\omega_H r + \kappa_{PH}^2\right]U_H(r) = 0 \tag{25}$$

with

$$\nu_{PH}(\nu_{PH}-1) = \left(\omega_H^2 L(L+1) + \alpha_{ML}(E+M_o)^2\frac{(V_H\omega_H)^2}{2}\right) \tag{26}$$

$$2q_{PH} = \left(\begin{array}{c}(E+M_o)\frac{(V_H\omega_H)}{2} + \alpha_{ML}(E+M_o)^2(V_H\omega_H)^2 \\ -2\alpha_{ML}(E^2-M_o{}^2)(E+M_o)(V_H\omega_H)\end{array}\right) \tag{27}$$

$$-\kappa_{PH}^2 = \left(\begin{array}{c}(E^2-M_o{}^2) - (E+M_o)\frac{(V_H\omega_H)}{2} \\ +2\alpha_{ML}(E^2-M_o{}^2)(E+M_o)(V_H\omega_H) \\ -\alpha_{ML}(E+M_o)^2(V_H\omega_H)^2 - 2\alpha_{ML}(E^2-M_o{}^2)^2\end{array}\right) \tag{28}$$

Equation (25) is second-order differential equation that will be reduced to hypergeometric differential equation type; by letting $\coth\omega_H r = 1-2z$, we get

$$z(1-z)\frac{d^2U_H(r)}{dz^2} + (1-2z)\frac{dU_H(r)}{dz} + \left[\nu'_{PH}(\nu'_{PH}-1) - \frac{-2q'_{PH}+\kappa'^2_{PH}}{4z} - \frac{2q'_{PH}+\kappa'^2_{PH}}{4(1-z)}\right]U_H(r) = 0 \tag{29}$$

and then we set

$$2q'_{PH} = \frac{2q_{PH}}{\omega_H^2};$$

$$\kappa'^2_{PH} = \frac{\kappa_{PH}^2}{\omega_H^2}$$

$$\nu'_{PH}(\nu'_{PH}-1) = \frac{\nu_{PH}(\nu_{PH}-1)}{\omega_H^2}; \tag{30}$$

$$-2q'_{PH} + \kappa'^2_{PH} = 4\alpha_{PH}^2;$$

$$2q'_{PH} + \kappa'^2_{PH} = 4\beta_{PH}^2$$

in (29), so we have

$$z(1-z)\frac{d^2U_H(r)}{dz^2} + (1-2z)\frac{dU_H(r)}{dz} + \left[\nu'_{PH}(\nu'_{PH}-1) - \frac{4\alpha_{PH}^2}{4z} - \frac{4\beta_{PH}^2}{4(1-z)}\right]U_H(r) = 0 \tag{31}$$

By setting $U_H(r) = z^{\alpha_{PH}}(1-z)^{\beta_{PH}}f_H(z)$, then (31) is reduced to AIM-type differential equation that is similar to (14).

$$f''_H(z) + \left[\frac{(2\alpha_{PH}+1) - (2\alpha_{PH}+2\beta_{PH}+2)y}{z(1-z)}\right]f'_H(z) + \left[\frac{\nu'_{PH}(\nu'_{PH}-1) - (\alpha_{PH}+\beta_{PH})(\alpha_{PH}+\beta_{PH}+1)}{z(1-z)}\right] \cdot f_H(z) = 0 \tag{32}$$

By comparing (14) and (32), we have

$$\lambda_{0_{PH}} = \frac{-(2\alpha_{PH}+1)}{z} + \frac{(2\beta_{PH}+1)}{1-z} \tag{33}$$

$$s_{0_{PH}} = \left[\begin{array}{c}\frac{(\alpha_{PH}+\beta_{PH})(\alpha_{PH}+\beta_{PH}+1) - \nu'_{PH}(\nu'_{PH}-1)}{z} \\ +\frac{(\alpha_{PH}+\beta_{PH})(\alpha_{PH}+\beta_{PH}+1) - \nu'_{PH}(\nu'_{PH}-1)}{(1-z)}\end{array}\right] \tag{34}$$

To obtain eigenvalue, we use (15)-(17) and (33)-(34), which yields

$$\nu'_{PH}(\nu'_{PH}-1) = (\alpha_{PH}+\beta_{PH}+n)(\alpha_{PH}+\beta_{PH}+(n+1)) \tag{35}$$

By inserting (27)-(28) and (30) into (35), we obtain the relativistic energy equation Klein-Gordon equation with minimal length for Hulthen potential, as follows:

$$(E^2-M_o{}^2) = \left\{\begin{array}{c}-\omega_H^2\left[\begin{array}{c}\left(\sqrt{\nu'_{PH}(\nu'_{PH}-1)+\frac{1}{4}}-n-\frac{1}{2}\right)^2 \\ +\frac{\zeta_{PH}}{4\left(\omega_H^2\sqrt{\nu'_{PH}(\nu'_{PH}-1)+1/4}-n-1/2\right)^2}\end{array}\right] \\ +(E+M_o)\frac{V_H}{2}+\tau_{PH}\end{array}\right\} \tag{36}$$

where

$$\nu'_{PH} = \frac{1}{2} + \sqrt{\frac{\left(\omega_H^2 L(L+1) + \alpha_{ML}(E+M_o)^2\left((V_H\omega_H)^2/2\right)\right)}{\omega_H^2} + \frac{1}{4}} \tag{37}$$

$$\zeta_{PH} = \left(\begin{array}{c}(E+M_o)\frac{(V_H\omega_H)}{2} + \alpha_{ML}(E+M_o)^2(V_H\omega_H)^2 \\ -2\alpha_{ML}(E^2-M_o{}^2)(E+M_o)(V_H\omega_H)\end{array}\right) \tag{38}$$

TABLE 1: Relativistic energy with $V_H = 7$, $\omega_H = 0.1$, $M_o = 1$, and $L = 0$.

n	$E(eV)$			
	$\alpha_{ML} = 0$	$\alpha_{ML} = 0.001$	$\alpha_{ML} = 0.01$	$\alpha_{ML} = 0.1$
0	-1.9770	-1.9654	-1.8759	-1.5259
1	-1.8629	-1.8537	-1.7801	-1.4725
2	-1.7473	-1.7402	-1.6818	-1.4171
3	-1.6298	-1.6248	-1.5808	-1.3594
4	-1.5102	-1.5071	-1.4767	-1.2990
5	-1.3879	-1.3865	-1.3690	-1.2357

$$\tau_{PH} = \begin{bmatrix} \alpha_{ML}(E+M_o)^2 (V_H\omega_H)^2 \\ +2\alpha_{ML}(E^2 - M_o{}^2)^2 \\ -2\alpha_{ML}(E^2 - M_o{}^2)(E+M_o)(V_H\omega_H) \end{bmatrix} \tag{39}$$

and n is quantum number. Equation (36) is relativistic energy equation of the minimal length formalism in Klein-Gordon equation with Hulthen potential. The relativistic energies were calculated numerically by using Matlab software. The results of relativistic energies are listed in Table 1.

Table 1 shows that the presence of minimal length and Hulthen potential in Klein-Gordon equation gives influence to the relativistic energy value. The relativistic energy value without minimal length parameter ($\alpha_{ML} = 0$) is lower than the relativistic energy value with minimal length parameter. Then, the relativistic energy value increases for the larger minimal length parameter and for the larger quantum number (n). The influence of Hulthen potential gives negative value in the relativistic energy value. If we set $\alpha_{ML} = 0$, without the presence of the minimal length parameter in relativistic energy equation (36), it was reduced to the relativistic energy equation which is in agreement with [5]. In [5] it was shown that the relativistic energy equation for the Klein-Gordon equation for Hulthen potential without the minimal length depends on the squared of quantum number (n).

To get the unnormalized wave function, we used (19)-(20), so we have

$$\begin{aligned} N_{PH} &= -1; \\ a_{PH} &= \left(\beta_{PH} + \frac{1}{2}\right); \\ b_{PH} &= 1; \\ t_{PH} &= \left(\alpha_{PH} - \frac{1}{2}\right); \\ \sigma_{PH} &= 2\alpha_{PH} + 1; \\ \rho_{PH} &= 2\alpha_{PH} + 2\beta_H + 1 \end{aligned} \tag{40}$$

Inserting (40) into (19) yields

$$f_H(z) = \begin{bmatrix} (-1)^n C'(1)^n (2\alpha_{PH}+1)_n \\ {}_2F_1(-n, 2\alpha_{PH} + 2\beta_{PH} + 1 + n, 2\alpha_{PH} + 1, z) \end{bmatrix} \tag{41}$$

Equation (41) is substituted into $U_H(r) = z^{\alpha_{PH}}(1-z)^{\beta_{PH}} f_H(z)$, so we have the unnormalized wave function given as

$$U_H(r) = \begin{bmatrix} z^{\alpha_{PH}}(1-z)^{\beta_{PH}} (-1)^n C'(1)^n (2\alpha_{PH}+1)_n \\ {}_2F_1(-n, 2\alpha_{PH} + 2\beta_{PH} + 1 + n, 2\alpha_{PH} + 1, z) \end{bmatrix} \tag{42}$$

By applying $\coth\omega_H r = 1 - 2z$ in (42), we obtain the unnormalized wave function of minimal length formalism in Klein-Gordon equation, given as

$$U_{H_0}(z) = C'\left(\frac{1-\coth\omega_H r}{2}\right)^{\alpha_{PH}} \left(\frac{1+\coth\omega_H r}{2}\right)^{\beta_{PH}} \tag{43}$$

$$U_{H_1}(z) = \begin{Bmatrix} -C'\left(\frac{1-\coth\omega_H r}{2}\right)^{\alpha_{PH}} \left(\frac{1+\coth\omega_H r}{2}\right)^{\beta_{PH}} (2\alpha_{PH}+1) \\ \left[1 + \frac{(-1)(2\alpha_{PH} + 2\beta_{PH} + 2)((1-\coth\omega_H r)/2)}{(2\alpha_{PH}+1)1!}\right] \end{Bmatrix} \tag{44}$$

$$U_{H_2}(r) = \begin{Bmatrix} -C'\left(\frac{1-\coth\omega_H r}{2}\right)^{\alpha_{PH}} \left(\frac{1+\coth\omega_H r}{2}\right)^{\beta_{PH}} \begin{pmatrix} (2\alpha_{PH}+1) \\ (2\alpha_{PH}+2) \end{pmatrix} \\ \left[\begin{matrix} 1 + \dfrac{\begin{pmatrix} (-1)(2\alpha_{PH}+2\beta_{PH}+2) \\ ((1-\coth\omega_H r)/2) \end{pmatrix}}{(2\alpha_{PH}+1)1!} \\ + \dfrac{\begin{pmatrix} (-2)(-1)(2\alpha_{PH}+2\beta_{PH}+3) \\ (2\alpha_{PH}+2\beta_{PH}+4)((1-\coth\omega_H r)/2)^2 \end{pmatrix}}{(2\alpha_{PH}+1)(2\alpha_{PH}+2)2!} \end{matrix}\right] \end{Bmatrix} \tag{45}$$

Equations (43), (44), and (45) are the unnormalized wave functions n=0 for ground state, n=1 for energy level 1, and n=2 for energy level 2, respectively. The visualization of the unnormalized wave functions is shown in Figures 3 and 4.

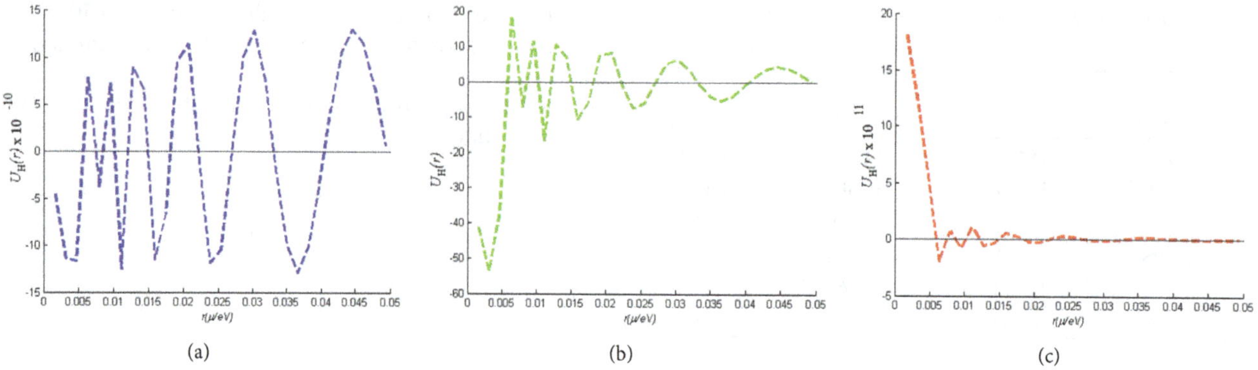

FIGURE 3: The unnormalized wave functions with $\alpha_{ML} = 0$ for (a) n=0, (b) n=1, and (c) n=2.

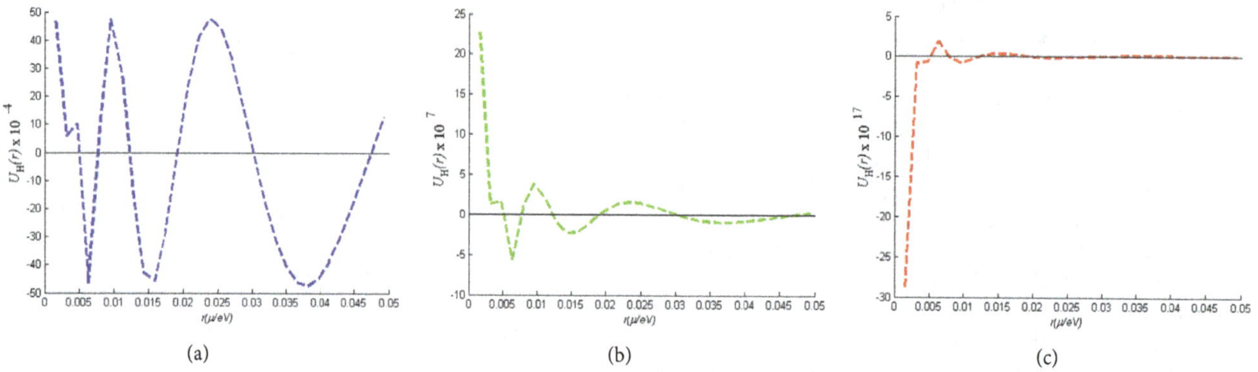

FIGURE 4: The unnormalized wave functions with $\alpha_{ML} = 0.1$ for (a) n=0, (b) n=1, and (c) n=2.

By inspecting Figures 3 and 4, we can see that the influence of minimal length parameter increases the amplitude of the unnormalized wave function.

6. Conclusion

The investigation of the minimal length formalism in the Klein-Gordon equation is obtained by approximate solution. The minimal length in Klein-Gordon equation for Hulthen potential is solved using Asymptotic Iteration Method. The Asymptotic Iteration Method is used to obtain the relativistic energy and unnormalized wave functions. The results show that the relativistic energies value increases for the larger minimal length parameter and for the larger quantum number (n). Then, the influence of minimal length parameter exerts effect in increasing the amplitude value of the unnormalized wave functions.

Conflicts of Interest

The authors declare that they have no conflicts of interest.

Acknowledgments

This research was partly supported by Ministry of Research, Technology, and Higher Education with Contract no. 089/SP2H/LT/DRPM/2018.

References

[1] A. D. Antia, A. N. Ikot, E. E. Ituen, and I. O. Akpan, "Bound state solutions of the Klein - Gordon equation for deformed Hulthen potential with position dependent mass," *Sri Lankan Journal of Physics*, vol. 13, no. 1, pp. 27–40, 2012.

[2] M. C. Zhang and Z. B. Wang, "Exact solutions of the Klein-Gordon equation with Makarov potential and a recurrence relation," *Chinese Physics*, vol. 16, no. 7, article 1863, 2007.

[3] S. Ortakaya, "Exact solutions of the klein—Gordon equation with ring-shaped oscillator potential by using the Laplace integral transform," *Chinese Physics B*, vol. 21, no. 7, 2012.

[4] H. Hassanabadi, S. Zarrinkamar, and H. Rahimov, "Approximate solution of D-dimensional Klein - Gordon equation with hulthén-type potential via SUSYQM," *Communications in Theoretical Physics*, vol. 56, no. 3, pp. 423–428, 2011.

[5] A. N. Ikot, L. E. Akpabiol, and E. J. Uwah, "Bound state solutions of the klein gordon equationwith the hulthen potential," *Electronic Journal of Theoretical Physics*, vol. 8, no. 25, pp. 225–232, 2011.

[6] D. A. Nugraha, A. Suparmi, C. Cari, and B. N. Pratiwi, "Asymptotic iteration method for solution of the Kratzer potential in D-dimensional Klein-Gordon equation," *Journal of Physics: Conference Series*, vol. 820, 2017.

[7] D. A. Nugraha, A. Suparmi, and C. Cari, "Asymptotic Iteration Method foranalytical solution of Klein Gordon equation for Trigonometric Pöschl-Teller potential in D dimensions," *Journal of Theoretical and Applied Physics*, vol. 795, no. 1, p. 42, 2017.

[8] A. Poszwa, "Relativistic generalizations of the quantum harmonic oscillator," *Acta Physica Polonica A*, vol. 126, no. 6, pp. 1226–1234, 2014.

[9] L. Garay, "Quantum gravity and minimum length," *International Journal of Modern Physics A*, vol. 10, no. 2, pp. 145–165, 1995.

[10] A. Tilbi, M. Merad, and T. Boudjedaa, "Particles of spin zero and 1/2 in electromagnetic field with confining scalar potential in modified heisenberg algebra," *Few-Body Systems*, vol. 56, no. 2-3, pp. 139–147, 2015.

[11] M. Chabab, A. El Batoul, A. Lahbas, and M. Oulne, "On γ-rigid regime of the Bohr-Mottelson Hamiltonian in the presence of a minimal length," *Physics Letters B*, vol. 758, pp. 212–218, 2016.

[12] M. Alimohammadi and H. Hassanabadi, "Alternative solution of the gamma-rigid Bohr Hamiltonian in minimal length formalism," *Nuclear Physics A*, vol. 957, pp. 439–449, 2017.

[13] T. K. Jana and P. Roy, "Exact solution of the Klein-Gordon equation in the presence of a minimal length," *Physics Letters A*, vol. 373, no. 14, pp. 1239–1241, 2009.

[14] M. Merad, F. Zeroual, and H. Benzair, "Spinless relativistic particle in the presence of a minimal length," *Electronic Journal of Theoretical Physics*, vol. 6, no. 23, pp. 41–56, 2010.

[15] I. L. Elviyanti, A. Suparmi, C. Cari, D. A. Nugraha, and B. N. Pratiwi, "Solution of Klein Gordon equation for hyperbolic cotangent potential in the presence of a minimal length using Hypergeometric method," *Journal of Physics: Conference Series*, vol. 909, no. 1, 2017.

[16] A. Suparmi, C. Cari, and I. L. Elviyanti, "Solution of Klein Gordon equation for trigonometric cotangent potential in the presence of a minimal length using Asymptotic Iteration Method," *Journal of Physics: Conference Series*, vol. 909, no. 1, 2017.

[17] C. Cari, A. Suparmi, and I. L. Elviyanti, "Approximate solution for the minimal length case of klein gordon equation for trigonometric cotangent potential using asymptotic iteration method," *Journal of Physics: Conf. Series*, vol. 909, 2017.

[18] S. M. Ikhdair and M. Hamzavi, "Effects of external fields on a two-dimensional Klein—Gordon particle under pseudo-harmonic oscillator interaction," *Chinese Physics B*, vol. 21, no. 11, p. 110302, 2012.

[19] H. Ciftci, R. L. Hall, and N. Saad, "Construction of exact solutions to eigenvalue problems by theasymptotic iteration method," *Journal of Physics A: Mathematical and General*, vol. 36, no. 47, pp. 11807–11816, 2003.

[20] B. N. Pratiwi, A. Suparmi, C. Cari, and A. S. Husein, "Asymptotic iteration method for the modified poschl-teller potential and trigonometric Scarf II non-central potential in the Dirac equation spin symmetry," *Pramana—Journal of Physics*, vol. 88, no. 2, 2017.

[21] S. Pramono, A. Suparmi, and C. Cari, "Relativistic energy analysis of five-dimensional q-deformed radial rosen-morse potential combined with q-deformed trigonometric scarf non-central potential using asymptotic iteration method," *Advances in High Energy Physics*, vol. 2016, 2016.

[22] Y. P. Varshni, "Eigenenergies and oscillator strengths for the Hulthén potential," *Physical Review A: Atomic, Molecular and Optical Physics*, vol. 41, no. 9, pp. 4682–4689, 1990.

[23] L. Naderi and H. Hassanabadi, "Bohr Hamiltonian with Eckart potential for triaxial nuclei," *The European Physical Journal Plus*, vol. 131, no. 5, 2016.

19

Rational Solutions of a Weakly Coupled Nonlocal Nonlinear Schrödinger Equation

Huijuan Zhou,[1] Chuanzhong Li,[1] and Yueh-Lung Lin[2]

[1] *Department of Mathematics, Ningbo University, Ningbo 315211, China*

[2] *School of Mathematical Sciences, University of Nottingham, Ningbo 315100, China*

Correspondence should be addressed to Chuanzhong Li; lichuanzhong@nbu.edu.cn

Academic Editor: Antonio Scarfone

In this article, we investigate an integrable weakly coupled nonlocal nonlinear Schrödinger (WCNNLS) equation including its Lax pair. Afterwards, Darboux transformation (DT) of the weakly coupled nonlocal NLS equation is constructed, and then the degenerated Darboux transformation can be got from Darboux transformation. Applying the degenerated Darboux transformation, the new solutions ($q^{[1]}$, $r^{[1]}$) and self-potential function ($V^{[1]}$) are created from the known solutions (q, r). The ($q^{[1]}$, $r^{[1]}$) satisfy the parity-time (PT) symmetry condition, and they are rational solutions with two free phase parameters of the weakly coupled nonlocal nonlinear Schrödinger equation. From the plots of solutions, the compression effects of the real refractive index profile and the gain-or-loss distribution are produced.

1. Introduction

A nonlocal nonlinear Schrödinger (NLS) equation was recently found and shown to be an integrable infinite dimensional Hamiltonian equation. Unlike the local case, the nonlinearly induced potential is PT-symmetric [1]; thus the nonlocal NLS equation is also PT-symmetric. According to the work of Bender and Boettcher [2], the PT-symmetry plays a vital role in the spectrum of the Hamiltonian. They proved that a broad class of non-Hermitian Hamiltons [3] with PT-symmetry have real and positive spectrum; the work drew attention for several researchers who study PT-symmetry in quantum mechanics [4–7]. For a non-Hermitian Hamiltonian $\mathbf{H} = d^2/dx^2 + V(x)$, it is PT-symmetric when $V(x) = V(-x)^*$. Under this condition, the Schrödinger equation $i\Psi_t = \mathbf{H}\Psi$ is PT-symmetric. It has been shown that optics can provide a good condition for testing the theory of PT-symmetry or observing the phenomenon when the PT-symmetry is broken [8–11].

As is well known, the nonlinear Schrödinger equation

$$q_t(x,t) - iq_{xx}(x,t) - 2i\left|q(x,t)\right|^2 q(x,t) = 0, \quad (1)$$

which is PT-symmetric, received an extensive study since the work by Shabat and Zakharov [12]. The NLS equation has many important physical applications, such as water wave [13], nonlinear optics [14], plasma physics [15], and so on.

Recently, Musslimani and Ablowitz [16] considered the solution of the nonlocal NLS equation,

$$\begin{aligned} &q_t(x,t) - iq_{xx}(x,t) \pm 2iV(x,t)\,q(x,t) = 0,\\ &V(x,t) = q(x,t)\,q^*(-x,t). \end{aligned} \quad (2)$$

We first recall that the nonlocal focusing NLS equation

$$q_t(x,t) - iq_{xx}(x,t) + 2iV(x,t)\,q(x,t) = 0, \quad (3)$$

from zero background by using the IST method, which implied the PT-symmetry, in other words, $x \longrightarrow -x$, $t \longrightarrow -t$ and the complex conjugation of the field envelope. Here $q(x,t)$ and $V(x,t)$ (called the PT-symmetric potential) imply the electric field envelope of the optical beam and complex refractive index distribution or an optical potential, respectively [7, 8], and the asterisk denotes the complex conjugation. In optics, the real part of $V(x,t)$ defined by $V_R(x,t)$

indicates the refractive index profile, and the imaginary part of $V(x,t)$ represented by $V_I(x,t)$ denotes the gain or loss [17] distribution. Based on $V_R(x,t)=V_R(-x,t)$ and $V_I(x,t) = -V_I(-x,t)$, a PT-symmetric system can be designed.

Nevertheless, the nonlocal property of the soliton equations is not new, for example, nonlocal symmetry [18]. In this purpose, we consider another new nonsymmetric coupled nonlocal NLS equation called weakly coupled nonlocal NLS equation which is as follows:

$$\begin{aligned} &q_t(x,t) - iq_{xx}(x,t) + 2iV(x,t)\,q(x,t) = 0, \\ &r_t(x,t) - ir_{xx}(x,t) + 2iq(x,t)^2 r^*(-x,t) \\ &\quad + 4iV(x,t)\,r(x,t) = 0, \\ &V(x,t) = q(x,t)\,q^*(-x,t). \end{aligned} \tag{4}$$

The concept of PT-symmetry, based on the non-Hermitian Hamiltonians [2, 19–22], has recently attracted much attention [23], in particular in the fields of optics and photonics [9, 24, 25]. This concept offers a fertile ground for PT-related notions and experiments. Furthermore, applying such idea for the design of photonic devices has opened up many new possibilities, which allow a controlled interplay between gain and loss. The loss is abundant in physical systems but is typically considered as a problem. The gain, however, as afforded by lasers, is valuable in optoelectronics because it provides means to overcome loss [26]. Several studies have shown that the PT-symmetric optical structures can exhibit particular properties that are otherwise unattainable in traditional Hermitian structures, for example, the possibility for breaking this symmetry through an abrupt phase transition [11, 27], the unidirectional invisibility [28], and so on. For a few overview papers in the area of linear and nonlinear PT-symmetric systems, see [29–31].

If a Hamiltonian is PT-symmetric, there are two possibilities: Either the eigenvalues are entirely real, in which case the Hamiltonian is said to be in an unbroken PT-symmetric phase, or else the eigenvalues are partly complex and partly real, in which case the Hamiltonian is said to be in a broken PT-symmetric phase. Moreover, there is a PT-symmetry-breaking threshold, where the transition is between broken and unbroken symmetry. At this point, the behavior of Hamiltonians becomes even more interesting. In this paper, the solutions of (4) at the threshold value of eigenvalues are derived by degenerated DT. It is found that the solutions possess a similar structure with the second-order soliton, but both of them are different because the asymptotic amplitudes at two orbits for our rational solutions depend on each other and there does not exist any phase shift after the interaction. Besides, we consider the analyticity related to the phase parameters and classify the solution according to its asymptotic amplitudes on the phase parameters. We have provided the rational solutions $(q^{[1]}, r^{[1]})$ from the weakly nonlocal NLS equation, and the analytic properties of $(q^{[1]}, r^{[1]})$ are proved in detail.

This paper is arranged as follows: In Section 2, we construct a weakly coupled nonlocal NLS equation and its Lax representation. In Section 3, we calculate the Darboux transformation of the weakly coupled nonlocal NLS equation. In Section 4, we consider the rational solutions of the weakly coupled nonlocal NLS equation. Please see Section 5 for the conclusion.

2. Weakly Coupled Nonlocal NLS Equation and Its Lax Representation

The weakly coupled nonlocal NLS equation (4) can be expressed by the compatibility condition of the following Lax pair:

$$\begin{aligned} \Phi_x(x,t;\lambda) &= M(x,t;\lambda)\,\Phi(x,t;\lambda), \\ \Phi_t(x,t;\lambda) &= N(x,t;\lambda)\,\Phi(x,t;\lambda), \end{aligned} \tag{5}$$

where $\Phi(x,t;\lambda)$ is a matrix valued function, $\lambda \in \mathbb{C}$ is the spectral parameter, and

$$\begin{aligned} &M(x,t;\lambda) \\ &= \begin{pmatrix} -i\lambda E & q(x,t)\,E + r(x,t)\,\Gamma \\ q^*(-x,t)\,E + r^*(-x,t)\,\Gamma & i\lambda E \end{pmatrix}, \\ &N(x,t;\lambda) = \begin{pmatrix} N_1 & N_2 \\ N_3 & N_4 \end{pmatrix}, \end{aligned} \tag{6}$$

where

$$\begin{aligned} N_1 &= \left(-2i\lambda^2 - iq(x,t)\,q^*(-x,t)\right)E \\ &\quad - i\left(q(x,t)\,r^*(-x,t) + q^*(-x,t)\,r(x,t)\right)\Gamma, \\ N_2 &= \left(2\lambda q(x,t) + iq_x(x,t)\right)E \\ &\quad + \left(2\lambda r(x,t) + ir_x(x,t)\right)\Gamma, \\ N_3 &= \left(2\lambda q^*(-x,t) - iq_x^*(-x,t)\right)E \\ &\quad + \left(2\lambda r^*(-x,t) - ir_x^*(-x,t)\right)\Gamma, \\ N_4 &= \left(2i\lambda^2 + iq(x,t)\,q^*(-x,t)\right)E \\ &\quad + i\left(q(x,t)\,r^*(-x,t) + q^*(-x,t)\,r(x,t)\right)\Gamma, \end{aligned} \tag{7}$$

$$\begin{aligned} E &= \begin{pmatrix} 1 & 0 \\ 0 & 1 \end{pmatrix}, \\ \Gamma &= \begin{pmatrix} 0 & 0 \\ 1 & 0 \end{pmatrix}. \end{aligned} \tag{8}$$

That is, $M_t - N_x + [M,N] = 0$ leads to a weakly coupled nonlocal NLS equation (4),

$$\begin{aligned} &q_t(x,t) - iq_{xx}(x,t) + 2iV(x,t)\,q(x,t) = 0, \\ &r_t(x,t) - ir_{xx}(x,t) + 2iq(x,t)^2 r^*(-x,t) \\ &\quad + 4iV(x,t)\,r(x,t) = 0, \\ &V(x,t) = q(x,t)\,q^*(-x,t). \end{aligned} \tag{9}$$

3. Darboux Transformation for the Weakly Coupled Nonlocal NLS Equation

It was pointed out that the Darboux transformation is a method proposed by J. G. Darboux more than 100 years ago and it is an efficient method to generate soliton solutions of integrable equations. In this section, we would like to introduce the Darboux transformation [32, 33] and degenerated Darboux transformation for the weakly coupled nonlocal NLS equation. First, we will introduce a simple gauge transformation of spectral problems (5) as follows:

$$\Phi^{[1]} = T\Phi. \tag{10}$$

It can transform linear problems (5) into a new one

$$\begin{aligned} \Phi_x^{[1]} &= M^{[1]}\Phi^{[1]}, \\ \Phi_t^{[1]} &= N^{[1]}\Phi^{[1]}, \end{aligned} \tag{11}$$

where $M^{[1]}$, $N^{[1]}$ have the same forms with M, N except that the $q(x,t)$, $r(x,t)$, $q^*(-x,t)$, $r^*(-x,t)$ in the matrices M, N are replaced with $q^{[1]}(x,t)$, $r^{[1]}(x,t)$, $q^{[1]*}(-x,t)$, $r^{[1]*}(-x,t)$ in the matrices $M^{[1]}$, $N^{[1]}$. It is easy to obtain the equations

$$\begin{aligned} M^{[1]}T &= T_x + TM, \\ N^{[1]}T &= T_t + TN, \end{aligned} \tag{12}$$

which lead to

$$\begin{aligned} M^{[1]} &= T_xT^{-1} + TMT^{-1}, \\ N^{[1]} &= T_zT^{-1} + TNT^{-1}. \end{aligned} \tag{13}$$

In general, if the transformation T is a polynomial of the parameter λ, we can start from

$$T = T_1\lambda + T_0, \tag{14}$$

where

$$\begin{aligned} T_1 &= \begin{pmatrix} a_{10}E + a_{11}\Gamma & b_{10}E + b_{11}\Gamma \\ c_{10}E + c_{11}\Gamma & d_{10}E + d_{11}\Gamma \end{pmatrix}, \\ T_0 &= \begin{pmatrix} a_0E + a_1\Gamma & b_0E + b_1\Gamma \\ c_0E + c_1\Gamma & d_0E + d_1\Gamma \end{pmatrix}. \end{aligned} \tag{15}$$

For (13), we can get

$$\begin{aligned} T_x &= T_{1x}\lambda + T_{0x} \\ &= \begin{pmatrix} a_{10x}E + a_{11x}\Gamma & b_{10x}E + b_{11x}\Gamma \\ c_{10x}E + c_{11x}\Gamma & d_{10x}E + d_{11x}\Gamma \end{pmatrix}\lambda \\ &\quad + \begin{pmatrix} a_{0x}E + a_{1x}\Gamma & b_{0x}E + b_{1x}\Gamma \\ c_{0x}E + c_{1x}\Gamma & d_{0x}E + d_{1x}\Gamma \end{pmatrix} \\ &= \begin{pmatrix} \mathbf{0} & -2ib_{10}E - 2ib_{11}\Gamma \\ 2ic_{10}E + 2ic_{11}\Gamma & \mathbf{0} \end{pmatrix}\lambda^2 \\ &\quad + \begin{pmatrix} H1 & H2 \\ H3 & H4 \end{pmatrix}\lambda + \begin{pmatrix} K1 & K2 \\ K3 & K4 \end{pmatrix}, \end{aligned} \tag{16}$$

where

$$\begin{aligned} H_1 &= \left(q^{[1]}(x,t)c_{10} - b_{10}q^*(-x,t)\right)E + \left(r^{[1]}(x,t)c_{10}\right. \\ &\quad \left. + q^{[1]}(x,t)c_{11} - b_{10}r^*(-x,t) - b_{11}q^*(-x,t)\right)\Gamma, \\ H_2 &= \left(-2ib_0 + q^{[1]}(x,t)d_{10} - a_{10}q(x,t)\right)E + \left(-2ib_1\right. \\ &\quad + q^{[1]}(x,t)d_{11} + r^{[1]}(x,t)d_{10} - a_{10}r(x,t) \\ &\quad \left. - a_{11}q(x,t)\right)\Gamma, \\ H_3 &= \left(2ic_0 + q^{[1]*}(-x,t)a_{10} - d_{10}q^*(-x,t)\right)E \\ &\quad + \left(2ic_1 + q^{[1]*}(-x,t)a_{11} + r^{[1]*}(-x,t)a_{10}\right. \\ &\quad \left. - d_{10}r^*(-x,t) - d_{11}q^*(-x,t)\right)\Gamma, \\ H_4 &= \left(q^{[1]*}(-x,t)b_{10} - c_{10}q(x,t)\right)E \\ &\quad + \left(q^{[1]*}(-x,t)b_{11} + r^{[1]*}(-x,t)b_{10} - c_{10}r(x,t)\right. \\ &\quad \left. - c_{11}q(x,t)\right)\Gamma, \\ K_1 &= \left(q^{[1]}c_0 - q^*(-x,t)b_0\right)E + \left(r^{[1]}(x,t)c_0\right. \\ &\quad \left. + q^{[1]}(x,t)c_1 - r^*(-x,t)b_0 - b_1q^*(-x,t)\right)\Gamma, \\ K_2 &= \left(q^{[1]}d_0 - a_0q(x,t)\right)E + \left(q^{[1]}(x,t)d_1\right. \\ &\quad \left. + r^{[1]}(x,t)d_0 - a_0r(x,t) - a_1q(x,t)\right)\Gamma, \\ K_3 &= \left(q^{[1]*}(-x,t)a_0 - d_0q^*(-x,t)\right)E \\ &\quad + \left(q^{[1]*}(-x,t)a_1 + r^{[1]*}(-x,t)a_0 - r^*(-x,t)d_0\right. \\ &\quad \left. - d_1q^*(-x,t)\right)\Gamma, \\ K_4 &= \left(q^{[1]*}(-x,t)b_0 - c_0q(x,t)\right)E + \left(q^{[1]*}(-x,t)b_1\right. \\ &\quad \left. + r^{[1]*}b_0 - c_0r(x,t) - c_1q(x,t)\right)\Gamma. \end{aligned} \tag{17}$$

By comparing the coefficient in terms of λ, we can see

$$\begin{aligned} &a_{10x} = a_{11x} = 0, \\ &b_{10} = b_{11} = 0, \\ &d_{10x} = d_{11x} = 0, \\ &c_{10} = c_{11} = 0, \\ &q^{[1]}(x,t)d_{10} - 2ib_0 - q(x,t)a_{10} = 0, \\ &r^{[1]}(x,t)d_{10} + q^{[1]}(x,t)d_{11} - 2ib_1 - q(x,t)a_{11} \\ &\quad - r(x,t)a_{10} = 0. \end{aligned} \tag{18}$$

So functions a_{10}, a_{11}, d_{10}, d_{11} are independent of x. For $T_t + TN = N^{[1]}T$, we can get that functions a_{10}, a_{11}, d_{10},

d_{11} should be independent of t using the same method. This suggests that a_{10}, a_{11}, d_{10}, and d_{11} are constants. Then the form of Darboux transformation operator is as follows:

$$\begin{aligned} T &= T_1(\lambda) \\ &= \begin{pmatrix} a_{10}E + a_{11}\Gamma & \mathbf{0} \\ \mathbf{0} & d_{10}E + d_{11}\Gamma \end{pmatrix} \lambda \\ &+ \begin{pmatrix} a_0E + a_1\Gamma & b_0E + b_1\Gamma \\ c_0E + c_1\Gamma & d_0E + d_1\Gamma \end{pmatrix}. \end{aligned} \tag{19}$$

Now we will try to determine the specific expression of the matrix T. In order to solve this problem, we must use eigenfunction which contains λ to determine their expressions through the parameterized method. Next, we will introduce the properties of eigenfunction in spectral problem, with lemma as follows.

Lemma 1. *The eigenfunctions possess the following property:*

$$\begin{aligned} &\begin{pmatrix} \phi_{110}(x,t;\lambda_1)E + \phi_{111}(x,t;\lambda_1)\Gamma \\ \phi_{120}(x,t;\lambda_1)E + \phi_{121}(x,t;\lambda_1)\Gamma \end{pmatrix} \\ &= \begin{pmatrix} \phi_{120}^*(-x,t;-\lambda_1^*)E + \phi_{121}^*(-x,t;-\lambda_1^*)\Gamma \\ -\phi_{110}^*(-x,t;-\lambda_1^*)E - \phi_{111}^*(-x,t;-\lambda_1^*)\Gamma \end{pmatrix}, \end{aligned} \tag{20}$$

where

$$\Phi_1(x,t;\lambda_1) = \begin{pmatrix} \phi_{110}(x,t;\lambda_1)E + \phi_{111}(x,t;\lambda_1)\Gamma \\ \phi_{120}(x,t;\lambda_1)E + \phi_{121}(x,t;\lambda_1)\Gamma \end{pmatrix} \tag{21}$$

is an eigenfunction associated with $\lambda = \lambda_1$.

Proof. If $\Phi_1(x,t;\lambda_1)$ is the solution of (5),

$$\begin{aligned} \phi_{110x}(x,t;\lambda_1) &= -i\lambda_1\phi_{110}(x,t;\lambda_1) \\ &\quad + q(x,t)\phi_{120}(x,t;\lambda_1), \\ \phi_{111x}(x,t;\lambda_1) &= -i\lambda_1\phi_{111}(x,t;\lambda_1) \\ &\quad + q(x,t)\phi_{121}(x,t;\lambda_1) \\ &\quad + r(x,t)\phi_{120}(x,t;\lambda_1), \\ \phi_{120x}(x,t;\lambda_1) &= i\lambda_1\phi_{120}(x,t;\lambda_1) \\ &\quad + q^*(-x,t)\phi_{110}(x,t;\lambda_1), \\ \phi_{121x}(x,t;\lambda_1) &= i\lambda_1\phi_{121}(x,t;\lambda_1) \\ &\quad + q^*(-x,t)\phi_{111}(x,t;\lambda_1) \\ &\quad + r^*(-x,t)\phi_{110}(x,t;\lambda_1). \end{aligned} \tag{22}$$

Take the conjugate for the above equation on both sides at the same time, and let $x = -x$:

$$\begin{aligned} \phi_{110x}^*(-x,t;\lambda_1) &= i\lambda_1^*\phi_{110}^*(-x,t;\lambda_1) \\ &\quad + q(-x,t)^*\phi_{120}^*(-x,t;\lambda_1), \\ \phi_{111x}^*(-x,t;\lambda_1) &= i\lambda_1^*\phi_{111}^*(-x,t;\lambda_1) \\ &\quad + q(-x,t)^*\phi_{121}^*(-x,t;\lambda_1) \\ &\quad + r(-x,t)^*\phi_{120}^*(-x,t;\lambda_1), \\ \phi_{120x}^*(-x,t;\lambda_1) &= -i\lambda_1^*\phi_{120}^*(-x,t;\lambda_1) \\ &\quad + q(x,t)\phi_{110}^*(-x,t;\lambda_1), \\ \phi_{121x}^*(-x,t;\lambda_1) &= -i\lambda_1^*\phi_{121}^*(-x,t;\lambda_1) \\ &\quad + q(x,t)\phi_{111}^*(-x,t;\lambda_1) \\ &\quad + r(x,t)\phi_{110}^*(-x,t;\lambda_1), \end{aligned} \tag{23}$$

so if $\lambda_1 = -\lambda_1^*$, then

$$\begin{aligned} \phi_{110x}^*(-x,t;-\lambda_1^*) &= -i\lambda_1\phi_{110}^*(-x,t;-\lambda_1^*) \\ &\quad + q(-x,t)^*\phi_{120}^*(-x,t;-\lambda_1^*), \\ \phi_{111x}^*(-x,t;-\lambda_1^*) &= -i\lambda_1\phi_{111}^*(-x,t;-\lambda_1^*) \\ &\quad + q(-x,t)^*\phi_{121}^*(-x,t;-\lambda_1^*) \\ &\quad + r(-x,t)^*\phi_{120}^*(-x,t;-\lambda_1^*), \\ \phi_{120x}^*(-x,t;-\lambda_1^*) &= i\lambda_1\phi_{120}^*(-x,t;-\lambda_1^*) \\ &\quad + q(x,t)\phi_{110}^*(-x,t;-\lambda_1^*), \\ \phi_{121x}^*(-x,t;-\lambda_1^*) &= i\lambda_1\phi_{121}^*(-x,t;-\lambda_1^*) \\ &\quad + q(x,t)\phi_{111}^*(-x,t;-\lambda_1^*) \\ &\quad + r(x,t)\phi_{110}^*(-x,t;-\lambda_1^*). \end{aligned} \tag{24}$$

Taking a similar procedure, the symmetry property also holds for the t-part of the Lax pair; that is, if $\lambda_1 = -\lambda_1^*$, the eigenfunction

$$\begin{pmatrix} \phi_{120}^*(-x,t;-\lambda_1^*)^*E + \phi_{121}^*(-x,t;-\lambda_1^*)\Gamma \\ -\phi_{110}^*(-x,t;-\lambda_1^*)E - \phi_{111}^*(-x,t;-\lambda_1^*)\Gamma \end{pmatrix} \tag{25}$$

is also the eigenfunction corresponding to λ_1, because it satisfies the same Lax equations. We can let

$$\begin{aligned} \phi_{210}(x,t;\lambda_1) &= \phi_{120}^*(-x,t;-\lambda_1^*), \\ \phi_{211}(x,t;\lambda_1) &= \phi_{121}^*(-x,t;-\lambda_1^*), \\ \phi_{220}(x,t;\lambda_1) &= -\phi_{110}^*(-x,t;-\lambda_1^*), \\ \phi_{221}(x,t;\lambda_1) &= -\phi_{111}^*(-x,t;-\lambda_1^*), \end{aligned} \tag{26}$$

and $\lambda_2 = -\lambda_1^*$ in the following. □

According to Lemma 1, by choosing $a_{10} = a_{11} = 1, d_{10} = d_{11} = 1$, we can get the onefold Darboux transformation for the weakly coupled NLS equation as follows:

$$\begin{aligned} T &= T_1(\lambda) \\ &= \begin{pmatrix} (\lambda + a_0) E + (\lambda + a_1)\Gamma & b_0 E + b_1 \Gamma \\ c_0 E + c_1 \Gamma & (\lambda + d_0) E + (\lambda + d_1)\Gamma \end{pmatrix} \\ &= \begin{pmatrix} E+\Gamma & 0 \\ 0 & E+\Gamma \end{pmatrix}\lambda + \begin{pmatrix} a_0 E + a_1\Gamma & b_0 E + b_1\Gamma \\ c_0 E + c_1\Gamma & d_0 E + d_1\Gamma \end{pmatrix}, \end{aligned} \tag{27}$$

and then the new solution $q^{[1]}$, $r^{[1]}$ can be obtained by the Darboux transformation

$$\begin{aligned} q^{[1]}(x,t) &= q(x,t) + 2ib_0, \\ r^{[1]}(x,t) &= r(x,t) + 2i(b_1 - b_0). \end{aligned} \tag{28}$$

The Darboux transformation matrix T_1 must satisfy the following equation:

$$T_1(\lambda, \lambda_j)\Phi_j = 0, \quad j = 1, 2. \tag{29}$$

From the above equation, we can get the expressions of functions b_0, b_1 easily by the eigenfunctions as follows:

$$\begin{aligned} b_0 &= \frac{(\lambda_1 - \lambda_2)\phi_{110}\phi_{210}}{\phi_{110}\phi_{220} - \phi_{120}\phi_{210}}, \\ b_1 &= \frac{(\lambda_1 - \lambda_2)\left(\phi_{110}^2(\phi_{211}\phi_{220} - \phi_{210}\phi_{221}) + \phi_{210}^2(\phi_{110}\phi_{121} - \phi_{111}\phi_{120})\right)}{(\phi_{110}\phi_{220} - \phi_{120}\phi_{210})^2}. \end{aligned} \tag{30}$$

Combining with (28), the following theorem can be derived.

Theorem 2. *The Darboux transformation of the weakly coupled NLS equation can be written:*

$$\begin{aligned} q^{[1]} &= q + 2i\frac{(\lambda_1 - \lambda_2)\phi_{110}\phi_{210}}{\phi_{110}\phi_{220} - \phi_{120}\phi_{210}}, \\ r^{[1]} &= r + 2i\frac{(\lambda_1 - \lambda_2)\left(\phi_{110}^2(\phi_{220}(\phi_{211} - \phi_{210}) - \phi_{210}\phi_{221}) + \phi_{210}^2(\phi_{110}(\phi_{120} + \phi_{121}) - \phi_{111}\phi_{120})\right)}{(\phi_{110}\phi_{220} - \phi_{120}\phi_{210})^2}, \end{aligned} \tag{31}$$

Then, degenerated Darboux transformation of the weakly coupled NLS equation is given explicitly by a series expansion of the $\phi_{ijk}(\lambda = \lambda_1 + \epsilon)$, $i, j = 1, 2$, $k = 0, 1$. Degenerated Darboux transformation of the weakly coupled nonlocal NLS equation will help us to find the rational solution in the next section.

4. The Rational Solutions of the Weakly Coupled Nonlocal NLS Equation

From the previous section, degenerated Darboux transformation of the weakly coupled nonlocal NLS equation has been discussed. In this section, we will be focused on a kind of solutions which starts from seed solutions by the degenerated Darboux transformation. The solutions are called rational solutions. The seed solutions can be assumed as $(q(x,t), r(x,t)) = (ce^{ibt}, ice^{ibt})$, b, c being arbitrary real constants. Substituting seed solution into (4), through a proper simplification, we can obtain a constraint

$$b = -2c^2. \tag{32}$$

Taking the transformation

$$\begin{aligned} \Phi_x(x,t;\lambda) &= \psi_x(x,t;\lambda)e^{ibt\sigma}, \\ \sigma &= \begin{pmatrix} 1 & 0 \\ 0 & -1 \end{pmatrix}, \end{aligned} \tag{33}$$

then

$$\begin{aligned} \psi_x(x,t;\lambda) &= \widehat{M}\psi(x,t;\lambda) = \begin{pmatrix} -i\lambda E & cE \\ cE & i\lambda E \end{pmatrix}, \\ \psi_t(x,t;\lambda) &= \widehat{N}\psi(x,t;\lambda) = \begin{pmatrix} -2i\lambda^2 E & 2c\lambda E \\ 2c\lambda E & 2i\lambda^2 E \end{pmatrix}. \end{aligned} \tag{34}$$

So solving the Lax pair equation (34) is equivalent to solving (5).

By defining

$$R = \sqrt{c^2 - \lambda^2}, \tag{35}$$

the following basic solutions ψ_{110}, ψ_{111}, ψ_{120}, ψ_{121} are obtained in terms of R from the the spectral problem of (5) as

$$\begin{aligned}
\psi_{110}(R) &= e^{2tR\lambda+xR},\\
\psi_{111}(R) &= e^{2tR\lambda+xR},\\
\psi_{120}(R) &= \frac{(R+i\lambda)}{c}e^{2tR\lambda+xR},\\
\psi_{121}(R) &= \frac{(R+i\lambda)}{c}e^{2tR\lambda+xR}.
\end{aligned} \tag{36}$$

Here Lemma 1 and the superposition principle are adopted to construct the nontrivial eigenfunctions ϕ_{110}, ϕ_{120}, ϕ_{111}, ϕ_{121}:

$$\begin{aligned}
\phi_{110} &= J\psi_{110}(R)(x,t,\lambda) + K\psi_{120}^{*}(R)(-x,t,-\lambda)\\
&= Je^{2tR\lambda+xR} + K\frac{(R+i\lambda)}{c}e^{-2tR\lambda-xR},\\
\phi_{111} &= J\psi_{111}(R)(x,t,\lambda) + K\psi_{121}^{*}(R)(-x,t,-\lambda)\\
&= Je^{2tR\lambda+xR} + K\frac{(R+i\lambda)}{c}e^{-2tR\lambda-xR},\\
\phi_{120} &= J\psi_{120}(R)(x,t,\lambda) - K\psi_{110}^{*}(R)(-x,t,-\lambda)\\
&= J\frac{(R+i\lambda)}{c}e^{2tR\lambda+xR} - Ke^{-2tR\lambda-xR},\\
\phi_{121} &= J\psi_{121}(R)(x,t,\lambda) - K\psi_{111}^{*}(R)(-x,t,-\lambda)\\
&= J\frac{(R+i\lambda)}{c}e^{2tR\lambda+xR} - Ke^{-2tR\lambda-xR},
\end{aligned} \tag{37}$$

where $J = J_1 + iJ_2$, $K = K_1 + iK_2$, J_1, J_2, K_1, K_2 are arbitrary real constants. It can be proved that ϕ_{110}, ϕ_{111}, ϕ_{120}, and ϕ_{121} are also the solutions of the Lax equation with $\lambda_1 = c + \epsilon$. Now applying the degenerated Darboux transformation, we can get one kind of rational solution of the nonlocal NLS equation at the threshold $\lambda = c$:

$$\begin{aligned}
q^{[1]} &= \frac{F_1}{G_1}e^{-2it},\\
r^{[1]} &= \frac{F_2+F_3+F_4}{G_1^2}e^{-2it},
\end{aligned} \tag{38}$$

where

$$\begin{aligned}
F_1 &= -8\left(J_1^2+J_2^2-2J_2+1\right)t^2 - 8\left(iJ_1^2+iJ_2^2-2iJ_2+i\right.\\
&\quad \left.+J_1\right)t - 2\left(2J_2-J_1^2-J_2^2-1\right)x^2 - 2\left(iJ_2^2+iJ_1^2-i\right)\\
&\quad \cdot x - 4iJ_1 + J_1^2 + J_2^2 - 4J_2 + 1,\\
G_1 &= 8\left(J_1^2+J_2^2-2J_2+1\right)t^2 + 8J_1t + 2\left(2J_2-J_1^2-J_2^2\right.\\
&\quad \left.-1\right)x^2 + 2i\left(J_1^2+J_2^2-1\right)x + J_1^2 + J_2^2 + 1,
\end{aligned}$$

$$\begin{aligned}
F_2 &= ((2+i)\\
&\quad \cdot \left(64\left(J_1^2+J_2^2\right)^2 - 256J_2\left(J_2^2+J_1^2+1\right) + J_2^2\right)\\
&\quad + 128iJ_1^2 + 64(6+i))t^4 + 64(i\left(J_2^2+J_1^2\right)^2\\
&\quad + 4\left(J_2-1\right)^2\left(J_1-iJ_2\right)\\
&\quad + 2J_1\left(i\left(\left(J_1+1\right)^2 - J_1\left(1+2J_2\right)\right) + 2\left(J_1^2-iJ_2\right)\right)\\
&\quad + 2iJ_2^2\left(J_1-1\right)+i)t^3 + (16(i+1)\\
&\quad \cdot \left(\left(J_1^2+J_2^2\right)^2+1\right) + 32J_1^2\left(3i\left(1+J_1\right)-iJ_2+5\right)\\
&\quad + 32J_2^2\left(3iJ_1-iJ_2+i-1\right) - 32iJ_2\\
&\quad + 96iJ_1\left(1-2J_2\right))t^2\\
&\quad + 8(2iJ_2\left(J_2-1-\left(\left(J_1^2-J_2\right)^2\right)\right)\\
&\quad + J_1\left(6iJ_1+2(i+1)+J_1^2\left(2(1+i)-iJ_1\right)\right)\\
&\quad + i\left(\left(J_1^2+J_2^2\right)^2+J_1^4+1\right))t,\\
F_3 &= 4(2+i)(\left(J_1^2+J_2^2\right)^2 - \left(4J_2^2-6J_2+8\right)J_2\\
&\quad + 2J_1^2\left(1-2J_2\right)+1)x^4 + (2i-1)(8-8\left(J_1^2+J_2^2\right)^2\\
&\quad + J_2\left(J_1^2+16\left(J_2^2-1\right)\right))x^3\\
&\quad + (J_1^2\left(8\left(1+iJ_2-iJ_1\right)-12J_1^2\right))\\
&\quad - 8i(J_2^2\left(24-12J_2^2+8iJ_2\right) - 12 + \left(J_1^2+J_2^2\right)^2\\
&\quad + J_1\left(1-J_2\right)^2)x^2 + (4(i-1)\left(\left(J_1^2+J_2^2\right)^2-1\right)\\
&\quad + 8\left(J_1\left(J_2\left(iJ_1-J_2\right)+1\right) - J_1^3 - iJ_2 + iJ_2^3\right))x,\\
F_4 &= (J_1^2\left(64(2+i)\left(2J_2-1\right)-24J_2^2\right) + (2+i)\\
&\quad \cdot \left(128J_2 - 32\left(J_1^2+J_2^2\right)^2 + J_2^3 - 192J_2^2 - 32\right))t^2x^2\\
&\quad + 32((2i-1)\left(J_1^2+J_2^2\right)^2 + J_1^2\left(1+2J_2(1-2i)\right)\\
&\quad - 2J_2(1-2i)\left(1-J_2^2\right) - 2i)t^2x\\
&\quad + (32J_2^2\left(2iJ_2-2i-J_1(2+i)\right) - 32i\left(J_1-J_2\right)^2\\
&\quad - 16i\left(J_1^2+J_2^2\right)^2 - 16i - 32J_1^3(i+2)\\
&\quad + 32J_1\left(4J_2-i-2\right) + 64iJ_2\left(J_1^2+1\right))tx^2 + 32(J_2^3\\
&\quad + J_2\left(J_1^2-1\right) - 16\left(J_1^2+J_2^2\right)^2 + (2i-1)
\end{aligned}$$

$$\cdot J_1\left(J_1^2+J_2^2-1\right)\Big)tx+i+2iJ_1^2\left(2J_2+1\right)+2iJ_2^2\left(1+2J_1\right)+i\left(J_1^2+J_2^2\right)^2+4i\left(J_1+J_1^3\right)+4\left(J_2+4J_2^3\right). \quad (39)$$

Here, for convenience, we have set $c = 1$. It can be verified that the solutions (38) satisfy the PT-symmetry. Furthermore, the solutions are analytic under certain condition.

Theorem 3. *If $J_1^2 + J_2^2 \neq 1$, then $|q^{[1]}|$ and $|r^{[1]}|$ are analytic. Otherwise, $|q^{[1]}|$ and $|r^{[1]}|$ have singular points.*

Proof. First, we display that $|q^{[1]}|$ and $|r^{[1]}|$ are analytic when $J_1^2 + J_2^2 \neq 1$. In (38) $|q^{[1]}|$ and $|r^{[1]}|$ are rational solutions, so the singular points just come from zero points of the denominator $|G_1|$; that is, it is enough to prove $|G_1| \neq 0$ when $J_1^2 + J_2^2 \neq 1$.

We know that $|G_1| = 0$ is equivalent to $G_{1R} = 0$ and $G_{1I} = 0$, where G_{1R} and G_{1I} denote the real part and imaginary part of G_1, respectively.

If $G_{1I} = 0$, we can get $x = 0$ in the case of $J_1^2 + J_2^2 \neq 1$. Substituting $x = 0$ into $G_{1R} = 0$, then

$$G_{1R} = \left(8J_1^2+8J_2^2-16J_2+8\right)t^2+8J_1t+J_1^2+J_2^2+1 = 0, \quad (40)$$

which implies

$$t = \frac{-2J_1 \pm \sqrt{\Delta_1}}{4\left(J_1^2+J_2^2-2J_2+1\right)}, \qquad \Delta_1 = -2\left(J_1^2+J_2^2-J_2\right)^2-2\left(J_2-1\right)^2. \quad (41)$$

Through simple calculation, we find that Δ_1 arrives at its maximum value 0 when $J_1 = 0$ and $J_1 = 1$. Since $J_1^2 + J_2^2 \neq 1$ is always negative, this implies there is no $t \in R$ making $G_{1R} = 0$. Therefore, the rational solutions (38) are analytic when $J_1^2 + J_2^2 \neq 1$.

Second, let $J_1^2 + J_2^2 = 1$ in (38). For convenience, we define $J_1 = \cos(\theta)$ and $J_2 = \sin(\theta)$. Then we can get

$$q^{[1]} = \frac{-8it\sin(\theta)-8t^2\sin(\theta)+2x^2\sin(\theta)+2i\cos(\theta)+8it+4t\cos(\theta)+8t^2-2x^2+2\sin(\theta)-1}{Q},$$
$$r^{[1]} = \frac{\left(i\left(\sin(\theta)\left(-16it^2+4ix^2+8t^2-2x^2+2i+8t\right)+\cos(\theta)\left(8it-2-4t\right)-8t^2+2x^2+16it^2-4ix^2-8t-1\right)\right)}{Q}, \quad (42)$$

where $Q = (2-2\sin(\theta))x^2 + 8\sin(\theta)t^2 - 4t\cos(\theta) - 8t^2 - 1$.

When $\sin(\theta) = 1$, $(q^{[1]}, r^{[1]})$ are analytic but trivial owing to $q^{[1]} = -1, r^{[1]} = 2 + I$ in this case. So when $\sin(\theta) \neq 1$ we need to prove that there always exist $x \in R$ and $t \in R$ making Q be zero.

Now we need to prove the discriminant of Q about x

$$\Delta_2 = 8\left(\sin(\theta)-1\right)\left(8\sin(\theta)t^2-4\cos(\theta)t-8t^2-1\right) \quad (43)$$

is positive. Owing to $\sin(\theta) - 1 < 0$, let $f(\theta) = 8t^2(\sin(\theta) - 1) - 4\cos(\theta)t - 1$; we need to prove $f(\theta) < 0$. Let $f(\theta_0)' = 0$; then $\theta_0 = -\arctan(2t)$, taking θ_0 into $f(\theta)$, leading to $f(\theta_0) = -4t\sqrt{4t^2+1} - 8t^2 - 1$. Obviously, there exists $t \in R$ making $f(\theta_0) < 0$ so that the discriminant $\Delta_2 > 0$. Thus, $q^{[1]}$ and $r^{[1]}$ have singular points when $J_1^2 + J_2^2 = 1$. □

The dynamical structures of $(|q_1|^2, |r_1|^2)$ under the condition $J_1^2 + J_2^2 \neq 1$ were shown in Figures 1 and 3. When $x \longrightarrow \infty, t \longrightarrow \infty$, we can know $|q^{[1]}|^2 \longrightarrow 1, |r^{[1]}|^2 \longrightarrow 5$ with c=1. In Figure 1, we find that $(|q_1|^2)$ has four types of patterns. The first and second cases, shown in Figures 1(a) and 1(b), display two wavefronts, a bright one and a dark one, moving from $t = -\infty$ to $t = \infty$ on two straight lines and intersecting with each other in the neighborhood of the origin. At the intersection point, amplitudes of the bright and the dark wavelet decrease rapidly, and then their amplitudes simultaneously increase rapidly to their original amplitudes before the intersection. The only difference between Figures 1(a) and 1(b) is that the bright and dark wavefronts exchange their locations. The third and fourth cases, shown in Figures 1(c) and 1(d), display the evolution of two bright wavefronts, and the amplitudes of these bright wavefronts have similar properties as the wavefronts in the first and second cases. In Figure 3, we find that $(|r_1|^2)$ has two types of patterns. The first case, shown in Figure 3(a), is similar to Figure 1(b). The second case, shown in Figure 3(b), is similar to Figure 1(a). Figure 2 and Figure 4 are density plots of Figure 1 and Figure 3, respectively.

In the end of the section, we study the self-induced potential function $V^{[1]}(x,t)$, the PT-symmetric potential $V^{[1]}(x,t) = V_R^{[1]} + iV_I^{[1]} = q^{[1]}(x,t)q^{[1]}(-x,t)$,

$$V^{[1]}(x,t) = -\frac{F_5}{F_6}, \quad (44)$$

$$F_5 = \left(8itJ_1^2+2ixJ_1^2+8it-2ix+8t^2J_1^2+8t^2J_2^2-2x^2J_1^2-2x^2J_2^2-16itJ_2-16t^2J_2+4x^2J_2+4iJ_1+8itJ_2^2+2ixJ_2^2+8t^2+8tJ_1-2x^2-J_1^2-J_2^2+4J_2-1\right)\left(1-8t^2-4J_2+8itJ_1^2+8it-2ixJ_1^2+2ix+J_1^2+J_2^2-(16i)tJ_2-8t^2J_1^2-8t^2J_2^2+16t^2J_2+2x^2J_1^2\right.$$

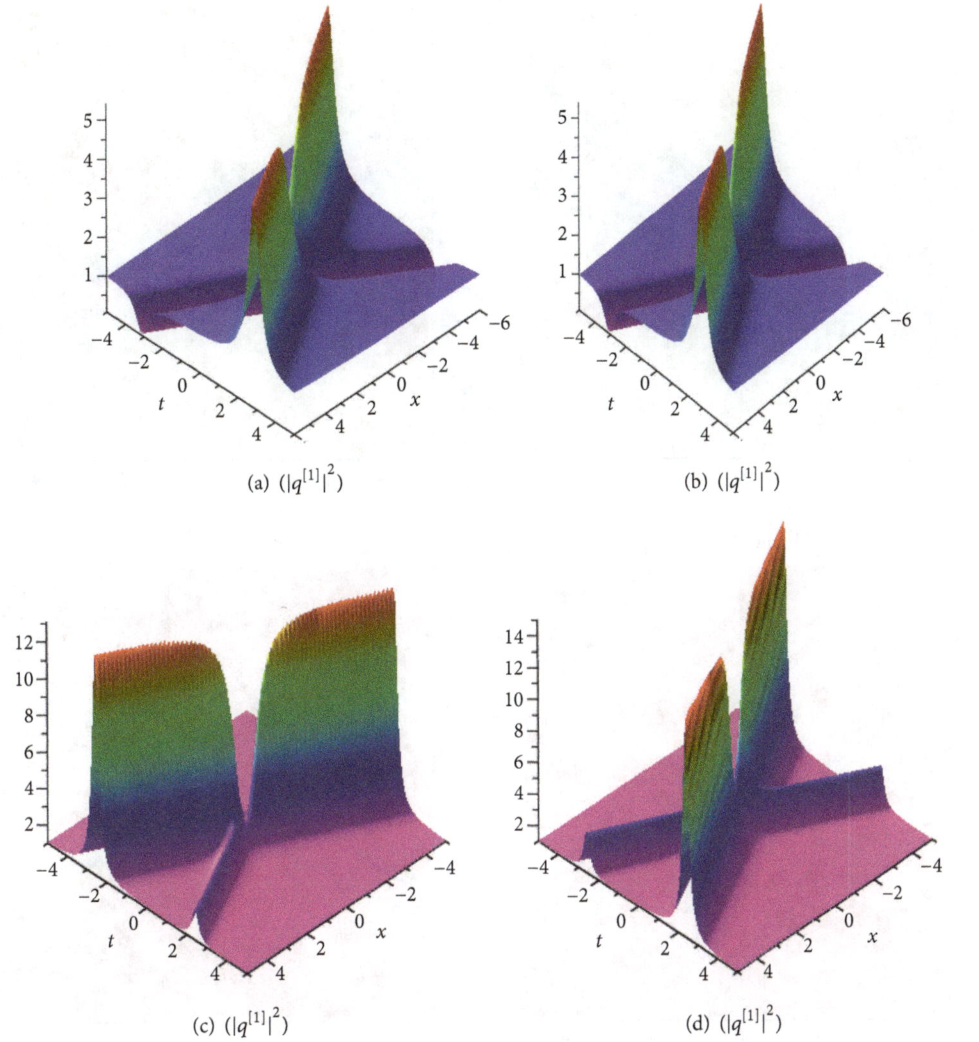

(a) $(|q^{[1]}|^2)$ (b) $(|q^{[1]}|^2)$

(c) $(|q^{[1]}|^2)$ (d) $(|q^{[1]}|^2)$

FIGURE 1: $|q^{[1]}|^2$ displays four structures with the height of background being 1. (a) The dark-bright structure: $(J_1, J_2) = (2, 3)$. (b) The bright-dark structure: $(J_1, J_2) = (0.2, 0.2)$. (c) The bright-bright 1 structure: $(J_1, J_2) = (0.2, -0.1)$. (d) The bright-bright 2 structure: $(J_1, J_2) = (3, -1)$.

$$
\begin{aligned}
&+ 2x^2J_2^2 - 4x^2J_2 - 8tJ_1 + 2x^2 + 4iJ_1 + 8itJ_2^2 \\
&- 2ixJ_2^2\big), \\
F_6 &= \big(8t^2J_1^2 + 8t^2J_2^2 - 16t^2J_2 + 8t^2 - 2x^2J_1^2 - 2x^2J_2^2 \\
&+ 2ixJ_1^2 + 2ixJ_2^2 + 4x^2J_2 + 8tJ_1 - 2x^2 - 2ix + J_1^2 \\
&+ J_2^2 + 1\big)^2.
\end{aligned} \tag{45}
$$

Figures 5 and 6 imply that in the PT-symmetric waveguide system that involved refractive index and gain or loss profiles, when a part of the system involves loss, an equal amount of gain is observed near this part, which may be applied to overcome the loss effects in the system. Thus, our results may guide the interplay between gain and loss and the manufacture of a new generation of multifunction optical devices and networks.

5. Conclusion

In this article, we study a weakly coupled nonlocal NLS equation. We have shown its integrability by proposing its Lax pair. Then, we have derived the Darboux transformation of the weakly coupled nonlocal NLS equation which implies the representations of $(q^{[1]}, r^{[1]})$ generated from known solutions (q, r). By choosing some special eigenvalues $\lambda_2 = -\lambda_1^*$, and eigenfunctions using the reduction conditions $\varphi_{210}(x,t,\lambda_2) = \varphi_{120}^*(-x,t,-\lambda_1^*)$, $\varphi_{211}(x,t,\lambda_2) = \varphi_{121}^*(-x,t,-\lambda_1^*)$, $\varphi_{220}(x,t,\lambda_2) = -\varphi_{110}^*(-x,t,-\lambda_1^*)$, $\varphi_{221}(x,t,\lambda_2) = -\varphi_{111}^*(-x,t,-\lambda_1^*)$, the representations of

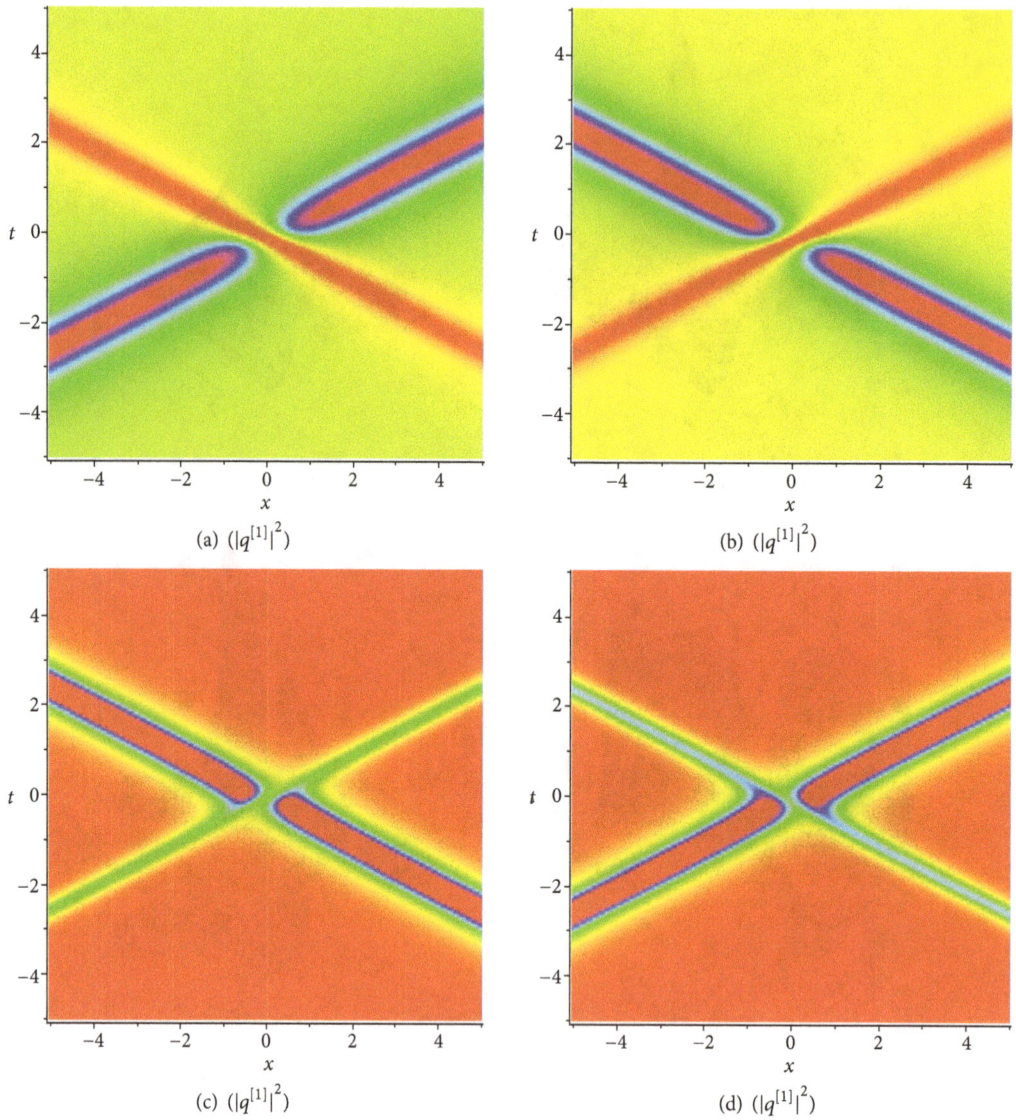

FIGURE 2: Density plots of $|q^{[1]}|^2$ on (x,t)-plane associated with Figure 1. (a) The dark-bright structure: $(J_1, J_2) = (2, 3)$. (b) The bright-dark structure: $(J_1, J_2) = (0.2, 0.2)$. (c) The bright-bright 1 structure: $(J_1, J_2) = (0.2, -0.1)$. (d) The bright-bright 2 structure: $(J_1, J_2) = (3, -1)$.

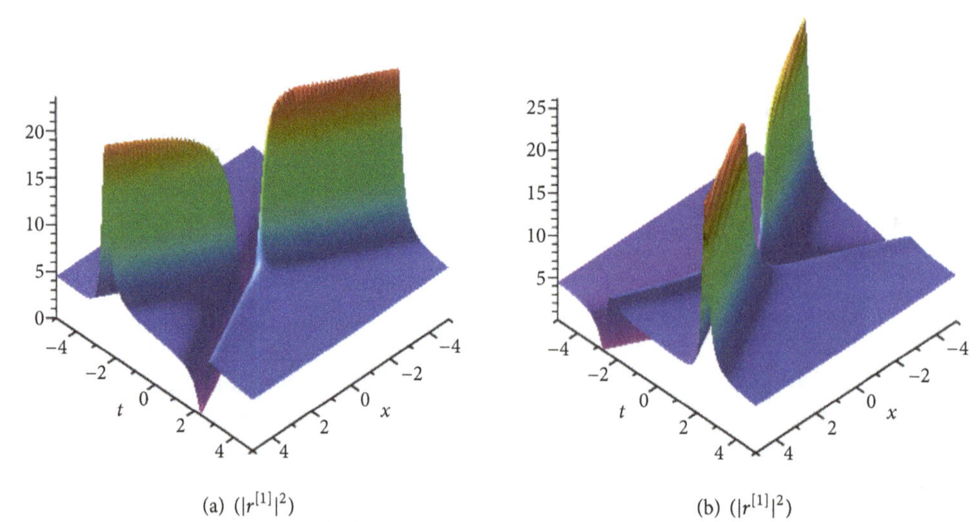

FIGURE 3: $|r^{[1]}|^2$ displays two structures with the height of background being 5. (a) The bright-bright 1 structure: $(J_1, J_2) = (0.2, -0.1)$. (b) The bright-bright 2 structure: $(J_1, J_2) = (3, -1)$.

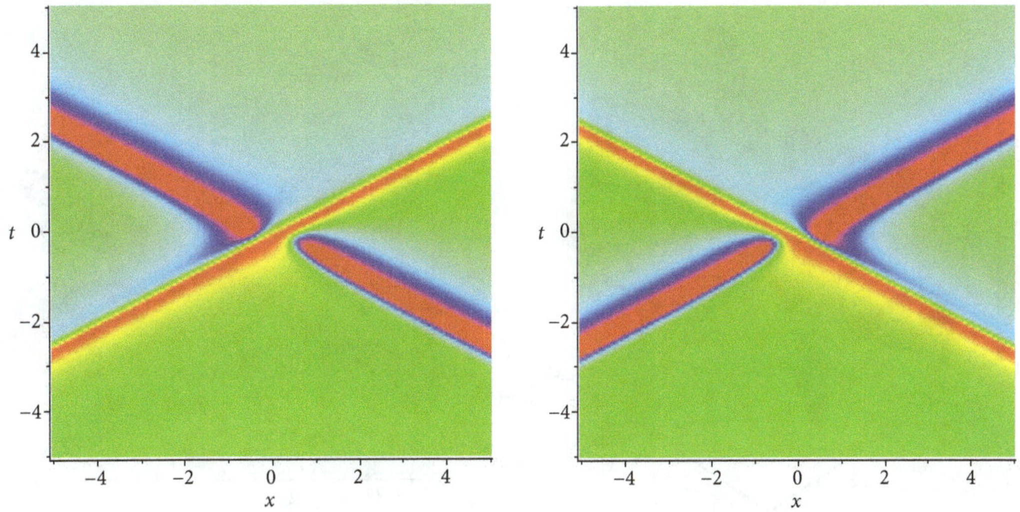

(a) $(|r^{[1]}|^2)$ (b) $(|r^{[1]}|^2)$

FIGURE 4: Density plots of $|r^{[1]}|^2$ on (x, t)-plane associated with Figure 1. (a) The bright-bright 1 structure: $(J_1, J_2) = (0.2, -0.1)$. (b) The bright-bright 2 structure: $(J_1, J_2) = (3, -1)$.

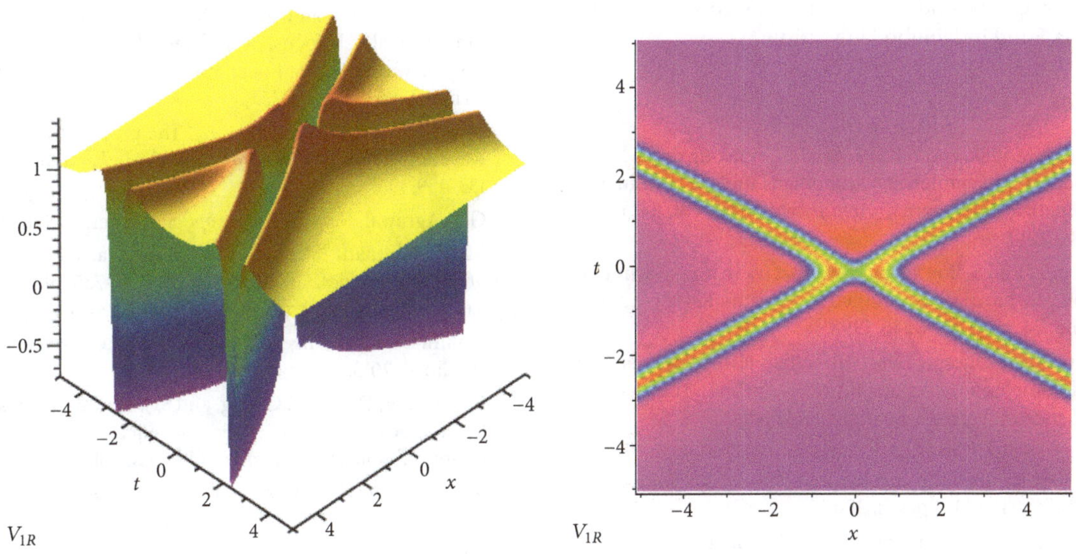

FIGURE 5: The real part of the PT-symmetric potential generated by the rational solution. The profile of V_{1I} with $(J_1, J_2) = (2, 3)$.

(31) provided some different solutions of the weakly coupled nonlocal NLS equation. By using degenerate Darboux transformation, we have derived rational solutions, the real refractive index profile $(V_R^{[1]})$, and the gain-or-loss distribution $(V_I^{[1]})$ of the weakly coupled nonlocal NLS equation under the PT-symmetry condition and we have proved that it is analytic if $J_1^2 + J_2^2 \neq 1$ [17]. We have divided the obtained solutions $q^{[1]}, r^{[1]}$ into four categories solutions and two categories solutions which are graphically illustrated in Figures 1 and 3, respectively. This observation may be useful for synthesizing new artificial optical structures and materials by mixing together the refractive index distribution and the gain-or-loss profile.

Conflicts of Interest

The authors declare that there are no conflicts of interest regarding the publication of this paper.

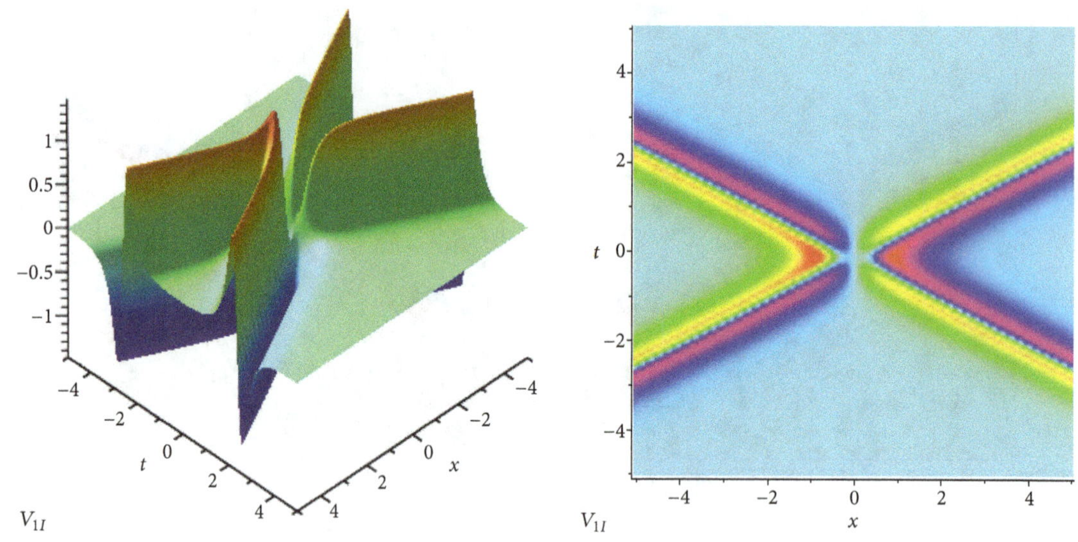

FIGURE 6: The imaginary part of the PT-symmetric potential generated by the rational solution. The profile of V_{1I} with $(J_1, J_2) = (2, 3)$.

Acknowledgments

Chuanzhong Li is supported by the National Natural Science Foundation of China under Grant no. 11571192 and K. C. Wong Magna Fund in Ningbo University.

References

[1] B. Bagchi, S. Modak, and P. K. Panigrahi, "Tracking down localized modes in pt-symmetric hamiltonians under the influence of a competing nonlinearity," *Acta Polytechnica*, vol. 54, no. 2, pp. 79–84, 2014.

[2] C. M. Bender and S. Boettcher, "Real spectra in non-Hermitian Hamiltonians having PT symmetry," *Physical Review Letters*, vol. 80, no. 24, pp. 5243–5246, 1998.

[3] B. Bagchi and C. Quesne, "Non-Hermitian HAMiltonians with real and complex eigenvalues in a Lie-algebraic framework," *Physics Letters A*, vol. 300, no. 1, pp. 18–26, 2002.

[4] C. M. Bender, D. C. Brody, H. F. Jones, and B. K. Meister, "Faster than Hermitian quantum mechanics," *Physical Review Letters*, vol. 98, no. 4, 040403, 4 pages, 2007.

[5] K. G. Makris, R. El-Ganainy, Z. H. Musslimani, and D. N. Christodoulides, "Theory of coupled optical PT-symmetric structures," *Optics Expresss*, vol. 32, no. 17, pp. 2632–2634, 2007.

[6] Z. H. Musslimani, K. G. Makris, R. El-Ganainy, and D. N. Christodoulides, " Optical Solitons in ," *Physical Review Letters*, vol. 100, no. 3, 2008.

[7] K. G. Makris, R. El-Ganainy, D. N. Christodoulides, and Z. H. Musslimani, "Beam dynamics in PT symmetric optical lattices," *Physical Review Letters*, vol. 100, no. 10, Article ID 103904, 2008.

[8] C. E. Rüter, K. G. Makris, R. El-Ganainy, D. N. Christodoulides, M. Segev, and D. Kip, "Observation of parity-time symmetry in optics," *Nature Physics*, vol. 6, no. 3, pp. 192–195, 2010.

[9] A. Regensburger, C. Bersch, M.-A. Miri, G. Onishchukov, D. N. Christodoulides, and U. Peschel, "Parity-time synthetic photonic lattices," *Nature*, vol. 488, no. 7410, pp. 167–171, 2012.

[10] A. Regensburger, M. Miri, C. Bersch et al., " Observation of Defect States in ," *Physical Review Letters*, vol. 110, no. 22, 2013.

[11] A. Guo, G. J. Salamo, D. Duchesne et al., "Observation of PT-symmetry breaking in complex optical potentials," *Physical Review Letters*, vol. 103, no. 9, Article ID 093902, 2009.

[12] V. E. Zakharov and A. B. Shabat, "Exact theory of two-dimensional self-focusing and one-dimensional self-modulation of waves in nonlinear media," *Sov. Phys. JETP*, vol. 34, Article ID 6353506, 1972.

[13] D. J. Benney and A. C. Newell, "The propagation of nonlinear wave envelopes," *Studies in Applied Mathematics*, vol. 46, pp. 133–139, 1967.

[14] G. P. Agrawal, *Nonlinear fiber optics*, Academic, San Diego, 1989.

[15] M. V. Goldman, "Strong turbulence of plasma waves," *Reviews of Modern Physics*, vol. 56, no. 4, pp. 709–735, 1984.

[16] M. J. Ablowitz and Z. H. Musslimani, "Integrable Nonlocal Nonlinear Schrödinger Equation," *Physical Review Letters*, vol. 110, no. 6, 2013.

[17] Y. S. Zhang, D. Qiu, Y. Cheng, and S. J. He, "Rational solution of the nonlocal nonlinear Schrödinger equation and its application in optics," *Romanian Journal of Physics*, vol. 61, p. 108, 2017.

[18] S.-Y. Lou and H.-Y. Ruan, "Similarity reductions and nonlocal symmetry of the KdV equation," *Communications in Theoretical Physics*, vol. 25, no. 2, pp. 241–244, 1996.

[19] G. Lévai and M. Znojil, "Systematic search for PT symmetric potentials with real energy spectra," *Journal of Physics A: Mathematical and General*, vol. 33, no. 40, pp. 7165–7180, 2000.

[20] A. Mostafazadeh, "Pseudo-Hermiticity versus PT symmetry: the necessary condition for the reality of the spectrum of a non-Hermitian Hamiltonian," *Journal of Mathematical Physics*, vol. 43, no. 1, pp. 205–214, 2002.

[21] C. M. Bender, "Making sense of non-Hermitian Hamiltonians," *Reports on Progress in Physics*, vol. 70, no. 6, pp. 947–1018, 2007.

[22] P. F. Li, D. Mihalache, and L. Li, "Asymmetric solitons in parity-time-symmetric double-hump Scarff-2 potentials," *Romanian Journal of Physics*, vol. 61, pp. 1028–1039, 2016.

[23] Z. Yan, Z. Wen, and V. V. Konotop, "Solitons in a nonlinear Schrödinger equation with PT-symmetric potentials and inhomogeneous nonlinearity: Stability and excitation of nonlinear modes," *Physical Review A*, vol. 92, Article ID 023821, 2015.

[24] M. I. Stockman, "Spaser action, loss compensation, and stability in plasmonic systems with gain," *Physical Review Letters*, vol. 106, no. 15, Article ID 156802, 2011.

[25] P. Berini and I. de Leon, "Surface plasmon-polariton amplifiers and lasers," *Nature Photonics*, vol. 6, no. 1, pp. 16–24, 2012.

[26] M. A. Noginov, G. Zhu, A. M. Belgrave et al., "Demonstration of a spaser-based nanolaser," *Nature*, vol. 460, no. 7259, pp. 1110–1112, 2009.

[27] Y. D. Chong, L. Ge, and A. D. Stone, " Publisher's Note: ," *Physical Review Letters*, vol. 108, no. 26, 2012.

[28] Z. Lin, H. Ramezani, T. Eichelkraut, T. Kottos, H. Cao, and D. N. Christodoulides, " Unidirectional Invisibility Induced by ," *Physical Review Letters*, vol. 106, no. 21, 2011.

[29] V. V. Konotop, J. Yang, and D. A. Zezyulin, "Nonlinear waves in PT-symmetric systems," *Reviews of Modern Physics*, vol. 88, no. 3, Article ID 035002, 2016.

[30] S. V. Suchkov, A. A. Sukhorukov, J. Huang, S. V. Dmitriev, C. Lee, and Yu. S. Kivshar, "Nonlinear switching and solitons in PT-symmetric photonic systems," *Laser & Photonics Reviews*, vol. 10, no. 2, pp. 177–213, 2016.

[31] Y. J. He, X. Zhu, and D. Mihalache, "Dynamics of spatial solitons in parity-time-symmetric optical latties: a selection of recent theoretical results," *Romanian Journal of Physics*, vol. 61, pp. 595–613, 2016.

[32] V. B. Matveev and M. A. Salle, *Darboux Transformations and Solitons*, Springer, Berlin, Gemany, 1991.

[33] C. H. Gu, *Darboux Transformation in Soliton Theory and its Geometric Applications Shanghai Sci.-Tech*, Edu, Publishing House, Shanghai, 2005.

20

Dimensional Regularization Approach to the Renormalization Group Theory of the Generalized Sine-Gordon Model

Takashi Yanagisawa

Electronics and Photonics Research Institute, National Institute of Advanced Industrial Science and Technology, 1-1-1 Umezono, Tsukuba, Ibaraki 305-8568, Japan

Correspondence should be addressed to Takashi Yanagisawa; t-yanagisawa@aist.go.jp

Academic Editor: Emmanuel Lorin

We present the dimensional regularization approach to the renormalization group theory of the generalized sine-Gordon model. The generalized sine-Gordon model means the sine-Gordon model with high frequency cosine modes. We derive renormalization group equations for the generalized sine-Gordon model by regularizing the divergence based on the dimensional method. We discuss the scaling property of renormalization group equations. The generalized model would present a new class of scaling property.

1. Introduction

The sine-Gordon model is an interesting model and plays an important role in physics [1–13]. There are many phenomena that are related to the sine-Gordon model. In this sense, the sine-Gordon model has universality. In the weak coupling phase the sine-Gordon model is perturbatively equivalent to the massive Thirring model [1, 14–16]. The two-dimensional (2D) sine-Gordon model describes a crossover between weak coupling region and strong coupling region. The renormalization equations are the same as those for the Kosterlitz-Thouless transition of the 2D classical XY model [17–19]. The 2D sine-Gordon model is mapped to the Coulomb gas model where particles interact with each other through logarithmic interaction [4, 20, 21]. The Kondo problem belongs to the same universality class where the renormalization group equations are given by the same equations for the 2D sine-Gordon model [20–27]. The renormalization group equations in the Kondo problem was derived before those by Kosterlitz and Thouless. The one-dimensional Hubbard model is mapped to the 2D sine-Gordon model by using a bosonization method [28–31], where the Hubbard model is an important model that describes the metal-insulator transition and high-temperature superconductivity [32–39]. The sine-Gordon model appears in a multiband superconductor where the Nambu-Goldstone modes become massive due to the Josephson couplings [40–47]. The Josephson plasma oscillation in layered high-temperature superconductors was analyzed based on the sine-Gordon model [48]. In a series of papers [41–43, 45, 46] we introduced the sine-Gordon model into the study of superconductivity and examined significant excitation modes in superconductors. A generalization from U(1) to a compact continuous group G for the sine-Gordon model was also investigated [49] where the sine-Gordon model considered in this paper and in references cited above is a model with U(1) group.

In this paper, we investigate the renormalization group theory for the 2D generalized sine-Gordon model by using the dimensional regularization method to regularize the divergence [50–52]. Here the generalized sine-Gordon model is a sine-Gordon model that includes high frequency cosine potential terms such as $\cos(n\phi)$ for an integer n. The renormalization of the generalized sine-Gordon model was investigated [53] by the Wegner-Houghton method [54] and by the functional renormalization group method [55]. We use the dimensional regularization method in deriving the renormalization group equation for the generalized sine-Gordon model. The divergence is regularized near two dimensions by putting the dimension $d = 2 + \epsilon$. The divergent part of integral is evaluated as a pole in the form $1/\epsilon$. This is called the minimal subtraction method. Then the beta function for the coupling constant is derived.

2. Lagrangian

Let us consider a real scalar field ϕ. The Lagrangian of the generalized sine-Gordon model is given by

$$\mathscr{L} = \frac{1}{2t_0}\left(\partial_\mu \phi_B\right)^2 + \frac{1}{t_0}\sum_n \alpha_{n0}\cos\left(n\phi_B\right), \quad (1)$$

where ϕ_B is a bare real scalar field and t_0 and α_{0n} are bare coupling constants. The second term indicates the potential energy of the scalar field ϕ. The generalized sine-Gordon model contains high frequency terms such as $\cos(n\phi)$ ($n = 1, 2, \cdots$). We write the renormalized coupling constants as t and α_n, respectively. We adopt that $t > 0$ and $\alpha_n \geq 0$. α_m for some m may be zero, but at least one α_n should be positive (nonzero). The dimensions of t and α_n are given as $[t] = \mu^{2-d}$ and $[\alpha_n] = \mu^2$ where μ is a parameter representing the energy scale. The scalar field ϕ is dimensionless. The relations between bare and renormalized quantities are given by

$$t_0 = t\mu^{2-d} Z_t, \qquad \alpha_{n0} = \alpha_n \mu^2 Z_{\alpha n}, \quad (2)$$

where Z_t and $Z_{\alpha n}$ are renormalization constants. t and α_n are dimensionless constants by virtue of the energy scale μ. We define the renormalized field ϕ_R by

$$\phi_B = \sqrt{Z_\phi}\phi_R, \quad (3)$$

where Z_ϕ is the renormalization constant for the field ϕ. The Lagrangian with renormalized quantities is written as

$$\mathscr{L} = \frac{\mu^{d-2} Z_\phi}{2tZ_t}\left(\partial_\mu \phi\right)^2 + \sum_n \frac{\mu^d \alpha_n Z_{\alpha n}}{tZ_t}\cos\left(n\sqrt{Z_\phi}\phi\right), \quad (4)$$

where ϕ denotes the renormalized field ϕ_R. The second term represents the interaction of the field ϕ as seen by expanding $\cos(n\sqrt{Z_\phi}\phi)$ as a power series. There is the other representation of interaction parameters. We can absorb the parameter t in the definition of the field ϕ and the parameter α_n. In this case, field ϕ in the interaction term includes the parameter in the form $\cos(n\sqrt{Z_\phi}\beta\phi)$ where $\beta = \sqrt{t}$. We will obtain the same result since it does not depend on the representation.

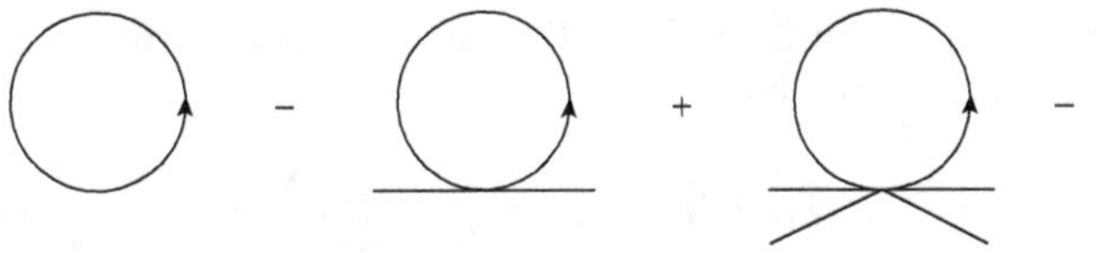

FIGURE 1: One-loop contributions to the renormalization of α_n.

3. Renormalization of α_n

We consider the renormalization of α_n up to the lowest order of α_n. By considering tadpole diagrams in Figure 1, the cosine function is renormalized to

$$\begin{aligned}&\cos\left(n\sqrt{Z_\phi}\phi\right)\\ &\longrightarrow \left(1 - \frac{1}{2}n^2 Z_\phi\left\langle\phi^2\right\rangle + \cdots\right)\cos\left(n\sqrt{Z_\phi}\phi\right) \\ &= \exp\left(-\frac{1}{2}n^2 Z_\phi\left\langle\phi^2\right\rangle\right)\cos\left(n\sqrt{Z_\phi}\phi\right).\end{aligned} \quad (5)$$

Since the expectation value $\langle\phi^2\rangle$ diverges, we regularize it using the dimensional regularization method:

$$Z_\phi\left\langle\phi^2\right\rangle = t\mu^{2-d} Z_t \int \frac{d^dk}{(2\pi)^d}\frac{1}{k^2 + m_0^2} = -\frac{t}{\epsilon}\frac{\Omega_d}{(2\pi)^d}, \quad (6)$$

for $d = 2 + \epsilon$ where m_0 is introduced to avoid infrared divergence, Ω_d is the solid angle in d dimensions and Z_t was put as 1. In order to remove the divergence, the constant $Z_{\alpha n}$ is determined as follows:

$$Z_{\alpha n} = 1 - \frac{n^2}{2}t\frac{1}{\epsilon}\frac{\Omega_d}{(2\pi)^d}. \quad (7)$$

Since the bare coupling constant α_{n0} is independent of μ, we have $\mu\, \partial\alpha_{n0}/\partial\mu = 0$. This results in

$$\mu\frac{\partial\alpha_n}{\partial\mu} = -2\alpha_n - \alpha_n\mu\frac{\partial \ln Z_{\alpha n}}{\partial\mu}. \quad (8)$$

We set $Z_t = 1$ up to the lowest order of α_n, so that we have $\mu\, \partial t/\partial\mu = (d-2)t$. The beta function for α_n at the lowest order in α_n is given by

$$\beta\left(\alpha_n\right) \equiv \mu\frac{\partial\alpha_n}{\partial\mu} = -2\alpha_n + n^2 t\alpha_n\frac{1}{2}\frac{\Omega_d}{(2\pi)^d}. \quad (9)$$

$\beta(\alpha_n)$ has a zero at $t = t_{cn}$:

$$t_{cn} = \frac{8\pi}{n^2}, \quad (10)$$

for $d = 2$. There is a fixed point of t for each n.

4. Renormalization of t

There is an effect of renormalization on the coupling constant t that is the correction to the kinetic term. Let us consider the two-point function $\Gamma(p)$. The bare lowest order two-point function is given by

$$\Gamma_B^{(0)}(p) = \frac{1}{t_0}p^2 = \frac{1}{t\mu^{2-d}Z_t}p^2, \quad (11)$$

This corresponds to the kinetic part of the bare Lagrangian:

$$\mathscr{L}_B^{(0)} = \frac{\mu^{d-2}Z_\phi}{2tZ_t}\left(\partial_\mu\phi\right)^2. \quad (12)$$

4.1. Real Space Formulation. The lowest order correction to the two-point function is given by a second-order term for α_ℓ ($\ell = 1, 2, \cdots$) such as $\alpha_n\alpha_m\cos(n\sqrt{Z_\phi}\phi(x))\cos(m\sqrt{Z_\phi}\phi(x'))$. From the formula $\cos\theta_1\cos\theta_2 = (1/2)(\cos(\theta_1+\theta_2)+\cos(\theta_1-\theta_2))$, the correction to the action comes from

$$I_{nm} \equiv -\frac{1}{2!}\frac{1}{2}\left(\frac{\mu^d}{tZ_t}\right)^2 Z_{\alpha n}Z_{\alpha m}\alpha_n\alpha_m\int d^dx\int d^dx' \cdot\cos\left(n\sqrt{Z_\phi}\phi(x) - m\sqrt{Z_\phi}\phi(x')\right), \quad (13)$$

where we consider connected contributions. By taking into account the contribution of tadpole diagrams, this reduces to

$$I_{nm} = -\frac{1}{4}Z_{\alpha n}Z_{\alpha m}\alpha_n\alpha_m\int d^dx\int d^dx'\left(1-\frac{1}{2}Z_\phi\left\langle n^2\phi(x)^2 + m^2\phi(x')^2 - 2nm\phi(x)\phi(x')\right\rangle + \cdots\right) \times\left(\frac{\mu^d}{tZ_t}\right)^2\cos\left(n\sqrt{Z_\phi}\phi(x) - m\sqrt{Z_\phi}\phi(x')\right) \cong -\frac{1}{4} \cdot Z_{\alpha n}Z_{\alpha m}\alpha_n\alpha_m\int d^dx\int d^dx'\exp\left(-\frac{1}{2}Z_\phi\left\langle n^2\phi(x)^2 + m^2\phi(x')^2\right\rangle + +nmZ_\phi\left\langle\phi(x)\phi(x')\right\rangle\right)\left(\frac{\mu^d}{tZ_t}\right)^2 \cdot\cos\left(n\sqrt{Z_\phi}\phi(x) - m\sqrt{Z_\phi}\phi(x')\right). \quad (14)$$

The expectation value $\langle\phi(x)\phi(y)\rangle$ is given by

$$\langle\phi(x)\phi(y)\rangle = \frac{t\mu^{2-d}Z_t}{Z_\phi}\int\frac{d^dp}{(2\pi)^d}\frac{e^{ip\cdot(x-y)}}{p^2+m_0^2} = \frac{t\mu^{2-d}Z_t}{Z_\phi}\frac{\Omega_d}{(2\pi)^d}K_0\left(m_0|x-y|\right), \quad (15)$$

where K_0 is the 0th modified Bessel function. Because $K_0(x)$ increases divergently as x approaches zero, I_{nm} is approximated as

$$I_{nm} \cong -\frac{1}{4}Z_{\alpha n}Z_{\alpha m}\alpha_n\alpha_m\int d^dx\int d^dr \cdot\exp\left(-\frac{1}{2}Z_\phi\left(n^2+m^2\right)\left\langle\phi^2\right\rangle + nmt\frac{\Omega_d}{(2\pi)^d}K_0\left(m_0r\right)\right) \cdot\left(\frac{\mu^d}{tZ_t}\right)^2\cos\left(\sqrt{Z_\phi}(n-m)\phi(x)\right)\left(1-\frac{1}{2}m^2Z_\phi\left(\nabla_x\phi \cdot \mathbf{r}\right)^2\right), \quad (16)$$

where we put $x' = x + \mathbf{r}$. The cosine function $\cos(\sqrt{Z_\phi}(n-m)\phi(x))$ would oscillate as a function of x, the contribution for $n \neq m$ will be small. Thus, we consider only the contributions with $n = m$:

$$I_{nm} \cong \frac{1}{4}\alpha_n^2Z_{\alpha n}^2\left(\frac{\mu^d}{tZ_t}\right)^2\int d^dxd^dr\frac{1}{2}n^2Z_\phi\left(\nabla_x\phi\right)^2\frac{1}{2}r^2 \cdot\exp\left(-Z_\phi n^2\left\langle\phi^2\right\rangle + n^2t\frac{\Omega_d}{(2\pi)^d}K_0\left(m_0r\right)\right). \quad (17)$$

We extract the divergent term in I_{nn}. There may be two ways to do this. We discuss these methods in the following.

(1) In the first method, we regularize $\langle\phi^2\rangle$ by introducing a cutoff a in the real space:

$$Z_\phi\left\langle\phi^2\right\rangle = t\mu^{2-d}Z_t\frac{\Omega_d}{(2\pi)^d}K_0\left(m_0a\right), \quad (18)$$

by replacing $K_0(m_0r)$ with $K_0(m_0\sqrt{r^2+a^2})$. By using the asymptotic relation $K_0(z) \approx -\gamma - \ln(z/2)$ with the Euler constant γ, the integral with respect to r is performed as follows [49]:

$$J_n \equiv \int d^drr^2\exp\left(n^2t\frac{\Omega_d}{(2\pi)^d}K_0\left(m_0\sqrt{r^2+a^2}\right)\right) \approx \Omega_d\int drr^{d+1}\frac{1}{\left[cm_0^2\left(r^2+a^2\right)\right]^{n^2t/4\pi}}, \quad (19)$$

near $d = 2$ where we set $c = (e^\gamma/2)^2$. We consider the case where t is close to the critical value $t_{cn} = 8\pi/n^2$:

$$\frac{n^2t}{8\pi} = 1 + \nu_n, \quad (20)$$

where v_n represents the deviation from the critical point. In the lowest order of v_n, we have

$$\begin{aligned} J_n &= \Omega_d \left(cm_0^2\right)^{-2} \frac{1}{2-d} + O\left(v_n\right) \\ &= -\Omega_d \left(cm_0^2\right)^{-2} \frac{1}{\epsilon} + O\left(v_n\right). \end{aligned} \tag{21}$$

Then we obtain

$$\begin{aligned} I_{nn} &\approx -\frac{1}{8}\alpha_n^2 Z_{\alpha n}^2 \left(\frac{\mu^d}{tZ_t}\right)^2 \exp\left(n^2 t Z_t \frac{1}{4\pi}\log\left(cm_0^2 a^2\right)\right) \\ &\cdot \Omega_d \left(cm_0^2\right)^{-2} \frac{1}{\epsilon} \int d^d x \frac{1}{2} n^2 Z_\phi \left(\partial_\mu \phi\right)^2 + O\left(v_n\right) \\ &\approx -\frac{n^2}{32}\alpha_n^2 \mu^{d+2} a^4 \frac{1}{\epsilon} \int d^d x \frac{\mu^{d-2} Z_\phi}{2tZ_t} n^2 \left(\partial_\mu \phi\right)^2 \\ &+ O\left(v_n\right). \end{aligned} \tag{22}$$

The constant $Z_{\alpha n}$ was absorbed for the renormalization of α_n. Then, by taking the sum from each term, the kinetic part $\mathscr{L}_B^{(0)}$ is renormalized to

$$\mathscr{L}_B^{(2)} = \frac{\mu^{d-2} Z_\phi}{2tZ_t} \left[1 - \sum_n \frac{n^4}{32}\alpha_n^2 \mu^{d+2} a^4 \frac{1}{\epsilon}\right] \left(\partial_\mu \phi\right)^2. \tag{23}$$

This indicates that we choose

$$\frac{Z_\phi}{Z_t} = 1 + \sum_n \frac{n^4}{32}\alpha_n^2 \mu^{d+2} a^4 \frac{1}{\epsilon}. \tag{24}$$

Z_ϕ and Z_t appear as a ratio Z_ϕ/Z_t in this order, and then the coupling constant t is renormalized as $t_0 = t\mu^{2-d} Z_t/Z_\phi$ or we can choose $Z_\phi = 1$. The equation $\mu\, \partial t_0/\partial\mu = 0$ results in

$$\begin{aligned} \beta(t) &\equiv \mu \frac{\partial t}{\partial \mu} \\ &= (d-2)\, t \\ &+ \sum_n \frac{n^4}{32} a^4 \frac{t}{\epsilon} \left(2\alpha_n \mu \frac{\partial \alpha_n}{\partial \mu} \mu^{d+2} + (d+2)\, \alpha_n^2 \mu^{d+2}\right) \\ &= (d-2)\, t + \sum_n \frac{n^4}{32} t \alpha_n^2 \mu^{d+2} a^4 + O\left(t^2\right). \end{aligned} \tag{25}$$

Lastly, we put $a = \mu^{-1}$ to obtain

$$\beta(t) = (d-2)\, t + \sum_n \frac{n^4}{32} t \alpha_n^2 + O\left(t^2\right). \tag{26}$$

The numerical coefficient is not important and this depends on the choice of the cutoff a.

(2) In the second way, the divergence comes from $\langle \phi^2 \rangle$ where we adopt that the integral with respect to r is finite. This treatment is similar to that in [31] where the Wilson renormalization group method was used. The correction I_{nn} is written as

$$\begin{aligned} I_{nn} &= \frac{1}{8}\alpha_n^2 Z_{\alpha n}^2 \left(\frac{\mu^d}{tZ_t}\right)^2 \exp\left(-n^2 Z_\phi \langle \phi^2 \rangle\right) \int d^d r r^2 \\ &\cdot \exp\left(n^2 t \frac{\Omega_d}{(2\pi)^d} K_0\left(m_0 r\right)\right) \int d^d x \frac{1}{2} n^2 Z_\phi \left(\partial_\mu \phi\right)^2 \\ &\approx -\frac{n^4}{8}\alpha_n^2 \mu^{d+2} \frac{1}{2\pi\epsilon} \int d^d r r^2 \exp\left(\frac{n^2 t}{2\pi} K_0\left(m_0 r\right)\right) \\ &\cdot \int d^d x \frac{\mu^{d-2} Z_\phi}{2tZ_t} \left(\partial_\mu \phi\right)^2. \end{aligned} \tag{27}$$

In order to let the integral for r be dimensionless, we change the variable $r = au$ and put

$$A_n = \int d^d u u^2 \exp\left(\frac{n^2 t}{2\pi} K_0\left(m_0 a \sqrt{u^2+1}\right)\right), \tag{28}$$

by introducing a cutoff in the integral. Then, we have

$$I_{nn} \cong -\frac{A_n}{16\pi} n^4 \alpha_n^2 \mu^{d+2} a^{d+2} \frac{1}{\epsilon} \int d^d x \frac{\mu^{d-2} Z_\phi}{2tZ_t} \left(\partial_\mu \phi\right)^2. \tag{29}$$

This results in the same beta function $\beta(t)$ with the numerical factor being slightly different:

$$\beta(t) = (d-2)\, t + \sum_n \frac{A_n}{16\pi} n^4 t \alpha_n^2. \tag{30}$$

4.2. Momentum Space Formulation. In the momentum space, we evaluate the two-point function by calculating the diagrams in Figure 2 [6]. This set of diagrams gives the self-energy $\Sigma(p)$. $\Sigma(p)$ is written as a sum of $\Sigma_{nm}(p)$ that comes from the interaction term $\cos(n\phi)\cos(m\phi)$. The diagrams in Figure 2 are summed up to give

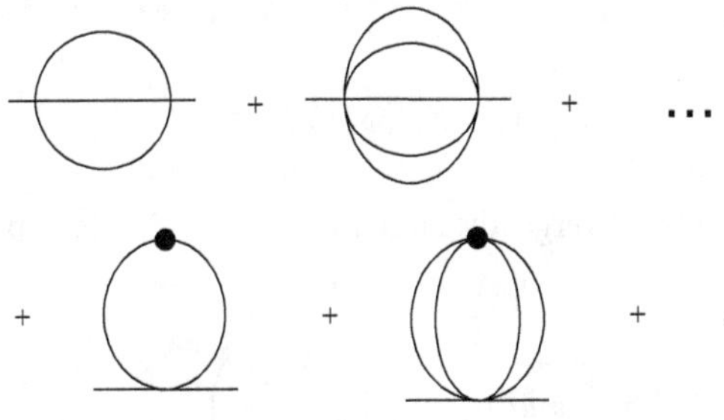

FIGURE 2: The contributions to the two-point function $\Gamma^{(2)}(p)$ up to the order of $\alpha_n \alpha_m$.

$$\begin{aligned} \Sigma_{nm}(p) &= nm\alpha_n \alpha_m Z_{\alpha n} Z_{\alpha m} \left(\frac{\mu^d}{tZ_t}\right)^2 \\ &\cdot \int d^d x \left[e^{ip\cdot x}\left(\sinh\left(nmG_0(x)\right) - nmG_0(x)\right)\right. \\ &\left. - \left(\cosh\left(nmG_0(x)\right) - 1\right)\right], \end{aligned} \tag{31}$$

where we put

$$G_0(x) = Z_\phi \langle \phi(x)\phi(0)\rangle. \quad (32)$$

Since $G_0(x)$ is divergently large as $x \longrightarrow 0$, Σ_{nm} is approximated as

$$\Sigma_{nm}(p) = \frac{1}{2} nm\alpha_n\alpha_m Z_{\alpha n} Z_{\alpha m}\left(\frac{\mu^d}{tZ_t}\right)^2 \cdot \int d^d x\left(e^{ip\cdot x}-1\right)\exp\left(nmG_0(x)\right). \quad (33)$$

Using the expansion $e^{ip\cdot x} = 1 + ip\cdot x - (1/2)(p\cdot x)^2 + \cdots$, we keep the p^2 term. By using the formula $K_0(x) \approx -\gamma - \log(x/2)$ for small x, $\Sigma_{nm}(p)$ is written as

$$\begin{aligned}\Sigma_{nm}(p) &= -\frac{1}{4} nm\alpha_n\alpha_m Z_{\alpha n} Z_{\alpha m}\left(\frac{\mu^d}{tZ_t}\right)^2 \cdot \int d^d x x^{d+1}\frac{1}{\left[cm_0^2\left(x^2+a^2\right)\right]^{nmt/4\pi}} = -\frac{1}{8} \\ &\cdot p^2 nm\alpha_n\alpha_m Z_{\alpha n} Z_{\alpha m}\left(\frac{\mu^d}{tZ_t}\right)^2 \Omega_d \int dx x^{d+1} \cdot \frac{1}{\left[cm_0^2\left(x^2+a^2\right)\right]^{nmt/4\pi}}.\end{aligned} \quad (34)$$

The integral diverges when $t = t_{cn} = 8\pi/n^2$ and $n = m$. Then we consider the case $n = m$, which gives the correction to the two-point function $\Gamma_B^{(2)n} = -\Sigma_{nn}$ when $t \approx t_{cn}$ as follows:

$$\begin{aligned}\Gamma_B^{(2)n}(p) &= -\frac{1}{8} p^2 n^2 \alpha_n^2 Z_{\alpha n}^2\left(\frac{\mu^d}{tZ_t}\right)^2 \Omega_d\left(cm_0^2\right)^{-2}\frac{1}{\epsilon} \\ &\cong -\frac{1}{t\mu^{2-d}Z_t} p^2 \frac{1}{32} n^4 \alpha_n^2 \mu^{d+2}\left(cm_0^2\right)^{-2}\frac{1}{\epsilon} \\ &\quad + O(\nu_n),\end{aligned} \quad (35)$$

where we set $t \approx (8\pi/n^2)\cdot(1+\nu_n)$. This term mainly comes from the region where $t \approx t_{cn}$. The two-point function up to this order is

$$\begin{aligned}&\Gamma_B^{(2)}(p) \\ &= \frac{1}{t\mu^{2-d}Z_t}\left[p^2 - p^2\sum_n \frac{n^4}{32}\alpha_n^2\mu^{d+2}\left(cm_0^2\right)^{-2}\frac{1}{\epsilon}\right].\end{aligned} \quad (36)$$

The renormalized two-point function is given as $\Gamma_R^{(2)}(p) = Z_\phi \Gamma_B^{(2)}(p)$. The renormalization constants are determined as shown above and thus we obtain the same renormalization group equation.

5. Renormalization Group Flow

Let us consider the case with two parameters α_1 and α_2. The renormalization group equations are

$$\mu\frac{\partial\alpha_1}{\partial\mu} = -2\alpha_1\left(1 - \frac{1}{8\pi}t\right), \quad (37)$$

$$\mu\frac{\partial\alpha_2}{\partial\mu} = -2\alpha_2\left(1 - \frac{1}{2\pi}t\right), \quad (38)$$

$$\mu\frac{\partial t}{\partial\mu} = (d-2)t + \frac{1}{32}t\alpha_1^2 + \frac{1}{8}t\alpha_2^2. \quad (39)$$

We have the critical value $t = t_{c1} = 8\pi$ for α_1 and $t = t_{c2} = 2\pi$ for α_2. The parameter t is an increasing function of μ in two dimensions $d = 2$. The space (t, α_1, α_2) may be divided into four regions which are classified by the values to which the pair (α_1, α_2) is renormalized as μ increases. We call them regions I, II, III, and IV:

$$\text{I: } (\alpha_1, \alpha_2) \longrightarrow (\infty, \infty), \quad (40)$$

$$\text{II: } (\alpha_1, \alpha_2) \longrightarrow (0, \infty), \quad (41)$$

$$\text{III: } (\alpha_1, \alpha_2) \longrightarrow (\infty, 0), \quad (42)$$

$$\text{IV: } (\alpha_1, \alpha_2) \longrightarrow (0, 0). \quad (43)$$

In region $t \approx 8\pi$, we put $t = 8\pi(1+\nu_1)$, $x_1 = 2\nu_1$, $y_1 = \alpha_1/4$and $y_2 = \alpha_2/2$. The equations read

$$\mu\frac{\partial x_1}{\partial\mu} = y_1^2 + y_2^2, \quad (44)$$

$$\mu\frac{\partial y_1}{\partial\mu} = x_1 y_1, \quad (45)$$

$$\mu\frac{\partial y_2}{\partial\mu} = 6y_2. \quad (46)$$

When y_2 is small, the equations reduce to those of the conventional sine-Gordon model (Kosterlitz-Thouless transition).

When $t \approx 2\pi$, we put $t = 2\pi(1+\nu_2)$ and $x_2 = 2\nu_2$ to obtain

$$\mu\frac{\partial x_2}{\partial\mu} = y_1^2 + y_2^2, \quad (47)$$

$$\mu\frac{\partial y_1}{\partial\mu} = -\frac{3}{2}y_1, \quad (48)$$

$$\mu\frac{\partial y_2}{\partial\mu} = x_2 y_2. \quad (49)$$

In this region, y_1 acts as a perturbation to the scaling equation of the conventional sine-Gordon model. We show the renormalization group flow as μ increases in Figure 3.

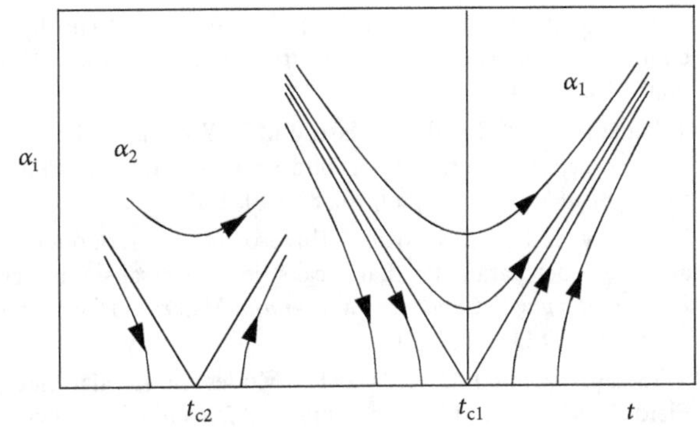

FIGURE 3: Renormalization group flow as $\mu \longrightarrow \infty$ for α_1 and α_2.

6. Summary

We have discussed the dimensional regularization approach to the renormalization group theory of the generalized sine-Gordon model. There are multiple critical points for the coupling constant t given as $t_{cn} = 8\pi/n^2$. In the case where t close to t_{cn} for some n, the renormalization group equations are approximated by those for the sine-Gordon model with single-cosine potential (conventional sine-Gordon model). A nontrivial simple generalized model is the sine-Gordon model with α_1 and α_2. When t is $t_{c1} = 8\pi$, α_2 acts as a perturbation for t and α_1. The renormalization flow as $\mu \longrightarrow \infty$ or $\mu \longrightarrow 0$ depends on an initial set of parameters t, α_1, and α_2. This can be viewed as a competition between two interactions α_1 and α_2. This may lead to a generalization of the Kosterlitz-Thouless transition, the crossover phenomenon in the Kondo effect, and other phenomena. In the Kondo problem, the appearance of logarithmic singularity [22, 56, 57] suggested the renormalizability of the model. In materials with many magnetic impurities, the interaction between magnetic impurities, called the RKKY interaction [58–60], should be considered. In this case the renormalization group flow is drawn on a two-dimensional plane of two parameters. There may be a relation to the generalized sine-Gordon model.

Conflicts of Interest

The author declares that there are no conflicts of interest.

Acknowledgments

The author expresses his sincere thanks to K. Odagiri for useful discussions. This work was supported in part by Grant-in-Aid from the Ministry of Education, Culture, Sports, Science and Technology (MEXT) of Japan (no. 17K05559).

References

[1] S. Coleman, "Quantum sine-Gordon equation as the massive Thirring model," *Physical Review*, vol. 11, p. 2088, 1975.

[2] R. F. Dashen, B. Hasslacher, and A. Neveu, "Particle spectrum in model field theories from semiclassical functional integral techniques," *Physical Review D: Particles, Fields, Gravitation and Cosmology*, vol. 11, no. 12, pp. 3424–3450, 1975.

[3] J. V. José, L. P. Kadanoff, S. Kirkpatrick, and D. R. Nelson, "Renormalization, vortices, and symmetry-breaking perturbations in the two-dimensional planar model," *Physical Review B: Condensed Matter and Materials Physics*, vol. 16, no. 3, pp. 1217–1241, 1977.

[4] S. Samuel, "Grand partition function in field theory with applications to sine-Gordon field theory," *Physical Review*, vol. 18, p. 1916, 1978.

[5] A. B. Zamolodchikov and A. B. Zamolodchikov, "Factorized S-matrices in two dimensions as the exact solutions of certain relativistic quantum field theory models," *Annals of Physics*, vol. 120, no. 2, pp. 253–291, 1979.

[6] D. J. Amit, Y. Y. Goldschmidt, and G. Grinstein, "Renormalisation group analysis of the phase transition in the 2D Coulomb gas, sine-Gordon theory and XY-model," *Journal of Physics A: Mathematical and General*, vol. 13, no. 2, pp. 585–620, 1980.

[7] P. Weigman, "One-dimensional Fermi system and plane xy model," *Journal of Physics C: Solid State Physics*, vol. 11, no. 8, p. 1583, 1987.

[8] J. Balog and A. Hegedus, "Two-loop beta functions of the Sine-Gordon model," *Journal of Physics A: Mathematical and General*, vol. 33, p. 6543, 2000.

[9] R. Rajaraman, *Solitons and Instantons*, North-Holland Publishing Company, Amsterdam, The Netherlands, 1982.

[10] N. Manton and P. Sutcliffe, *Topological Solitons*, Cambridge University Press, Cambridge, UK, 2004.

[11] S. Coleman, *Aspects of Symmetry*, Cambridge University Press, Cambridge, UK, 1985.

[12] E. C. Marino, *Quantum Field Theory Approach to Condensed Matter Physics*, Cambridge University Press, Cambridge, UK, 2017.

[13] E. Weinberg, *Classical Solutions in Quantum Field Theory*, Cambridge University Press, Cambbridge, UK, 2015.

[14] S. Mandelstam, "Soliton operators for the quantized sine-Gordon equation," *Physical Review*, vol. 11, p. 3026, 1975.

[15] B. Schroer and T. Truong, "Equivalence of the sine-Gordon and Thirring models and cumulative mass effects," *Physical Review*, vol. 15, p. 1684, 1977.

[16] M. Faber and A. N. Ivanov, "On the equivalence between sine-Gordon model and Thirring model in the chirally broken phase of the Thirring model," *European Physical Journal*, vol. 20, p. 723, 2001.

[17] V. I. Berezinski, "Destruction of Long-range Order in One-dimensional and Two-dimensional Systems Possessing a Continuous Symmetry Group. II. Quantum Systems," *Journal of Experimental and Theoretical Physics*, vol. 34, p. 610, 1972.

[18] J. M. Kosterlitz and D. Thouless, "Ordering, metastability and phase transitions in two-dimensional systems," *Journal of Physics*, vol. 6, p. 1181, 1973.

[19] J. M. Kosterlitz, "Critical Exponents of the Two-Dimensional XY Model," *Journal of Physics*, vol. 7, p. 1046, 1974.

[20] J. Jose, "Sine-Gordon Theory and the Classical Two-Dimensional xy Model," *Physical Review*, vol. 14, p. 2826, 1976.

[21] J. Zinn-Justin, *Quantum Field Theory and Critical Phenomena*, Oxford University Press, Oxford, UK, 1989.

[22] J. Kondo, "Resistance Minimum in Dilute Magnetic Alloys," *Progress of Theoretical Physics*, vol. 32, no. 1, p. 34, 1964.

[23] J Kondo, *The Physics of Dilute Magnetic Alloys*, Cambridge University Press, Cambridge, UK, 2012.

[24] P. W. Anderson, "A poor man's derivation of scaling laws for the Kondo problem," *Journal of Physics C: Solid State Physics*, vol. 3, p. 2436, 1970.

[25] P. W. Anderson and G. Yuval, *Physical Review Letters*, vol. 23, p. 89, 1969.

[26] G. Yuval and P. W. Anderson, "Exact Results for the Kondo Problem: One-Body Theory and Extension to Finite Temperature," *Physical Review*, vol. 1, p. 1522, 1970.

[27] P. W. Anderson, G. Yuval, and D. R. Hamann, "Exact Results in the Kondo Problem. II. Scaling Theory, Qualitatively Correct Solution, and Some New Results on One-Dimensional Classical Statistical Models," *Physical Review B*, vol. 1, p. 4464, 1970.

[28] J. Sólyom, "The Fermi gas model of one-dimensional conductors," *Advances in Physics*, vol. 28, no. 2, pp. 201–303, 1979.

[29] F. D. N. Haldane, "'Luttinger liquid theory' of one-dimensional quantum fluids. I. Properties of the Luttinger model and their extension to the general 1D interacting spinless Fermi gas," *Journal of Physics*, vol. 14, pp. 2585–2609, 1981.

[30] S.-T. Chui and P. A. Lee, "Equivalence of a One-Dimensional Fermion Model and the Two-Dimensional Coulomb Gas," *Physical Review Letters*, vol. 35, p. 315, 1975.

[31] J. B. Kogut, "An introduction to lattice gauge theory and spin systems," *Reviews of Modern Physics*, vol. 51, no. 4, pp. 659–713, 1979.

[32] J. Hubbard, "Electron correlations in narrow energy bands," *Proceedings of the Royal Society of London Series A: Mathematical and Physical*, vol. 276, no. 1365, pp. 238–257, 1963.

[33] M. C. Gutzwiller, " Correlation of Electrons in a Narrow ," *Physical Review A: Atomic, Molecular and Optical Physics*, vol. 137, no. 6A, pp. A1726–A1735, 1965.

[34] J. E. Hirsch, "Monte Carlo Study of the Two-Dimensional Hubbard Model," *Physical Review Letters*, vol. 51, p. 1900, 1983.

[35] S. Sorella, S. Baroni, R. Car, and M. Parrinello, "A novel technique for the simulation of interacting fermion systems," *Europhysics Letters*, vol. 8, p. 663, 1989.

[36] S. R. White, D. J. Scalapino, R. L. Sugar, E. Y. Loh, J. E. Gubernatis, and R. T. Scalettar, "Numerical study of the two-dimensional Hubbard model," *Physical Review*, vol. 40, p. 506, 1989.

[37] K. Yamaji, T. Yanagisawa, T. Nakanishi, and S. Koike, "Variational Monte Carlo study on the superconductivity in the two-dimensional Hubbard model," *Physica C: Superconductivity*, vol. 304, p. 225, 1988.

[38] T. Yanagisawa and Y. Shimoi, "Exact results in strongly correlated electrons - Spin-reflection positivity and the Perron-Frobenius theorem," *International Journal of Modern Physics*, vol. 10, p. 3383, 1996.

[39] S. Koikegami, "Variational Monte Carlo study on the superconductivity in the two-dimensional Hubbard model," *Physical Review*, vol. 67, p. 134517, 2003.

[40] A. J. Leggett, "Number-Phase Fluctuations in Two-Band Superconductors," *Progress of Theoretical Physics*, vol. 36, p. 901, 1966.

[41] Y. Tanaka and T. Yanagisawa, "Chiral Ground State in Three-Band Superconductors," *Journal of the Physical Society of Japan*, vol. 79, p. 114706, 2010.

[42] Y. Tanaka and T. Yanagisawa, "Chiral state in three-gap superconductors," *Solid State Communications*, vol. 150, no. 41-42, pp. 1980–1982, 2010.

[43] T. Yanagisawa, Y. Tanaka, I. Hase, and K. Yamaji, "Chiral state in three-gap superconductors, Solid State Commun," *Journal of the Physical Society of Japan*, vol. 81, p. 024712, 2012.

[44] V. Stanev and Z. Tesanovic, "Three-band superconductivity and the order parameter that breaks time-reversal symmetry," *Physical Review B: Condensed Matter and Materials Physics*, vol. 81, Article ID 134522, 2010.

[45] T. Yanagisawa and I. Hase, "Massless Modes and Abelian Gauge Fields in Multi-Band Superconductors," *Journal of the Physical Society of Japan*, vol. 82, p. 124704, 2013.

[46] T. Yanagisawa and Y. Tanaka, "Fluctuation-induced Nambu–Goldstone bosons in a Higgs–Josephson model," *New Journal of Physics*, vol. 16, p. 123014, 2014.

[47] T. Yanagisawa, "Nambu–Goldstone Bosons Characterized by the Order Parameter in Spontaneous Symmetry Breaking," *Journal of the Physical Society of Japan*, vol. 86, p. 104711, 2017.

[48] T. Koyama and M. Tachiki, *Physical Review B: Condensed Matter and Materials Physics*, vol. 54, no. 22, pp. 16183–16191, 1996.

[49] T. Yanagisawa, "Chiral sine-Gordon model," *Europhysics Letters*, vol. 113, p. 41001, 2016.

[50] G. 'tHooft and M. Veltman, "Regularization and renormalization of gauge fields," *Nuclear Physics*, vol. 44, p. 189, 1972.

[51] D. Gross, "Methods in field theory," in *Les Houches Lecture Notes*, R. Balian and J. Zinn-Justin, Eds., North-Holland Publishing Company, Amsterdam, Netherlands, 1976.

[52] T. Yanagisawa, *Recent Studies in Perturbation Theory*, D. I. Uzunov, Ed., InTech Open Publisher, 2017.

[53] I. Nándori, U. Jentschura, K. Sailer, and G. Soff, "Renormalization-group analysis of the generalized sine-Gordon model and of the Coulomb gas for d>~3 dimensions," *Physical Review D: Particles, Fields, Gravitation and Cosmology*, vol. 69, no. 2, 2004.

[54] F. J. Wegner and A. Houghton, "Renormalization Group Equation for Critical Phenomena," *Physical Review A: Atomic, Molecular and Optical Physics*, vol. 8, no. 1, pp. 401–412, 1973.

[55] S. Nagy, I. Nándori, J. Polonyi, and K. Sailer, "Functional Renormalization Group Approach to the Sine-Gordon Model," *Physical Review Letters*, vol. 102, no. 24, 2009.

[56] Y. Nagaoka, "Self-Consistent Treatment of Kondo's Effect in Dilute Alloys," *Physical Review A: Atomic, Molecular and Optical Physics*, vol. 138, no. 4A, pp. A1112–A1120, 1965.

[57] D. R. Hamann, "New Solution for Exchange Scattering in Dilute Alloys," *Physical Review*, vol. 158, p. 570, 1967.

[58] M. A. Ruderman and C. Kittel, "Indirect Exchange Coupling of Nuclear Magnetic Moments by Conduction Electrons," *Physical Review A: Atomic, Molecular and Optical Physics*, vol. 96, no. 1, pp. 99–102, 1954.

[59] T. Kasuya, "A Theory of Metallic Ferro- and Antiferromagnetism on Zener's Model," *Progress of Theoretical Physics*, vol. 16, p. 45, 1956.

[60] K. Yosida, "Magnetic Properties of Cu-Mn Alloys," *Physical Review A: Atomic, Molecular and Optical Physics*, vol. 106, no. 5, pp. 893–898, 1957.

21

Expanded (G/G^2) Expansion Method to Solve Separated Variables for the 2+1-Dimensional NNV Equation

Yong Meng

College of Science, Ningbo University, Ningbo 315211, China

Correspondence should be addressed to Yong Meng; 1611071420@nbu.edu.cn

Academic Editor: Ping Li

The traditional (G/G^2) expansion method is modified to extend the symmetric extension to the negative power term in the solution to the positive power term. The general traveling wave solution is extended to a generalized solution that can separate variables. By using this method, the solution to the detached variables of the symmetric extended form of the 2+1-dimensional NNV equation can be solved, also the soliton structure and fractal structure of Dromion can be studied well.

1. Introduction

The nature of this world is nonlinear, and linearity is only an approximation under certain conditions. And the reason why this large world is so colorful and varied is that there is a nonlinear relationship between things. In nonlinear science we start from the actual nonlinear phenomena; according to the laws of physics, the corresponding mathematical models are established under reasonable approximations, and then the corresponding laws are obtained by studying these mathematical models. A large part of these equations is nonlinear partial differential equations. It is not easy to solve these partial differential equations. For many years people have done a lot of work for this and have achieved a series of important results, such as the Inverse Scattering Transformation(IST) [1], Darboux transform [2, 3], Lie group and nonclassical Lie group method [4, 5], Hirota bilinear method [6–8], and other important methods. Among them, the function expansion [9–19] has the advantages of simplicity, directness, and ease of operation. So there is a lot of use in solving nonlinear partial differential equations.

Li proposed an expansion method in [20] to solve the new traveling wave solution for the Vakhnenko equation. However, for a large number of nonlinear systems, it is difficult to fully describe only the traveling wave solution, and it will inevitably lose other solutions with a rich structure. This paper reconstructs this method and extends the expansion of the solution symmetrically to negative power. Moreover, the variable is separated by the solution, and more forms of solutions are obtained. The symmetric extension of the 2+1-dimensional NNV equation is solved by using this method. A solution of this equation is studied to get soliton solution and fractal structure solution.

2. Separation Variable Solutions to Symmetric Extensions of 2+1-Dimensional NNV Equations

The 2+1-dimensional NNV equation

$$u_t + au_{xxx} + bu_{yyy} + cu_x + du_y - 3a(uv)_x - 3b(uw)_y = 0 \tag{1}$$

$$u_x = v_y \tag{2}$$

$$u_y = w_x \tag{3}$$

is a Lax integrable generalization of all isotropic properties of the 1+1-dimensional KdV equation. We first determine the maximum power of the expansion term in u, v, w expansion formula by the principle of homogeneous equilibrium [21–23] to be 2 and extend the symmetry extension of the solution to the negative power term (-2). Then set the expression for u, v, w to

$$u = a_{-2}\left(\frac{G'(\phi)}{G(\phi)^2}\right)^{-2} + a_{-1}\left(\frac{G'(\phi)}{G(\phi)^2}\right)^{-1} + a_0 + a_1\frac{G'(\phi)}{G(\phi)^2} + a_2\left(\frac{G'(\phi)}{G(\phi)^2}\right)^2 \tag{4}$$

$$v = b_{-2}\left(\frac{G'(\phi)}{G(\phi)^2}\right)^{-2} + b_{-1}\left(\frac{G'(\phi)}{G(\phi)^2}\right)^{-1} + b_0 + b_1\frac{G'(\phi)}{G(\phi)^2} + b_2\left(\frac{G'(\phi)}{G(\phi)^2}\right)^2 \tag{5}$$

$$w = c_{-2}\left(\frac{G'(\phi)}{G(\phi)^2}\right)^{-2} + c_{-1}\left(\frac{G'(\phi)}{G(\phi)^2}\right)^{-1} + c_0 + c_1\frac{G'(\phi)}{G(\phi)^2} + c_2\left(\frac{G'(\phi)}{G(\phi)^2}\right)^2 \tag{6}$$

where $\phi, a_i, b_j, c_k, (i, j, k = -2..2)$ are functions of x, y, t.

At the same time $G(\xi)$ achieves an ordinary differential equation

$$\left(\frac{G'(\xi)}{G(\xi)^2}\right)' = \mu + \lambda\left(\frac{G'(\xi)}{G(\xi)^2}\right)^2 \tag{7}$$

It can be obtained from (7)

$$\mu\lambda < 0 : \frac{G'}{G^2} = -\frac{\sqrt{|\mu\lambda|}}{\lambda} + \frac{\sqrt{|\mu\lambda|}}{2}\frac{C_1\sinh\left(\sqrt{|\mu\lambda|}\xi\right) + C_2\cosh\left(\sqrt{|\mu\lambda|}\xi\right)}{C_1\cosh\left(\sqrt{|\mu\lambda|}\xi\right) + C_2\sinh\left(\sqrt{|\mu\lambda|}\xi\right)} \tag{8}$$

$$\mu\lambda > 0 : \frac{G'}{G^2} = \frac{C_1\cos\left(\sqrt{\mu\lambda}\xi\right) + C_2\sin\left(\sqrt{\mu\lambda}\xi\right)}{C_1\sin\left(\sqrt{\mu\lambda}\xi\right) - C_2\sin\left(\sqrt{\mu\lambda}\xi\right)} \tag{9}$$

$$\mu = 0, \lambda \neq 0 : \frac{G'}{G^2} = \frac{C_1}{\lambda(C_1\xi + C_2)} \tag{10}$$

In order to achieve the separation of variables, we set

$$\phi(x, y, t) = p(x, t) + q(y, t) \tag{11}$$

Substituting (4)~(6) into (1)~(3), if there is a high-order derivative of $G(\phi)$, then use the ordinary differential equation (7) to replace the first derivative of G. That is, That is, (1)~(3) become three power polynomials about (G'/G^2). Let the coefficient of each power of (G'/G^2) of this polynomial be zero, resulting in a set containing a_i, b_j, c_k partial differential equations. Then, (11) is substituted into the partial differential equation group, and finally the partial differential equation group is solved to obtain three sets of solutions.

First group is

$$\begin{aligned}
&a_{-2} = b_{-2} = c_{-2} = a_{-1} = b_{-1} = c_{-1} = a_1 = 0,\\
&a_0 = 2\lambda\left(\frac{\partial}{\partial y}q\right)\left(\frac{\partial}{\partial x}p\right)\mu,\\
&b_0 = \frac{2}{3}\left(\frac{\partial}{\partial x}p\right)^2\lambda\mu - \frac{F_1(t)\,b}{((\partial/\partial x)\,p)\,a} + \frac{1}{3}\left(\frac{c}{a} + \frac{(\partial^3/\partial x^3)\,p}{(\partial/\partial x)\,p} + \frac{(\partial/\partial t)\,p}{((\partial/\partial x)\,p)\,a}\right),\\
&c_0 = \frac{2}{3}\left(\frac{\partial}{\partial y}q\right)^2\lambda\mu + \frac{1}{3}\left(\frac{d}{b} + \frac{(\partial^3/\partial y^3)\,q}{(\partial/\partial y)\,q} + \frac{(\partial/\partial t)\,q}{((\partial/\partial y)\,q)\,b}\right) + \frac{F_1(t)}{(\partial/\partial y)\,q},\\
&b_1 = 2\left(\frac{\partial^2}{\partial x^2}p\right)\lambda,\\
&c_1 = 2\lambda\frac{\partial^2}{\partial y^2}q,\\
&a_2 = 2\left(\frac{\partial}{\partial y}q\right)\left(\frac{\partial}{\partial x}p\right)\lambda^2,\\
&b_2 = 2\left(\frac{\partial}{\partial x}p\right)^2\lambda^2,\\
&c_2 = 2\lambda^2\left(\frac{\partial}{\partial y}q\right)^2
\end{aligned} \tag{12}$$

Second Group is

$$a_{-1} = a_1 = b_1 = c_1 = a_2 = b_2 = c_2 = 0,$$

$$a_{-2} = 2\mu^2\left(\frac{\partial}{\partial x}p\right)\frac{\partial}{\partial y}q,$$

$$b_{-2} = 2\left(\frac{\partial}{\partial x}p\right)^2\mu^2,$$

$$c_{-2} = 2\left(\frac{\partial}{\partial y}q\right)^2\mu^2,$$

$$b_{-1} = -2\mu\frac{\partial^2}{\partial x^2}p,$$

$$c_{-1} = -2\mu\frac{\partial^2}{\partial y^2}q,$$

$$a_0 = 2\lambda\left(\frac{\partial}{\partial y}q\right)\left(\frac{\partial}{\partial x}p\right)\mu,$$

$$b_0 = \frac{2}{3}\left(\frac{\partial}{\partial x}p\right)^2\lambda\mu + \frac{1}{3}\left(\frac{c}{a} + \frac{(\partial^3/\partial x^3)p}{(\partial/\partial x)p} + \frac{(\partial/\partial t)p}{((\partial/\partial x)p)a}\right) - \frac{F_2(t)b}{((\partial/\partial x)p)a},$$

$$c_0 = \frac{2}{3}\left(\frac{\partial}{\partial y}q\right)^2\lambda\mu + \frac{1}{3}\left(\frac{d}{b} + \frac{(\partial^3/\partial y^3)q}{(\partial/\partial y)q} + \frac{(\partial/\partial t)q}{((\partial/\partial y)q)b}\right) + \frac{F_2(t)}{(\partial/\partial y)q} \tag{13}$$

The third group is

$$a_{-2} = 2\mu^2\left(\frac{\partial}{\partial x}p\right)\frac{\partial}{\partial y}q,$$

$$b_{-2} = 2\left(\frac{\partial}{\partial x}p\right)^2\mu^2,$$

$$c_{-2} = 2\left(\frac{\partial}{\partial y}q\right)^2\mu^2,$$

$$a_{-1} = a_1 = 0,$$

$$b_{-1} = -2\mu\frac{\partial^2}{\partial x^2}p,$$

$$c_{-1} = -2\mu\frac{\partial^2}{\partial y^2}q,$$

$$a_0 = 4\lambda\left(\frac{\partial}{\partial y}q\right)\left(\frac{\partial}{\partial x}p\right)\mu,$$

$$b_0 = \frac{1}{3}\left(\frac{c}{a} + \frac{(\partial^3/\partial x^3)p}{(\partial/\partial x)p} + \frac{(\partial/\partial t)p}{((\partial/\partial x)p)a}\right) - \frac{4}{3}\left(\frac{\partial}{\partial x}p\right)^2\lambda\mu - \frac{F_3(t)b}{((\partial/\partial x)p)a},$$

$$c_0 = \frac{1}{3}\left(\frac{d}{b} + \frac{(\partial^3/\partial y^3)q}{(\partial/\partial y)q} + \frac{(\partial/\partial t)q}{((\partial/\partial y)q)b}\right) - \frac{4}{3}\left(\frac{\partial}{\partial y}q\right)^2\lambda\mu + \frac{F_3(t)}{(\partial/\partial y)q},$$

$$b_1 = 2\lambda\frac{\partial^2}{\partial x^2}p,$$

$$c_1 = 2\lambda\frac{\partial^2}{\partial y^2}q,$$

$$a_2 = 2\left(\frac{\partial}{\partial y}q\right)\left(\frac{\partial}{\partial x}p\right)\lambda^2,$$

$$b_2 = 2\left(\frac{\partial}{\partial x}p\right)^2\lambda^2,$$

$$c_2 = 2\lambda^2\left(\frac{\partial}{\partial y}q\right)^2 \tag{14}$$

where $F_1(t), F_2(t), F_3(t)$ is any function about t.Then (12)~(14) are substituted into (4)~(5) and three solutions to the 2+1-dimensional NNV equation are obtained.

$$u_1 = 2\lambda\left(\frac{\partial}{\partial y}q\right)\left(\frac{\partial}{\partial x}p\right)\left(\mu + \lambda\left(\frac{G'}{G^2}\right)^2\right) \tag{15}$$

$$v_1 = \frac{2}{3}\left(\frac{\partial}{\partial x}p\right)^2\lambda\mu + \frac{1}{3}\left(\frac{c}{a} + \frac{(\partial^3/\partial x^3)p}{(\partial/\partial x)p} + \frac{(\partial/\partial t)p}{((\partial/\partial x)p)a}\right) - \frac{F_1(t)b}{((\partial/\partial x)p)a} + 2\left(\frac{\partial^2}{\partial x^2}p\right)\lambda \cdot \frac{G'}{G^2} + 2\left(\frac{\partial}{\partial x}p\right)^2\lambda^2\left(\frac{G'}{G^2}\right)^2 \tag{16}$$

$$w_1 = \frac{2}{3}\left(\frac{\partial}{\partial y}q\right)^2\lambda\mu + \frac{1}{3}\left(\frac{d}{b} + \frac{(\partial^3/\partial y^3)q}{(\partial/\partial y)q} + \frac{(\partial/\partial t)q}{((\partial/\partial y)q)b}\right) + \frac{F_1(t)}{(\partial/\partial y)q} + 2\lambda\left(\frac{\partial^2}{\partial y^2}q\right)\frac{G'}{G^2} + 2\lambda^2\left(\frac{\partial}{\partial y}q\right)^2\left(\frac{G'}{G^2}\right)^2 \tag{17}$$

$$u_2 = 2\left(\frac{\partial}{\partial y}q\right)\left(\frac{\partial}{\partial x}p\right)\mu\left(\lambda + \mu\left(\frac{G'}{G^2}\right)^{-2}\right) \tag{18}$$

$$v_2 = \frac{2}{3}\left(\frac{\partial}{\partial x}p\right)^2\lambda\mu + \frac{1}{3}\left(\frac{c}{a} + \frac{(\partial^3/\partial x^3)p}{(\partial/\partial x)p} + \frac{(\partial/\partial t)p}{((\partial/\partial x)p)a}\right) - \frac{F_2(t)b}{((\partial/\partial x)p)a} - 2\mu\left(\frac{\partial^2}{\partial x^2}p\right) \cdot \left(\frac{G'}{G^2}\right)^{-1} + 2\left(\frac{\partial}{\partial x}p\right)^2\mu^2\left(\frac{G'}{G^2}\right)^{-2} \tag{19}$$

$$w_2 = \frac{2}{3}\left(\frac{\partial}{\partial y}q\right)^2\lambda\mu + \frac{1}{3}\left(\frac{d}{b} + \frac{(\partial^3/\partial y^3)q}{(\partial/\partial y)q} + \frac{(\partial/\partial t)q}{((\partial/\partial y)q)b}\right) + \frac{F_2(t)}{(d/dy)q} - 2\mu\frac{d^2}{dy^2}q\left(\frac{G'}{G^2}\right)^{-1} + 2\left(\frac{\partial}{\partial y}q\right)^2\mu^2\left(\frac{G'}{G^2}\right)^{-2} \tag{20}$$

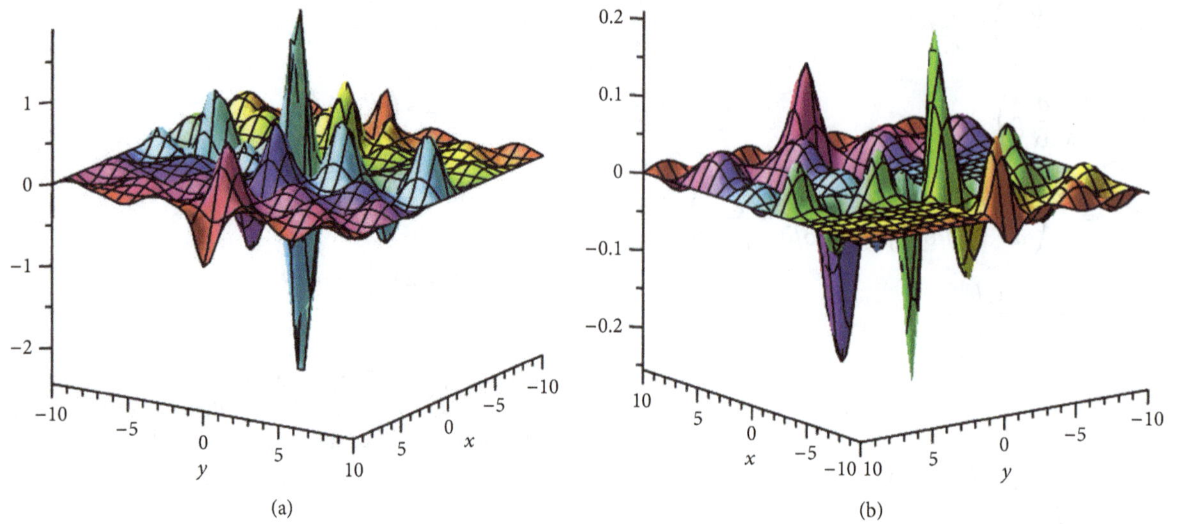

FIGURE 1: Figure (a) is a three-dimensional view of Dromion single soliton at t=-0.5 and figure (b) is a three-dimensional view of Dromion double soliton at t=2.9.

$$u_3 = 4\lambda\left(\frac{\partial}{\partial x}p\right)\left(\frac{\partial}{\partial y}q\right)\mu + 2\left(\frac{\partial}{\partial x}p\right)\left(\frac{\partial}{\partial y}q\right) \cdot \lambda^2\left(\frac{G'}{G^2}\right)^2 + 2\left(\frac{\partial}{\partial x}p\right)\left(\frac{\partial}{\partial y}q\right)\mu^2\left(\frac{G'}{G^2}\right)^{-2} \tag{21}$$

$$\begin{aligned} v_3 = {} & \frac{1}{3}\left(\frac{c}{a} + \frac{\left(\partial^3/\partial x^3\right)p}{(\partial/\partial x)\,p} + \frac{(\partial/\partial t)\,p}{((\partial/\partial x)\,p)\,a} - 4\left(\frac{\partial}{\partial x}p\right)^2\lambda\mu\right) - \frac{F_3(t)\,b}{((\partial/\partial x)\,p)\,a} + 2\left(\frac{\partial^2}{\partial x^2}p\right)\lambda \\ & \cdot \frac{G'}{G^2} + 2\left(\frac{\partial}{\partial x}p\right)^2\lambda^2\left(\frac{G'}{G^2}\right)^2 - 2\mu\left(\frac{\partial^2}{\partial x^2}p\right) \\ & \cdot \left(\frac{G'}{G^2}\right)^{-1} + 2\left(\frac{\mathrm{d}}{\mathrm{d}x}p\right)^2\mu^2\left(\frac{G'}{G^2}\right)^{-2} \end{aligned} \tag{22}$$

$$\begin{aligned} w_3 = {} & \frac{1}{3}\left(\frac{d}{b} + \frac{\left(\partial^3/\partial y^3\right)q}{(\partial/\partial y)\,q}\frac{d}{b} + \frac{(\partial/\partial t)\,q}{((\partial/\partial y)\,q)\,b} - 4\left(\frac{\partial}{\partial y}q\right)^2\lambda\mu\right) + \frac{F_3(t)}{(\partial/\partial y)\,q} + 2\lambda\frac{\partial^2}{\partial y^2}q\frac{G'}{G^2} \\ & + 2\lambda^2\left(\frac{\partial}{\partial y}q\right)^2\left(\frac{G'}{G^2}\right)^2 - 2\mu\frac{\partial^2}{\partial y^2}q\left(\frac{G'}{G^2}\right)^{-1} \\ & + 2\left(\frac{\partial}{\partial y}q\right)^2\mu^2\left(\frac{G'}{G^2}\right)^{-2} \end{aligned} \tag{23}$$

where $G = G(p+q)$ is the expression in (8)~(10).

3. Soliton Structure and Fractal Structure of the 2+1 Dimensional NNV Equation

Substituting (8) into (15) gives

$$u = 2\lambda\left(\frac{\partial}{\partial x}p\right)\left(\frac{\partial}{\partial y}q\right)\mu + 2\left(\frac{\partial}{\partial x}p\right)\left(\frac{\partial}{\partial y}q\right)\lambda^2\left(-\frac{\sqrt{|\mu\lambda|}}{\lambda} + \frac{\sqrt{|\mu\lambda|}}{2}B(p,q)\right)^2 \tag{24}$$

$$B(p,q) = \frac{C_1\sinh\left(\sqrt{|\mu\lambda|}\,(p+q)\right) + C_2\cosh\left(\sqrt{|\mu\lambda|}\,(p+q)\right)}{C_1\cosh\left(\sqrt{|\mu\lambda|}\,(p+q)\right) + C_2\sinh\left(\sqrt{|\mu\lambda|}\,(p+q)\right)} \tag{25}$$

Then select the parameters

$$C_1 = 4, \quad C_2 = 2, \quad \lambda = 1, \quad \mu = -1 \tag{26}$$

p, q selected as

$$p = \sin(x - 0.5t) + \tanh(x - 0.5t) \tag{27}$$

$$q = 0.4\sin(y) + \tanh(y) \tag{28}$$

as well as

$$p = 0.1\sin(x - 0.5t) + 0.5\tanh(x + t) + 0.5\tanh(0.5x - t) \tag{29}$$

$$q = 0.1\sin(0.9y) + 0.5\tanh(0.9y) \tag{30}$$

The periodic wave background Dromion soliton and double Dromion soliton can be obtained as shown in Figure 1.

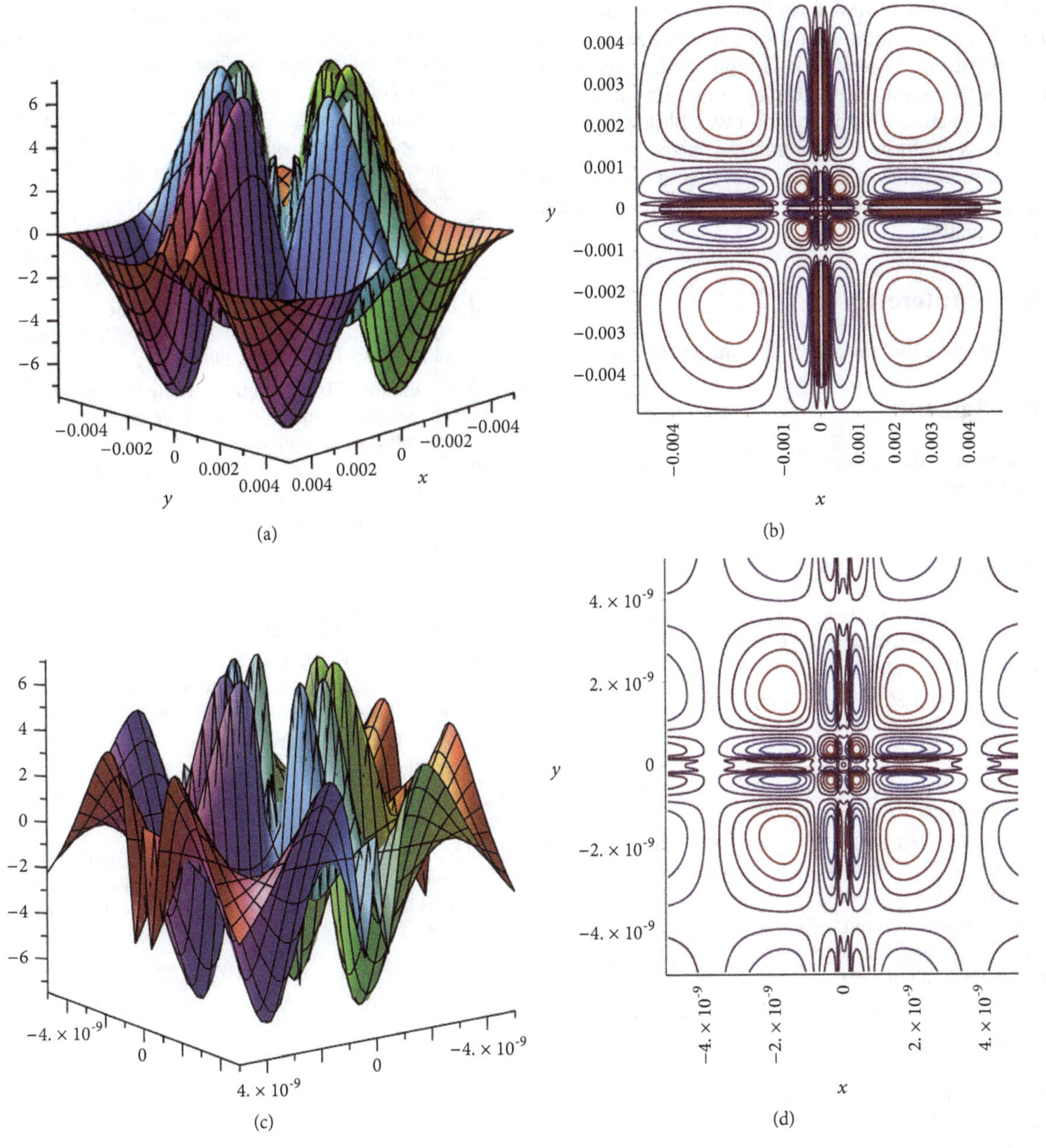

FIGURE 2: (a) Three-dimensional map, $x, y \in (-5\times10^{-3}, 5\times10^{-3})$. (b) Contour map $x, y \in (-5\times10^{-3}, 5\times10^{-3})$. (c) Three-dimensional map, $x, y \in (-5\times10^{-9}, 5\times10^{-9})$. (d) Contour map, $x, y \in (-5\times10^{-9}, 5\times10^{-9})$.

When we select the following parameters,

$$\begin{aligned} C_1 &= 3, \\ C_2 &= 2, \\ \lambda &= 1, \\ \mu &= -1 \end{aligned} \tag{31}$$

p, q selected as

$$p = 1 + e^{x(x+\sin(\ln(x^2)))} \tag{32}$$

$$q = 1 + e^{y(y+\sin(\ln(y^2)))} \tag{33}$$

we get Dromion's fractal structure.

It can be seen from Figure 2 that the Dromion structure remains basically unchanged at different scales and has self-similar properties, so it is a fractal structure.

4. Conclusion

In this paper, in order to obtain a richer solution of the nonlinear partial differential equations, improved the (G'/G^2) expansion method, starting from the solution to the positive exponential expansion symmetric extension to the negative power term and comparing the expressions (15)~(23) of the solution of the 2+1-dimensional NNV equation u, v, w, it can be found that symmetric extension to the negative power can indeed obtain more new formal solutions. At the same time, traveling wave solutions used for solving the equations are transformed into generalized solutions being capable of separating variables. Equation (11) plays an important role in achieving the separation of the equation variables, which also greatly expands the form of the nonlinear partial differential equation solution. The method is simple and direct. The advantages can be used to solve other nonlinear partial differential equations.

Conflicts of Interest

The authors declare that they have no conflicts of interest.

Acknowledgments

The author would like to thank Yuan Qun and Cai Jiaqi for their support and help.

References

[1] M. J. Ablowitz and P. A. Clarkson, *Solitons, Nonlinear Evolution Equations and Inverse Scattering*, Cambridge University Press, New York, NY, USA, 1991.

[2] V. B. Matveev and M. A. Salle, "Darbour transformations for certain nonliear evolution equation-s," *Journal of Mathematical Physics*, vol. 15, pp. 2157–2165, 1974.

[3] V. B. Matveev and M. A. Salle, *Darboux Transformations and Solitons*, Springer, Berlin, Gemany, 1991.

[4] G. W. Bluman and S. Kumei, *Symmetries and Differential Equations*, vol. 81 of *Applied Mathematical Sciences*, Springer, New York, NY, USA, 1989.

[5] P. J. Olver, *Applications of Lie Groups to Differential Equations*, vol. 107 of *Graduate Texts in Mathematics*, Springer, Berlin, Germany, 2nd edition, 1993.

[6] R. Hirota, "Exact envelope-soliton solutions of a nonlinear wave equation," *Journal of Mathematical Physics*, vol. 14, no. 7, pp. 805–809, 1973.

[7] R. Hirota, "Exact solution of the korteweg—de vries equation for multiple Collisions of solitons," *Physical Review Letters*, vol. 27, no. 18, pp. 1192–1194, 1971.

[8] R. Hirota, *The Direct Method in Soliton Theory*, Cambridge University Press, Cambridge, UK, 2004.

[9] E. G. Fan, "Extended tanh-function method and its applications to nonlinear equations," *Physics Letters A*, vol. 277, no. 4-5, pp. 212–218, 2000.

[10] Y.-T. Gao and B. Tian, "Deformed Korteweg-de Vries equation with symbolic computation," *International Journal of Modern Physics C*, vol. 12, no. 9, pp. 1335–1344, 2001.

[11] Y.-T. Gao and B. Tian, "Generalized hyperbolic-function method with computerized symbolic computation to construct the solitonic solutions to nonlinear equations of mathematical physics," *Computer Physics Communications*, vol. 133, no. 2-3, pp. 158–164, 2001.

[12] M. Wang and Y. Zhou, "The periodic wave solutions for the Klein-Gordon-Schrodinger equations," *Physics Letters A*, vol. 318, no. 1-2, pp. 84–92, 2003.

[13] M. Wang, X. Li, and J. Zhang, "The G'/G-expansion method and travelling wave solutions of nonlinear evolution equations in mathematical physics," *Physics Letters A*, vol. 372, no. 4, pp. 417–423, 2008.

[14] Y. Zhou, M. Wang, and T. Miao, "The periodic wave solutions and solitary wave solutions for a class of nonlinear partial differential equations," *Physics Letters A*, vol. 323, no. 1-2, pp. 77–88, 2004.

[15] K. S. Liu, T. Z. Fu, D. S. Liu, and Q. Zhao, "Jacobi elliptic function expansion method and periodic wave solutions of nonlinear wave equations," *Physics Letters A*, vol. 289, p. 69, 2001.

[16] J. Weiss, M. Tabor, and G. Carnevale, "The Painlevé property for partial differential equations," *Journal of Mathematical Physics*, vol. 24, no. 3, pp. 522–526, 1983.

[17] M. Eslami, "Trial solution technique to chiral nonlinear Schrodinger's equation in (1+2)-dimensions," *Nonlinear Dynamics*, vol. 85, no. 2, pp. 813–816, 2016.

[18] A. Neirameh and M. Eslami, "An analytical method for finding exact solitary wave solutions of the coupled $(2+1)$-dimensional Painlevé Burgers equation," *Scientia Iranica*, vol. 24, no. 2, pp. 715–726, 2017.

[19] H. Triki, T. Ak, M. Ekici et al., "Some new exact wave solutions and conservation laws of potential Korteweg–de Vries equation," *Nonlinear Dynamics*, vol. 89, no. 1, pp. 501–508, 2017.

[20] L. Wen-An, C. Hao, and Z. Guo-Cai, "The (ω/g)-expansion method and its application to Vakhnenko equation," *Chinese Physics B*, vol. 18, no. 2, pp. 400–404, 2009.

[21] M. L. Wang, "Solitary wave solutions for variant Boussinesq equations," *Physics Letters A*, vol. 199, no. 3-4, pp. 169–172, 1995.

[22] M. Wang, "Exact solutions for a compound KdV-Burgers equation," *Physics Letters A*, vol. 213, no. 5-6, pp. 279–287, 1996.

[23] M. L. Wang, Y. B. Zhou, and Z. B. Li, "Application of a homogeneous balance method to exact solutions of nonlinear equations in mathematical physics," *Physics Letters A*, vol. 216, p. 67, 1996.

22

Abundant Lump-Type Solutions and Interaction Solutions for a Nonlinear (3+1) Dimensional Model

R. Sadat,[1] **M. Kassem**,[1] **and Wen-Xiu Ma**[2,3,4,5]

[1]*Department of Mathematics and Physics, Faculty of Engineering, Zagazig University, Egypt*
[2]*Department of Mathematics and Statistics, University of South Florida, Tampa, FL 33620, USA*
[3]*Department of Mathematics, King Abdulaziz University, Jeddah, Saudi Arabia*
[4]*College of Mathematics and Systems Science, Shandong University of Science and Technology, Qingdao 266590, Shandong, China*
[5]*International Institute for Symmetry Analysis and Mathematical Modelling, Department of Mathematical Sciences, North-West University, Mafikeng Campus, Mmabatho 2735, South Africa*

Correspondence should be addressed to R. Sadat; eng_rahmaelsadat@yahoo.com

Academic Editor: Qin Zhou

We explore dynamical features of lump solutions as diversion and propagation in the space. Through the Hirota bilinear method and the Cole-Hopf transformation, lump-type solutions and their interaction solutions with one- or two-stripe solutions have been generated for a generalized (3+1) shallow water-like (SWL) equation, via symbolic computations associated with three different ansatzes. The analyticity and localization of the resulting solutions in the $(x, y, z,$ and $t)$ space have been analyzed. Three-dimensional plots and contour plots are made for some special cases of the solutions to illustrate physical motions and peak dynamics of lump soliton waves in higher dimensions. The study of lump-type solutions moderates the visuality of optics media and oceanography waves.

1. Introduction

It is very important to control the physical mechanisms of rough waves and interaction waves specially with lump-type waves. The significance of nonlinear waves of these types appears from natural disasters. Many physical phenomena need analytical approaches to classify the physical dynamics of nonlinear evaluation equations. The Darboux transformation (DT) and the Lie symmetry (LS) method [1–3] are efficient approaches to obtaining closed-form solutions. However, some problems occur in applying those methods, such as how to find Lax pairs in the DT method and how to carry out the back-substitution procedure in the LS method. There are also new types of closed-form solutions, for example, positions and complexions [4–8], and even new collision phenomena including fissions and fusions [9–14]. The Hirota bilinear method plays an influential role in discovering all the mentioned types of solutions to overcome a lot of analytic problems. Most studies apply the Hirota method to completely integrability nonlinear problems as in [10, 15–23]. We would like to demonstrate that the Hirota method can be used to explore various types of closed-form solutions: interaction solutions of lumps with solitons, kinks, line-solitons, resonance solutions, and one- or two-stripe solitons; and two classes of breather solutions (time periodic or space periodic solutions). Our analysis will show that those solutions can predicate the characteristics and physical significance of nonlinear problems.

Consider the following generalized (3+1) SWL equation [24, 25]:

$$u_{xxxy} + 3u_{xx}u_y + 3u_xu_{xy} - u_{yt} - u_{xz} = 0. \tag{1}$$

There are a few studies on this equation. For example, Tian et al. in [26] generated a traveling wave solution via the tanh method. In 2010, Zayed [27] used *the* G'/G method to obtain some traveling wave solutions by reducing the independent variables using the linear D'lambert transformation.

In what follows, we investigate lump soliton solutions and their dynamics and the susceptibility of their interactions with other types of solutions using the Hirota method for (1). By using the singular manifold method (SMM) with two-term truncated series, one derives the same ansatz in [24, 25]

$$u(x,y,t,z)=2(\ln(\psi(x,y,t,z))_x. \tag{2}$$

This is called the Cole-Hopf transformation, where ψ is an auxiliary or test function that will be determined later. Starting by substituting (2) into (1), one gets

$$\begin{aligned}&8\psi_{xxxy}\psi_x\psi-4\psi_{xxx}\psi_{xy}\psi+2\psi_{xxxx}\psi_y\psi-12\psi_{xxy}\psi_x{}^2\\&-2\psi_{xt}\psi_y\psi-2\psi_{xy}\psi_t\psi-2\psi_{yt}\psi_x\psi-2\psi_{xx}\psi_z\psi\\&-4\psi_{xz}\psi_x\psi+4\psi_x{}^2\psi_z+2\psi_{xxz}\psi^2-2\psi_{xxxxy}\psi^2\\&+2\psi_{xyt}\psi^2-4\psi_{xxx}\psi_x\psi_y+12\psi_{xx}\psi_x\psi_{xy}\\&+4\psi_x\psi_t\psi_y=0.\end{aligned}\tag{3}$$

The transformation increases the nonlinearity but allows us to work with the test function. In [24], Zhang used Bell polynomial theories to generate lump-kink solutions, lumps with one-stripe solitons and lumps with two-stripe solitons for (1), but he supposed that $z=x$ to minimize the number of independent variables and so studied the equation in a (2+1)-dimensional domain.

2. Lump Soliton Solutions

To generate single-lump solutions, we suppose that

$$\begin{aligned}\psi&=\beta^2+\gamma^2+\alpha_{11},\\\beta&=\alpha_1x+\alpha_2y+\alpha_3t+\alpha_4z+\alpha_5,\\\gamma&=\alpha_6x+\alpha_7y+\alpha_8t+\alpha_9z+\alpha_{10}.\end{aligned}\tag{4}$$

where $\alpha_i, i=1\ldots.11$, are real unknowns that will be found subsequently. We carry out a direct substitution of (4) into (3) and gather the coefficients of the resulting polynomial in x, y, t, and z, to obtain a nonlinear algebraic system in α_k. By solving this system of nonlinear algebraic equations with the aid of Maple, we acquire some sets of solutions for the parameters. Avoiding the redundancy, we surpass one studying case as follows:

$$\begin{aligned}\alpha_1&=\alpha_1,\\\alpha_2&=\alpha_2,\\\alpha_3&=-\frac{1}{\alpha_6\alpha_{11}(\alpha_1\alpha_7-\alpha_2\alpha_6)}\left(-\alpha_1^2\alpha_7\alpha_8\alpha_{11}\right.\\&\quad+\alpha_2\alpha_1\alpha_6\alpha_8\alpha_{11}+3\alpha_6^5\alpha_7+3\alpha_1^5\alpha_2+6\alpha_1^3\alpha_6^2\alpha_2\\&\quad\left.+3\alpha_1^4\alpha_6\alpha_7+6\alpha_1^2\alpha_6^3\alpha_7+3\alpha_6^4\alpha_1\alpha_2\right),\\\alpha_4&=\frac{1}{\alpha_6\alpha_{11}(\alpha_1\alpha_7-\alpha_2\alpha_6)}\left(3\alpha_1^4\alpha_2^2+6\alpha_1^3\alpha_2\alpha_6\alpha_7\right.\\&\quad+3\alpha_6^2\alpha_1^2\alpha_2^2+3\alpha_6^2\alpha_1^2\alpha_7^2-\alpha_2\alpha_1\alpha_7\alpha_8\alpha_{11}+6\alpha_6^3\alpha_1\alpha_2\alpha_7\\&\quad\left.+\alpha_2^2\alpha_{11}\alpha_6\alpha_8+3\alpha_6^4\alpha_7^2\right),\\\alpha_5&=\alpha_5,\\\alpha_6&=\alpha_6,\\\alpha_7&=\alpha_7,\\\alpha_8&=\alpha_8,\\\alpha_9&=\frac{\left(3\alpha_1^3\alpha_2+3\alpha_1^2\alpha_6\alpha_7+3\alpha_6^2\alpha_1\alpha_2-\alpha_7\alpha_8\alpha_{11}+3\alpha_6^3\alpha_7\right)}{\alpha_6\alpha_{11}},\\\alpha_{10}&=0,\\\alpha_{11}&=\alpha_{11}.\end{aligned}\tag{5}$$

Using the aggregation equation (4), one can represent the auxiliary function as

$$\begin{aligned}\psi&=\left(\alpha_1x+\alpha_2y-\frac{1}{\alpha_6\alpha_{11}(\alpha_1\alpha_7-\alpha_2\alpha_6)}\left(-\alpha_1^2\alpha_7\alpha_8\alpha_{11}\right.\right.\\&\quad+\alpha_2\alpha_1\alpha_6\alpha_8\alpha_{11}+3\alpha_6^5\alpha_7+3\alpha_1^5\alpha_2+6\alpha_1^3\alpha_6^2\alpha_2\\&\quad\left.+3\alpha_1^4\alpha_6\alpha_7+6\alpha_1^2\alpha_6^3\alpha_7+3\alpha_6^4\alpha_1\alpha_2\right)t\\&\quad+\frac{1}{\alpha_6\alpha_{11}(\alpha_1\alpha_7-\alpha_2\alpha_6)}\left(3\alpha_1^4\alpha_2^2+6\alpha_1^3\alpha_2\alpha_6\alpha_7\right.\\&\quad+3\alpha_6^2\alpha_1^2\alpha_2^2+3\alpha_6^2\alpha_1^2\alpha_7^2-\alpha_2\alpha_1\alpha_7\alpha_8\alpha_{11}+6\alpha_6^3\alpha_1\alpha_2\alpha_7\\&\quad\left.\left.+\alpha_2^2\alpha_{11}\alpha_6\alpha_8+3\alpha_6^4\alpha_7^2\right)z+\alpha_5\right)^2+\left(\alpha_6x+\alpha_7y+\alpha_8t\right.\\&\quad+\frac{\left(3\alpha_1^3\alpha_2+3\alpha_1^2\alpha_6\alpha_7+3\alpha_6^2\alpha_1\alpha_2-\alpha_7\alpha_8\alpha_{11}+3\alpha_6^3\alpha_7\right)}{\alpha_6\alpha_{11}}\\&\quad\left.\cdot z\right)^2+\alpha_{11},\end{aligned}\tag{6}$$

where

$$\Delta=\begin{vmatrix}\alpha_1&\alpha_2\\\alpha_6&\alpha_7\end{vmatrix}\neq0,\quad\alpha_6\alpha_{11}\neq0.\tag{7}$$

By using (2), the solution of (1) has the form

$$u=4\frac{\alpha_1\beta+\alpha_6\gamma}{\psi}\tag{8}$$

Incorporating (6) and (5) into (8), one gets a class of lump solutions of (1) depicted in Figure 1.

3. Interaction Solutions

3.1. Lump Solitons with One-Stripe Waves. Suppose that the test function is a confederation of a quadratic function withan

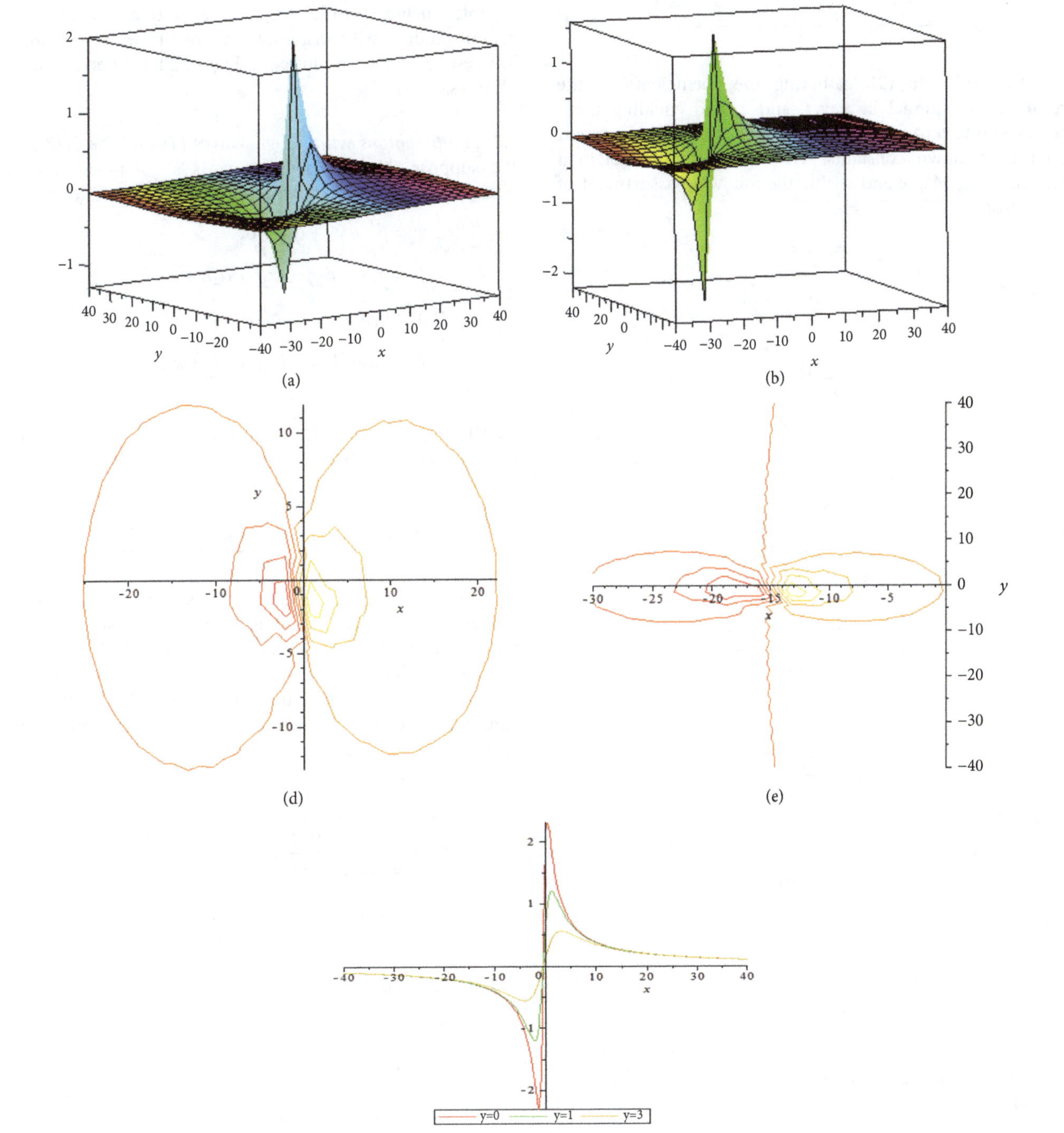

FIGURE 1: Proportion sight of the solution equation (8) with (6) and (5). For arbitrary constant values, $\alpha_2 = \alpha_1 = 1, \alpha_7 = -1, \alpha_6 = 1, \alpha_{10} = 0, \alpha_{11} = 1$, and $\alpha_8 = 5$. (a-b) 3D plots for $t = 0$, and 3, respectively. (d-e) Consistent contour plot of (a), (b). 2D plot present in (f) for various values of y.

exponential function as follows:

$$\psi = \beta^2 + \gamma^2 + \alpha_{11} + \lambda,$$
$$\beta = \alpha_1 x + \alpha_2 y + \alpha_3 t + \alpha_4 z + \alpha_5,$$
$$\gamma = \alpha_6 x + \alpha_7 y + \alpha_8 t + \alpha_9 z + \alpha_{10}$$
$$\lambda = e^{k_1 x + k_2 y + k_3 t + k_4 z + k_5}. \tag{9}$$

where $\alpha_i, i = 1 \ldots .11$ and $k_j, j = 1..5$, are real unknown constants that will be determined subsequently. Using the ansatz in (2),

$$u = 2\frac{2\alpha_1\beta + 2\alpha_6\gamma + k_1e^{k_1x+k_2y+k_3t+k_4z+k_5}}{\psi}. \quad (10)$$

Inserting (9) into (3), gathering the coefficients of the resulting polynomial in $x, y, t,$ and z, and equaling these coefficients to zero, we explore a complicated algebraic system on the unknown constants. We then solve the obtained system using Maple and snaffle the following assortment of solutions:

$$\begin{aligned}
\alpha_1 &= \alpha_1,\\
\alpha_2 &= \frac{k_2\left(\alpha_1^2+\alpha_6^2\right)}{k_1\alpha_1},\\
\alpha_3 &= 3k_1^2\alpha_1,\\
\alpha_5 &= \alpha_5,\\
\alpha_6 &= \alpha_6,\\
\alpha_7 &= 0,\\
\alpha_8 &= -3\frac{k_2\left(\alpha_1^2\right)}{\alpha_6},\\
\alpha_9 &= \frac{3k_1k_2\left(\alpha_1^2+\alpha_6^2\right)}{\alpha_6},\\
\alpha_{10} &= 0,\\
\alpha_{11} &= \frac{\left(\alpha_1^2+\alpha_6^2\right)}{k_1^2},\\
k_1 &= k_1,\\
k_2 &= k_2,\\
k_3 &= k_1^3,\\
k_4 &= 0,\\
k_5 &= k_5.
\end{aligned} \quad (11)$$

To avoid the singularity and promote the wave to localize in all directions, the following stipulation must be taken into consideration:

$$k_1a_1a_6 \neq 0. \quad (12)$$

Substituting (11) into (9), we obtain

$$\begin{aligned}
\psi &= \left(\alpha_1x + \frac{k_2\left(\alpha_1^2+\alpha_6^2\right)}{k_1\alpha_1}y + 3k_1^2\alpha_1t + \alpha_5\right)^2\\
&+ \left(\alpha_6x - 3\frac{k_2\left(\alpha_1^2\right)}{\alpha_6}t + \frac{3k_1k_2\left(\alpha_1^2+\alpha_6^2\right)}{\alpha_6}z\right)^2\\
&+ \frac{\left(\alpha_1^2+\alpha_6^2\right)}{k_1^2} + e^{k_1x+k_2y+k_1^3t+k_5}.
\end{aligned} \quad (13)$$

Introducing (13) into (10), we generate a class of interaction solutions with stripe soliton (solitary wave) solutions. The results have been plotted in Figure 2 for different values of times.

3.2. Lump Solitons with Tough Waves (Two-Stripe Solitons). We suppose that the new ansatz is a combination of a quadratic function and a hyperbolic function as follows:

$$\begin{aligned}
\psi &= \beta^2 + \gamma^2 + \alpha_{11} + \delta,\\
\beta &= \alpha_1x + \alpha_2y + \alpha_3t + \alpha_4z + \alpha_5,\\
\gamma &= \alpha_6x + \alpha_7y + \alpha_8t + \alpha_9z + \alpha_{10},\\
\delta &= \cosh\left(k_1x + k_2y + k_3t + k_4z + k_5\right).
\end{aligned} \quad (14)$$

Substituting (17) into (2), we snaffle an assortment of solutions for (1) as follows:

$$u = 2\frac{2\alpha_1\beta + 2\alpha_6\gamma + k_1\sinh\left(k_1x + k_2y + k_3t + k_4z + k_5\right)}{\psi}. \quad (15)$$

More complicated calculations have been done using Maple, to acquire the unidentified constants. Substituting (14) into (3), equaling the coefficients of $x, y, t,$ and z to zero, and solving the resulting nonlinear algebraic system (up to 150 equations), we explore the following solution cases of the constant parameters. In each case, we do back substitution in (14).

$$\begin{aligned}
\alpha_1 &= 0,\\
\alpha_2 &= \frac{12\left(k_2^2\alpha_4\alpha_6^2\right)}{9k_2^2\alpha_6^2k_1^2 + 4\alpha_4^2},\\
\alpha_3 &= \frac{-k_1\alpha_4}{k_2},\\
\alpha_4 &= \alpha_4,\\
\alpha_5 &= 0,\\
\alpha_6 &= \alpha_6,\\
\alpha_7 &= \frac{\alpha_6k_2\left(4\alpha_4^2 - 9k_2^2\alpha_6^2k_1^2\right)}{k_1\left(4\alpha_4^2 + 9k_2^2\alpha_6^2k_1^2\right)},\\
\alpha_8 &= \frac{3\left(k_1^2\alpha_6\right)}{2},\\
\alpha_9 &= \frac{3\left(k_1k_2\alpha_6\right)}{2},\\
\alpha_{10} &= 0,\\
\alpha_{11} &= \frac{9k_2^2\alpha_6^2k_1^2 + 4k_1^4\alpha_4^2 + 16\alpha_4^2\alpha_6^2 - 36\alpha_6^2k_1^2k_2^2}{16\alpha_4^2\alpha_6^2k_1^2},\\
k_1 &= k_1,
\end{aligned}$$

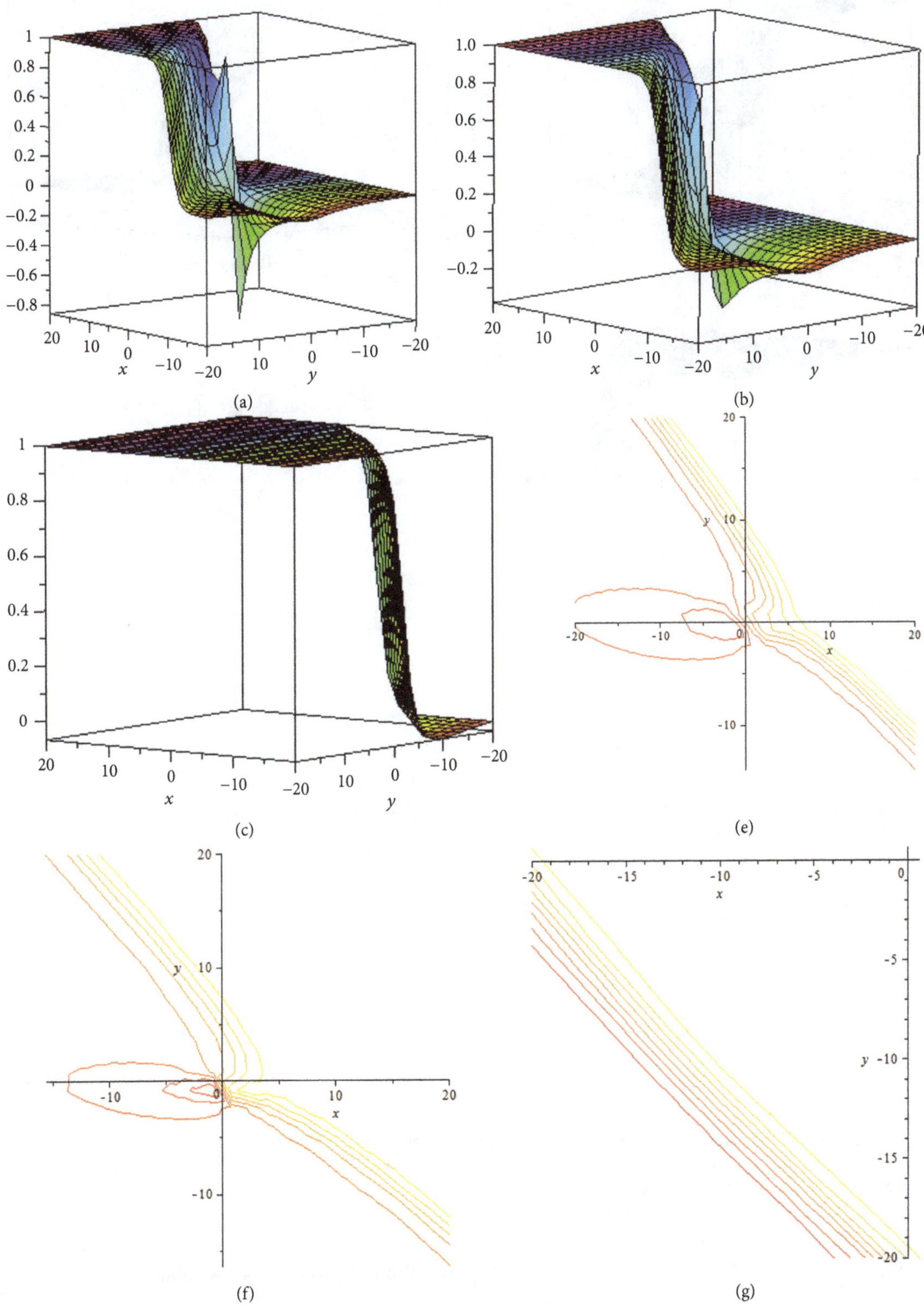

FIGURE 2: Proportion scene of the solution equation (10) with (13) and (11) for the values of arbitrary constants is $\alpha_1 = 1, \alpha_4 = 0, \alpha_6 = 3, \alpha_5 = 1, \alpha_{10} = 0$, and $k_1 = k_2 = k_5 = 1$. (a-c) represent 3D plots for (10) at $t = 0, 2$, and 30, $z = 0$, respectively. (e-g) Consistent contour plot of (a, b, c), respectively.

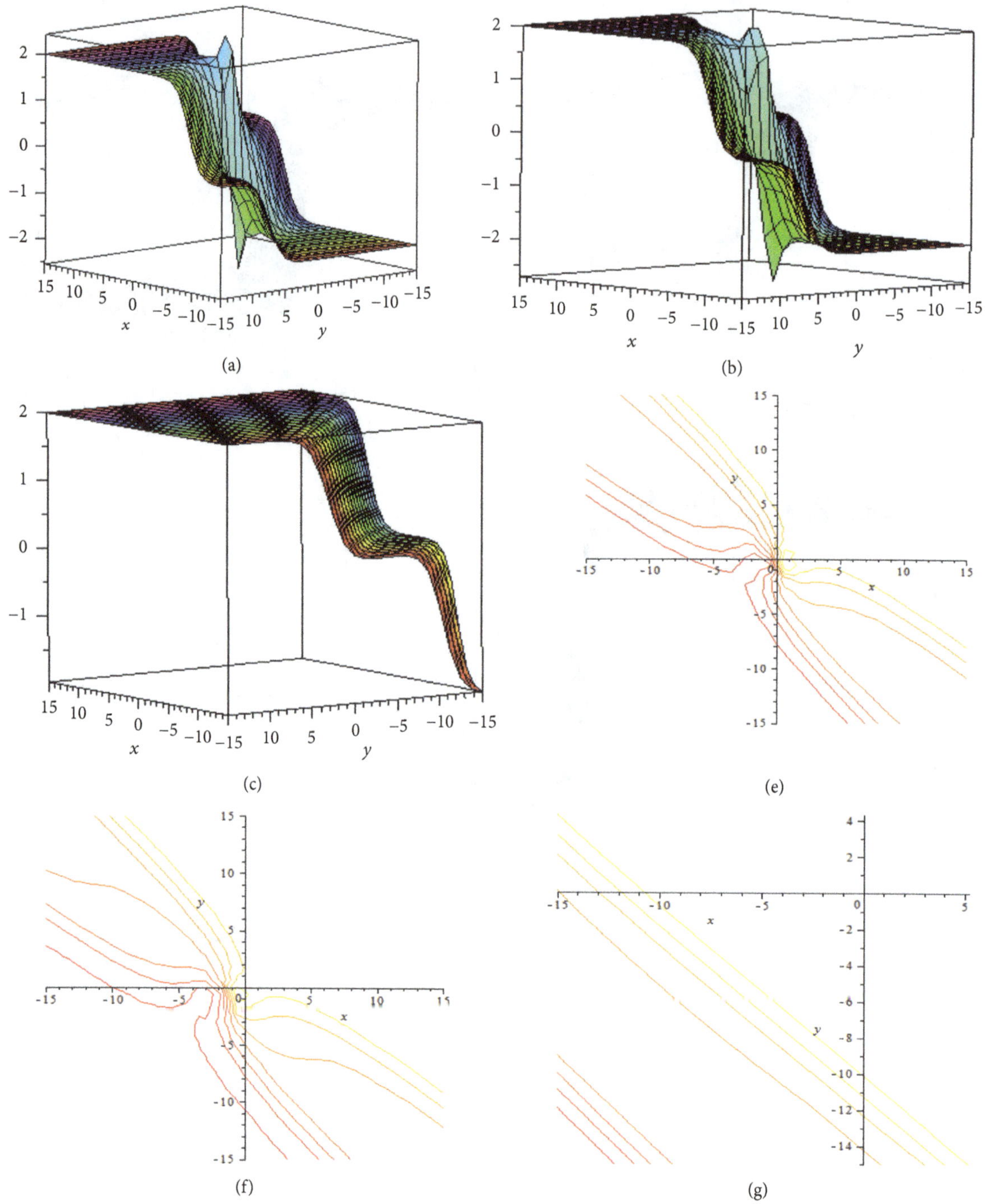

FIGURE 3: Proportion scenes of the solution equation (15) with (16) and (17) for the values of arbitrary constants are for $\alpha_2 = -1, \alpha_6 = 1, \alpha_7 = 2, \alpha_5 = 1, \alpha_{10} = 0, \alpha_8 = 1, \alpha_{11} = 1, k_1 = 1, k_2 = 1, k_5 = 1$. (a), (b), and (c) represent 3D plots for (10) at $t = 0, 5, 18$ and $z = 0$, respectively. (e), (f), and (g) Consistent contour plot of (a), (b), and (c).

$$
\begin{aligned}
k_2 &= k_2,\\
k_3 &= k_1^3,\\
k_4 &= 0,\\
k_5 &= k_5.
\end{aligned}
\tag{16}
$$

Substituting (16) into (14), we obtain

$$
\psi = \left(\alpha_1 x + \frac{12\left(k_2^2\alpha_4\alpha_6^2\right)}{9k_2^2\alpha_6^2k_1^2 + 4\alpha_4^2}y - \frac{k_1\alpha_4}{k_2}t + \alpha_4 z\right)^2 + \left(\alpha_6 x + \frac{\alpha_6 k_2\left(4\alpha_4^2 - 9k_2^2\alpha_6^2k_1^2\right)}{k_1\left(4\alpha_4^2 + 9k_2^2\alpha_6^2k_1^2\right)}y + \frac{3\left(k_1^2\alpha_6\right)}{2}t\right.
$$

$$
\begin{aligned}
&+ \frac{3\left(k_1 k_2 \alpha_6\right)}{2} z \Bigg)^2 \\
&+ \frac{9k_2^2\alpha_6^2k_1^2 + 4k_1^4\alpha_4^2 + 16\alpha_4^2\alpha_6^2 - 36\alpha_6^2k_1^2k_2^2}{16\alpha_4^2\alpha_6^2k_1^2} \\
&+ \cosh\left(k_1 x + k_2 y + k_1^3 t + k_5\right).
\end{aligned}
\tag{17}
$$

Through the same procedure, we get a class of solutions of (1) and plot a special solution in Figure 3.

4. Conclusions

Starting from the Cole-Hopf transformation, investigated in the SMM with a two-term truncated series, we derived novel lump solitons, lump-kinks, interacted lumps with one-stripe solitons or kinks, and interacted lumps with two-stripe solitons or kink waves, after some complicated calculations using the Maple software. The presented three-dimensional plots of the interaction solutions show that the lump solitons are coalesced or spliced up by the stripe solitons. To the best of our knowledge, those types of solutions for (1) are presented for the first time. Our solutions are localized in the four dimensional space (x, y, z, and t), but in [24], the authors assumed that $z = x$ and generated only one lump solution.

Conflicts of Interest

The authors declare that they have no conflicts of interest.

References

[1] R. Sadat and A. Halim, "Explicit Solutions of a Generalized Hirota-Satsuma Equation Using Darboux Transformation," *International Journal of Modern Mathematical Sciences*, vol. 14, no. 3, pp. 325–334, 2016.

[2] R. Sadat and A. Halim, "New Soliton Solutions for the Kadomtsev-Petviashvili equation Using Darboux Transformation," *International Journal of Modern Mathematical Sciences*, vol. 15, pp. 112–122, 2017.

[3] R. Sadat and M. Kassem, "Explicit Solutions for the (2+ 1)-Dimensional Jaulent–Miodek Equation Using the Integrating Factors Method in an Unbounded Domain," *Mathematical & Computational Applications*, vol. 23, no. 1, Paper No. 15, 9 pages, 2018.

[4] Y. Zhou, *Lump, Complexiton and Algebro-Geometric SOLutions to SOLiton Equations*, 2017.

[5] P. G. Varlashkin and E. F. Zganjar, "Shifts of annihilation radiation energy from the electron rest mass energy for positons annihilating in copper and sodium," *Nuclear Physics A*, vol. 130, no. 1, pp. 182–194, 1969.

[6] W. Liu and Y. Zhang, "Resonant multiple wave solutions, complexiton solutions and rogue waves of a generalized (3+1)-dimensional nonlinear wave in liquid with gas bubbles," *Waves in Random and Complex Media*, pp. 1–11, 2018.

[7] Z. Sun, Y. Gao, X. Yu, W. Liu, and Y. Liu, "Bound vector solitons and soliton complexes for the coupled nonlinear Schrödinger equations," *Physical Review E: Statistical, Nonlinear, and Soft Matter Physics*, vol. 80, no. 6, 2009.

[8] B. Nawaz, S. T. Rizvi, K. Ali, and M. Younis, "Optical soliton for perturbed nonlinear fractional Schrödinger equation by extended trial function method," *Optical and Quantum Electronics*, vol. 50, no. 5, 2018.

[9] S. Wang, X.-Y. Tang, and S.-Y. Lou, "Soliton fission and fusion: Burgers equation and Sharma-Tasso-Olver equation," *Chaos, Solitons & Fractals*, vol. 21, no. 1, pp. 231–239, 2004.

[10] Y. Zhou, S. Manukure, and W. X. Ma, "Lump and lump-soliton solutions to the Hirota–Satsuma–Ito equation," *Communications in Nonlinear Science and Numerical Simulation*, vol. 68, pp. 56–62, 2019.

[11] Z. Ma, J. Chen, and J. Fei, "Lump and line soliton pairs to a (2+ 1)-dimensional integrable KadomtsevPetviashvili equation," *Computers & Mathematics with Applications*, 2018.

[12] X. Liu, H. Triki, Q. Zhou, W. Liu, and A. Biswas, "Analytic study on interactions between periodic solitons with controllable parameters," *Nonlinear Dynamics*, vol. 94, no. 1, pp. 703–709, 2018.

[13] C. Yang, W. Liu, Q. Zhou, D. Mihalache, and B. A. Malomed, "One-soliton shaping and two-soliton interaction in the fifth-order variable-coefficient nonlinear Schrödinger equation," *Nonlinear Dynamics*, pp. 1–12, 2018.

[14] K. Ali, S. T. R. Rizvi, S. Ahmad, S. Bashir, and M. Younis, "Bell and kink type soliton solutions in birefringent nano-fibers," *Optik - International Journal for Light and Electron Optics*, vol. 142, pp. 327–333, 2017.

[15] S. Batwa and W.-X. Ma, "A study of lump-type and interaction solutions to a (3+ 1)-dimensional Jimbo–Miwa-like equation," *Computers & Mathematics with Applications. An International Journal*, vol. 76, no. 7, pp. 1576–1582, 2018.

[16] X. Yong, W.-X. Ma, Y. Huang, and Y. Liu, "Lump solutions to the Kadomtsev-Petviashvili I equation with a self-consistent source," *Computers & Mathematics with Applications*, 2018.

[17] W.-X. Ma, "Lump solutions to the Kadomtsev-Petviashvili equation," *Physics Letters A*, vol. 379, no. 36, Article ID 23311, pp. 1975–1978, 2015.

[18] L. Cheng and Y. Zhang, "Lump-type solutions for the (4+ 1)-dimensional Fokas equation via symbolic computations," *Modern Physics Letters B. Condensed Matter Physics, Statistical Physics, Applied Physics*, vol. 31, no. 25, 1750224, 9 pages, 2017.

[19] J. Pu and H. Hu, "Construction of lump soliton and mixed lump stripe solutions of (3+ 1)-dimensional soliton equation," *Applied Mathematics Letters*, vol. 85, pp. 77–81, 2018.

[20] J. Yang, W. Ma, and Z. Qin, "Lump and lump-soliton solutions to the (2+1)-dimensional Ito equation," *Analysis and Mathematical Physics*, vol. 1, pp. 1–10, 2017.

[21] W. Tan and Z. Dai, "Spatiotemporal dynamics of lump solution to the (1+ 1)-dimensional Benjamin–Ono equation," *Nonlinear Dynamics*, vol. 89, no. 4, pp. 2723–2728, 2017.

[22] Y. Zhang, C. Yang, W. Yu, M. Mirzazadeh, Q. Zhou, and W. Liu, "Interactions of vector anti-dark solitons for the coupled nonlinear Schrödinger equation in inhomogeneous fibers," *Nonlinear Dynamics*, vol. 94, no. 2, pp. 1351–1360, 2018.

[23] M. R. Ali and A. R. Hadhoud, "Hybrid Orthonormal Bernstein and Block-Pulse functions wavelet scheme for solving the 2D Bratu problem," *Results in Physics*, 2018.

[24] Y. Zhang, H. Dong, X. Zhang, and H. Yang, "Rational solutions and lump solutions to the generalized (3+1)-dimensional Shallow Water-like equation," *Computers & Mathematics with Applications*, vol. 73, no. 2, pp. 246–252, 2017.

[25] Y.-N. Tang, W.-X. Ma, and W. Xu, "Grammian and Pfaffian solutions as well as Pfaffianization for a (3+1)-dimensional generalized shallow water equation," *Chinese Physics B*, vol. 21, no. 7, 2012.

[26] B. Tian and Y.-T. Gao, "Beyond travelling waves: A new algorithm for solving nonlinear evolution equations," *Computer Physics Communications*, vol. 95, no. 2-3, pp. 139–142, 1996.

[27] E. Zayed, "travelling wave solutions for higher dimensional nonlinear evaluation equations using G'/G expansion method," *Journal of Applied Mathematics & Informatics*, vol. 28, no. 1, 2, pp. 383–395, 2010.

Permissions

All chapters in this book were first published in AMP, by Hindawi Publishing Corporation; hereby published with permission under the Creative Commons Attribution License or equivalent. Every chapter published in this book has been scrutinized by our experts. Their significance has been extensively debated. The topics covered herein carry significant findings which will fuel the growth of the discipline. They may even be implemented as practical applications or may be referred to as a beginning point for another development.

The contributors of this book come from diverse backgrounds, making this book a truly international effort. This book will bring forth new frontiers with its revolutionizing research information and detailed analysis of the nascent developments around the world.

We would like to thank all the contributing authors for lending their expertise to make the book truly unique. They have played a crucial role in the development of this book. Without their invaluable contributions this book wouldn't have been possible. They have made vital efforts to compile up to date information on the varied aspects of this subject to make this book a valuable addition to the collection of many professionals and students.

This book was conceptualized with the vision of imparting up-to-date information and advanced data in this field. To ensure the same, a matchless editorial board was set up. Every individual on the board went through rigorous rounds of assessment to prove their worth. After which they invested a large part of their time researching and compiling the most relevant data for our readers.

The editorial board has been involved in producing this book since its inception. They have spent rigorous hours researching and exploring the diverse topics which have resulted in the successful publishing of this book. They have passed on their knowledge of decades through this book. To expedite this challenging task, the publisher supported the team at every step. A small team of assistant editors was also appointed to further simplify the editing procedure and attain best results for the readers.

Apart from the editorial board, the designing team has also invested a significant amount of their time in understanding the subject and creating the most relevant covers. They scrutinized every image to scout for the most suitable representation of the subject and create an appropriate cover for the book.

The publishing team has been an ardent support to the editorial, designing and production team. Their endless efforts to recruit the best for this project, has resulted in the accomplishment of this book. They are a veteran in the field of academics and their pool of knowledge is as vast as their experience in printing. Their expertise and guidance has proved useful at every step. Their uncompromising quality standards have made this book an exceptional effort. Their encouragement from time to time has been an inspiration for everyone.

The publisher and the editorial board hope that this book will prove to be a valuable piece of knowledge for researchers, students, practitioners and scholars across the globe.

List of Contributors

Yukun Song, Heyuan Wang and Fengming Liu
College of Science, Liaoning University of Technology, Jinzhou 121001, China

Jianmei Peng
School of Science, East China University of Technology, Nanchang, Jiangxi 330013, China

Shufang Qiu and Wen Zhang
School of Science, East China University of Technology, Nanchang, Jiangxi 330013, China
Institute of Science and Engineering Computing, East China University of Technology, Nanchang 330013, China

Eunji Jeong, Junha Kim and Jihoon Lee
Department of Mathematics, Chung-Ang University, Seoul 06974, Republic of Korea

Xiaona Li, Minjie Xu and Xianqing Lv
Physical Oceanography Laboratory, Ocean University of China, Qingdao 266100, China

Kai Fu
School of Mathematical Sciences, Ocean University of China, Qingdao 266100, China

Abeer A. Mahmoud, Essam M. Abulwafa, Abd-alrahman F. Al-Araby and Atalla M. Elhanbaly
Theoretical Physics Research Group, Physics Department, Faculty of Science, Mansoura University, Mansoura 35516, Egypt

Hichem Eleuch
Institute for Quantum Science and Engineering, Texas A&M University, College Station, Texas 77843, USA
Department of Applied Science and Mathematics, College of Arts and Sciences, Abu Dhabi University, Abu Dhabi, UAE

Ingrid Rotter
Max Planck Institute for the Physics of Complex Systems, 01187 Dresden, Germany

Anna Perelomova
Gdansk University of Technology, Faculty of Applied Physics and Mathematics, Ul. Narutowicza 11/12, 80-233 Gdansk, Poland

Xifang Cao
School of Mathematical Sciences, Yangzhou University, China

A. Ludu
Department of Mathematics, Embry-Riddle Aeronautical University, Daytona Beach, FL 32114, USA

M. E. K. Fuziki
Departamento de Engenharia Química, Universidade Tecnologica Federal do Paraná, Ponta Grossa, PR 84016-210, Brazil

M. K. Lenzi
Departamento de Engenharia Química, Universidade Federal do Paraná, Curitiba, PR 81531-990, Brazil

M. A. Ribeiro and A. Novatski
Departamento de Física, Universidade Estadual de Ponta Grossa, Ponta Grossa, PR 87030-900, Brazil

E. K. Lenzi
Departamento de Física, Universidade Estadual de Ponta Grossa, Ponta Grossa, PR 87030-900, Brazil
National Institute of Science and Technology for Complex Systems, Centro Brasileiro de Pesquisas Física, Rio de Janeiro, RJ 22290-180, Brazil

Yafei Nie and Xianqing Lv
Physical Oceanography Laboratory, Ocean University of China, Qingdao 266100, China

Kai Fu
School of Mathematical Sciences, Ocean University of China, Qingdao 266100, China

Lu Xiao and Haiqin Qian
School of Management, Jiangsu University, Zhenjiang, Jiangsu, 212013, China

Mengmeng Qu and JunWang
Faculty of Science, Jiangsu University, Zhenjiang, Jiangsu, 212013, China

Samuel O. Adesanya
Department of Mathematical Sciences, Redeemer's University, Ede, Nigeria
Department of Education, Vaal University of Technology, Vanderbijlpark, South Africa

Ramosheuw S. Lebelo and K. C. Moloi
Department of Education, Vaal University of Technology, Vanderbijlpark, South Africa

Cong Sun
School of Applied Mathematics, Jilin University of Finance and Economics, Changchun, Jilin 130117, China

Lixia Li
College of Mathematics and Physics, Bohai University, Liaoning 121013, China

Vagif M. Mirsalimov
Azerbaijan Technical University, H. Javid 25, AZ1073 Baku, Azerbaijan
Institute of Mathematics and Mechanics of the Academy of Sciences of Azerbaijan, B. Vahabzade 9, AZ1141 Baku, Azerbaijan

Parvana E. Akhundova
Azerbaijan Technical University, H. Javid 25, AZ1073 Baku, Azerbaijan
Azerbaijan State University of Economics, Istiglaliyyat 6, AZ1001 Baku, Azerbaijan

H. Eleuch
Institute for Quantum Science and Engineering, Texas A&M University, College Station, TX 77843, USA
Department of Applied Sciences and Mathematics, College of Arts and Sciences, Abu Dhabi University, Abu Dhabi, UAE

Chanon Promsakon, Nattapong Kamsrisuk and Jessada Tariboon
Intelligent and Nonlinear Dynamic Innovations Research Center, Department of Mathematics, Faculty of Applied Science, King Mongkut's University of Technology North Bangkok, Bangkok 10800, Thailand

Sotiris K. Ntouyas
Department of Mathematics, University of Ioannina, 451 10 Ioannina, Greece
Nonlinear Analysis and Applied Mathematics (NAAM)-Research Group, Department of Mathematics, Faculty of Science, King Abdulaziz University, Jeddah 21589, Saudi Arabia

Isnaini Lilis Elviyanti, Beta Nur Pratiwi, A. Suparmi and C. Cari
Physics Department, Graduate Program, Sebelas Maret University, Indonesia

Huijuan Zhou and Chuanzhong Li
Department of Mathematics, Ningbo University, Ningbo 315211, China

Yueh-Lung Lin
School of Mathematical Sciences, University of Nottingham, Ningbo 315100, China

Takashi Yanagisawa
Electronics and Photonics Research Institute, National Institute of Advanced Industrial Science and Technology, 1-1-1 Umezono, Tsukuba, Ibaraki 305-8568, Japan

Yong Meng
College of Science, Ningbo University, Ningbo 315211, China

R. Sadat and M. Kassem
Department of Mathematics and Physics, Faculty of Engineering, Zagazig University, Egypt

Wen-Xiu Ma
Department of Mathematics and Statistics, University of South Florida, Tampa, FL 33620, USA
Department of Mathematics, King Abdulaziz University, Jeddah, Saudi Arabia
College of Mathematics and Systems Science, Shandong University of Science and Technology, Qingdao 266590, Shandong, China
International Institute for Symmetry Analysis and Mathematical Modelling, Department of Mathematical Sciences, North-West University, Mafikeng Campus, Mmabatho 2735, South Africa

Index

A

Adjoint Method, 26, 43, 49, 51, 53-56
Asymptotic Iteration Method, 186, 192-193
Axisymmetric Inviscid Hall-mhd Equations, 27

C

Caputo-fabrizio Fractional Time Operator, 120
Central Difference Scheme, 43, 49-51, 54
Cole-hopf Transformation, 219-220, 225
Conservative Semi-lagrangian Method, 133
Couple Stress Fluid, 144, 149

D

Darboux Transformation, 194, 205, 225
Dimensional Regularization Approach, 206, 211
Dipole Equation, 96, 98, 100-102, 106, 108, 110
Dirichlet Boundary Value Problem, 12
Double Confluent Heun Equation, 100, 111
Dust Acoustic Solitary Waves, 57, 64

F

Fluid-particle Interaction Model, 1, 10-11
Fourier Coefficients, 12, 167

G

G'/g-expansion Method, 218
Generalized Diffusion Equations, 120
Generation Rate, 144, 149

H

Heat Equation, 12, 26
Heating-cooling Function, 77-78, 80, 86
Hirota Bilinear Method, 219
Hulthen Potential, 186, 192

K

Kadomtsev-petviashvili Equation, 93, 174, 176, 225
Kdv Equations, 57
Klein-gordon Equation, 142, 176, 186, 192-193
Klein-gordon-zakharov Equations, 159

L

Larger Minimal Length Parameter, 186
Lie Symmetry, 144, 219
Linear D'lambert Transformation, 219
Linear Dispersion Relation, 59
Lump Soliton Solutions, 219-220
Lump-type Solutions, 93, 219, 225

M

Magnetic Dipole, 94, 96, 98, 106, 108
Magnetoacoustic Heating, 77-78, 86-87
Magnetoacoustic Waves, 77, 82, 86-87
Magnetohydrodynamic, 77, 149
Markovian Processes, 113
Maxwell Equations, 27, 142
Mean Absolute Error, 50, 53-55
Mesoscopic Superconductor, 94
Minimal Length Formalism, 186, 192-193
Momentum Operators, 186

N

Navier-stokes Equations, 1, 11, 27-28
Negative Power Term, 213
Non-hermitian Quantum, 68-69, 72, 75
Non-newtonian Fluid, 1, 11
Nonlinear Phenomena, 77, 82, 87, 93, 159, 213
Nonlinear Schrödinger Equation, 194

O

Oldroyd 6-constant Fluid, 144, 149
Open Quantum Systems, 68-69, 74-75

P

Periodic Initial Value Problem, 159
Plasma Parameters, 57, 59, 64
Plasma System, 57, 64
Poisson's Equation, 12, 59, 64
Positive Ground State Solutions, 134, 142

Q

Q-deformed Functions, 170
Q-deformed Sinh-gordon Equation, 170, 174
Quantum Calculus, 177, 185
Quantum Mechanics, 75, 186, 194, 204
Quantum Vortices, 94

R

Reductive Perturbation Theory, 59

Regularization Method, 12, 25-26, 206-207
Renormalization Group Theory, 206, 211

S
Sawada-kotera Equation, 93
Scalar Potential, 186, 193
Schrodinger-like Equation, 186
Second-order Quantum, 177
Separated Boundary Conditions, 177
Sine-gordon Model, 206-207, 211-212
Singular Manifold Method, 220
Spintronics, 94

T
Thin Film Flow, 144, 147, 149
Tide Wave Equations, 124-125, 132

www.ingramcontent.com/pod-product-compliance
Lightning Source LLC
LaVergne TN
LVHW080711210926
848485LV00003B/448